W0091081

Progress in Mathematics
Volume 191

Physical Combinatorics

Masaki Kashiwara
Tetsuji Miwa
Editors

Birkhäuser
Boston • Basel • Berlin

Masaki Kashiwara
Research Institute for
Mathematical Sciences (RIMS)
Kyoto University
606-01 Kyoto, Japan

Tetsuji Miwa
Research Institute for
Mathematical Sciences (RIMS)
Kyoto University
606-01 Kyoto, Japan

Library of Congress Cataloging-in-Publication Data

Physical combinatorics / Masaki Kashiwara, Tetsuji Miwa.
p. cm. — (Progress in mathematics ; v. 191)
Includes bibliographical references.
ISBN 0-8176-4175-0 (alk. paper). – ISBN 3-7643-4175-0 (alk. paper)
1. Combinatorial analysis–Congresses. 2. Representations of algebras–Congresses. 3. Integral equations–Congresses. I. Kashiwara, Masaki, 1947- II. Miwa, T. (Tetsuji) III. Progress in mathematics (Boston, Mass.) ; vol. 191.
QA164.P48 2000
511'.6–dc21

00-037945
CIP

AMS Subject Classifications: 81Qxx, 05A10, 05A30, 33D15

Printed on acid-free paper.

Birkhäuser 

ISBN 0-8176-4175-0 SPIN 10754198
ISBN 3-7643-4175-0

Reformatted from editors' files by John Spiegelman, Philadelphia, PA.
Printed and bound by Edwards Brothers, Ann Arbor, MI.
Printed in the United States of America.

9 8 7 6 5 4 3 2 1

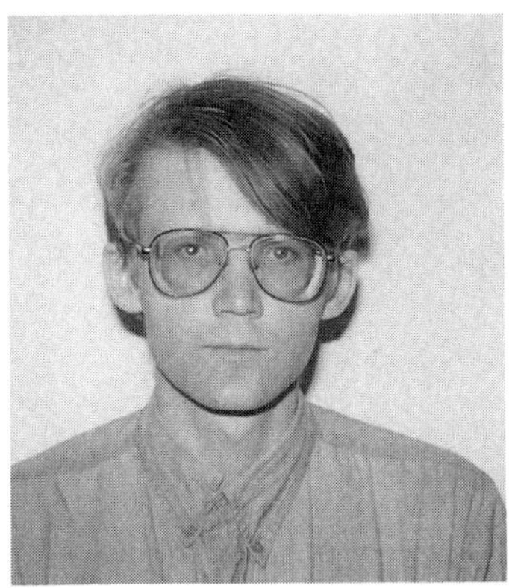

Dedicated to the memory of our friend and colleague
Denis Uglov (January 12, 1968–October 4, 1999)

Contents

Preface

Masaki Kashiwara and Tetsuji Miwa

This is the proceedings of a workshop entitled "Physical Combinatorics" held in 1999 during January 29–30 at the International Institute for Advanced Studies and February 1–2 at Research Institute for Mathematical Sciences, Kyoto University.

This conference was concerned with the combinatorial aspects arising in the theory of exactly solvable models and representation theory. Recent developments in integrable models reveal an unexpected link between representation theory and statistical mechanics through combinatorics. For example, Young tableaux, which describe the basis of irreducible representations, appear in Bethe Ansatz method in quantum spin chains as labels for the eigenstates of Hamiltonians. This connection brought new ideas both in representation theory and integrable models in statistical mechanics and also in combinatorics. For example, the classical Rogers–Ramanujan identities were generalized extensively, and their interpretations are given from both representation theory and integrable models.

In this workshop "Physical Combinatorics," the most recent developments were discussed by researchers from three different areas: representation theory, integrable models, and combinatorics.

All of the papers in this volume were refereed.

We thank Kumiko Matsumura and Natsuko Morino for their help for the organization of the workshop and the preparation of these proceedings. We also thank the International Institute for Advanced Studies for their hospitality during the workshop.

Denis Uglov, our friend and one of the contributors to this volume, passed away on October 4, 1999. We feel much sorrow at his passing. This volume is dedicated to his memory.

Masaki Kashiwara and Tetsuji Miwa
Kyoto, Japan
January 2000

An Insertion Scheme for C_n Crystals

T. H. Baker

Abstract. We define bumping and sliding algorithms for the C_n tableaux of Kashiwara and Nakashima. Together, they are used to define for such tableaux an insertion scheme which turns out to be a crystal isomorphism.

1 Introduction

Young tableaux play a pivotal role in the interplay between the fields of representation theory—of quantized universal enveloping algebras of (affine) Lie algebras—and combinatorics. Indeed, since representation theory itself is known to be a vital tool [6] in the solution of certain kinds of two-dimensional lattice models in statistical mechanics, one might say that Young tableaux are a central ingredient in *physical combinatorics*. Given a partition $\lambda = (\lambda_1, \lambda_2, \dots, \lambda_p)$ of a positive integer n (i.e., $n = \lambda_1 + \cdots + \lambda_p$ with $\lambda_1 \geq \lambda_2 \geq \cdots \geq \lambda_p > 0$), a Young diagram is a left justified array of square boxes, with the top row consisting of λ_1 boxes, the second row consisting of λ_2 boxes and so on (here we adopt the *English* convention for diagrams and tableaux). A Young tableau T of weight (or content) $wt(T) = (m_1, \dots, m_n)$ is a filling of a Young diagram with entries from the set $\{1, 2, \dots, n\}$ using m_1 numbers 1, m_2 numbers 2, etc. If the filling is such that the entries are weakly increasing along the rows (reading from left to right) and strictly increasing down the columns (reading from top to bottom), such a tableau is called *semistandard*.

The usefulness of semistandard tableaux in representation theory of the symmetric group S_n and the Lie algebra $sl(n)$ has long been demonstrated (see, e.g., [5]). More recently, the theory of crystal bases [7] of irreducible representations (irreps) of quantized Lie algebras has provided a new setting for semistandard tableaux and operations upon them. In that setting, the elements of a crystal base $B(\lambda)$ of an irrep $V(\lambda)$ of $U_q(sl(n))$ are labeled by semistandard tableaux, and the modified Chevalley generators (*Kashiwara* or *crystal* operators) $\tilde{e}_i$, $\tilde{f}_i$ act by changing one particular entry in the tableau. Such operators were known to Robinson in his proof of the Littlewood–Richardson rule [24]. This was extended to crystal bases of the other classical Lie algebras by Kashiwara and Nakashima [8], who gave analogues of semistandard tableaux for such cases as well as described the action of the operators $\tilde{e}_i$, $\tilde{f}_i$.

One of the most important features of crystal bases is the action of the Kashiwara operators $\tilde{e}_i$, $\tilde{f}_i$ on the tensor product of two crystals [7]

$$\tilde{f}_i(b_1 \otimes b_2) = \begin{cases} \tilde{f}_i(b_1) \otimes b_2 & \phi_i(b_1) > \epsilon_i(b_2), \\ b_1 \otimes \tilde{f}_i(b_2) & \phi_i(b_1) \leq \epsilon_i(b_2), \end{cases}$$

$$\tilde{e}_i(b_1 \otimes b_2) = \begin{cases} \tilde{e}_i(b_1) \otimes b_2 & \phi_i(b_1) \geq \epsilon_i(b_2), \\ b_1 \otimes \tilde{e}_i(b_2) & \phi_i(b_1) < \epsilon_i(b_2), \end{cases} \tag{1.1}$$

where $\phi_i(b) = \max\{k|(\tilde{f}_i)^k b \neq 0\}$, $\varepsilon_i(b) = \max\{k|(\tilde{e}_i)^k b \neq 0\}$. This rule is one of the main computational tools in the theory of crystal bases.

In the case of semistandard tableaux, there is an important connection between the A_{n-1} crystal action and the Schensted insertion algorithm [25], which we now briefly recall. The *column* insertion procedure inserts a letter x into a tableau b as follows: let y be the smallest integer in the first column $\geq x$. Then x replaces y in the first column, while y is then "bumped" into the next column where the process is repeated. This procedure stops when the entry which is bumped is larger than all the entries in the next column, in which case it is placed at the end of that column. The result of such a procedure is a tableau which is one box larger than b, call it $b \leftarrow x$ say (Note: this notation is usually reserved for row bumping (e.g., in [5]) but we need it for compatibility with the Japanese order of word reading). Using this insertion procedure, one can define the "product tableaux"

$$b * c := (\cdots((b \leftarrow c_1) \leftarrow c_2) \leftarrow \cdots) \leftarrow c_p), \tag{1.2}$$

where $w(c) = c_1 c_2 \cdots c_p$ is the *Japanese* reading of c, i.e., reading from top to bottom and leftwards down successive columns. This product is associative and gives rise to a monoid structure on semistandard tableaux, called the *plactic monoid* [15].

Now, suppose we have two irreducible $U_q(A_{n-1})$ crystals $B(\mu)$, and $B(\nu)$. Then it is well known that their tensor product has a finite decomposition

$$B(\mu) \otimes B(\nu) \cong \bigoplus_j B(\lambda_j),$$

which can be described exactly via the Littlewood–Richardson rule; see [19] or [18, p. 143]. In fact, the above isomorphism is given explicitly by the following.

Proposition 1.1. *Let* $\psi_A : B(\mu) \otimes B(\nu) \longrightarrow \bigoplus_j B(\lambda_j)$ *be defined by*

$$\psi_A(b_1 \otimes b_2) = b_1 * b_2.$$

Then ψ_A is the unique crystal isomorphism describing the above tensor product decomposition.

The row and column insertion algorithms have a convenient characterization in terms of *Knuth* equivalence classes. Namely, if two (column) words w_1, w_2 are connected by a series of elementary Knuth transformations [12]

$$x\,z\,y \to z\,x\,y \qquad x < y \leq z, \tag{K}$$

$$y\,z\,x \to y\,x\,z \qquad x \leq y < z, \tag{K$'$}$$

(in which case we write $w_1 \sim w_2$), then inserting the letters of the words w_1 and w_2 into the empty tableau results in the same *insertion* tableau $P(w_1) = P(w_2)$. The key to proving Proposition 1.1 is the following fact: the Knuth equivalence classes are stable under $\tilde{e}_i$, $\tilde{f}_i$, i.e., if $w_1 \sim w_2$, then $\tilde{e}_i w_1 \sim \tilde{e}_i w_2$ and similarly for $\tilde{f}_i$. Again, this is a classical fact, used in the proof of the LR rule [15].

Another important operation on tableaux is the *jeu-de-taquin* of Schützenberger. Here, one is given a tableau with a "hole" in it (this may appear in the operations of *evacuation* or *promotion*, for example [26, 27]). The hole slides in a southerly or easterly direction until it reaches the boundary of the tableau, thereby producing a new tableau containing one fewer box. The movement of the hole is uniquely determined to ensure that the tableau row and column conditions are preserved. Although it is not required in the proof of Proposition 1.1, it is nonetheless remarkable that *sliding preserves the Knuth equivalence classes.*

The aim of the present article is to define an insertion scheme for the C_n tableaux of Kashiwara and Nakashima which will provide the analogue of the above proposition for the case of C_n crystals. Apart from the tableaux of Kashiwara and Nakashima, analogues of semistandard Young tableaux which describe irreducible finite-dimensional representations of the classical Lie algebras B_n, C_n, and D_n have been introduced by King [9], De Concini [3], King and El-Sharkaway [10], Koike and Terada [13], Sundaram [30], King and Welsh [11], and Proctor [23]. For a good overview of the various bijections existing between them, consult the article of Fulmek and Krattenthaler [4]. For some of these types of tableaux, insertion schemes have been generated, such as the one of Berele [2] (for the C_n tableaux of King), Sundaram [30], Proctor [22], Benkart and Stroomer [1], and Okada [20, 21].

As in the case of Berele's insertion algorithm, we shall require analogues of both bumping *and* sliding to describe this scheme. Moreover, the proof of the C_n analogue of Proposition 1.1 will require both the fact that the equivalence classes of the C_n Knuth relations are stable under the Kashiwara operators, and that the C_n analogue of sliding preserves the C_n Knuth classes. Thus, what appears to be a happy accident in the A_{n-1} case is an essential property in the C_n case.

The plan of our presentation is as follows: In Section 2 we introduce the necessary notations and basic results about crystals of irreducible finite-dimensional representations of $U_q(C_n)$, in particular their description in terms of tableaux and the action of the Kashiwara operators $\tilde{e}_i$, $\tilde{f}_i$ on such tableaux. In Section 3, the (column) bumping procedure is defined, along with the reverse bumping procedure. In Section 4, the sliding procedure is defined, along with the reverse sliding procedure. In Section 5, these two procedures are combined to define an insertion scheme for the C_n tableaux. In Section 6, the elementary Knuth transformations governing the insertion scheme are deduced. In particular, we show that the C_n sliding procedure preserves Knuth equivalence, and that the Knuth equivalence classes are stable under the action of the Kashiwara operators. In Section 7 we give the main result, that the C_n insertion scheme realizes an isomorphism of C_n crystals. Finally, in Section 8 we present proofs that the bumping and sliding introduced earlier is indeed well-defined.

2 C_n tableaux and crystal action

2.1 Tableaux. Let Λ_i, $i = 1, \ldots, n$ denote the fundamental weights of the root system of C_n. The representation $V(\Lambda)$ with $\Lambda = \Lambda_{m_1} + \Lambda_{m_2} + \cdots + \Lambda_{m_p}$, $m_1 \leq m_2 \leq \cdots \leq m_p$ can be represented by a Young diagram consisting of columns of length $m_1, m_2, \ldots, m_p$ (going from right to left). Such representations (resp., crystal bases) can be embedded into the tensor product of multiple copies of the fundamental irreducible C_n representation (resp., crystal base) with highest weight Λ_1.

The representation $V(\Lambda_1)$ has a basis $\{\boxed{i}, \boxed{\overline{i}}\,; 1 \leq i \leq n\}$ and the action of the $U_q(C_n)$ generators is given as follows

$$\begin{array}{lll} q^h\boxed{i} = q^{\langle h,\epsilon_i\rangle}\boxed{i} & q^h\boxed{\overline{i}} = q^{-\langle h,\epsilon_i\rangle}\boxed{\overline{i}} & \\ e_j\boxed{i} = \delta_{i,j+1}\boxed{i-1} & e_j\boxed{\overline{i}} = \delta_{i,j}\boxed{\overline{i+1}} & 1 \leq j \leq n-1 \\ f_j\boxed{i} = \delta_{i,j}\boxed{i+1} & f_j\boxed{\overline{i}} = \delta_{i,j+1}\boxed{\overline{i-1}} & 1 \leq j \leq n-1 \\ e_n\boxed{i} = 0 & e_n\boxed{\overline{i}} = \delta_{i,n}\boxed{n} & \\ f_n\boxed{i} = \delta_{i,n}\boxed{\overline{n}} & f_n\boxed{\overline{i}} = 0. & \end{array} \tag{2.1}$$

As a result, the crystal graph of $B(\Lambda_1)$ has the following structure:

$$\boxed{1} \xrightarrow{1} \boxed{2} \xrightarrow{2} \cdots \xrightarrow{n-1} \boxed{n} \xrightarrow{n} \boxed{\overline{n}} \xrightarrow{n-1} \boxed{\overline{n-1}} \xrightarrow{n-2} \cdots \xrightarrow{2} \boxed{\overline{2}} \xrightarrow{1} \boxed{\overline{1}}.$$

The crystal structure on $B(\Lambda)$ for general $\Lambda = \Lambda_{m_1} + \cdots + \Lambda_{m_p}$ can be described by first embedding $B(\Lambda_{m_k})$ into $B(\Lambda_1)^{\otimes m_k}$ for each $k = 1, \ldots, p$ and then embedding $B(\Lambda)$ into $B(\Lambda_{m_1}) \otimes \cdots \otimes B(\Lambda_{m_p})$. As a result, the crystal base $B(\Lambda)$ can be described by a set of tableaux of shape Λ on the alphabet $\mathcal{X} = \mathcal{A} \cup \overline{\mathcal{A}}$, $\mathcal{A} := \{1, 2, \ldots, n\}$ with the (total) order $1 < 2 < \cdots n-1 < n < \overline{n} < \overline{n-1} < \cdots \overline{2} < \overline{1}$. The entries of such tableaux obey certain conditions that we now describe.

For the alphabet $\mathcal{X}$, we follow the convention that greek letters (α, β, etc.) belong to $\mathcal{A} \cup \overline{\mathcal{A}}$ while latin letters x, y, etc. (resp., $\overline{x}$, $\overline{y}$, etc.) belong to $\mathcal{A}$ (resp., $\overline{\mathcal{A}}$). Given a column C of length N containing the entries (reading from top to bottom) $\alpha_1, \alpha_2, \ldots, \alpha_N$, define the following function on C:

$$\mathrm{pos}_C(\alpha_k) = \begin{cases} k & \text{if } \alpha_k \in \mathcal{A}, \\ N+1-k & \text{if } \alpha_k \in \overline{\mathcal{A}}. \end{cases}$$

In other words, $\mathrm{pos}_C(a)$ (resp., $\mathrm{pos}_C(\overline{a})$) is the position of the entry a (resp., $\overline{a}$) w.r.t. the top (resp., bottom) of the column C. Sometimes we shall drop the subscript C when the meaning is clear. We shall also extend the domain of $\mathrm{pos}(\cdot)$ to include empty boxes, in which case $\mathrm{pos}(\square)$ will be the distance of the empty box from the top or the bottom of the column, the choice being clear from the context. Say a column C satisfies the *one-column condition* (1CC) if for all pairs $a, \overline{a}$ in C,

$$\mathrm{pos}_C(a) + \mathrm{pos}_C(\overline{a}) \leq a. \tag{2.2}$$

Given two adjacent columns C, C' consisting of the entries (reading from top to bottom) $\alpha_1, \ldots, \alpha_N$ and $\beta_1, \ldots, \beta_M$, respectively ($M \leq N$), say there is an (a, b) configuration if there exists $i \leq j < k \leq l$ such that either $(\alpha_i, \alpha_j, \alpha_k, \beta_l) = (a, b, \overline{b}, \overline{a})$ or $(\alpha_i, \beta_j, \beta_k, \beta_l) = (a, b, \overline{b}, \overline{a})$ (note that we include the possibility that $a = b$). Sometimes we shall distinguish the two cases and call the former a *left* (a, b) configuration and the latter a *right* (a, b) configuration. Given an (a, b) configuration, define a function $p(a, b|\overline{b}, \overline{a}) := (j-i)+(l-k)$. Say a pair of adjacent columns satisfies the *two-column condition* (2CC) if for every (a, b) configuration

$$p(a, b|\overline{b}, \overline{a}) < b - a. \tag{2.3}$$

More generally, suppose there exist integers $i \leq j < k \leq l$ such that $(\alpha_i, \alpha_j, \alpha_k, \beta_l) = (a, b, \overline{c}, \overline{d})$ or $(\alpha_i, \beta_j, \beta_k, \beta_l) = (a, b, \overline{c}, \overline{d})$. We call such a configuration a (left/right) $(a, b|\overline{c}, \overline{d})$ configuration.

The crystal base $B(\Lambda)$ then consists of the set of tableaux T on the shape Λ with entries in $\mathcal{X}$ satisfying the following conditions:

1. The entries of T increase weakly along the rows.
2. The entries of T increase strictly down the columns.
3. For each column C, the 1CC holds.
4. For each pair of adjacent columns C, C', the 2CC holds.

Note that (2.3) implies that a C_n tableau T has no (a, a) configurations for any a.

The following lemmas given by Kashiwara and Nakashima [8, Lemmas 4.3.1 and 4.4.2], will be essential tools in the proofs of our results.

Lemma 2.1. *Suppose a and $\overline{b}$ belong to a column C in a C_n tableau T. Then*

$$\mathrm{pos}_C(a) + \mathrm{pos}_C(\overline{b}) \leq \max(a, b).$$

Lemma 2.2. *For any $(a, b|\overline{c}, \overline{d})$-configuration, define $p(a, b|\overline{c}, \overline{d}) := (j-i) + (l-k)$. Then we have*

$$p(a, b|\overline{c}, \overline{d}) < \max(b, c) - \min(a, d).$$

Remark. During the proofs of our results, we will also need to consider situations where there is a tableau containing an empty box. A careful analysis of the proofs of Lemmas 2.1 and 2.2 show that the upper bounds given there hold when the empty box lies between a and $\overline{b}$ (for Lemma 2.1) or between b and $\overline{c}$ or above a or below $\overline{d}$ (for Lemma 2.2). Moreover, if the empty box lies above a or below $\overline{b}$ (for Lemma 2.1) or between a and b or between $\overline{c}$ and $\overline{d}$ (for Lemma 2.2), then the upper bound increases by 1.

2.2 Crystal action on $B(\Lambda)$. Given a tableau T, define the (Japanese) reading of T, $w(T)$ to be the word obtained by reading the columns of T from top to bottom and right to left (such a reading is necessary to be compatible with the tensor product rule (1.1)). For a fixed i, to compute the action of the Kashiwara operators $\tilde{e}_i$, $\tilde{f}_i$ on a tableau T, perform the following steps (if $i = n$ ignore all expressions containing $i+1$ in what follows).

1. Write down $w(T)$ and ignore all letters different from $i, i+1, \overline{i}, \overline{i+1}$.
2. Ignore adjacent letters $(i, i+1)$, $(i, \overline{i})$, $(\overline{i+1}, i+1)$, $(\overline{i+1}, \overline{i})$.
3. Repeat step 2 until the result does not change.

The resulting word will be of the form $u_-^s u_+^t$ where u_- (resp., u_+) is one of $\{i+1, \overline{i}\}$ (resp., $\{i, \overline{i+1}\}$). If $s = 0$ (resp., $t = 0$), then $\tilde{e}_i\, T = 0$ (resp., $\tilde{f}_i\, T = 0$). Otherwise $\tilde{e}_i$ changes the rightmost u_- to the corresponding u_+, i.e., $i \mapsto i+1$ and $\overline{i+1} \mapsto \overline{i}$ (resp., $\tilde{f}_i$ changes the leftmost u_+ to the corresponding u_-).

Example 2.3.

$$T \quad = \quad \begin{array}{|c|c|c|c|}\hline 1 & 2 & 3 & \overline{5} \\ \hline 2 & 4 & 4 & \overline{4} \\ \hline 4 & 5 & \overline{5} \\ \cline{1-3} \overline{4} & \overline{3} & \overline{2} \\ \cline{1-3} \end{array} \qquad\qquad \tilde{f}_4\, T \ = \ \begin{array}{|c|c|c|c|}\hline 1 & 2 & 3 & \overline{5} \\ \hline 2 & 4 & 5 & \overline{4} \\ \hline 4 & 5 & \overline{5} \\ \cline{1-3} \overline{4} & \overline{3} & \overline{2} \\ \cline{1-3} \end{array}$$

$$w(T) \quad = \quad \overline{5}\,\overline{4}\,3\,\mathbf{4}\,\overline{5}\,\overline{2}\,2\,4\,5\,\overline{3}\,1\,2\,4\,\overline{4}$$

3 Bumping

3.1 Forward bumping. The (column) bumping process for type A_{n-1} tableaux is simply described as follows: to insert an element x into a column C, if x is strictly greater than all entries in C, place x at the bottom of the column C. Otherwise select the entry $y \in C$ which is the smallest entry $\geq x$. The entry y is replaced by x, and y is "bumped" into the next column to the right of C.

In the case of the C_n tableaux of Kashiwara and Nakashima, to insert an element α into a column C, proceed as in the type A case unless one of the following two situations arise:

I. if $\overline{y} \in C$ is the smallest letter $\geq \alpha$ and $y \in C$ as well;

II. if $\alpha = x$ and $\overline{x} \in C$.

It can be seen that such conditions are mutually exclusive. If such conditions arise, one must carry out a "special" bump, which we now describe.

Type I special bump

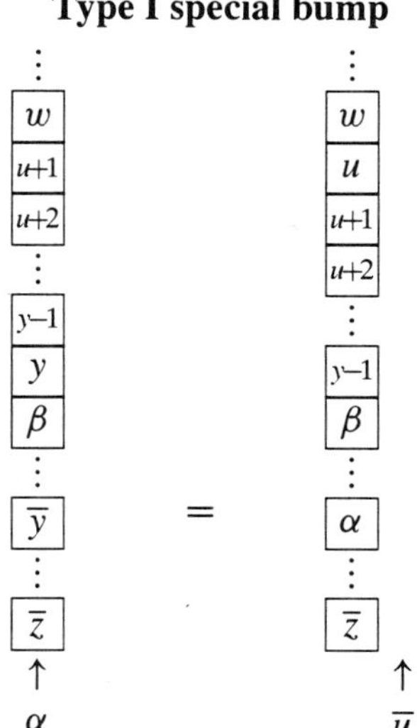

In the situation above, suppose there is a string (possibly empty) of consecutive integers $y-1, y-2, \ldots, u+2, u+1$ lying above the entry y in the column C, with the entry w immediately above $u+1$ being $\leq u-1$. Then the type I bump consists of replacing the entry $\overline{y}$ with α, deleting the entry y and shifting the entries $y-1, \ldots, u+1$ down one position, and inserting an entry u between w and $u+1$ in the column C. The entry $\overline{u}$ is then bumped into the next column to the right.

In the second case, suppose that there is a string (possibly empty) of consecutive entries $\overline{x+1}, \overline{x+2}, \ldots, \overline{v-2}, \overline{v-1}$ above the entry $\overline{x} \in C$. Let β be the next entry above $\overline{v-1}$. Then we have two subcases depending on whether (a) $v \leq \beta \leq \overline{v+1}$ or (b) $\beta \leq v-1$ or no such β exists.

Type IIa special bump

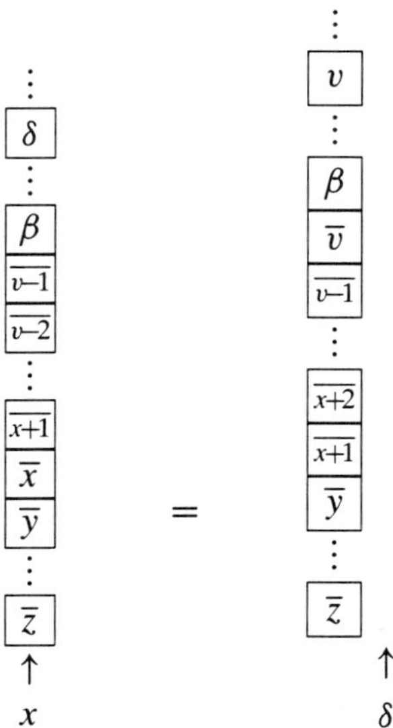

In the case above, suppose δ is the smallest entry in C which is $\geq v$ where $\overline{v}$ is the first letter absent from the string of consecutive letters above $\overline{x}$. Then the type IIa bump consists of deleting the entry $\overline{x}$, shifting the entries $\overline{x+1}, \ldots, \overline{v-1}$ down one position, inserting entry $\overline{v}$ between $\overline{v-1}$ and β, and replacing δ with v. The entry δ is then bumped into the next column.

Type IIb special bump

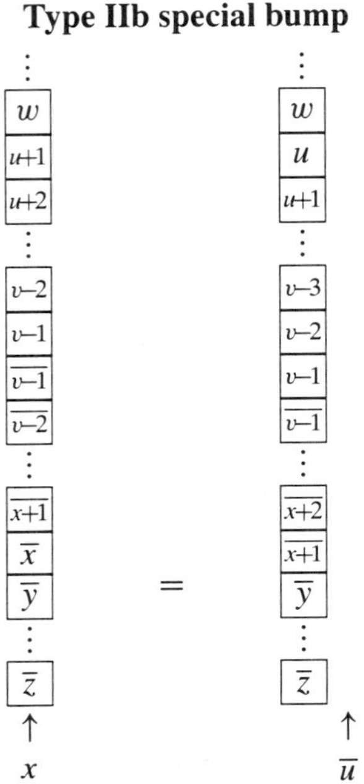

In the case above, suppose that there is a string of consecutive integers $v-1, v-2, \ldots, u+1$ above the entry $\overline{v-1}$ and let w be the entry immediately above $u+1$, which is $< u$. This only occurs when $\beta = v-1$ in case (b); if $\beta < v-1$, such a string is empty and $u := v-1$. The type IIb bump consists of deleting the entry $\overline{x}$, shifting the entries $\overline{x+1}, \ldots, \overline{v-1}, v-1, \ldots, u+1$ down one position, and inserting an entry u between $u+1$ and w. The entry $\overline{u}$ is then bumped into the next column.

The fact that the above bumping process is well defined (i.e., the result of such a bump is a valid C_n tableau) is not obvious. The proofs will be provided in Section 8. Moreover, although these special bumps appear foreboding, they actually have a very simple description in terms of elementary Knuth transformations, which we describe in Section 6.

3.2 Reverse bumping. The bumping procedures given in the last section can be reversed. To reverse insert an element ψ into a column C, proceed as in the type A case unless one of the following two situations arise:

I^R. if $v \in C$ is the largest letter $\leq \psi$ and $\overline{v} \in C$ as well;

II^R. if $\psi = \overline{u}$ and $u \in C$.

Again, it can be seen that such conditions are mutually exclusive. If such conditions arise, one must carry out a "special" reverse bump.

Type I^R reverse bump

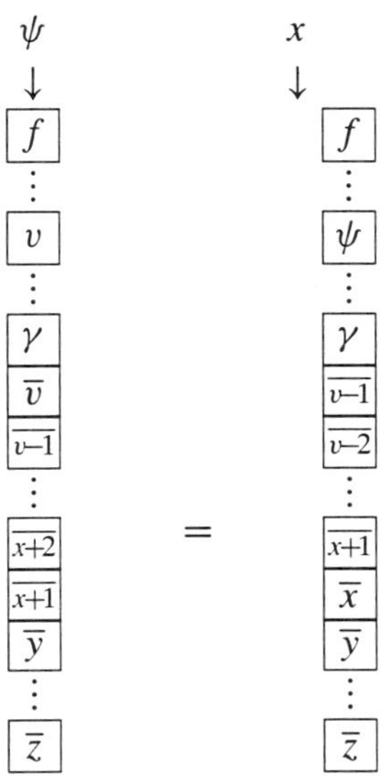

In the case above, suppose that there is a string of consecutive letters $\overline{v-1}$, $\overline{v-2}$, $\ldots, \overline{x+1}$ underneath $\overline{v}$ with the next letter $\overline{y} > \overline{x}$. The I^R reverse bump consists of replacing the entry v with ψ, deleting the entry $\overline{v}$, moving the string of entries $\overline{v-1}, \ldots, \overline{x+1}$ up one position and inserting $\overline{x}$ into the vacant cell. An entry x is then reverse bumped into the next column.

In case II^R, suppose there is a string of consecutive entries $u+1, u+2, \ldots, v-1$ below the entry $u \in C$. Let $\alpha > v$ be the next entry below $v-1$. Then there are two subcases, depending on whether (a) $v+1 \leq \alpha \leq \overline{v}$ or (b) $\alpha \geq \overline{v-1}$ or no such α exists.

Type II^Ra reverse bump

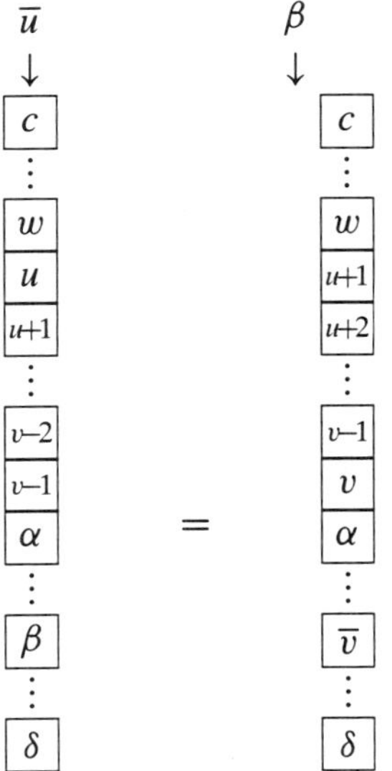

In the case above, β is the largest letter $\leq \overline{v}$ occuring in column C. The II^Ra reverse bump consists of deleting the entry u, shifting the consecutive letters $u+1, u+2, \ldots, v-1$ up one position and inserting the entry v into the vacant cell. The entry β is then replaced by $\overline{v}$ and β is reverse bumped into the next column (on the left).

Type II^{R}b reverse bump

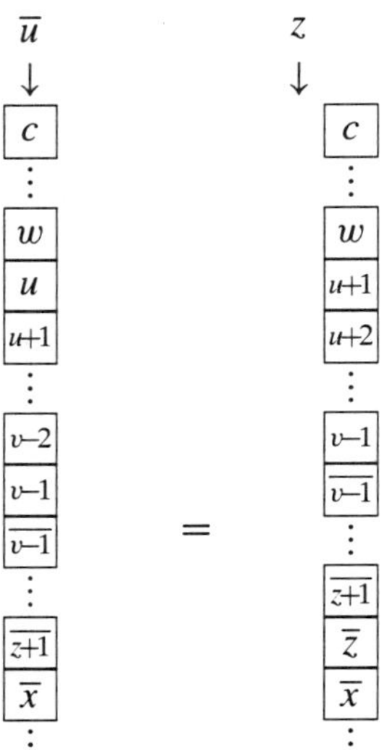

In the case above, suppose $\alpha = \overline{v-1}$ and there is a consecutive string $\overline{v-2}, \ldots, \overline{z+2}, \overline{z+1}$ below α (if $\alpha > \overline{v-1}$ then such a string is empty and $z := v-1$). The II^Rb reverse bump consists of deleting the entry u, moving the consecutive string of letters $u+1, \ldots, v-1, \overline{v-1}, \ldots, \overline{z+1}$ up one position, and inserting the entry $\overline{z}$ in the vacant cell. An entry z is then reverse bumped into the next column on the left.

From the definitions, it is clear that the reverse bumps I^R, II^Ra, II^Rb are the inverse processes of the forward bumps IIa, I, and IIb, respectively.

Example. We give an example of the bumping procedures described earlier. In this example, the sequence of bumps is IIa, IIb, and then I. By reading the sequence from right to left, one can examine the reverse bumps also.

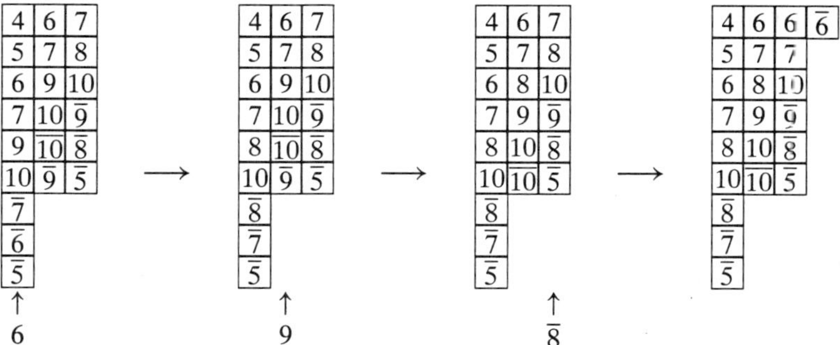

4 Sliding

4.1 Forward sliding. We define forward sliding for C_n tableaux by proceeding as in the (usual) A case unless one of the following four situations is encountered.

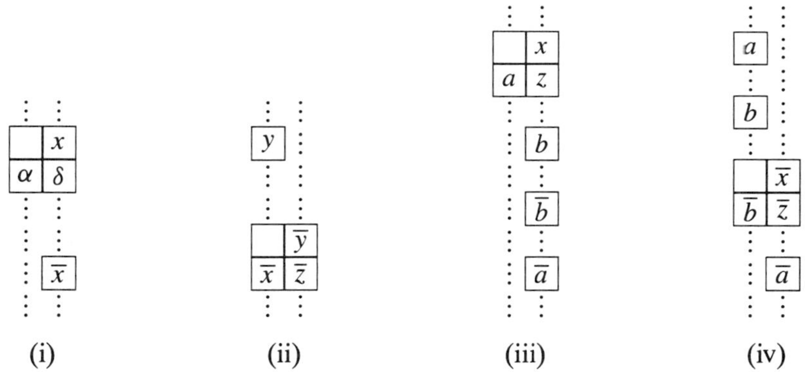

The relevant conditions on the entries are as follows.

Case (i). $x < \alpha$.

Case (ii). $\overline{y} < \overline{x}$.

Case (iii). $a \leq x$, $p(a, b|\overline{b}, \overline{a}) = b - a - 1$, and b is the smallest integer satisfying such a condition.

Case (iv). $\overline{b} \leq \overline{x}$, $p(a, b|\overline{b}, \overline{a}) = b - a - 1$, and a is the largest integer satisfying such a condition.

Note that all these cases are mutually exclusive. Thus in the A_n case, cases (i) and (ii) would correspond to horizontal slides while cases (ii) and (iv) would correspond to vertical slides. In the C_n case, it turns out that all these special slides correspond to horizontal slides (i.e., the empty box moves horizontally), but with some modifications, which we now describe.

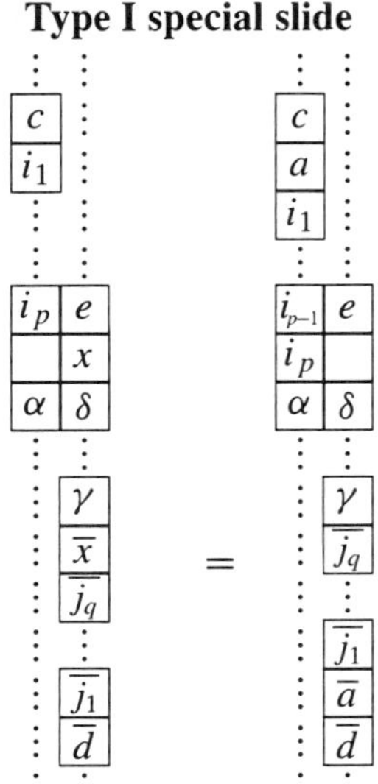

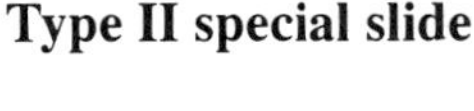

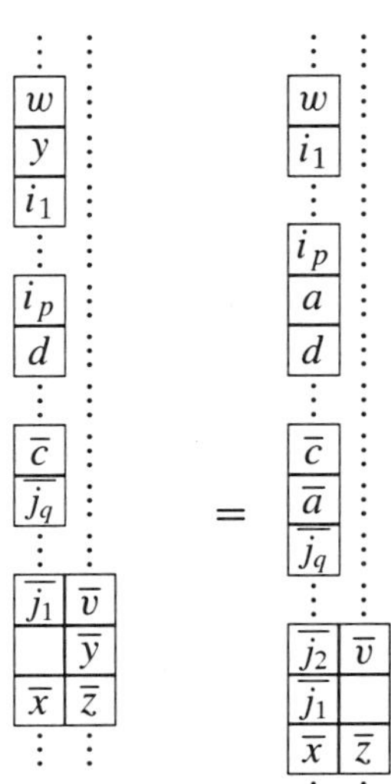

In the diagram for the type I slide, $c \leq a-1$, $\overline{d} \geq \overline{a-1}$, and $\{i_1, \ldots, i_p\} \sqcup \{j_1, \ldots, j_q\} = \{a+1, a+2, \ldots, x-1\}$. In fact, this union is disjoint as if it were not, it is not difficult to show that such a column would violate the 2CC. The letter a is thus the largest letter $< x$ which does not appear in the column above the empty box nor (as a barred entry) below the entry $\overline{x}$. The type I slide then consists of moving the entries $i_1, \ldots, i_p$ down one position and inserting the entry a in the vacant cell; deleting the entry x and replacing it with an empty box; deleting the entry $\overline{x}$, moving the entries $\overline{j_1}, \ldots, \overline{j_q}$ up one position, and inserting the entry $\overline{a}$ into the vacant cell.

In the diagram for the type II slide, $d \geq a+1$, $\overline{c} \leq \overline{a+1}$, and $\{i_1, \ldots, i_p\} \sqcup \{j_1, \ldots, j_q\} = \{y+1, y+2, \ldots, a-1\}$. Again, this union is disjoint. Thus a is the smallest letter $> y$ which is not below the entry y, nor (as a barred letter) above the empty box. The type II slide then consists of deleting the entry y, moving the entries $i_1, \ldots, i_p$ up one position, and inserting a into the vacant cell; moving the entries $\overline{j_1}, \ldots, \overline{j_q}$ down one position and inserting an $\overline{a}$ into the vacated cell; deleting $\overline{y}$ in the next column and replacing it with an empty box.

In the diagram for type III slides, $\{i_1, \ldots, i_p\} \sqcup \{j_1, \ldots, j_q\} = \{c+1, c+2, \ldots, a-1\}$. Thus c is the largest integer $< a$ which does not appear above the empty box nor below the entry $\overline{a}$. The sliding process is as follows: move the entries $i_1, \ldots, i_p$ down one position and insert c into the vacated cell; delete the entry b and move all entries beginning from the cell to the right of the empty box to the cell immediately above b down one position, leaving an empty box in the vacated cell; delete the entry $\overline{b}$, move all the entries from the cell below $\overline{b}$ to the entry $\overline{j_q}$ up one position and insert the entry $\overline{c}$ into the vacated cell.

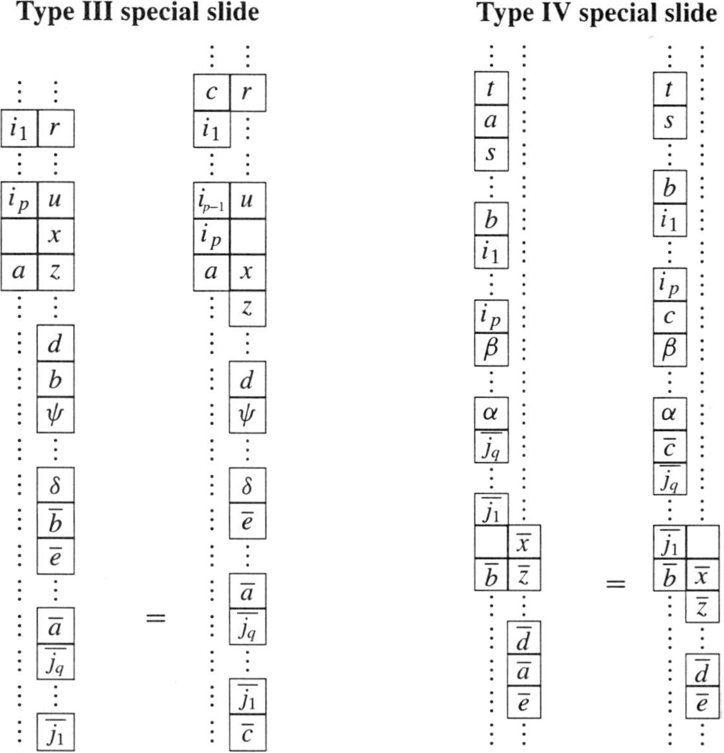

In the diagram for type IV slides, $\{i_1, \ldots, i_p\} \sqcup \{j_1, \ldots, j_q\} = \{b+1, b+2, \ldots, c-1\}$. Hence c here is the smallest integer $> b$ which does not appear below b nor above the empty box. The sliding process is described as follows: delete the entry a, move the entries beginning from the entry below a to the entry i_p up one position and insert an entry c into the vacated cell; move the entries $\overline{j_1}, \ldots, \overline{j_q}$ down one position and insert the entry $\overline{c}$ into the vacated call; delete the entry $\overline{a}$, move the entries beginning from the cell to the right of the empty box to the cell immediately above $\overline{a}$ down one position and leave an empty box in the vacated cell.

4.2 Reverse sliding. Just as for the bumping procedure, the sliding procedure can be reversed. A type C_n reverse slide consists of an ordinary (type A_n) reverse slide unless one of the following four situations is encountered:

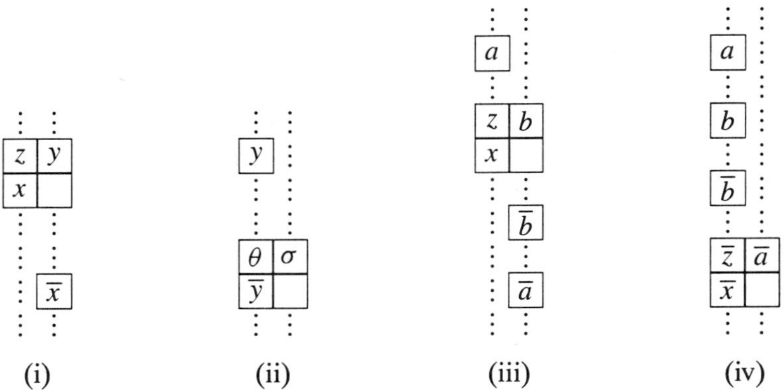

The relevant conditions on the entries are as follows.

Case (i). $x > y$.

Case (ii). $\overline{y} > \sigma$.

Case (iii). $x \leq b$, $p(a, b|\overline{b}, \overline{a}) = b - a - 1$ and a is the largest integer satisfying such a condition.

Case (iv). $\overline{x} \leq \overline{b}$, $p(a, b|\overline{b}, \overline{a}) = b - a - 1$ and b is the smallest integer satisfying such a condition.

Again, note that all these cases are mutually exclusive and that in the A_n case, cases (i) and (ii) would correspond to horizontal slides while cases (iii) and (iv) would correspond to vertical slides. In the C_n case, all these special reverse slides correspond to horizontal reverse slides with some modifications.

In the left of Figure 4.1, $\{i_1, \ldots, i_p\} \sqcup \{j_1, \ldots, j_q\} = \{x+1, x+2, \ldots, a-1\}$, $\gamma \geq a+1$, $\delta \leq \overline{a-1}$. The I^R reverse slide thus proceeds as follows: move the entries $i_1, \ldots, i_p$ up one position into the empty cell, and insert a into the cell vacated; delete the entry $\overline{x}$ and move the entries $\overline{j_1}, \ldots, \overline{j_q}$ down one position, inserting $\overline{a}$ into the cell vacated; replace the entry x with an empty box.

Similarly for the II^R reverse slide, $\{i_1, \ldots, i_p\} \sqcup \{j_1, \ldots, j_q\} = \{a+1, a+2, \ldots, y-1\}$, $c \leq a-1$, $\overline{d} \geq \overline{a-1}$ and the slide proceeds as follows: delete the entry y, move the entries $i_1, \ldots, i_p$ down one position and place a into the vacated cell; delete the entry $\overline{y}$ and replace by an empty box; move the entries $\overline{j_1}, \ldots, \overline{j_q}$ up one position and insert $\overline{a}$ into the vacated cell.

In Figure 4.2, for the type III^R slide, $\{i_1, \ldots, i_p\} \sqcup \{j_1, \ldots, j_q\} = \{b+1, b+2, \ldots, c-1\}$, $\alpha \geq c+1$ and $\beta \leq \overline{c+1}$. The procedure for the III^R slide is: delete the entry $\overline{a}$ and move all entries from the cell immediately above $\overline{a}$ to $\overline{j_q}$ down one position, inserting the entry $\overline{c}$ into the vacated cell; move the entries $i_1, \ldots, i_p$ up one position, inserting the entry c into the vacated cell; delete the entry a and move all the entries from the cell immediately below a to the cell immediately to the left of the empty box up one position, leaving the vacated cell as the new empty box.

Similarly, in the diagram for the type IV^R slide, $\{i_1, \ldots, i_p\} \sqcup \{j_1, \ldots, j_q\} = \{c+1, c+2, \ldots, a-1\}$, $f \leq c-1$ and $\overline{g} \geq \overline{c-1}$. The procedure for the IV^R slide is: move the entries $\overline{j_1}, \ldots, \overline{j_q}$ up one position, placing the entry $\overline{c}$ into the vacated

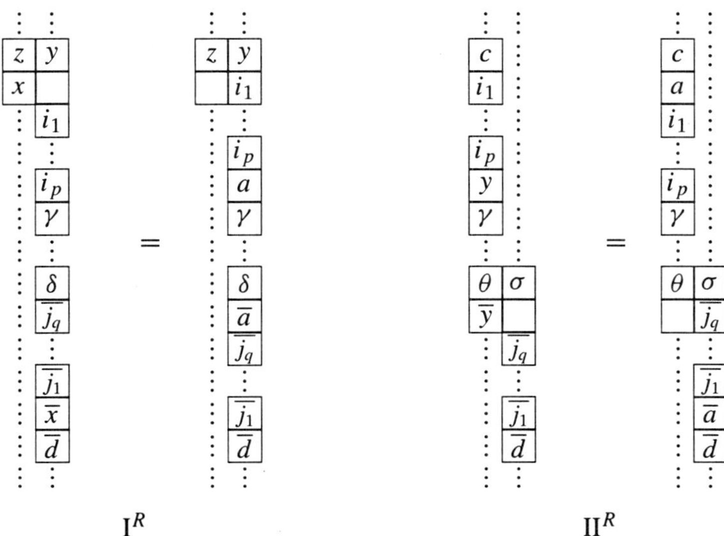

Figure 4.1. Type I^R and II^R reverse slides.

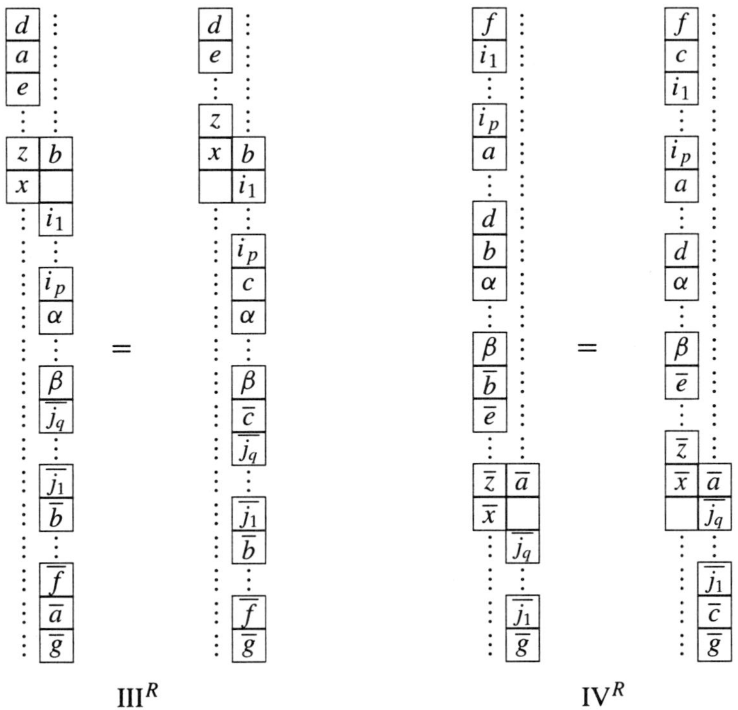

Figure 4.2. Type III^R and IV^R reverse slides.

cell; delete the entry $\overline{b}$ and move all the entries from the cell immediately to the left of the empty box to the cell immediately below b up one position, leaving the vacated cell as the new empty box; delete the entry b and move all the entries from i_1 to the cell immediately above b down one position, inserting the entry c into the vacated cell.

Example. In this example, we give a sequence of forward slides (omitting the vertical slides).

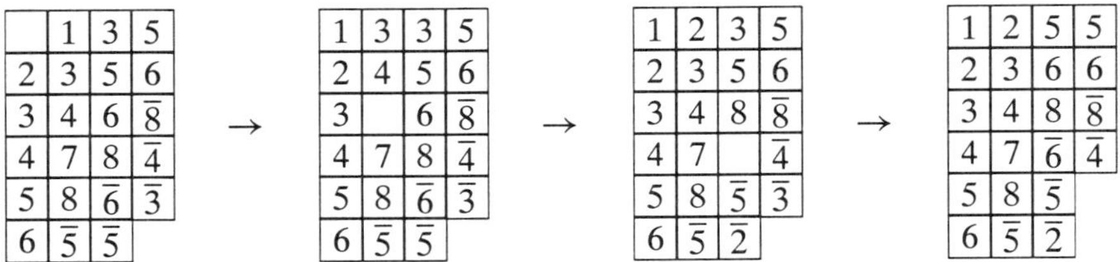

5 The insertion scheme

Having defined bumping and sliding procedures for C_n tableaux, we can now combine them into a procedure for inserting a letter α into a tableau b, the result being a new C_n tableau whose size is one box greater or less than b.

Before doing so, it is instructive to look at the corresponding insertion scheme of Berele [2] when inserting letters into the symplectic tableaux of King [9]. The symplectic tableaux in that case are tableaux on the letters $\{1, 2, \ldots, n\} \cup \{\overline{1}, \overline{2}, \ldots, \overline{n}\}$ with the order $1 < \overline{1} < 2 < \overline{2} < \cdots < n < \overline{n}$. The conditions on the tableaux are threefold: the rows are weakly increasing with respect to the aforementioned order; the columns are strictly increasing; in row i, all the entries are $\geq i$. The Berele insertion consists of *row* bumping, and the bumping procedure is exactly as it is for ordinary (type A) tableaux. However, it may happen that an entry $\overline{i}$ in row i is bumped (by an i) into row $i + 1$. This would cause a violation of the symplectic row condition. Berele's solution was that, instead of the $\overline{i}$ being bumped into the next row, it gets annihilated by the i, thus creating an empty box which is then slid out of the tableau using the usual sliding procedure. Let us call the stage in Berele's insertion scheme where bumping becomes sliding, the *bumping–sliding transition*.

Our insertion scheme also has a stage where a bump would cause a violation of the tableau conditions of Kashiwara and Nakashima, and hence we also have an annihilation process and subsequent sliding. This occurs when an element $\overline{x}$ is inserted into a column and is greater than all the elements in the column, and hence placed at the end of the column. It may happen that a 1CC violation occurs: that is, before the insertion, there were pairs a–$\overline{a}$, b–$\overline{b}$, etc. such that $\text{pos}(a) + \text{pos}(\overline{a}) = a$, etc. and the addition of the extra box at the end of the column has increased $\text{pos}(\overline{a})$ etc by one. Our procedure is as follows: locate the smallest a such that $\text{pos}(a)+\text{pos}(\overline{a}) = a+1$, and replace the cells containing both a and $\overline{a}$ by empty boxes. Now slide these boxes out using the sliding rules described in Section 4. We then have the following result.

Proposition 5.1.

a. *The bumping–sliding transition can only occur in the first column.*

b. *The lower box always slides out (with ordinary slides) vertically.*

Before proceeding to the proof, let us first prove a technical result.

Claim. *Assume the pair a–$\overline{a}$ in the diagram below satisfies* $\mathrm{pos}(a) + \mathrm{pos}(\overline{a}) = a$, *and let v_1 be the bottommost element in the column containing such a pair. Then* $\{1, 2, \ldots, v_1 - 1\} \subseteq \{w_1, w_2, \ldots, w_p\}$.

PROOF. Suppose not and suppose $x < v_1$ is the smallest letter not in $\{w_1, \ldots, w_p\}$. That is, $w_i = i$ for $i = 1, \ldots, x-1$ and $w_x \geq x+1$. We consider two cases depending on whether $w_x \leq v_1$ or $w_x > v_1$.

In the former case, Lemma 2.2 implies $p(w_x, a|\overline{a}, \overline{v_1}) \leq a - w_x - 1 \leq a - x - 2$. Now, by the definition of a, $p + q + 2 = a$, and thus from the diagram below, we have $p(w_x, a|\overline{a}, \overline{v_1}) = q + p + 1 - x = a - x - 1$, which is a contradiction.

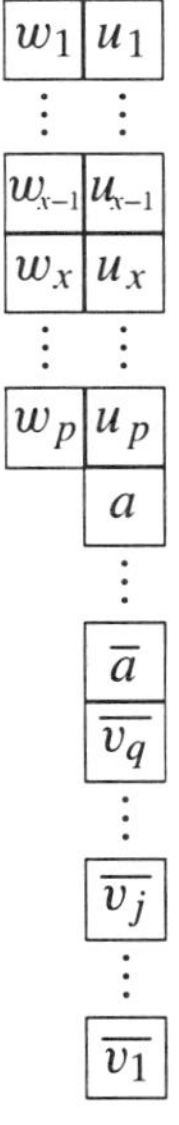

In the latter case, there exists a j such that $v_{j-1} < w_x \leq v_j$ (we include the case where $j = q+1$, i.e., $v_q < w_x \leq v_{q+1} := a$). From the diagram we see that $p(w_x, a|\overline{a}, \overline{v_j}) = (p + 1 - x) + (q + 1 - j) = a - x - j$. Also, note that the strictly increasing column condition implies $v_{j-1} \geq v_1 + j - 2 > x + j - 2$, so that $v_{j-1} \geq x + j - 1$. Now the inequality $v_{j-1} < w_x \leq v_j$ and Lemma 2.2 implies $p(w_x, a|\overline{a}, \overline{v_j}) \leq a - 1 - \min(w_x, v_j) = a - 1 - w_x < a - 1 - v_{j-1} \leq a - 1 - (x + j - 1) = a - x - j$, which is a contradiction. □

PROOF OF PROPOSITION 5.1A. Suppose $\overline{z}$ gets appended below $\overline{v_1}$ at some stage in the bumping procedure. That is $\overline{z} > \overline{v_1} \Leftrightarrow z < v_1$. How did $\overline{z}$ arise from the previous column? Certainly not from an ordinary bump, as the above claim implies that z is in the previous column. Thus, $\overline{z}$ must have arisen as a result of a type I or IIb bump. But before such a special bump, z must have been absent from the column, again in contradiction with the above claim. □

PROOF OF PROPOSITION 5.1B. We first show that $\overline{v_j} \leq \overline{c_{j+1}}$ for all j such that $\overline{c_{j+1}}$ exists (we include the possibility that $\overline{c_1}$ exists, so let us define $\overline{v_0} := \overline{x}$ for ease of notation). Suppose for a contradiction, that $\overline{v_j} > \overline{c_{j+1}} \Leftrightarrow v_j < c_{j+1}$. Now, the fact that the a–$\overline{a}$ pair satisfies $\mathrm{pos}(a) + \mathrm{pos}(\overline{a}) = a + 1$ and a is minimal means that the entries above a and below $\overline{a}$ have a disjoint union equal to the set $\{1, 2, \ldots, a-1\}$. Suppose w is the smallest integer $> v_j$ appearing above a. Then the above remark implies that there is a consecutive string $\overline{v_j}, \overline{v_j+1}, \ldots, \overline{w-1}$ appearing below $\overline{a}$. Below, this is depicted in the diagram on the left, with entries $\overline{c_j}, \ldots, \overline{c_{j+r}}$ to their immediate right, where $r := w - 1 - v_j$.

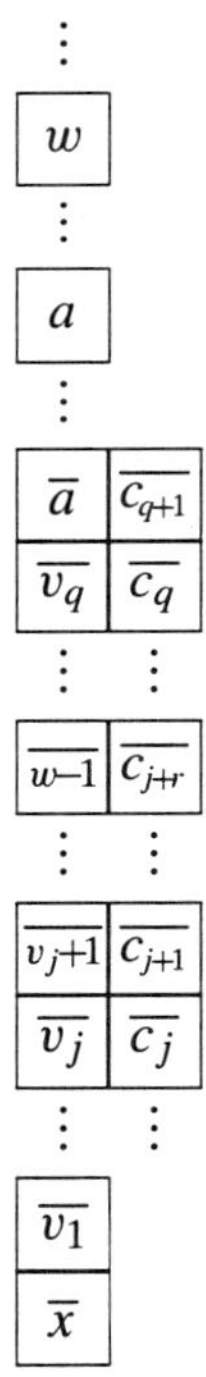

Note that $c_{j+r} \geq c_{j+1} + r - 1 \geq v_j + r = w - 1$. Certainly the entries between w and a, and between $\overline{w-1}$ and a form a disjoint union of $\{w+1, w+2, \ldots, a-1\}$. Thus

$$p(w, a|\overline{a}, \overline{c_{j+r}}) = \#\{w-1, w, \ldots, a-1\} = a - w + 1$$

But Lemma 2.2 implies $p(w, a|\overline{a}, \overline{c_{j+r}}) \leq a - 1 - \min(w, c_{j+r}) = a - w$ which is a contradiction.

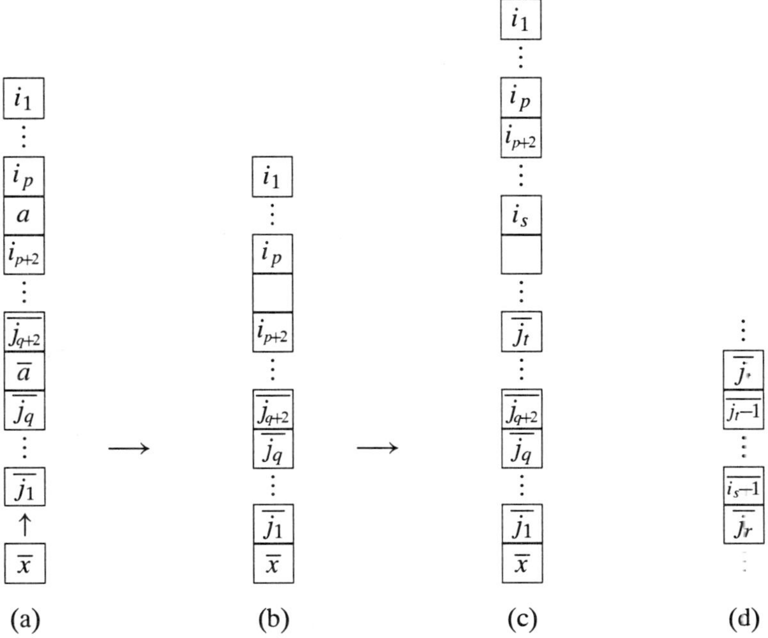

Figure 5.1. Reversibility of bumping–sliding transition.

Thus when the $\overline{a}$ is replaced by an empty box and such a box is slid out, the slides are ordinary vertical slides (and in particular never a type IV slide). □

Finally, we show that the bumping–sliding transition can be reversed. That is, after a sequence of reverse slides produces an empty box in the first column, we can recreate the elements a–$\overline{a}$ and their positions in the first column from whence they vanished.

In Figure 5.1(a), suppose there is a pair $i_{p+1} = a, \overline{j_{q+1}} = \overline{a}$ such that $p+q+2 = a$, and we insert an $\overline{x}$ such that $\overline{x} > \overline{j_1}$. After we replace the a–$\overline{a}$ pair with a pair of empty boxes, we slide out the bottom box vertically, in accordance with Proposition 5.1 to produce Figure 5.1(b). In this diagram, $\mathrm{pos}(i_p) = p$ $\mathrm{pos}(\overline{j_q}) = q + 1$ and i_p (resp., j_q) is the largest (resp., smallest) letter $< a$ (resp., $> \overline{a}$) appearing in the first column.

The condition for the reverse transition is now clear: suppose after a sequence of reverse slides, the empty box is in the first column and let i be the entry in the cell immediately above it. Let a be the smallest integer $> i$ which does not appear in the barred section of the column and let $\overline{j}$ be the smallest integer $> \overline{a}$ such that $\overline{j}$ *does* appear in the barred section of the column. Then if $\mathrm{pos}(i) + \mathrm{pos}(\overline{j}) = a - 1$, insert a in the empty box, shift the entries below (and including) $\overline{j}$ down one position, inserting an $\overline{a}$ in the vacated cell, and finally bump the lowest cell in the column out of the tableau.

We must check that the above conditions specify a and the positions of a and $\overline{a}$ in the first column *uniquely*. Suppose they did not, i.e., suppose we continue *forward* sliding vertically down until we reach the situation in Figure 5.1(c) where the above conditions hold (note that we are implicitly using the fact that all vertical slides are ordinary slides, so the entries in the first column do not change). Thus, suppose b is the smallest integer $> i_s$ which does not appear among the barred entries, and suppose $\overline{j_t}$ is the smallest integer $> \overline{b}$ such that $\overline{j_t}$ *does* appear and that

$$\mathrm{pos}_A(i_s) + \mathrm{pos}_A(\overline{j_t}) = b - 1.$$

Here, A denotes the column as it appears *after* the bumping–sliding transition. Now, from the definition of i_s and $\overline{j_t}$, there must be a consecutive string below the entry $\overline{j_t}$ ending at $\overline{i_s + 1}$ so that the next entry $\overline{j_r}$ is such that $\overline{j_r} \geq \overline{i_s} \Leftrightarrow j_r \leq i_s$ (see Figure 5.1(d)). Let B denote the column as it appears *before* the bump (i.e., in Figure 5.1(a)). Then

$$\mathrm{pos}_B(i_s) = \mathrm{pos}_A(i_s) + 1, \quad \mathrm{pos}_B(\overline{j_r}) = \mathrm{pos}_A(\overline{j_r}), \quad \mathrm{pos}_B(i_s) + \mathrm{pos}_B(\overline{j_r}) \leq i_s, \tag{5.1}$$

the last inequality following from Lemma 2.1. Since $\mathrm{pos}_A(\overline{j_t}) = \mathrm{pos}_A(\overline{j_r}) + j_t - i_s$ we have from (5.1)

$$\begin{aligned} \mathrm{pos}_A(i_s) + \mathrm{pos}_A(\overline{j_t}) &= \mathrm{pos}_A(i_s) + \mathrm{pos}_A(\overline{j_r}) + j_t - i_s \\ &\leq (j_t - i_s) + i_s - 1 = j_t - 1 < b - 1, \end{aligned}$$

which is a contradiction. □

Thus inserting a letter α into a tableau b results in a new C_n tableau $b \leftarrow \alpha$ whose shape is one box larger or smaller than the shape of b. Thus we can define a correspondence $w \leftrightarrow (P(w), Q(w))$ where $P(w) := (\cdots((\emptyset \leftarrow w_1) \leftarrow w_2) \cdots \leftarrow w_p)$ and $Q(w)$ is an *oscillating* tableau, i.e., a sequence of Young diagrams whose shape differs by one [29].

Example. With $w = 3\overline{3}2\overline{2}2\overline{2}123\overline{1}$, we have the following sequence of insertions:

$$\emptyset \to \begin{array}{|c|}\hline 3 \\ \hline\end{array} \to \begin{array}{|c|}\hline 3 \\ \hline \overline{3} \\ \hline\end{array} \to \begin{array}{|c|c|}\hline 2 & 3 \\ \hline \overline{3} \\ \cline{1-1}\end{array} \to \begin{array}{|c|c|}\hline 2 & 3 \\ \hline \overline{3} \\ \cline{1-1} \overline{2} \\ \cline{1-1}\end{array} \to \begin{array}{|c|c|}\hline 2 & 3 \\ \hline 3 & \overline{3} \\ \hline \overline{3} \\ \cline{1-1}\end{array} \to \begin{array}{|c|c|}\hline 2 & 3 \\ \hline \overline{3} \\ \cline{1-1} \overline{2} \\ \cline{1-1}\end{array} \to$$

$$\to \begin{array}{|c|c|c|}\hline 1 & 2 & 3 \\ \hline \overline{3} \\ \cline{1-1} \overline{2} \\ \cline{1-1}\end{array} \to \begin{array}{|c|c|c|}\hline 1 & 2 & 3 \\ \hline 3 & \overline{3} \\ \cline{1-2} \overline{3} \\ \cline{1-1}\end{array} \to \begin{array}{|c|c|c|}\hline 1 & 2 & 3 \\ \hline 2 & \overline{3} \\ \cline{1-2} 3 & \overline{2} \\ \cline{1-2}\end{array} \to \begin{array}{|c|c|}\hline 1 & 3 \\ \hline 2 & \overline{3} \\ \hline \overline{3} & \overline{1} \\ \hline\end{array}$$

Thus we can associate the word $w = (3\bar{3}2\bar{2}2\bar{2}123\bar{1})$ with the pair

$$\left(\begin{array}{|c|c|} \hline 1 & 3 \\ \hline 2 & \bar{3} \\ \hline \bar{3} & \bar{1} \\ \hline \end{array} \;\; ; \;\; \ydiagram{1} \; \ydiagram{1,1} \; \ydiagram{2,1} \; \ydiagram{2,1,1} \; \ydiagram{2,2,1} \; \ydiagram{2,1,1} \; \ydiagram{3,1,1} \; \ydiagram{3,2,1} \; \ydiagram{3,2,2} \; \ydiagram{2,2,2} \right)$$

Note that both w and $P(w)$ have symplectic weight (011).

6 Knuth transformations

6.1 Bumping and Knuth transformations. The bumps and slides we defined in the previous two sections may appear complicated, but their appearance belies the fact that they have a very simple description in terms of sequences of elementary Knuth-like transformations. Indeed, the ordinary column bumps that we use are generated by a sequence of the following Knuth transformations:

$$\gamma\,\beta\,\alpha \to \beta\,\gamma\,\alpha, \qquad \gamma < \alpha \le \beta, \qquad (\beta, \gamma) \neq (\overline{x}, x), \tag{K1}$$

$$\alpha\,\beta\,\gamma \to \alpha\,\gamma\,\beta, \qquad \gamma \le \alpha < \beta, \qquad (\beta, \gamma) \neq (\overline{x}, x). \tag{K2}$$

These are the usual elementary Knuth relations, but with an extra condition on the pair of letters which swap positions. Just as in the type A case, an ordinary bump can be realized as a sequence of K2 transformations followed by a sequence of K1 transformations on the word of the column [12, 5]. Symbolically, let us write

$$\text{ordinary bump} \;\leftrightarrow\; \text{K1}^r\ \text{K2}^s,$$

where the transformations K1 and K2 act on the right.

Let us now show that the special bumps I, IIa and IIb can also be realized as a sequence of elementary transformations involving not only K1 and K2, but also

$$y+1\;\overline{y+1}\;\beta \to \overline{y}\;y\;\beta, \qquad y < \beta < \overline{y}, \qquad y < n, \tag{K3}$$

$$\alpha\;\overline{x}\;x \to \alpha\;x+1\;\overline{x+1}, \qquad x < \alpha < \overline{x}, \qquad x < n. \tag{K4}$$

Let us take, for example, the type I bump whereby we insert α into a column C which has a pair y–$\overline{y}$ with $\overline{y}$ the smallest letter $\ge \alpha$.

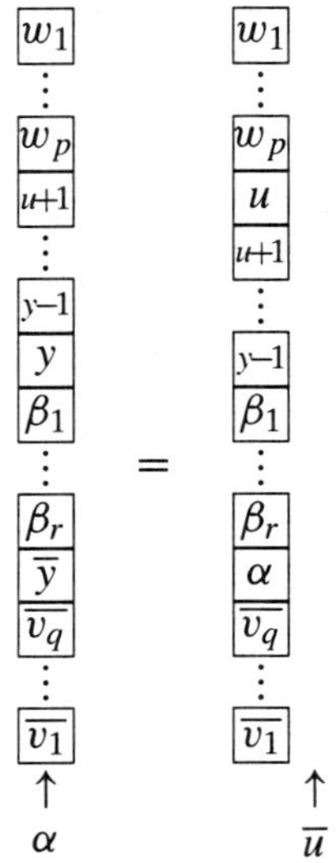

$$
\begin{array}{c}
\boxed{w_1}\\ \vdots \\ \boxed{w_p}\\ \boxed{u{+}1}\\ \vdots \\ \boxed{y{-}1}\\ \boxed{y}\\ \boxed{\beta_1}\\ \vdots \\ \boxed{\beta_r}\\ \boxed{\overline{y}}\\ \boxed{\overline{v_q}}\\ \vdots \\ \boxed{\overline{v_1}}\\ \uparrow \\ \alpha
\end{array}
\quad = \quad
\begin{array}{c}
\boxed{w_1}\\ \vdots \\ \boxed{w_p}\\ \boxed{u}\\ \boxed{u{+}1}\\ \vdots \\ \boxed{y{-}1}\\ \boxed{\beta_1}\\ \vdots \\ \boxed{\beta_r}\\ \boxed{\alpha}\\ \boxed{\overline{v_q}}\\ \vdots \\ \boxed{\overline{v_1}}
\end{array}
\;\;
\begin{array}{c}
\uparrow \\ \overline{u}
\end{array}
$$

The sequence of transformations on the column reading is as follows:

$$
\begin{aligned}
& w_1 \cdots w_p\, u{+}1 \cdots y{-}1\; y\, \beta_1 \cdots \beta_r\, \overline{y}\, \overline{v_q} \cdots \overline{v_1}\, \alpha \\
\overset{K2^q}{\sim}\quad & w_1 \cdots w_p\, u{+}1 \cdots y{-}1\; y\, \beta_1 \cdots \beta_r\, \overline{y}\, \alpha\, \overline{v_q} \cdots \overline{v_1} \\
\overset{K1^r}{\sim}\quad & w_1 \cdots w_p\, u{+}1 \cdots y{-}1\; y\, \overline{y}\, \beta_1 \cdots \beta_r\, \alpha\, \overline{v_q} \cdots \overline{v_1} \\
\overset{K3^{y-u}}{\sim}\quad & w_1 \cdots w_p\, \overline{u}\, u\, u{+}1 \cdots y{-}1\, \beta_1 \cdots \beta_r\, \alpha\, \overline{v_q} \cdots \overline{v_1} \\
\overset{K1^p}{\sim}\quad & \overline{u}\, w_1 \cdots w_p\, u\, u{+}1 \cdots y{-}1\, \beta_1 \cdots \beta_r\, \alpha\, \overline{v_q} \cdots \overline{v_1}.
\end{aligned}
$$

Thus, symbolically, we have

$$\text{type I bump} \;\leftrightarrow\; \mathrm{K1}^d\, \mathrm{K3}^c\, \mathrm{K1}^b\, \mathrm{K2}^a.$$

A similar analysis reveals that the IIa and IIb bumps can also be analyzed in terms of sequences of elementary Knuth relations, so that

$$
\begin{aligned}
\text{type IIa bump} &\;\leftrightarrow\; \mathrm{K1}^d\, \mathrm{K2}^c\, \mathrm{K4}^b\, \mathrm{K2}^a, \\
\text{type IIb bump} &\;\leftrightarrow\; \mathrm{K1}^d\, \mathrm{K3}^c\, \mathrm{K4}^b\, \mathrm{K2}^a.
\end{aligned}
$$

There is one final Knuth transformation which governs the transformation of the column reading of the tableaux in the insertion scheme, namely when there is a bumping–sliding transition in the first column. This Knuth transformation reads

$$w_1\, w_2 \cdots w_p\, a\, \beta_1 \cdots \beta_r\, \overline{a}\, \overline{v_q} \cdots \overline{v_1} \;\rightarrow\; w_1\, w_2 \cdots w_p\, \beta_1 \cdots \beta_r\, \overline{v_q} \cdots \overline{v_1}, \qquad \text{(K5)}$$

where $p+q = a-1$, $w_1 < \cdots < w_p < a < \beta_1 < \cdots < \beta_r < \overline{a} < \overline{v_q} < \cdots < \overline{v_1}$ and $\{w_1, \ldots, w_p\} \cap \{v_1, \ldots, v_q\} = \emptyset$.

We note here that the C_n Knuth relations (K1)–(K5) were first derived by Lascoux, Leclerc, and Thibon [14] (see also [17]).

6.2 Sliding and Knuth transformations. In this section we aim to show that the sliding procedures introduced in Section 4 preserve the Knuth relations (K1)–(K5) (the fact that the ordinary slides preserve K1 and K2 is well known). To begin with, let us first consider the type I slide. To show that they are related by a sequence of Knuth transformations, it is sufficient to take the readings of the two columns that change under the slide, and insert them into the empty tableau: if they produce the same (two-column) tableau, then we are done.

Note that such a process is not trivial, even in the case of inserting the column reading of a two-column tableaux into the empty tableau. In particular given a two-column tableau with columns C_1 and C_2, it may happen that there are pairs a–$\overline{a}$ where $a \in C_1$ and $\overline{a} \in C_2$. Let us denote the set of such elements as $C_1^+ \cap C_2^-$. In this case for example, after having inserted the letters of C_2 into the empty tableau, there will exist an $\overline{a}$ in the first column, and so when we come to insert the letter a, we will require a type IIa or IIb bump. In general we can group the barred elements of C_2 into disjoint groups of consecutive strings of letters containing elements of $C_1^+ \cap C_2^-$. Clearly any such elements must lie consecutively in C_1. As an example, let us consider the case where there are two elements in $C_1^+ \cap C_2^-$ so that the column reading of such a tableau reads thus

$$w = \begin{pmatrix} c_1 & \cdots & c_p & c_{p+1} & c_{p+2} & c_{p+3} & \cdots & c_t & \overline{d_1} & \cdots & \overline{d_v} & \overline{u-1} & \cdots & \overline{f} & \cdots & \overline{e} \\ a_1 & \cdots & a_p & e & f & \alpha_1 & \cdots & \cdots & \cdots & \cdots & \alpha_q & \beta_{u-1} & \cdots & \beta_f & \cdots & \beta_e \end{pmatrix}.$$

Here for convenience, we write the word of the tableau in such a way as to reflect the relative positions of the elements in the tableau. We assume for simplicity that $C_1^+ \cap C_2^- = \{e, f\}$ (and hence $\overline{u-1} \cdots \overline{f} \cdots \overline{e}$ are consecutive integers) and that $\overline{e}$ lies at the bottom of the column. Note that Lemma 2.2 applied to the $(e, c_{p+1} | \overline{u-1}, \overline{e})$ configuration implies that that $c_{p+1} \geq u$ (and so $c_{p+j} \geq u+j-1$ for $1 \leq j \leq t-p$). Similar considerations show that $\beta_j \leq \overline{j+1}$ for $u-1 \leq j \leq e$.

With these relations in mind the sequence of insertions is as follows (inserting the letters of C_2 as well as $a_1, \ldots, a_p$ is clear and hence omitted):

$$\begin{pmatrix} c_1 & \cdots & c_p & & & & & & & & & & & & \\ a_1 & \cdots & a_p & c_{p+1} & c_{p+2} & c_{p+3} & \cdots & c_t & \overline{d_1} & \cdots & \overline{d_v} & \overline{u-1} & \cdots & \overline{f} & \cdots & \overline{e} \end{pmatrix}$$

$$\xrightarrow[\text{IIa}]{e} \begin{pmatrix} c_1 & \cdots & c_p & c_{p+1} & & & & & & & & & & & \\ a_1 & \cdots & a_p & u & c_{p+2} & c_{p+3} & \cdots & c_t & \overline{d_1} & \cdots & \overline{d_v} & \overline{u} & \cdots & \overline{f} & \cdots & \overline{e+1} \end{pmatrix}$$

$$\xrightarrow[\text{IIa}]{f} \begin{pmatrix} c_1 & \cdots & c_p & c_{p+1} & c_{p+2} & & & & & & & & & & & \\ a_1 & \cdots & a_p & u & u+1 & c_{p+3} & \cdots & c_t & \overline{d_1} & \cdots & \overline{d_v} & \overline{u+1} & \cdots & \overline{f+1} & \overline{f-1} & \cdots & \overline{e+1} \end{pmatrix}$$

$$\xrightarrow{\{\alpha_1, \ldots, \alpha_q\}} \begin{pmatrix} c_1 & \cdots & c_p & c_{p+1} & c_{p+2} & c_{p+3} & \cdots & c_t & \overline{d_1} & \cdots & \overline{d_v} & & & & & \\ a_1 & \cdots & a_p & u & u+1 & \alpha_1 & \cdots & \cdots & \cdots & \cdots & \alpha_q & \overline{u+1} & \cdots & \overline{f+1} & \overline{f-1} \cdots & \overline{e+1} \end{pmatrix}$$

$$\xrightarrow[\text{I}]{\beta_{u-1}} \begin{pmatrix} c_1 & \cdots & c_p & c_{p+1} & c_{p+2} & c_{p+3} & \cdots & c_t & \overline{d_1} & \cdots & \overline{d_v} & \overline{u-1} & & & & \\ a_1 & \cdots & a_p & u-1 & u & \alpha_1 & \cdots & \cdots & \cdots & \cdots & \alpha_q & \beta_{u-1} & \cdots & \overline{f+1} & \overline{f-1} \cdots & \overline{e+1} \end{pmatrix}$$

$$\vdots$$

$$
\begin{aligned}
&\xrightarrow[\mathrm{I}]{\beta_{f-1}} \begin{pmatrix} c_1 & \cdots & c_p & c_{p+1} & c_{p+2} & c_{p+3} & \cdots & c_t & \overline{d_1} & \cdots & \overline{d_v} & \overline{u-1} & \cdots & \overline{f-1} & & \\ a_1 & \cdots & a_p & f-1 & f & \alpha_1 & \cdots & \cdots & \cdots & \cdots & \alpha_q & \beta_{u-1} & \cdots & \beta_{f-1} & \overline{f-1}\cdots & \overline{e+1} \end{pmatrix} \\
&\xrightarrow[\mathrm{I}]{\beta_{f-2}} \begin{pmatrix} c_1 & \cdots & c_p & c_{p+1} & c_{p+2} & c_{p+3} & \cdots & c_t & \overline{d_1} & \cdots & \overline{d_v} & \overline{u-1} & \cdots & \overline{f-1} & \overline{f-2} & \\ a_1 & \cdots & a_p & f-2 & f & \alpha_1 & \cdots & \cdots & \cdots & \cdots & \alpha_q & \beta_{u-1} & \cdots & \beta_{f-1} & \beta_{f-2}\cdots & \overline{e+1} \end{pmatrix} \\
&\quad\vdots \\
&\xrightarrow[\mathrm{I}]{\beta_e} \begin{pmatrix} c_1 & \cdots & c_p & c_{p+1} & c_{p+2} & c_{p+3} & \cdots & c_t & \overline{d_1} & \cdots & \overline{d_v} & \overline{u-1} & \cdots & \overline{f} & \cdots & \overline{e} \\ a_1 & \cdots & a_p & e & f & \alpha_1 & \cdots & \cdots & \cdots & \cdots & \alpha_q & \beta_{u-1} & \cdots & \beta_f & \cdots & \beta_e \end{pmatrix}.
\end{aligned}
\tag{6.1}
$$

The cases where the consecutive string in C_2^- contains more than two elements follow a similar pattern as above. We can also consider the case where there are several *disjoint* consecutive strings containing elements of C_1^+, in which case the highest block could give rise to type IIb bumps instead of IIa bumps, but the details are similar.

Turning to the type I slide, let us consider the case where there are no entries above $i_1, \ldots, i_p$ nor below $\overline{j_q}, \ldots, \overline{j_1}$ nor between x and $\overline{x}$. The column reading of such a configuration looks as follows:

$$
w = \begin{pmatrix} c_1 & \cdots & \cdots & \cdots & \cdots & \cdots & c_t & x & \overline{x} & \overline{j_q} & \cdots & \overline{j_1} \\ a+1 & \cdots & \langle j_1 \rangle & \cdots & \langle j_q \rangle & \cdots & x-1 & \square & \alpha_{q+1} & \alpha_q & \cdots & \alpha_1 \end{pmatrix}.
$$

Here $t = x - a - q - 1$ and we consider the entries $i_1, \ldots, i_p$ as the consecutive string $a+1, a+2, \ldots, x-2, x-1$ punctuated by the absences of $j_1, \ldots, j_q$. If we now start inserting this word into the empty tableau we get the following sequence of bumps (again, inserting the upper row and the letters up to $x - 1$ in the bottom row is clear and hence omitted):

$$
\begin{aligned}
&\begin{pmatrix} c_1 & \cdots & c_{j_q-a-q} & c_{j_q-a-q+1} & \cdots & c_t & & & & & \\ a+1 & \cdots & j_q-1 & j_q+1 & \cdots & x-1 & x & \overline{x} & \overline{j_q} & \cdots & \overline{j_1} \end{pmatrix} \\
&\xrightarrow[\mathrm{I}]{\alpha_{q+1}} \begin{pmatrix} c_1 & \cdots & c_{j_{q-1}-a-q+1} & c_{j_{q-1}-a-q+2} & \cdots & c_{j_q-a-q} & c_{j_q-a-q+1} & \cdots & c_t & \overline{j_q} & & & \\ a+1 & \cdots & j_{q-1}-1 & j_{q-1}+1 & \cdots & j_q-1 & j_q & \cdots & x-2 & x-1 & \alpha_{q+1} & \overline{j_q} & \cdots & \overline{j_1} \end{pmatrix} \\
&\xrightarrow[\mathrm{I}]{\alpha_q} \begin{pmatrix} c_1 & \cdots & c_{j_{q-1}-a-q+1} & c_{j_{q-1}-a-q+2} & \cdots & c_{j_q-a-q+1} & c_{j_q-a-q+2} & \cdots & c_t & \overline{j_q} & \overline{j_{q-1}} & & & \\ a+1 & \cdots & j_{q-1}-1 & j_{q-1} & \cdots & j_q-1 & j_q+1 & \cdots & x-2 & x-1 & \alpha_{q+1} & \alpha_q & \overline{j_{q-1}}\cdots & \overline{j_1} \end{pmatrix} \\
&\quad\vdots \\
&\xrightarrow[\mathrm{I}]{\alpha_1} \begin{pmatrix} c_1 & \cdots & c_{j_k-a-k+1} & c_{j_k-a-k+2} & \cdots & c_t & \overline{j_q} & \overline{j_{q-1}} & \cdots & \overline{j_1} & \overline{a} & \\ a & \cdots & j_k-1 & j_k+1 & \cdots & x-2 & x-1 & \alpha_{q+1} & \cdots & \alpha_3 & \alpha_2 & \alpha_1 \end{pmatrix}, \qquad 1 \le k \le q,
\end{aligned}
$$

which is precisely the column reading of the end product of the type I slide. For the general case of a type I slide, one simply combines the two types of insertions shown above. This is possible since there is always a "gap" between $\overline{j_1}$ and the entry immediately below it, as well as between i_1 and the entry immediately above it, so the above two types of sequences of insertions are "disjoint" (i.e., do not interfere with each other).

The case of type II slides is a bit more complicated. Let us consider the simplest case where there is only one i. That is, we want to show the following Knuth equivalence

$$\begin{pmatrix} \gamma_0 & \gamma_1 & \overline{a-1} & \cdots & \overline{k+1} & \overline{k-1} & \cdots & \bar{i} & \overline{i-1} & \cdots & \overline{y+1} & \bar{y} \\ y & i & \overline{a-1} & \cdots & \overline{k+1} & \bar{k} & \cdots & \overline{i+1} & \overline{i-1} & \cdots & \overline{y+1} & \square \end{pmatrix}$$
$$\sim \begin{pmatrix} \gamma_0 & \gamma_1 & \overline{a-1} & \cdots & \overline{k+1} & \overline{k-1} & \cdots & \overline{i-1} & \overline{i-2} & \cdots & \overline{y+1} & \\ i & a & \bar{a} & \cdots & \overline{k+2} & \overline{k+1} & \cdots & \overline{i+1} & \overline{i-1} & \cdots & \overline{y+2} & \overline{y+1} \end{pmatrix}. \tag{6.2}$$

Here, we consider the string $\overline{j_1}, \dots, \overline{j_q}$ as the consecutive string $y+1, y+2, \dots, a-1$ punctuated by the absence of i. We also consider the entries to the immediate right of the js as the string $y+1, y+2, \dots, a-1$ punctuated by the absence of k, and column strictness implies that $k \geq i$. Note also that Lemma 2.2 implies (using an argument by contradiction) that $\gamma_0 \geq k$ and $\gamma_1 \geq a$. With these facts in mind, we can insert the word on the left-hand side of (6.2) into the empty tableau, the result of which is the sequence of insertions

$$(\gamma_0\ \gamma_1\ \overline{a-1} \cdots \overline{k+1}\ \overline{k-1} \cdots \bar{i}\ \overline{i-1} \cdots \overline{y+1}\ \bar{y})$$
$$\xrightarrow[\text{IIa}]{y} \begin{pmatrix} \gamma_0 & & & & & & & & & & \\ k & \gamma_1 & \overline{a-1} & \cdots & \overline{k+1} & \bar{k} & \overline{k-1} & \cdots & \bar{i} & \overline{i-1} & \cdots & \overline{y+1} \end{pmatrix}$$
$$\xrightarrow[\text{IIa}]{i} \begin{pmatrix} \gamma_0 & \gamma_1 & & & & & & & & & & \\ k & a & \bar{a} & \overline{a-1} & \cdots & \overline{k+1} & \bar{k} & \cdots & \overline{i+1} & \overline{i-1} & \cdots & \overline{y+1} \end{pmatrix}$$
$$\xrightarrow{\{\overline{a-1},\dots,\overline{k+1}\}} \begin{pmatrix} \gamma_0 & \gamma_1 & \overline{a-1} & \overline{a-2} & \cdots & \overline{k+1} & & & & & & \\ k & a & \bar{a} & \overline{a-1} & \cdots & \overline{k+2} & \overline{k+1} & \cdots & \overline{i+1} & \overline{i-1} & \cdots & \overline{y+1} \end{pmatrix}$$
$$\xrightarrow[\text{I}]{\bar{k}} \begin{pmatrix} \gamma_0 & \gamma_1 & \overline{a-1} & \cdots & \overline{k+1} & \overline{k-1} & & & & & \\ k-1 & a & \bar{a} & \cdots & \overline{k+2} & \overline{k+1} & \cdots & \overline{i+1} & \overline{i-1} & \cdots & \overline{y+1} \end{pmatrix}$$
$$\xrightarrow[\text{I}]{\{\overline{k-1},\dots,\overline{i+1}\}} \begin{pmatrix} \gamma_0 & \gamma_1 & \overline{a-1} & \cdots & \overline{k+1} & \overline{k-1} & \cdots & \bar{i} & & & \\ i & a & \bar{a} & \cdots & \overline{k+2} & \overline{k+1} & \cdots & \overline{i+1} & \overline{i-1} & \cdots & \overline{y+1} \end{pmatrix}$$
$$\xrightarrow{\{\overline{i-1},\dots,\overline{y+1}\}} \begin{pmatrix} \gamma_0 & \gamma_1 & \overline{a-1} & \cdots & \overline{k+1} & \overline{k-1} & \cdots & \overline{i-1} & \overline{i-2} & \cdots & \overline{y+1} & \\ i & a & \bar{a} & \cdots & \overline{k+2} & \overline{k+1} & \cdots & \overline{i+1} & \overline{i-1} & \cdots & \overline{y+2} & \overline{y+1} \end{pmatrix},$$

and the proof of (6.2) is complete. The situation for the more general case of $i_1, \dots, i_p$ is similar (although more complicated). Again, we must combine such a sequence of insertions with the sequence (6.1) for elements in $C_1^+ \cap C_2^-$ (apart from y), but again, such sequences are "disjoint" for reasons mentioned above, so there is no problem.

For type III and IV slides, the arguments are similar to those for the type I and II slides, respectively. Thus, since both the bumping *and* sliding processes preserve Knuth equivalence, we have in fact proved the following.

Proposition 6.1. *If $b \leftarrow \alpha$ denotes the tableau which results after inserting α into the tableau b, then*

$$w(b \leftarrow \alpha) \sim w(b).\alpha.$$

6.3 Stability of Knuth classes under $\tilde{e}_i$, $\tilde{f}_i$. In this subsection, we shall show that the Knuth classes are stable under the action of the Kashiwara operators. We consider only the case for the lowering operators $\tilde{f}_i$ with $1 \le i \le n-1$, as the proofs for the case $i = n$ and for $\tilde{e}_i$ are similar. Thus, we must show that if $w \sim w'$, then $\tilde{f}_i w \sim \tilde{f}_i w'$. It suffices to consider the cases when w and w' are related by an elementary Knuth relation. We first consider the transformations (K1)–(K4) and treat (K5) separately.

Recalling the action of $\tilde{f}_i$ given in Section 2.2, let us call the letter upon which $\tilde{f}_i$ acts, the *pivot* letter. For the transformations (K1)–(K4), it can be seen that whenever the pivot letter is *not* part of the three letters comprising the transformation, the pivot letter remains unchanged and hence the result follows immediately. Thus we can restrict ourselves to the cases when the pivot letter *is* one of the three letters comprising the transformation.

In the case of a K1 or K2 transformation, the pivot letter can only change when the two letters which get swapped are of the form u_+u_- or u_-u_+ in the notation of Section 2.2. In this case, the inequalities governing the transformations leave us with the following two situations where the pivot letter changes:

$$\begin{array}{ccc} i\ i{+}1\ \mathbf{i} & \xrightarrow{\text{K2}} & \mathbf{i}\ i\ i{+}1 \\ \downarrow \tilde{f}_i & & \downarrow \tilde{f}_i \\ i\ i{+}1\ i{+}1 & \xrightarrow{\text{K1}} & i{+}1\ i\ i{+}1 \end{array} \qquad \begin{array}{ccc} \overline{i+1}\ \bar{i}\ \overline{\mathbf{i+1}} & \xrightarrow{\text{K2}} & \overline{\mathbf{i+1}}\ \overline{i+1}\ \bar{i} \\ \downarrow \tilde{f}_i & & \downarrow \tilde{f}_i \\ \overline{i+1}\ \bar{i}\ \bar{i} & \xrightarrow{\text{K1}} & \bar{i}\ \overline{i+1}\ \bar{i}, \end{array}$$

where the pivot letter is shown in bold. Hence in these cases, the images under $\tilde{f}_i$ are also related by an elementary Knuth transformation.

For K1 and K2 transformations, we must also consider the case when the letters getting swapped are of the form u_+u_+ but are of different type, i.e., $i\ \overline{i+1}$ or $\overline{i+1}\ i$, since in these cases, we have the following situation:

$$\begin{array}{ccc} \alpha\ \overline{\mathbf{i+1}}\ i & \xrightarrow{\text{K2}} & \alpha\ \mathbf{i}\ \overline{i+1} \\ \downarrow \tilde{f}_i & & \downarrow \tilde{f}_i \\ \alpha\ \bar{i}\ i & \xrightarrow{\text{K4}} & \alpha\ i{+}1\ \overline{i+1} \\ & i < \alpha < \overline{i+1} & \end{array} \qquad \begin{array}{ccc} \mathbf{i}\ \overline{i+1}\ \beta & \xrightarrow{\text{K1}} & \overline{\mathbf{i+1}}\ i\ \beta \\ \downarrow \tilde{f}_i & & \downarrow \tilde{f}_i \\ i{+}1\ \overline{i+1}\ \beta & \xrightarrow{\text{K3}} & \bar{i}\ i\ \beta \\ & i < \beta \le \overline{i+1}. & \end{array}$$

Note that in the first case, the K2 inequalities allows for $\alpha = i$, but it is easy to show that in such a case, α becomes the pivot letter and the images under $\tilde{f}_i$ remain related by a K2 transformation.

For K3 and K4 transformations, a similar situation occurs, namely,

$$\begin{array}{ccc} \alpha\,\bar{i}\,\mathbf{i} & \xrightarrow{\text{K4}} & \alpha\; i{+}1\; \overline{\mathbf{i+1}} \\ \big\downarrow \tilde{f}_i & & \big\downarrow \tilde{f}_i \\ \alpha\,\bar{i}\; i{+}1 & \xrightarrow{\text{K2}} & \alpha\; i{+}1\; \bar{i} \end{array} \qquad\qquad \begin{array}{ccc} i{+}1\; \overline{\mathbf{i+1}}\; \beta & \xrightarrow{\text{K3}} & \bar{i}\; \mathbf{i}\; \beta \\ \big\downarrow \tilde{f}_i & & \big\downarrow \tilde{f}_i \\ i{+}1\; \bar{i}\; \beta & \xrightarrow{\text{K1}} & \bar{i}\; i{+}1\; \beta \end{array}$$

$$i < \alpha < \bar{i} \qquad\qquad\qquad i-1 < \beta < \bar{i}.$$

Again, note that the K3 inequalities allow for $\beta = i + 1$, but in such a case the middle letter is never the pivot letter, so there is never a problem.

Finally, turning to the K5 transformation, it suffices to consider only the cases where the pivot letter is annihilated. Hence for each i, there are two subcases: (a) the pivot letter is i; (b) the pivot letter is $\overline{i+1}$.

In subcase (a), suppose $w = w_1 \cdots w_p\, \mathbf{i}\, \beta_1 \cdots \beta_r\, \bar{i}\, \overline{v_q} \cdots \overline{v_1}$ with $p+q = i-1$. Then the only way i can be a pivot letter is if $\beta_1 \geq i+2$ and $\beta_r = \overline{i+1}$. In such a situation we have the following diagram:

$$\begin{array}{ccc} w_1 \cdots w_p\, \mathbf{i}\, \beta_1 \cdots \overline{i+1}\, \bar{i}\, \overline{v_q} \cdots \overline{v_1} & \xrightarrow{\text{K5}} & w_1 \cdots w_p\, \beta_1 \cdots \overline{\mathbf{i+1}}\, \overline{v_q} \cdots \overline{v_1} \\ \big\downarrow \tilde{f}_i & & \big\downarrow \tilde{f}_i \\ w_1 \cdots w_p\, i{+}1\, \beta_1 \cdots \overline{i+1}\, \bar{i}\, \overline{v_q} \cdots \overline{v_1} & \xrightarrow{\text{K5}} & w_1 \cdots w_p\, \beta_1 \cdots \bar{i}\, \overline{v_q} \cdots \overline{v_1}. \end{array}$$

Note that the condition $p + q = i - 1$ is the necessary K5 condition for both w and $\tilde{f}_i w$. In case (b), a similar argument holds so that the only nontrivial case is the following:

$$\begin{array}{ccc} w_1 \cdots w_{p-1}\, i\; i{+}1\, \beta_1 \cdots \beta_r\, \overline{\mathbf{i+1}}\, \overline{v_q} \cdots \overline{v_1} & \xrightarrow{\text{K5}} & w_1 \cdots w_{p-1}\, \mathbf{i}\, \beta_1 \cdots \beta_r\, \overline{v_q} \cdots \overline{v_1} \\ \big\downarrow \tilde{f}_i & & \big\downarrow \tilde{f}_i \\ w_1 \cdots w_{p-1}\, i\; i{+}1\, \beta_1 \cdots \beta_r\, \bar{i}\, \overline{v_q} \cdots \overline{v_1} & \xrightarrow{\text{K5}} & w_1 \cdots w_{p-1}\, i{+}1\, \beta_1 \cdots \beta_r\, \overline{v_q} \cdots \overline{v_1}, \end{array}$$

where $p + q = i$ and $\overline{v_q} \geq \overline{i-1}$.

Summarizing, we have proven the following.

Proposition 6.2. *The Kashiwara operators are compatible with the Knuth equivalence classes*

$$\tilde{e}_j\, w_1 \sim \tilde{e}_j\, w_2 \;\Leftrightarrow\; w_1 \sim w_2 \;\Leftrightarrow\; \tilde{f}_j\, w_1 \sim \tilde{f}_j\, w_2$$

for all $1 \leq j \leq n$.

Remark. It follows from the above proposition that the Knuth equivalence classes have a unique representative which is the word of a C_n tableau. This is because, from Proposition 6.2 and the fact that every word is connected to a highest-weight (Yamanouchi) word by a sequence $\tilde{e}_{i_1} \cdots \tilde{e}_{i_p}$ of raising operators, it is only necessary to consider the case where the Knuth class contains highest-weight words, in which case the result follows from the A_{n-1} case. It would be nice to have a direct (i.e., combinatorial) argument of this statement in terms of analyzing longest increasing subsequences and an induction argument, à la [5, Chapter 3].

7 Crystal isomorphism

Recall that we use the same symbols $\tilde{f}_i$, $\tilde{e}_i$ to denote the action of the Kashiwara operators on tableaux or on words. Thus if w denotes the operator which takes the column reading of a tableau, we have symbolically, $w\tilde{f}_i = \tilde{f}_i w$.

A very important consequence of Propositions 6.1 and 6.2 is that

$$w(\tilde{f}_j(b \leftarrow \alpha)) \sim \tilde{f}_j(w(b).\alpha). \tag{7.1}$$

In particular, this implies that for $1 \leq j < n$,

$$\phi_j(b \leftarrow i) = \begin{cases} \max(\phi_j(b) - 1, 0) & j = i - 1, \\ \phi_j(b) + 1 & j = i, \\ \phi_j(b) & \text{otherwise}, \end{cases}$$
$$\phi_j(b \leftarrow \bar{i}) = \begin{cases} \phi_j(b) + 1 & j = i - 1, \\ \max(\phi_j(b) - 1, 0) & j = i, \\ \phi_j(b) & \text{otherwise}, \end{cases} \tag{7.2}$$

while for $j = n$,

$$\phi_n(b \leftarrow i) = \begin{cases} \phi_n(b) + 1 & i = n, \\ \phi_n(b) & \text{otherwise}, \end{cases}$$
$$\phi_n(b \leftarrow \bar{i}) = \begin{cases} \max(\phi_n(b) - 1, 0) & i = n, \\ \phi_n(b) & \text{otherwise}, \end{cases} \tag{7.3}$$

along with similar expressions for $\epsilon_j(b \leftarrow i)$, etc. These are the key ingredients that allow us to prove the C_n analogue of Proposition 1.1.

Proposition 7.1. *Let $\psi_C : B(\mu) \otimes B(\nu) \longrightarrow \bigoplus_j B(\lambda_j)$ be defined by*

$$\psi_C(b_1 \otimes b_2) = b_1 * b_2$$

where $*$ is defined as in (1.2) *but with A_n insertion replaced by C_n insertion. Then ψ_C is the unique crystal isomorphism describing the above tensor product decomposition.*

Proof. By induction, it suffices to consider the case where b_2 consists of a single box. Thus one must show for all $1 \leq i, j \leq n$,

$$\tilde{f}_j\, \psi_C\left(b \otimes \boxed{i}\right) = \psi_C\, \tilde{f}_j\left(b \otimes \boxed{i}\right), \qquad \tilde{f}_j\, \psi_C\left(b \otimes \boxed{\bar{i}}\right) = \psi_C\, \tilde{f}_j\left(b \otimes \boxed{\bar{i}}\right),$$

and similarly for $\tilde{e}_j$. We consider only the case of $\tilde{f}_j$ with the unbarred box $\boxed{i}$, as the other cases can be proved in a similar manner.

Case $j \neq i-1, i$. From the tensor product rule (1.1), we have

$$\tilde{f}_j(b \otimes \boxed{i}) = \begin{cases} (\tilde{f}_j b) \otimes \boxed{i}, & \phi_j(b) > 0, \\ 0, & \phi_j(b) = 0. \end{cases}$$

In the latter case, when $\phi_j(b) = 0$, then certainly $\psi_C \tilde{f}_j(b \otimes \boxed{i}) = 0$. Also (7.2) implies $\tilde{f}_j(b \leftarrow i) = 0$ so $\tilde{f}\psi_C(b \otimes \boxed{i}) = 0$. Hence the result is true in this case.

In the former case, when $\phi_j(b) > 0$ we have from Proposition 6.1

$$w(\psi_C \tilde{f}_j(b \otimes \boxed{i})) = w((\tilde{f}_j b) \leftarrow i) \sim w(\tilde{f}_j b) \,.\, i,$$

while from (7.1) we have

$$w(\tilde{f}_j \psi_C(b \otimes \boxed{i})) = w(\tilde{f}_j(b \rightarrow i)) \sim \tilde{f}_j(w(b) \,.\, i) = w(\tilde{f}_j b) \,.\, i,$$

Since the words of $\psi_C \tilde{f}_j(b \otimes \boxed{i})$ and $\tilde{f}_j \psi_C(b \otimes \boxed{i})$ are Knuth equivalent, and they are both the words of a tableau, they must in fact be the same tableau (see remark below Proposition 6.2). Hence the result is proved in this case.

Case $j = i$. From the tensor product rule (1.1) we have

$$\tilde{f}_i(b \otimes \boxed{i}) = \begin{cases} (\tilde{f}_i b) \otimes \boxed{i}, & \phi_i(b) > 0, \\ b \otimes \boxed{i+1}, & \phi_i(b) = 0, \end{cases}$$

In the case where $\phi_i(b) = 0$,

$$w(\psi_C \tilde{f}_i(b \otimes \boxed{i}) = w(b \leftarrow (i+1)) \sim w(b) \,.\, i+1,$$

while

$$w(\tilde{f}_i \psi_C(b \otimes \boxed{i})) \sim \tilde{f}_i(w(b) \,.\, i) = w(b) \,.\, i+1,$$

where we have successively used (7.1) and the fact that $\phi_i(b) = 0$.

In the case where $\phi_i(b) > 0$, we have similarly $w(\psi_C \tilde{f}_i(b \otimes \boxed{i})) = w((\tilde{f}_i b) \leftarrow i) \sim w(\tilde{f}_i b) \,.\, i$ (using Proposition 6.1), while $w(\tilde{f}_i \psi_C(b \otimes \boxed{i})) \sim \tilde{f}_i(w(b) \,.\, i) = w(\tilde{f}_i b) \,.\, i$ using first (7.1) and then the fact that $\phi_i(b) > 0$. Hence the result is proved in this case.

Case $j = i-1$. The arguments are similar in this case and will not be repeated, the only thing special to note is that when $\phi_{i-1}(b) \leq 1$, (7.2) implies $\tilde{f}_{i-1}(b \leftarrow i) = 0$. □

8 Proofs

In this section, we give the proofs of the claim that the bumping and sliding procedures defined in Sections 3 and 4 yield valid C_n tableaux. We first remark that

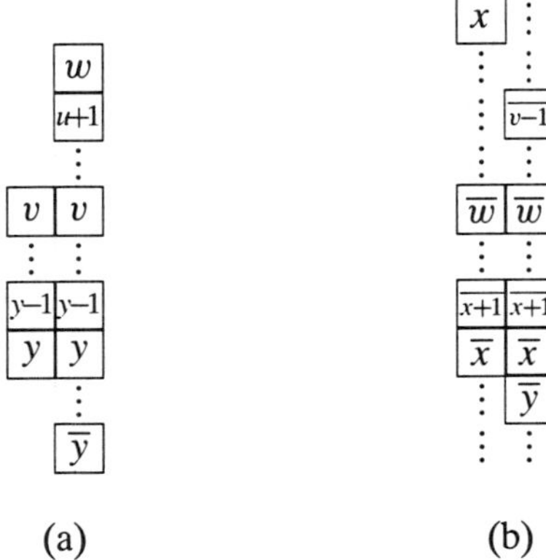

Figure 8.1. Bumping and the horizontal condition.

the bumping procedure is unambiguously defined and the elements u, v in the type I, IIb, and IIa bumps really exist (i.e., $1 \leq u, v \leq n$). For example, in the type I bump there would be a problem if $u + 1 = 1$. But in that case, we would have $\text{pos}(y) + \text{pos}(\overline{y}) \geq y + 1$ which is invalid to begin with. A similar remark applies to the IIb bump. For the IIa bump, we must always have $v \leq n - 1$ for β to satisfy the conditions for a IIa bump, so there will never be a problem. Similar remarks apply to the reverse bumps.

With regard to the sliding procedures, we can make similar observations. Thus in the case of the forward type I slide, we would be in trouble if $a < 1$, i.e., $\{i_1, \ldots, i_p\} \sqcup \{j_1, \ldots, j_q\} = \{1, 2, \ldots, x - 1\}$. But in that case, noting that the empty box must have slid down vertically from above the entry i_1 (see the first claim in section 8.5), Lemma 2.2 applied to the $(i_1, x|\overline{x}, \overline{j_1})$ configuration would yield a contradiction. Similar remarks apply to the remaining forward and reverse slides.

Given that we can carry out one of these bumps or slides, we must check that the result is a valid C_n tableaux. Thus there are four conditions we must check: strictly increasing columns, weakly increasing rows, the 1CC and the 2CC. For both bumping and sliding, the first condition is clear by the construction. In the following subsections, we check the remaining conditions.

8.1 Bumping and horizontal condition. There are four different types of bumps that we must consider: ordinary and types I, IIa, and IIb.

For the ordinary bump, the fact that bumping preserves the horizontal condition is well known.

For the type I bump, because the string of consecutive letters $u+1, u+2, \ldots, y-1$ moves down one position, a possible horizontal violation can only occur if before the bump, there is some v in such a string that has a cell on its left which also contains a v. But the horizontal and vertical conditions then force the cells containing $v+1, \ldots, y$ to have the same entries to their immediate left (see Figure 8.1(a)). This would mean that those two columns are in a (y, y) configuration, which contradicts the fact that we began with a valid C_n tableau.

For a type IIa bump, a similar situation occurs. Suppose we are bumping in x, and $\overline{x}$ already appears in the column. Then the only possible horizontal violation can occur when there exists a $\overline{w}$ in the consecutive string $\overline{v-1}, \ldots, \overline{x}$ and again the horizontal and vertical conditions before the bump force the letters $\overline{w}, \overline{w-1}, \ldots, \overline{x}$ to have the same entries in the cells to their immediate left. Now, since we are bumping in an x, such an x must have been bumped out of the previous column by either an ordinary or IIa bump. If it was an ordinary bump, then the situation before this bump must have been as appears in Figure 8.1(b). If it was a type IIa bump, then a consecutive string of barred letters may have moved during such a bump. By definition, the top of this string $\overline{v}$ is such that $v \leq x \Leftrightarrow \overline{x} \geq \overline{v}$. Thus the consecutive string $\overline{w}, \ldots, \overline{x+1}, \overline{x}$ does not move and the situation is also as appears in Figure 8.1(b). Thus, before the bump, there was an (x, x) configuration, which is a contradiction. The argument for the type IIb bump runs along similar lines.

8.2 Bumping and the 1CC. For an ordinary bump, a 1CC violation can occur in two possible cases: (i) a bumps an x and there exists an $\overline{a}$ such that $\text{pos}(x)+\text{pos}(\overline{a}) > a$ (ii) $\overline{a}$ bumps an $\overline{x}$ and there exists an a such that $\text{pos}(a) + \text{pos}(\overline{x}) > a$. In case (i), this cannot happen as such a bump must be, by definition, a type II bump. In case (ii), we must have $a \geq x$ and so by Lemma 2.1 $\text{pos}(a) + \text{pos}(\overline{x}) \leq \max(a, x) = a$.

We must also examine the case where the element inserted into a column is strictly greater than all the elements in the column, in which case it gets placed at the end of that column (this clearly only occurs in the case of an ordinary bump). We claim that, *apart from the case when we are inserting into the first column*, such a situation never gives rise to a 1CC violation. In the case of a 1CC violation in the first column, this signals that we must delete some boxes and start sliding (recall Section 5).

For a type I bump, we must check the 1CC for any s–$\overline{s}$ pair, where $s \in \{u+1, u+2, \ldots, y-1\}$, for a possible u–$\overline{u}$ pair, where u is a new element appearing at the top of the string, and for any x–$\overline{x}$ pair in the case where $\alpha = \overline{x}$. Let B (resp., A) be the column as it appears before (resp., after) the bump. We shall use this convention from now on. In the first case, we have by the definition of the 1CC that

$$\text{pos}_B(s) + \text{pos}_B(\overline{y}) = \text{pos}_B(y) + s - y + \text{pos}_B(\overline{y}) \leq s. \tag{8.1}$$

Since $s \leq y-1$, we must have $\text{pos}_B(\overline{s}) \leq \text{pos}_B(\overline{y}) - 1$, and hence from (8.1) $\text{pos}_B(s) + \text{pos}_B(\overline{s}) \leq s - 1$ and thus $\text{pos}_A(s) + \text{pos}_A(\overline{s}) \leq s$. In the second case, we know $\text{pos}_A(u) = \text{pos}_B(u+1)$ and $\text{pos}_A(\overline{u}) = \text{pos}_B(\overline{u}) < \text{pos}_B(\overline{y})$. Thus

$$\text{pos}_A(u) + \text{pos}_A(\overline{u}) < \text{pos}_B(u+1) + \text{pos}_B(\overline{y}) \leq u+1,$$

where, in the last inequality, we have used (8.1). Hence $\text{pos}_A(u) + \text{pos}_A(\overline{u}) \leq u$. Finally, if $\alpha = \overline{x}$ and there exists an $x \in A$, then since $x \geq y$, Lemma 2.1 implies $\text{pos}_B(x) + \text{pos}_B(\overline{y}) \leq x$ and so $\text{pos}_A(x) + \text{pos}_A(\overline{x}) \leq x$.

For a type IIa bump, the only letters which move downwards are barred letters, so the 1CC will always remain valid for such letters. However, we must check the 1CC for the new v–$\overline{v}$ pair which appears. Suppose that δ is the smallest letter

$\geq v$ in the column B, which gets replaced during the bump by a v. Then either $\text{pos}_B(\delta) = 1$ or there exists a $w < v$ in the cell immediately above δ. In the former case, $\text{pos}_B(\delta) + \text{pos}_B(\overline{v-1}) \leq 1 + v - 1 = v$ and thus $\text{pos}_A(v) + \text{pos}_A(\overline{v}) \leq v$. In the latter case, Lemma 2.1 implies $\text{pos}_B(w) + \text{pos}_B(\overline{v-1}) \leq v - 1$. Thus $\text{pos}_A(v) + \text{pos}_A(\overline{v}) \leq v$ as required.

For a type IIb bump, first note that since $\text{pos}_B(v-1) + \text{pos}_B(\overline{v-1}) \leq v - 1$, for any other pair e–$\overline{e}$ in the string $u+1, \ldots, v-1, \overline{v-1}, \ldots, \overline{x}$ we have

$$\text{pos}_B(e) + \text{pos}_B(\overline{e}) \leq e - (v - 1 - e), \tag{8.2}$$

We consider the two cases when $u \geq x$ and $u < x$. In the former case, consider all pairs s–$\overline{s}$ with $u < s \leq v-1$. Since the entire string $u+1, \ldots, v-1, \overline{v-1}, \ldots, \overline{x}$ moves down, the net effect on $\text{pos}(s) + \text{pos}(\overline{s})$ is zero. Similarly, we must check the 1CC for the newly created u–$\overline{u}$ pair. There, Lemma 2.1 implies that $\text{pos}_B(u+1) + \text{pos}_B(\overline{u}) \leq u + 1$, hence the result follows from noting that $\text{pos}_B(u+1) = \text{pos}_A(u)$ and $\text{pos}_B(\overline{u}) = \text{pos}_A(\overline{u}) + 1$.

In the latter case, there may be a pair t–$\overline{t}$ with $u < t < x$ so that $\text{pos}(t)$ increases by one and $\text{pos}(\overline{t})$ remains unchanged. But (8.2) implies that

$$\text{pos}_B(t) + \text{pos}_B(\overline{t}) < \text{pos}_B(x) + t - x + \text{pos}_B(\overline{x}) \leq t + x - (v-1) < t,$$

which gives the required result. We must also check the 1CC if there is an entry $\overline{u}$ below $\overline{x}$. By definition, we have $\text{pos}_B(x) + \text{pos}_B(\overline{x}) \leq x$ and so $\text{pos}_B(u+1) + \text{pos}_B(\overline{u}) \leq x - (x - u - 1) = u + 1$ and the result now follows.

8.3 Bumping and the 2CC. First, consider the case of an ordinary bump. From Lemma 2.2, the only possible 2CC violations could occur when we bump into the second or fourth entry of an $(a, \delta|\overline{b}, \overline{a})$ or $(a, b|\overline{b}, \overline{x})$ configuration, respectively. To see why this is so, suppose for example that a bumps an x in an $(x, b|\overline{b}, \overline{a})$ configuration. Then by definition $a \leq x$ and Lemma 2.2 implies that $p(x, b|\overline{b}, \overline{a}) < b - \min(x, a) = b - a$, and so the 2CC will always hold in such a case. Similarly for the case when $\overline{b}$ gets bumped into the third position of an $(a, b|\overline{x}, \overline{a})$ configuration.

In the case when b bumps an x in a (left or right) $(a, x|\overline{b}, \overline{a})$ configuration, this is by definition a type II bump, so we do not consider that here.

In the case where $\overline{a}$ bumps an $\overline{x}$ in an $(a, b|\overline{b}, \overline{x})$ configuration, then (calling the left column C and the right column C') the entry $\overline{a}$ being inserted into the column C' must have resulted from a type I or IIb bump in column C (but *neither* from an ordinary bump, *nor* a IIa bump). Before such bumps, there must have been a string of consecutive letters $a+1, a+2, \ldots, c$ and a letter $\overline{c}$ in column C. Moreover, the letter in the cell above $\overline{x}$ must have been a barred letter, $\overline{y}$ say, so that $\overline{y} < \overline{a} < \overline{x}$. Thus for the $(a+1, b|\overline{b}, \overline{y})$ configuration, Lemma 2.2 implies $p(a+1, b|\overline{b}, \overline{y}) \leq b - a - 1$. Hence $p(a+1, b|\overline{b}, \overline{x}) \leq b - a - 1$, and since a replaces $a+1$ after such bumps, the 2CC holds (note that in the case of a left configuration, b (and possibly $\overline{b}$) will move up one position, so the result will still hold).

For a type I bump, the only possible 2CC violations can come from the following three cases (see Figure 8.2):

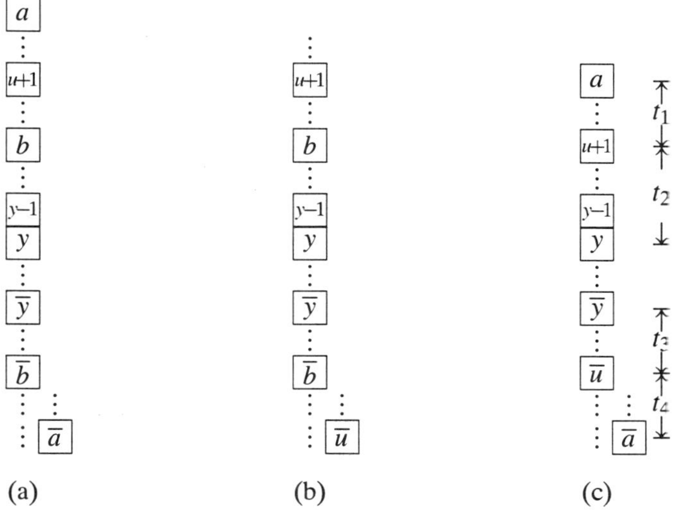

Figure 8.2. I bumps and 2CC violations.

(a) The entry b in a (left or right) $(a, b|\overline{b}, \overline{a})$ configuration belongs to the consecutive string which moves down one position in the type I bump (and a does not).

(b) There is a $(u + 1, b|\overline{b}, \overline{u})$ configuration where b belongs to the consecutive string below $u + 1$ and u gets created at the top of the string.

(c) There is an $(a, u + 1|\overline{u}, \overline{a})$ configuration and u gets created at the top of the string $u + 1, \dots, y$.

The proofs that such 2CC violations can never occur are similar, so we give only one of them, case (c). In that case, first note that Lemma 2.2 implies that $p(a, u + 1|\overline{u}, \overline{a}) \leq u - a$, so a problem can only occur if equality holds there. Let us assume therefore that $t_1 + t_4 = u - a$. Since $u + 1$ and y belong to a string of consecutive integers, $t_2 = y - u - 1$. Also, from the validity of the (a, y) configuration, we have $t_1 + t_2 + t_3 + t_4 \leq y - a - 1$. All of these facts imply that $t_3 \leq 0$, which is impossible.

Note that in cases (a) and (c), we must also consider the cases of right configurations, but the arguments are identical to the cases of left configurations. Also, the letter α which replaces the entry $\overline{y}$ as part of the type I bump might also cause a 2CC violation, but the reason that it does not is the same as the reasons given above for the validity of the ordinary bump.

For a type IIa bump there are also three situations that might result in a 2CC violation:

(a) The entry $\overline{a}$ in an (a, b) configuration moves down one position as part of a consecutive string.

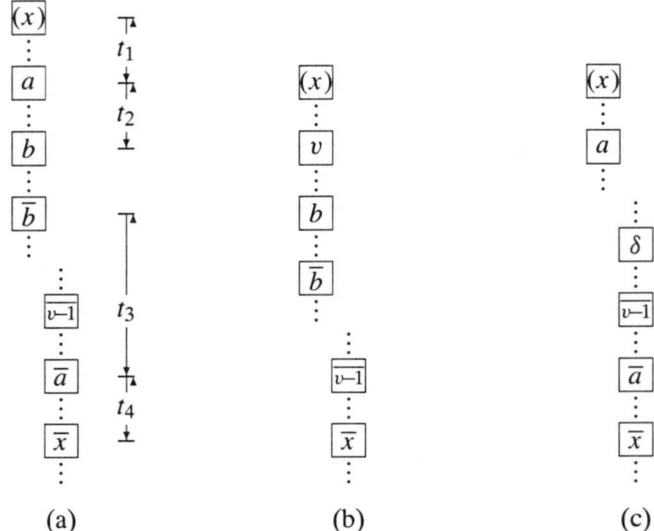

Figure 8.3. IIa bumps and 2CC violations.

(b) There is a $(v, b|\overline{b}, \overline{v-1})$ configuration and $\overline{v}$ appears at the top of the consecutive string $\overline{v-1}, \ldots, \overline{x+1}, \overline{x}$.

(c) There is an $(a, \delta|\overline{v-1}, \overline{a})$ where $\overline{v-1}$ is at the top of a consecutive string, $\overline{a}$ appears in the string, and δ is bumped by the v which is created.

See Figure 8.3. Since all the proofs are similar, we prove only (a).

First, note that since x is being bumped into a IIa situation, it must have been bumped from a previous column, either from an ordinary or a type IIa bump. Let us assume first of all, that it came from an ordinary bump. Figure 8.3(a) shows this prebump situation, where x is still in the left-hand column. Now, by the definition of an (a, b) configuration, we must have $p(a, b|\overline{b}, \overline{a}) \leq b - a - 1$. Indeed, we can assume that equality holds; otherwise, we will never have a problem. Thus $t_2 + t_3 = b - a - 1$. Also, since $\overline{a}$ is part of a consecutive string, $t_4 = a - x$. Finally, the valid $(x, b|\overline{b}, \overline{x})$ configuration tells us that $t_1 + t_2 + t_3 + t_4 \leq b - x - 1$. Hence $t_1 \leq 0$ which is not possible. If, on the other hand, x came from a type IIa bump, then either $\overline{b}$ was part of a consecutive string which moved down one cell (in which case no 2CC violation could have resulted anyway) or it was not, in which case the above argument also holds.

Finally, we must consider type IIb bumps. In this case, there are four situations in which a 2CC violation might arise:

(a) In an (a, b) configuration, the entry b moves down one cell (but all other entries do not).

(b) In an (a, b) configuration, the entry $\overline{a}$ moves down one cell (but all other entries do not).

(c) There is an $(a+1, b|\overline{b}, \overline{a})$ configuration with $a+1$ being at the top of the consecutive string, and b part of the consecutive string (but all other entries are not).

(d) There is an $(a, b+1|\overline{b}, \overline{a})$ configuration with $b+1$ being at the top of the consecutive string, and no other entries are part of it.

The proofs that such situations lead to a contradiction is similar to the type IIa case and will not be given.

8.4 Ordinary sliding. In this subsection, we show that ordinary slides can never lead to a violation of the four conditions of C_n tableaux. For the weakly increasing row conditions and the strictly increasing column conditions, this is a classical fact [5].

8.4.1 1*CC*. There are three possibilities where an ordinary slide might cause a violation of the 1CC:

(i) The entry $\overline{a}$ slides left one position and creates an a–$\overline{a}$ pair whose positions violate the 1CC.

(ii) The entry a slides left one position and creates an a–$\overline{a}$ pair whose positions violate the 1CC.

(iii) The entry $\overline{a}$ of an a–$\overline{a}$ pair slides up one position.

Case (i). This cannot happen, as such a slide must be a type II special slide, not an ordinary slide.

Case (ii). We must consider how the empty box arrived in the column containing $\overline{a}$. If it came from an ordinary horizontal slide or a type I slide, then the entry $\overline{a}$ does not move during such a slide. Thus if x occupied the cell to the left of a before such a slide, then $x \leq a$ and Lemma 2.1 implies that $\text{pos}(x) + \text{pos}(\overline{a}) \leq a$, so there can never be a 1CC violation in such a case. On the other hand, if the empty box came from a type III slide, then $\overline{a}$ may or may not move up one position during such a slide. If it does not move, the previous argument applies. If it does move, then before the type III slide, the entry, z say, to the left of a also moves down one position, so we must have $z \leq a$. But when the empty box slides down to the left of a, z is below the empty box, and a horizontal slide then takes place, so we must have $z > a$, which is a contradiction.

Case (iii). First, note that the only way for a 1CC violation to occur is if the path of the box is as appears in Figure 8.4(a). (If the empty box continues down vertically below $\overline{a}$ and disappears from the end of the column, no 1CC violation can ever occur.) Lemma 2.1 tells us that in Figure 8.4(a) $\text{pos}(a) + \text{pos}(\overline{a}) \leq a$. Moreover, the empty box must leave the column horizontally via either an ordinary, type II or IV slide. (Note that for a II or IV slide, it is possible that a or $\overline{a}$ might move, but this will only decrease $\text{pos}(\cdot)$, so for any violation to occur we assume they do not move.)

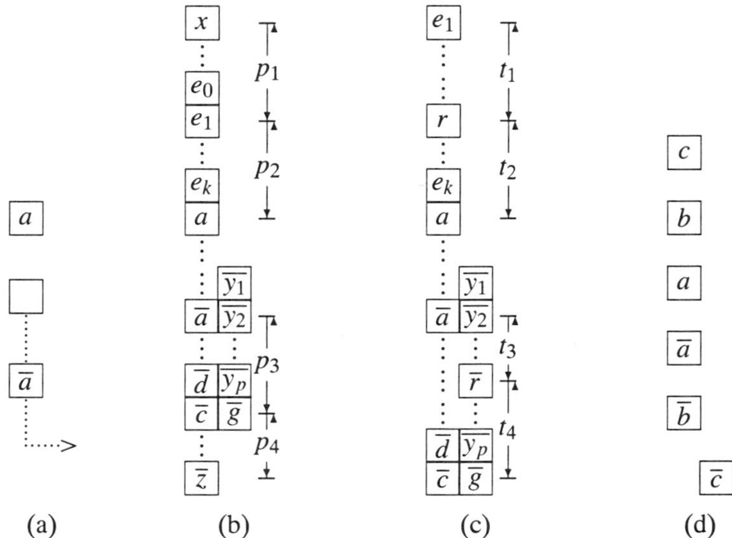

Figure 8.4. 1CC violation and ordinary vertical slides.

Consider first the case when the box leaves the column via an ordinary or type II slide. In this case we will show that the a–$\overline{a}$ pair is part of some left (r, a) configuration and that when the entry $\overline{a}$ moves up one position during the journey of the empty box, a type IV special slide must occur, contrary to assumption. First, note that a possible 1CC violation can occur only if $\mathrm{pos}(a) + \mathrm{pos}(\overline{a}) = a$. So let us assume such a relation holds. Referring to Figure 8.4(b), let us assume that the empty box moves out of the column at the position occupied by $\overline{d}$, so that $y_p > c$. (This must be true for both an ordinary slide and a type II slide.) Note that since the entry $\overline{a}$ has to move vertically, we must have $\overline{a} \leq \overline{y_1} \Leftrightarrow y_1 \leq a$. In fact, since we are assuming such a slide is an ordinary slide, we must have $y_1 < a$. (Otherwise, a type IV slide will occur.) Also, let e_1 be the smallest letter $> c$ lying above a. (We include the possibility that $e_1 = a$.)

We now claim that $\{e_1, \ldots, e_k\} \cap \{y_1, \ldots, y_p\} \neq \emptyset$. For a contradiction, suppose this is not so. Then since both of these sets are subsets of $\{c+1, c+2, \ldots, a\}$, we must have $p_2 + p_3 + 1 \leq a - c - 1$. Let us now assume that there exists an entry e_0 $(\leq c)$ lying immediately above e_1. Then Lemma 2.1 applied to the entries e_0 and c implies

$$p_1 + p_4 + 1 \leq c. \tag{8.3}$$

If such an e_0 does not exist, then $\mathrm{pos}_B(e_1) = 1$ (i.e., $p_1 = 0$) and since $\mathrm{pos}_B(\overline{c}) \leq c$, (8.3) still holds. Together these imply that $p_1 + p_2 + p_3 + p_4 + 2 \leq a - 1$, contrary to the assumption that $\mathrm{pos}(a) + \mathrm{pos}(\overline{a}) = p_1 + p_2 + p_3 + p_4 + 2 = a$.

Given that this intersection is nonempty, choose the smallest letter in such an intersection, call it r say (see Figure 8.4(c)). From the diagram, we have $p_2 = t_1 + t_2$, $p_3 = t_3 + t_4$. The fact that r is minimal implies $t_1 + t_4 - 1 \leq r - c - 1$. Also, (8.3)

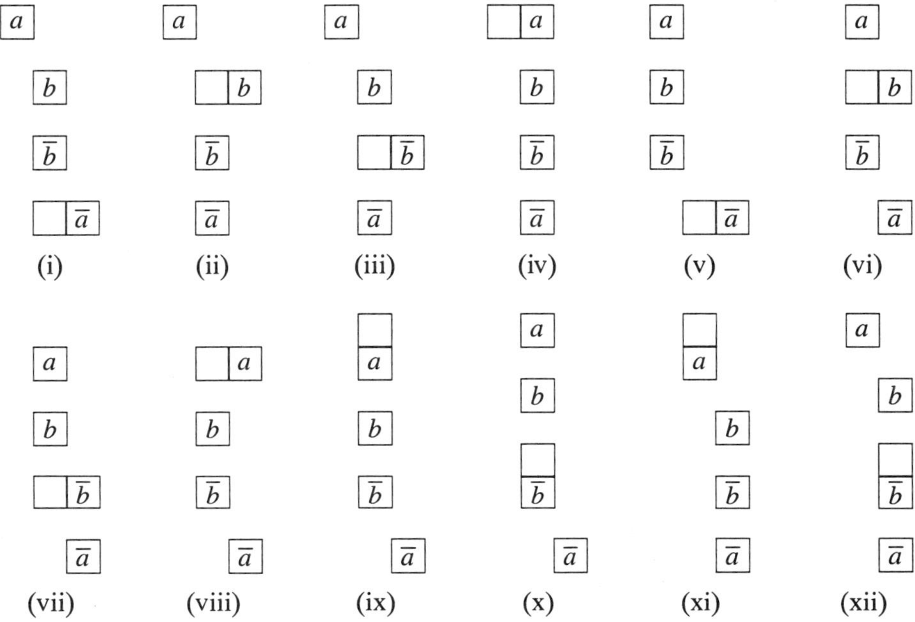

Figure 8.5. Ordinary slides and the 2CC.

implies $p_2 + p_3 \geq a - c - 1$. Thus

$$t_2 + t_3 = (p_2 + p_3) - (t_1 + t_4) \geq a - c - 1 - (r - c) = a - r - 1.$$

But from the definition of the (r, a) configuration, we know that $t_2 + t_3 \leq a - r - 1$. Hence $t_2 + t_3 = a - r - 1$ and so when the entry $\overline{a}$ slides up, a type IV special slide must occur, as promised.

Finally, suppose the empty box moves out of the column via a type IV slide. Then we have the situation as it appears in Figure 8.4(d) with $p(c, b|\overline{b}, \overline{c}) = b - c - 1$, $\text{pos}(a) + \text{pos}(\overline{a}) = a$. Then

$$p(c, a|\overline{a}, \overline{c}) = p(c, b|\overline{b}, \overline{c}) + \text{pos}(a) - \text{pos}(b) + \text{pos}(\overline{a}) - \text{pos}(\overline{b}) \geq a - c - 1.$$

However, by definition $p(c, a|\overline{a}, \overline{c}) \leq a - c - 1$ and so we must have $p(c, a|\overline{a}, \overline{c}) = a - c - 1$. Thus, when $\overline{a}$ moves up vertically, a type IV slide must occur, contrary to assumption.

8.4.2 *2CC.* Turning to ordinary slides which might invalidate the 2CC, we divide them into two categories: those that occur due to an ordinary horizontal slide and those due to an ordinary vertical slide. There are eight instances of the former and four of the latter, pictured schematically in Figure 8.5. Let us consider each of these cases in turn, showing that such violations never arise.

Case (i). Assume for a contradiction that $p(a, b|\overline{b}, \square) \geq b - a$. Let c denote the entry to the immediate right of the entry a. Then we have $c \geq a + 1$. (Certainly $c \geq a$ due to the row condition, but if $c = a$, we would have an illegal left (a, b)

configuration.) Let $\overline{e}$ be the entry in the cell immediately above $\overline{a}$. (Note that this must indeed be barred.) Certainly $e > a$. Applying Lemma 2.2 to the $(c, b|\overline{b}, \overline{e})$ configuration, we have $p(c, b|\overline{b}, \overline{e}) < b-a-1$ and so $p(a, b|\overline{b}, \Box) = p(c, b|\overline{b}, \overline{e})+1 < b - a$, which is a contradiction.

Case (ii). We must analyze how the empty box arrived at its position in the cell immediately to the left of b. Let us call the column to the left of a, the column containing a, and the column containing $\overline{b}$, $\overline{a}$, C_0, C_1, and C_2, respectively. First, note that if the box entered C_2 from (weakly) above a (via an ordinary, a type I, or a type III slide) then applying Lemma 2.2 immediately after such a slide to the $(a, e|\overline{b}, \overline{a})$ configuration (where e is the entry immediately to the left of b) shows that $p(a, e|\overline{b}, \overline{a}) < b - a$ and so a 2CC violation will never arise. Thus we only consider the case where the box enters below the row containing a, and above the row containing b. We consider two subcases:

(a) The box enters C_2 via an ordinary or type I slide.

(a) The box enters C_2 via a type III slide.

In subcase (a), note that if the box enters via a type I slide, then either the a–$\overline{a}$ pair in columns C_1, C_2 were created during the slide (in which case the entries in those cells were both $> a$ and an application of Lemma 2.2 shows we have no problem), or one or both of a, $\overline{a}$ moved down/up, respectively (and again, application of Lemma 2.2 shows there can never be a violation), or nothing moved. Thus we can assume the latter.

We next claim that the empty box must have entered column C_1 from C_0 from *above a* (and hence a must have slid up through an ordinary vertical slide). If this is not the case, then it either entered via an ordinary, type I, or type III slide and in all cases a did not move, so an application of Lemma 2.2 to the $(a, e|\overline{b}, \overline{a})$ configuration shows that there will not be a 2CC violation.

Given that the box passes through a vertically, assume the situation appears as in Figure 8.6(a). Note that since a slides vertically, we must have $x_0 > a$ (certainly $x_0 \geq a$ but if $x_0 = a$, a type III slide would occur), and that $x_p < b$. (Certainly $x_p \leq b$, but if $x_p = b$, then the empty box would slide out of C_2 in the row *above b*.) For a 2CC violation to occur, we assume $p(a, x_p|\overline{b}, \overline{a}) = b - a - 1$, i.e., $(p-1) + (q+1) = b-a-1$. We claim that $\{x_0, x_1, \ldots, x_p\} \cap \{y_1, \ldots, y_q\} \neq \emptyset$. Indeed, if the intersection were empty, then since both sets are subsets of $\{a+1, a+2, \ldots, b-1\}$ we would have $p + 1 + q \leq b - a - 1$, in contradiction to the above assumption. Let r be the maximal element of such an intersection. Referring to Figure 8.6(b), it follows from the maximality of r that $t_2 + t_3 - 1 \leq b - r - 1$. Thus

$$t_1 + t_4 = p(a, x_p|\overline{b}, \overline{a}) - (t_2 + t_3) \geq r - a - 1.$$

But by definition of the (a, r) configuration, we must have $t_1 + t_4 \leq r - a - 1$. Hence $t_1 + t_4 = r - a - 1$ and so when the entry a slides up we must have a type III slide, not an ordinary vertical slide, contrary to the assumption.

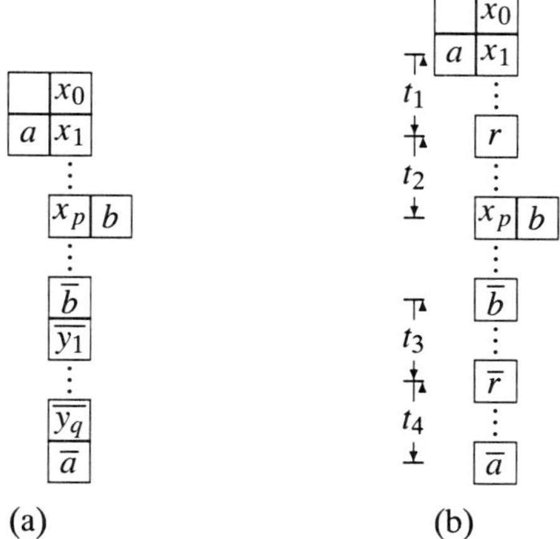

Figure 8.6. Case (ii) of the 2CC violations.

Finally, turning to subcase (b), there are several things which may occur when the box slides from column C_1 to column C_2 via a type III slide: the diagonal pair a–$\overline{a}$ might have been created, or a (resp., $\overline{a}$) might have moved down (resp., up) or both may have moved. Also, $\overline{b}$ and the entry to the left of b might move up or down, respectively. Analysis of all of these situations with Lemma 2.2 shows that strife can only occur when nothing is created and nothing moves—in this case, we can repeat the analysis of subcase (a) and show that the entry a must have moved down vertically as part of some (a, r) configuration, giving a type III slide there, contrary to the assumption.

Case (iii). This is a type II slide, contrary to assumption.

Case (iv). This is a type I slide, contrary to the assumption.

Case (v). This is similar to Case (ii), except with type II and IV slides taking the place of I and III slides. The arguments will not be repeated.

Case (vi). This is similar to Case (i): Let $\overline{c}$ be the entry immediately to the right of $\overline{b}$ in the diagram. Then $c < b$ and the entry immediately above b, e say, must satisfy $e < b$ also. Applying Lemma 2.2 to this $(a, e|\overline{c}, \overline{a})$ configuration gives the required result.

Case (vii). This is a type II slide, contrary to the assumption.

Case (viii). This is a type I slide, contrary to the assumption.

Case (ix). The argument here is similar to that presented for the vertical slides which might violate the 1CC. Namely, one can show that in the only situation which might cause a problem, there exists some r such that the entries a and $\overline{a}$ are part of some right (a, r) configuration and that moving a up one position will cause a type III slide, not an ordinary vertical slide, contrary to the assumption. The details are omitted.

Case (x). This is a type IV slide.

Case (xi). This is a type III slide.

Case (xii). A similar argument to Case (ix).

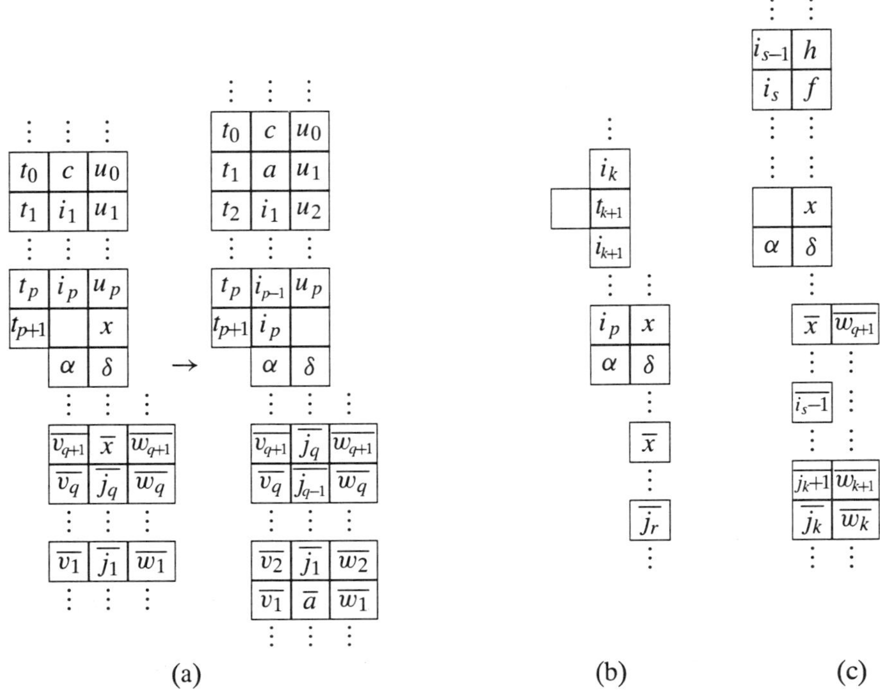

Figure 8.7. Type I slide.

8.5 Type I, II sliding. Let us begin with the row conditions which must be satisfied by the tableau after a type I slide. We recall the situation in Figure 8.7(a). It is clear that before the slide

$$t_k \le i_k \le u_k, \quad k = 1, \ldots, p, \qquad v_k \ge j_k \ge w_k, \quad k = 1, \ldots, q.$$

In particular, $a < i_1 \le u_1$ and $i_k < u_{k+1}$ for $1 \le k < p$. Similarly $v_1 \ge j_1 > a$ and $v_{k+1} > j_k$ for $1 < k \le q$. It thus remains to check that $t_1 \le a$, $t_{k+1} \le i_k$ for $1 \le k \le p$ and $a \ge w_1$, $j_k \ge w_{k+1}$ for $1 \le k \le q$. First, we make the following claim.

Claim. *The empty box has slid vertically from above i_1 by a sequence of ordinary vertical slides.*

Proof. Suppose not, and that at some stage, the box is located between the entries i_k and i_{k+1}, having entered that position from the column on the left. Then the following are the only possible scenarios:

(a) The box came from an ordinary horizontal slide, and the situation is as it appears in Figure 8.7(b).

(b) The box came from a type I slide, and so there was an entry p as part of a p–$\overline{p}$ pair which was located between the entries i_k and i_{k+1}, i.e., $i_k < p < i_{k+1}$.

(c) The box came from a type III slide.

In case (a), since $\{i_1, \ldots, i_p\} \sqcup \{j_1, \ldots, j_q\} = \{a+1, a+2, \ldots, x-1\}$, there exists some r such that $j_r = t_{k+1}$. Thus $\{i_{k+1}, \ldots, i_p\} \sqcup \{j_{r+1}, \ldots, j_q\} = \{t_{k+1}+1, t_{k+1}+2, \ldots, x-1\}$. Moreover, from the diagram, $p(t_{k+1}, x|\overline{x}, \overline{j_r}) = (p-k)+(q+1-r) = x - t_{k+1}$. But by definition of the 2CC, we must have $p(t_{k+1}, x|\overline{x}, \overline{j_r}) < x - t_{k+1}$, so we have a contradiction.

In case (b), a similar argument holds, since the entries $\{i_1, \ldots, i_p\}$ and $\{\overline{j_1}, \ldots, \overline{j_q}\}$ do not move in that particular slide.

In case (c), there was a b, part of an (a, b) configuration that disappears as part of the type III slide, and we must consider two subcases as to where the entry b lay before the III slide. If b was located among the string $i_{k+1}, \ldots, i_p, \alpha$, then a similar argument to cases (a) and (b) shows there was an r such that $j_r = b$ which was in an invalid (b, x) configuration. Alternatively, b was located below $(\alpha =)\ y$, in which case y must have been in the cell immediately to the left of x, which is also a contradiction as we are assuming that $x < y (= \alpha)$. □

It follows from the above claim that $t_{k+1} \leq i_k$ for all $k = 1, \ldots, p$. In addition, we must show that $t_1 \leq a$. To see this, note that when the empty box was at the top of the column of $i_1, \ldots, i_p$, then the previous move must have been an (ordinary) vertical slide or a horizontal slide (either ordinary, type I, or type III). In the first case, since $t_1 \leq c \leq a - 1$, we have $t_1 < a$ as required. If the slide was an ordinary horizontal slide, then t_1 must have been located between c and i_1 before such a slide. If $t_1 > a$ there, then an argument similar to that in the claim above yields a contradiction. If the slide was a type I or III slide, then t_1 had moved down one position during the slide, i.e., before it was immediately to the left of c, so that $t_1 \leq c < a$ and we are done.

It remains to check the inequalities $j_k \geq w_{k+1}, 1 \leq k \leq q$ and $a \geq w_1$. For this refer to Figure 8.7(c), where the entry i_s is the smallest integer $> j_k$ lying above a. Certainly the entries between i_s and the empty box and between $\overline{j_k}$ and $\overline{x}$ must form the disjoint union of the set $\overline{j_k + 1}, \overline{j_k + 2}, \ldots, \overline{x - 1}$. Thus

$$p(f, x|\overline{x}, \overline{w}_{k+1}) = x - j_k - 1. \tag{8.4}$$

But Lemma 2.2 tells us that $p(f, x|\overline{x}, \overline{w_{k+1}}) \leq x - 1 - \min(f, w_{k+1})$. Thus we have $\min(f, w_{k+1}) \leq j_k$. But $f \geq i_s \geq j_k + 1$, and so we can conclude that $w_{k+1} \leq j_k$. A similar argument shows that $a \geq w_1$.

Turning to the 1CC, let us first consider the column containing the is. If any $i_r = p$ was part of a p–$\overline{p}$ pair, then the above claim implies that the 1CC will still hold after the type I slide. Also, if there exists an entry $\overline{a}$ in the barred part of that column, then the above arguments show that before the empty box entered the column above the is, the entry which occupied the cell now occupied by a was always $< a$, so Lemma 2.1 tells us that the 1CC will hold for the new a–$\overline{a}$ pair.

Similarly, let us consider the column containing the js. Suppose there is an i and k such that $u_i = j_k$ say. Vertical constraints imply that such a u_i must occur in a row above that containing i_s in Figure 8.7(c). That is, $\mathrm{pos}(u_i) \leq \mathrm{pos}(h)$. Thus using (8.4), we have

$$\begin{aligned} \mathrm{pos}(u_i) + \mathrm{pos}(\overline{j_k}) &\leq \mathrm{pos}(h) + \mathrm{pos}(\overline{j_k}) \\ &= \mathrm{pos}(x) + \mathrm{pos}(\overline{x}) - p(f, x|\overline{x}, \overline{w_{k+1}}) - 2 \\ &\leq x - (x - j_k - 1) - 2 = j_k - 1, \end{aligned} \tag{8.5}$$

and hence the 1CC will hold when $\overline{j_k}$ moves up one position. We must also check the case when there already exists an entry a in the unbarred section of the column, as the creation of $\overline{a}$ may produce an a–$\overline{a}$ pair which fails the 1CC. The argument here is the same as in the case of the column containing the is.

Finally, we examine the 2CC. There are three possibilities for a 2CC violation to occur:

(a) there exist i, k such that $u_i = j_k$ which is part of an (e, u_i) configuration, where $e < a$.

(b) there exist r, s such that $i_r = v_s$ which is part of an (e, i_r) configuration, where $e < a$.

(c) there exist r, s (resp., i, k) such that $i_r = v_s$ (resp., $u_i = j_k$) which occur as part of an (a, i_r) (resp., (a, u_i)) configuration.

In case (a), the argument is similar to the 1CC case described above. From the definition of the 2CC, we know $p(e, x|\overline{x}, \overline{e}) < x - e$, and so it follows from this and (8.5) that $p(e, u_i|\overline{j_k}, \overline{e}) < j_k - e - 1$. Thus when j_s moves up one position during the type I slide, the 2CC will remain valid. The argument for case (b) is similar. In case (c), the argument is even simpler: since both i_1 and j_1 are $\geq a + 1$, Lemma 2.2 implies that for any $i_r = v_s$ for example, $p(i_1, i_r|\overline{v_s}, \overline{j_1}) < i_r - (a + 1)$ and so moving i_r down by one during the slide will not invalidate the 2CC.

To close this subsection, we remark that the proof of the validity of the type II slide follows a similar analysis as the type I slide and will not be repeated.

8.6 Type III, IV sliding. Let us consider the case of a type III slide. We must check that the four conditions for symplectic tableaux are preserved. Again, the vertical conditions clearly hold. For the horizontal conditions, consider the diagram below. During the type III slide, the entries $i_1, \ldots, i_p, k_0, \ldots, k_r$ move down one position so we need only concern ourselves with the entries to their immediate left, which could give rise to a violation of the horizontal condition. Similarly, the entries $\overline{j_1}, \ldots, \overline{j_q}, \overline{a}, \overline{l_1}, \ldots, \overline{l_s}$ move up one position, so we only need to look at the entries to their immediate right.

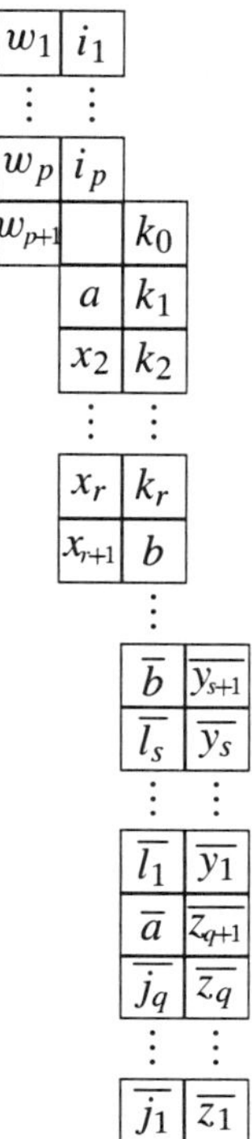

To prove that $w_{m+1} \le i_m$, for $1 \le m \le p$, we argue as in the previous section, that the empty box has slid vertically down from above i_1. The reasoning is almost identical: for example, if it enters between i_k and i_{k+1} via an ordinary slide, one can show there is some u lying above a, and $\overline{u}$ lying below $\overline{a}$ such that the union of the entries between u and a and between $\overline{u}$ and $\overline{a}$ is disjoint and equal to $\{u+1, u+2, \ldots, a-1\}$. This implies that $p(u, b|\overline{b}, \overline{u}) = p(a, b|\overline{b}, \overline{a}) + a - u + 1 = b - u$, using the fact that $p(a, b|\overline{b}, \overline{a}) = b - a - 1$. This is clearly impossible. The arguments for entering via a type I or III slide are identical to the previous section.

Similarly, we must prove $x_{m+1} \le k_m$, for $1 \le m \le r$ (certainly $a \le k_0$ as part of the conditions for a type III slide). The crucial point here is that since $p(a, b|\overline{b}, \overline{a}) = b - a - 1$, and b is the smallest integer satisfying that condition, we must have $\{k_0, k_1, \ldots, k_r\} \sqcup \{l_1, \ldots, l_s\} = \{a+1, a+2, \ldots, b-1\}$. In other words, we can consider the sequence $\overline{l_1}, \ldots, \overline{l_s}$ to be the consecutive sequence $a+1, a+2, \ldots, b-1$ interrupted by the absences of $k_0, \ldots, k_r$. In this setting, consider the entries x_{m+1} and $\overline{k_m + 1}$. It follows that

$$p(x_{m+1}, b|\overline{b}, \overline{k_m + 1}) = (r - m) + (b - (k_m + 1) - (r - m)) = b - k_m - 1.$$

However, Lemma 2.2 implies $p(x_{m+1}, b|\overline{b}, \overline{k_m + 1}) \le b - 1 - \min(x_{m+1}, k_m + 1)$. Hence we must have $x_{m+1} \le k_m$ as promised.

A similar argument to that presented in the above paragraph is used to show that $l_m \ge y_{m+1}$ for $1 \le m \le s$, while to show $j_m \ge z_{m+1}$ for $1 \le m \le q$, we can use a similar argument as was used in the previous section (used to show $j_k \geq w_{k+1}$ in Figure 8.7(a)).

Turning to the 1CC, the fact that this continues to hold in the column containing the is follows the same argument as used in the previous section. For the column containing the js, note that none of the ks can ever be part of some p–$\overline{p}$ pair, likewise with the ls. It remains to check the js themselves, and here we can follow the same argument as in the previous section (for the type I slide).

For the 2CC, the arguments are similar to the previous section and will not be repeated. Finally, we remark that the case of type IV slides can be analyzed in a similar manner.

8.7 Reverse slides. In this subsection, we will describe in what manner the reverse slides described in Section 4.2 really are "reverse." Indeed, we shall see that if one performs a sequence of forward slides on an empty box in a tableau, and then performs a sequence of reverse slides on the result, the paths taken by the empty box may not be the same. However the paths, and the corresponding tableaux, coincide at several points, in particular the initial point (after the bumping–sliding transition) and the final point (when the empty box moves to the edge of the tableau).

Consider first the case of the type I forward slide (refer to Figure 8.7(a)). There are two possible scenarios:

(a) $\{u_1, \ldots, u_p\} \cap \{j_1, \ldots, j_q\} \neq \emptyset$.

(b) $\{u_1, \ldots, u_p\} \cap \{j_1, \ldots, j_q\} = \emptyset$, i.e., $i_k = u_k$ for all $1 \leq k \leq p$.

In case (a), let s be the largest integer in $\{u_1, \ldots, u_p\} \cap \{j_1, \ldots, j_q\}$. If $u_r = s$, then the maximality of s implies that $u_m = i_m$ for $r + 1 \leq m \leq p$. If we now start to reverse slide the empty box in the right-hand diagram in Figure 8.7(a), it is clear that we will undertake a sequence of ordinary reverse vertical slides until the empty box is immediately below $u_r = s$. Using the maximality of s, one can show that we have $p(a, s|\overline{s}, \overline{a}) = s - a - 1$ and hence the next slide is a type III^R reverse slide. Since the smallest integer which does *not* appear below the empty box nor above $\overline{s}$ is x, the pair x–$\overline{x}$ gets created during the III^R slide and the pair a–$\overline{a}$ gets deleted. Moreover, the empty box gets inserted between i_r and i_{r+1}. From the previously derived inequalities $t_{k+1} \leq i_k$, the empty box now slides vertically up to the top of the column of is, and the situation now appears as it did before the type I forward slide.

In case (b), if we start reverse sliding the empty box in the right-hand diagram of Figure 8.7(a), then we will perform a sequence of ordinary vertical slides until the empty box is immediately below u_0. At that stage, we then have two subcases depending on whether $u_0 \geq a$ or $u_0 < a$. In the former case, it is clear that we have the necessary conditions for a III^R slide (since $u_0 < u_1 = i_1$ and so all the elements $\overline{a+1}, \overline{a+2}, \ldots, \overline{u_0 - 1}, \overline{u_0}$ appear immediately above $\overline{a}$). After performing such a slide, the empty box appears at the top of the column of is and the entries are precisely as they appeared before the forward slide. In the latter case, the situation is ripe for an I^R reverse slide, and again using the fact that $u_m = i_m$ for $1 \le m \le p$, after such a reverse slide the situation will be as at the beginning of the forward slide.

Next, consider a type III forward slide, the aftermath of which appears in the diagram below. We recall that $\{i_1, \ldots, i_p\} \sqcup \{j_1, \ldots, j_q\} = \{c+1, c+2, \ldots, a-1\}$ and $\{k_0, \ldots, k_r\} \sqcup \{l_1, \ldots, l_s\} = \{a+1, a+2, \ldots, b-1\}$. There are three possible situations to consider:

(a) There is an m, n such that $u_m = j_n$.

(b) There is an m, n such that $u_m = l_n$. (We include the case $u_m = l_0 := a$.)

(c) The conditions in (a) and (b) do not hold, so that $i_m = u_m$ for $1 \le m \le p$.

In case (a), the argument is the same as case (a) above for the type I slide, namely we will slide up vertically until the empty box is just below $u_m := t$ where t is the maximal element in $\{u_1, \ldots, u_p\} \cap \{j_1, \ldots, j_q\}$ at which point a type III^R reverse slide takes place. Note that the entries below t are $i_{m+1}, \ldots, i_p, k_0, \ldots, k_r$ while the entries above $\overline{t} = \overline{j_n}$ are $\overline{j_{n+1}}, \ldots, \overline{j_q}, \overline{a}, \overline{l_1}, \ldots, \overline{l_s}$. From the remarks above, the smallest (barred or unbarred) integer not appearing in the union of these entries is b. Hence after carrying out the reverse slide, the pair b–$\overline{b}$ gets created as required, while the empty box then slides vertically to the top of the column of is, and the situation is as it appeared before the initial forward III slide.

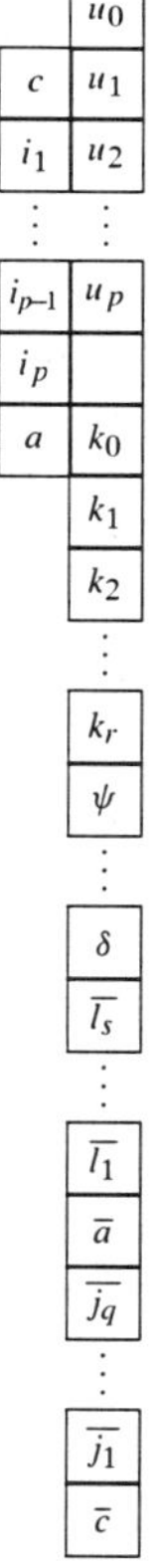

In case (b), if $u_m = l_n$ for some m, n, then we must have $u_p \geq a$ and hence a similar argument as above shows that the *very first* reverse slide must be a type III^R slide, the effect of which is the same as in the (a) case.

Finally, in case (c), as in case (b) for the forward type I slide, we will vertically slide the box until just below the entry u_0, at which stage we must have a reverse I^R or III^R slide according as $u_0 < a$ or $u_0 \geq a$ and the final result is as appears before the forward III slide.

We remark that the cases of reversing the forward II and IV slides are similar.

Acknowledgments. The author thanks the following people for helpful discussions or correspondence: O. Foda, P. Jarvis, M. Kashiwara, T. Miwa, S. Okada, M. Okado, Y. Pugai, A. Schilling, D. Uglov, and O. Warnaar. Special thanks are due to T. Welsh for many helpful discussions and comments and for reading through a previous draft. The financial support of the Japan Society for the Promotion of Science is acknowledged.

Note added. After completing this work, we became aware of [16] where a similar C_n insertion scheme is considered, and its relation to Sheats's "symplectic jeu-de-taquin" [28] is discussed. The author thanks B. Leclerc for informing him of this related work and C. Lecouvey for providing it.

References

[1] G. Benkart and J. Stroomer, Tableaux and insertion schemes for spinor representations of the orthogonal Lie algebra $so(2n+1, C)$, *J. Combin. Theory Ser. A*, **57** (1991), 211–237.

[2] A. Berele, A Schensted-type correspondence for the symplectic group, *J. Combin. Theory Ser. A*, **43** (1986), 320–328.

[3] C. De Concini, Symplectic standard tableaux, *Adv. Math.*, **34** (1979), 1–27.

[4] M. Fulmek and C. Krattenthaler, A bijection between Proctor's and Sundaram's odd orthogonal tableaux, *Discrete Math.*, **161** (1996), 101–120.

[5] W. Fulton, *Young Tableaux*, first ed., Cambridge University Press, Cambridge, 1997.

[6] M. Jimbo and T. Miwa, *Algebraic Analysis of Solvable Lattice Models*, CBMS Regional Conference Series in Mathematics 85, AMS, Providence, 1995, 1–152.

[7] M. Kashiwara, Crystallizing the q-analogue of universal enveloping algebra, *Comm. Math. Phys.*, **133** (1990), 249–260.

[8] M. Kashiwara and T. Nakashima, Crystal graphs for representations of the q-analogue of classical Lie algebras, *J. Algebra*, **165** (1994), 295–345.

[9] R. C. King, Weight multiplicities for the classical groups, in *Lecture Notes in Physics* 50, Springer-Verlag, New York, 1975, pp. 490–499.

[10] R. C. King and N. G. I. El-Sharkaway, Standard Young tableaux and weight multiplicities of the classical Lie groups, *J. Phys. A*, **16** (1983), 3153–3177.

[11] R. C. King and T. A. Welsh, Construction of orthogonal group modules using tableaux, *Linear Multlinear Algebra*, **33** (1993), 251–283.

[12] D. E. Knuth, Permutations, matrices and generalized Young tableaux, *Pacific J. Math.*, **34** (1970), 709–727.

[13] K. Koike and I. Terada, Young diagrammatic methods for the restriction of representations of complex classical Lie groups to reductive subgroups of maximal rank, *Adv. Math.*, **79** (1990), 104–135.

[14] A. Lascoux, B. Leclerc, and J-Y. Thibon, Crystal graphs and q-analogs of weight multiplicities for the root system A_n, *Lett. Math. Phys.*, **35** (1995), 359–374.

[15] A. Lascoux, B. Leclerc, and J-Y. Thibon, The plactic monoid, in *Algebraic Combinatorics on Words*, M. Lothaire, ed., pp. 146–174, to appear; `www-igm.univ-mlv.fr/~berstel/Lothaire/index.html`.

[16] C. Lecouvey, Schensted-type correspondence, plactic monoid and jeu-de-taquin for type C_n, preprint, Université Caen, Caen, France, 1999.

[17] P. Littelmann, A plactic algebra for semi-simple Lie algebras, *Adv. Math.*, **124** (1996), 312–331.

[18] I. G. Macdonald, *Symmetric Functions and Hall Polynomials*, second ed., Oxford University Press, Oxford, 1995.

[19] T. Nakashima, Crystal base and a generalization of Littlewood-Richardson rule for the classical Lie algebras, *Comm. Math. Phys.*, **154** (1993), 215–243.

[20] S. Okada, A Robinson-Schensted type algorithm for $so(2n, C)$, *J. Algebra*, **143** (1991), 334–372.

[21] S. Okada, Robinson-Schensted type algorithms for the spinor representations of the orthogonal Lie algebra $so(2n, C)$, *J. Algebra*, **158** (1993), 155–200.

[22] R. A. Proctor, A Schensted algorithm which models tensor representations of the orthogonal group, *Canad. J. Math.*, **42** (1990), 28–49.

[23] R. A. Proctor, Young tableaux, Gelfand patterns and branching rules for classical Lie groups, *J. Algebra*, **164** (1994), 299–360.

[24] G. de B. Robinson, On the representations of the symmetric group, *Amer. J. Math.*, **60** (1938), 745–760.

[25] C. Schensted, Longest increasing and decreasing subsequences, *Canad. J. Math.*, 13:179–191, 1961.

[26] M-P. Schützenberger, Promotion des morphisms d'ensembles ordonnés, *Discrete Math.*, **2** (1972), 73–94.

[27] M-P. Schützenberger, Las correspondance de Robinson, in *Lecture Notes in Mathematics* 579, Springer-Verlag, Berlin, New York, Heidelberg, 1977, pp. 59–135.

[28] J. T. Sheats, A symplectic jeu-de-taquin bijection between the tableaux of King and of De-Concini, preprint, University of North Carolina, Chapel Hill, NC, 1995.

[29] S. Sundaram, The Cauchy identity for $sp(2n)$, *J. Combin. Theory Ser. A*, **53** (1990), 209–238.

[30] S. Sundaram, Orthogonal tableaux and an insertion algorithm for $SO(2n+1)$, *J. Combin. Theory Ser. A*, **53** (1990), 239–256.

Research Institute for Mathematical Sciences
Kyoto University
Kyoto 606-8502
Japan
tbaker@kurims.kyoto-u.ac.jp

On the Combinatorics of Forrester–Baxter Models

Omar Foda and Trevor A. Welsh

Abstract. We provide further boson–fermion q-polynomial identities for the "finitized" Virasoro characters $\chi_{r,s}^{p,p'}$ of the Forrester–Baxter minimal models $M(p, p')$ for certain values of r and s. The construction is based on a detailed analysis of the combinatorics of the set $\mathcal{P}_{a,b,c}^{p,p'}(L)$ of q-weighted, length-L Forrester–Baxter paths, whose generating function $\chi_{a,b,c}^{p,p'}(L)$ provides a finitization of $\chi_{r,s}^{p,p'}$. In this paper, we restrict our attention to the case where the startpoint a and endpoint b of each path both belong to the set of *Takahashi lengths*. In the limit $L \to \infty$, these polynomial identities reduce to q-series identities for the corresponding characters.

We obtain two closely related fermionic polynomial forms for each (finitized) character. The first of these forms uses the classical definition of the Gaussian polynomials and includes a term that is a (finitized) character of a certain $M(\hat{p}, \hat{p}')$, where $\hat{p}' < p'$. We provide a combinatorial interpretation for this form using the concept of *particles*. The second form, which was first obtained using different methods by the Stony Brook group, requires a modified definition of the Gaussian polynomials, and its combinatorial interpretation requires not only the concept of particles, but also the additional concept of *particle annihilation*.

0 Introduction

0.1 Motivation. The physical spectrum of exactly solvable lattice models can be described in the language of highest-weight infinite-dimensional representations of affine and Virasoro algebras [16]. The characters of these representations are q-series that contain detailed information on the structure and symmetries of the corresponding models. In the following discussion, we wish to restrict attention to the characters of Virasoro highest-weight representations.

The earliest known expressions for these characters are due to Feigen and Fuchs [9] and Rocha-Caridi [17]. These expressions have alternating-signs. A number of years ago, the Stony Brook group discovered completely new expressions for the character formulae.[1] These expressions have constant signs.[2]

For physical reasons that are beyond the scope of this work, the original alternating-sign expressions are also known as *bosonic characters*. Correspondingly, the constant-sign expressions are also known as *fermionic characters*.[3]

The structure of these new character formulae hints at the presence of a completely new formulation of exactly solvable models.[4] This possibility has attracted attention for a number of reasons. One of these reasons is the fact that certain physical problems, such as the long-distance asymptotics of the correlation functions, are too difficult to handle in the current formulation. Further, there are reasons to believe that the new formulation could be the right starting point to tackle them. (See [6] and references therein.) At a more technical level, the availability of two distinct formulations is mathematically enriching, as we can use one to learn about the other.

However, although the bosonic characters are technically simple to write down, and are completely known for all Virasoro representations, the structure of the fermionic characters is strictly speaking known explicitly only in special cases, and generally only conceptually. In particular, the characters of the *nonunitary* Virasoro representations have turned out to be rather resistant to a complete formulation in fermionic form.[5]

This work is part of a series of papers that aim at a complete and explicit derivation of the fermionic characters of a certain class of models first discussed by Forrester and Baxter [14]. The characters of the Forrester–Baxter models correspond to the complete set of Virasoro characters of the discrete, though not necessarily unitary, Virasoro algebras with central charge $c < 1$, first discussed in [4]. As such, they form the largest class of Virasoro characters with no W-symmetries.

As in previous works, our approach is purely combinatorial. Further, the exposition is self-contained, in the sense that we have included all concepts required in the derivations. Our main result is a combinatorial derivation of two related finitized fermionic forms for the characters of a certain class of Forrester–Baxter models. The first of these requires the use of the classical form of Gaussian polynomials and can be interpreted combinatorially using the concept of *particles*. The second has already

---

[1]For references to the original Stony Brook papers, please refer to [6].

[2]The characterization of the different expressions of the characters as "alternating-sign" and "constant-sign" q-series is valid only for Virasoro but not for affine characters.

[3]For a complete discussion of the physical motivation of the terms "bosonic" and "fermionic," please refer to the original literature on the subject as cited in [6].

[4]Analogous developments in the context of highest-weight representations of affine algebras also took place. They are outside the scope of this work.

[5]The reason for that may of course eventually turn out to be the fact that we are not using the most efficient tools to tackle this problem.

appeared in the works of Berkovich, McCoy, and Schilling [7], requires the use of a modified form of Gaussian polynomials, and has a combinatorial interpretation in terms of particles and *particle annihilation*.

In a forthcoming paper, we further extend and refine the techniques of this work to obtain a complete and explicit derivation of the fermionic characters of the complete set of Forrester–Baxter models [13].

0.2 Overview of paper. The aim of this paper is to obtain fermionic expressions for $\chi^{p,p'}_{a,b,c}(L)$, the generating function for the set $\mathcal{P}^{p,p'}_{a,b,c}(L)$ of restricted length-L paths that have startpoint a and endpoint b.

These functions[6] first arose in the calculation of one-point functions of the Forrester–Baxter models [14]. The weighting originally assigned in [14] to the paths is significantly different from that used here. The weighting described in the current paper arose by obtaining a "weight-preserving" bijection between partitions with prescribed hook differences that were considered in [3] and the paths of [14]. This bijection is described in [10].

The paths in $\mathcal{P}^{p,p'}_{a,b,c}(L)$ may be depicted on a $(p'-2)\times L$ grid that we refer to as the (p,p')-model, as described in Section 1.1. Of particular importance is the shading of the (p,p')-model, which determines the weights $wt(h)$ that we assign to the paths $h\in\mathcal{P}^{p,p'}_{a,b,c}(L)$.

A bosonic expression for $\chi^{p,p'}_{a,b,c}(L)$ is given in Section 1.3. This expression is readily proved using L-recurrence relations [14], or by using the generating function for partitions with prescribed hook differences given in [3], and the bijection of [10]. The polynomial $\chi^{p,p'}_{a,b,c}(L)$ is seen to be a finitization of a Virasoro character.

In this paper, we tackle the particular cases where a and b are each one of the Takahashi lengths $\mathcal{T}$, or one of $\mathcal{T}'=\{p'-s : s\in\mathcal{T}\}$. These values depend on p and p', and are defined in Section 5.1. Our methods and results are a common generalization of those of [10, 12].

On equating the bosonic expression for $\chi^{p,p'}_{a,b,c}(L)$ with either of the fermionic expressions, we obtain boson–fermion polynomial identities. Taking the $L\to\infty$ limit (using, for example, the variable change employed in [10, 11]), these become q-series identities. Among them, in particular, are the Rogers–Ramanujan identities, and their generalizations by Andrews and Gordon [2]. In fact, the techniques employed by Agarwal and Bressoud [1, 8] in their combinatorial proof of the Andrews–Gordon identities provided the genesis of the techniques employed here.

Before we develop a generalization of Agarwal and Bressoud's 'Volcanic activity', we define in Section 2, a slightly different set $\mathcal{P}^{p,p'}_{a,b,e,f}(L)$ of paths, which have assigned presegments and postsegments that are determined by $e,f\in\{0,1\}$. Their generating function $\tilde{\chi}^{p,p'}_{a,b,e,f}(L)$ is defined in terms of a path weighting that differs slightly from that defined earlier.

[6]To be precise, a certain renormalization thereof.

The $\mathcal{B}$-transform, which is described in Section 3, enables $\tilde{\chi}^{p,p'+p}_{a',b',e,f}(L')$, for certain a', b' to be expressed in terms of $\tilde{\chi}^{p,p'+p}_{a,b,e,f}(L)$. We derive this transform combinatorially in three steps. The first step is known as the $\mathcal{B}_1$-transform and enlarges the features of a path, so that the resultant path resides in a larger model. The second step, referred to as a $\mathcal{B}_2(k)$-transform, lengthens a path by appending k pairs of segments to the path. Each of these pairs is known as a particle. The third step, the $\mathcal{B}_3(\lambda)$-transform deforms the path in a particular way. This process may be viewed as the particles *moving* through the path. The resulting transformation of generating functions is given in Corollary 3.14.

In Section 4, we see that $\tilde{\chi}^{p'-p,p'}_{a,b,1-e,1-f}(L)$ may be obtained from $\tilde{\chi}^{p,p'}_{a,b,e,f}(L)$ in a combinatorially trivial way. This process is referred to as a $\mathcal{D}$-transform. In fact, it is more convenient to use the $\mathcal{D}$-transform combined with the $\mathcal{B}$-transform. The resulting transformation of generating functions is given in Corollary 4.6.

To obtain a particular generating function $\tilde{\chi}^{p,p'}_{a,b,e,f}(L)$, where p and p' are coprime, we begin with one of the trivial generating functions $\tilde{\chi}^{1,3}_{a',b',e',f'}(L)$ given in Lemma 2.5, and perform a sequence of $\mathcal{B}$- and $\mathcal{BD}$-transforms. This sequence is determined by the continued fraction of p'/p which is described in Section 5.1.

In fact, a basic application of the transforms does not generate all elements of $\mathcal{P}^{p,p'}_{a,b,e,f}(L)$ in some cases. In these instances, the set generated is deficient in the full set of paths that do not rise above (or below) a certain height. Various results obtained in Section 6 enable us to keep track of this height. Lemma 6.4 shows that this height bounds a portion of the (p, p')-model which is identical to a smaller $(\hat{p}, \hat{p}')$-model. This property enables (in one case), the final generating function to be expressed using the generating function for paths in the $(\hat{p}, \hat{p}')$-model.

Section 7 provides one further ingredient for the final construction. There it is shown how appending or removing the first segment of the path affects the generating function.

Everything is now in place to carry out the proof of the main results. These results are stated in Section 8.1. We provide two similar expressions for $\chi^{p,p'}_{a,b,c}(L)$. These are Theorems 8.1 and 8.2. The first of these makes use of the classical definition of the Gaussian polynomial:

$$\begin{bmatrix} A \\ B \end{bmatrix}_q = \begin{cases} \dfrac{(q)_A}{(q)_{A-B}(q)_B} & \text{if } 0 \le B \le A, \\ 0 & \text{otherwise,} \end{cases} \tag{1}$$

where $(q)_0 = 1$ and $(q)_n = \prod_{i=1}^{n}(1-q^i)$ for $n > 0$. In some cases, the expression also includes a term $\chi^{\hat{p},\hat{p}'}_{\hat{a},\hat{b},\hat{c}}(L)$ for $\hat{p}' < p'$. Thus this expression may be viewed as a recursive fermionic expression for $\chi^{p,p'}_{a,b,c}(L)$. In the cases where this additional term is not present (for a and b further restricted in a certain way), the expressions were first stated in [5].

The expression of Theorem 8.2 makes use of a modified definition of the Gaussian polynomial [15]:

$$\begin{bmatrix} A \\ B \end{bmatrix}_q' = \begin{cases} \dfrac{(q^{A-B+1})_B}{(q)_B} & \text{if } 0 \le B, \\ 0 & \text{otherwise,} \end{cases} \tag{2}$$

where $(z)_0 = 1$ and $(z)_n = \prod_{i=0}^{n-1}(1 - zq^i)$ for $n > 0$. These expressions were first presented and proved in [7]. In fact, invoking the definition (2) is somewhat overkill, since the only value of $\left[{A \atop B}\right]_q'$ that we require that differs from $\left[{A \atop B}\right]_q$ is $\left[{-1 \atop 0}\right]_q' = 1$.

In [7], expressions for $\chi^{p,p'}_{a,b,c}(L)$ are presented, where b is now any value with $1 \le b \le p' - 1$. However, only $a \in \mathcal{T} \cup \mathcal{T}'$ is still permitted. In [13], we show that it is Theorem 8.1, and not Theorem 8.2, that generalizes to provide fermionic expressions for the most general $\chi^{p,p'}_{a,b,c}(L)$.

The remainder of Section 8 is concerned with the detailed derivation of the expression for first $\tilde{\chi}^{p,p'}_{a,b,e,f}(L)$, and then converting it to $\chi^{p,p'}_{a,b,c}(L)$. Section 8.3 describes the ***mn***-system which aids the actual evaluation of the fermionic expressions obtained. Section 8.4 describes how the proof for Theorem 8.1 modifies to provide a proof for Theorem 8.2. Here we see that the appearance of $\left[{-1 \atop 0}\right]_q'$ may be viewed in terms of "particle annihilation."

1 Paths

1.1 Paths and the (p, p')-model. Let p and p' be positive coprime integers for which $0 < p < p'$. Then, given $a, b, c, L \in \mathbb{Z}_{\ge 0}$ such that $1 \le a, b, c \le p' - 1$, $b = c \pm 1$, $L + a - b \equiv 0 \pmod 2$, a path $h \in \mathcal{P}^{p,p'}_{a,b,c}(L)$ is a sequence $h_0, h_1, h_2, \ldots, h_L$, of integers such that

1. $1 \le h_i \le p' - 1$ for $0 \le i \le L$;
2. $h_{i+1} = h_i \pm 1$ for $0 \le i < L$;
3. $h_0 = a$, $h_L = b$.

Note that the values of p and c do not feature in the above restrictions. As described below, they specify how the elements of $\mathcal{P}^{p,p'}_{a,b,c}(L)$ are weighted.

The integers $h_0, h_1, h_2, \ldots, h_L$, are readily depicted as a sequence of *heights* on a two-dimensional $L \times (p' - 2)$ grid. Adjacent heights are connected by *line segments* passing from (i, h_i) to $(i + 1, h_{i+1})$ for $0 \le i < L$.

Scanning the path from left to right, each of these line segments points either in the NE direction or in the SE direction. Figure 1 shows a typical path in the set $\mathcal{P}^{3,8}_{2,4,3}(14)$. The shadings in Figure 1 are explained below.

In the grid introduced above, the horizontal strip between adjacent heights is referred to as a *band*. There are $p' - 2$ bands. The hth band lies between heights h and $h + 1$.

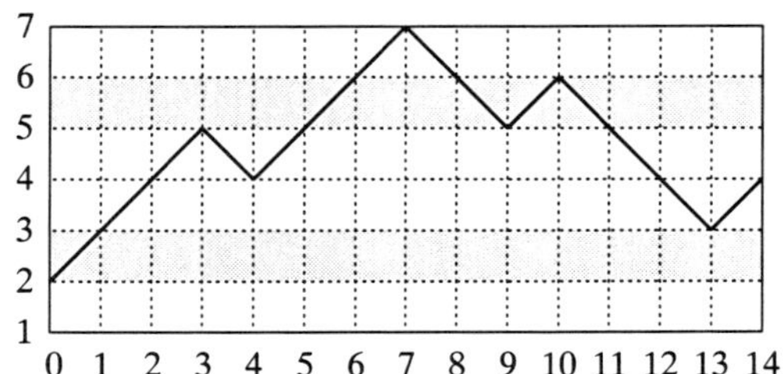

Figure 1. Typical path.

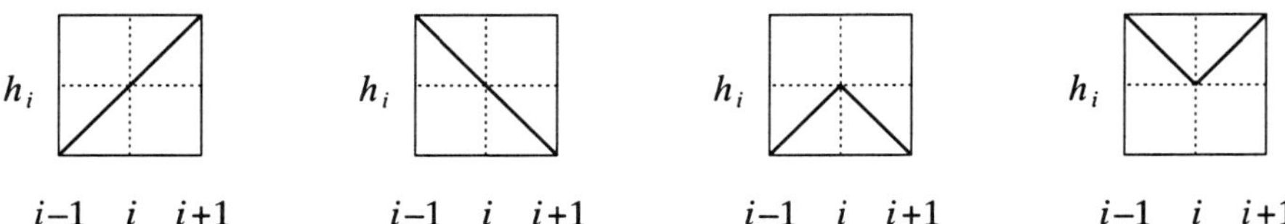

Figure 2. Vertex shapes.

We now assign a parity to each band: the hth band is said to be an *even* band if $\lfloor hp/p' \rfloor = \lfloor (h+1)p/p' \rfloor$; and an *odd* band if $\lfloor hp/p' \rfloor \neq \lfloor (h+1)p/p' \rfloor$. The array of odd and even bands so obtained will be referred to as the (p, p')-model. It may immediately be deduced that the (p, p')-model has $p' - p - 1$ even bands and $p - 1$ odd bands. In addition, it is easily shown that for $1 \leq r < p$, the band lying between heights $\lfloor rp'/p \rfloor$ and $\lfloor rp'/p \rfloor + 1$ is odd: it will be referred to as the rth odd band.

When drawing the (p, p')-model, we distinguish the bands by shading the odd bands. This was done in Figure 1 for the (3, 8)-model.

We note that the band structure of the (p, p')-model is up–down symmetrical and that if $p' > 2p$, then the first band and the $(p' - 2)$th band are both even, and there are no two adjacent odd bands.

For $2 \leq a \leq p' - 2$, we say that a is *interfacial* if $\lfloor (a+1)p/p' \rfloor = \lfloor (a-1)p/p' \rfloor + 1$. Thus a is interfacial if and only if a lies between an odd and even band in the (p, p')-model. Thus for the case of the (3, 8)-model depicted in Figure 1, a is interfacial for $a = 2, 3, 5, 6$. Note that if a is interfacial, the odd band that it borders is the $\lfloor (a+1)p/p' \rfloor$th.

As is easily seen, the $(p' - p, p')$-model differs from the (p, p')-model in that each band has changed parity. It follows that if a is interfacial in the (p, p')-model then a is also interfacial in the $(p' - p, p')$-model.

1.2 Weighting the paths. Given a path h of length L, for $1 \leq i < L$, the values of h_{i-1}, h_i and h_{i+1} determine the shape of the vertex at the point i. The four possible shapes are given in Figure 2.

The four types of vertices shown in Figure 2 are referred to as a *straight-up* vertex, a *straight-down* vertex, a *peak-up* vertex, and a *peak-down* vertex, respectively. Each

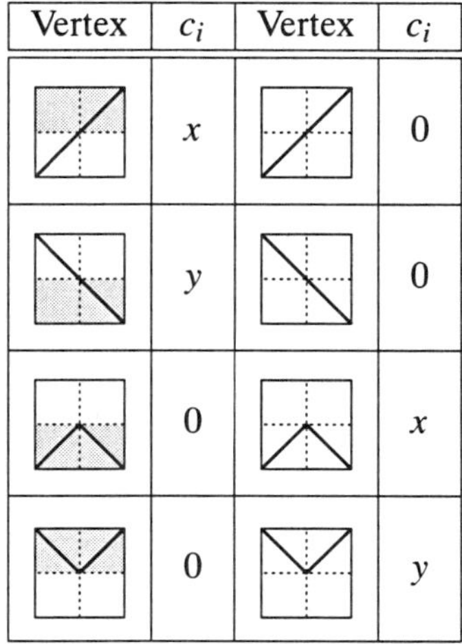

Vertex	c_i	Vertex	c_i
	x		0
	y		0
	0		x
	0		y

Table 1. Vertex weights.

vertex is also assigned a parity: this is the parity of the band in which the segment between (i, h_i) and $(i+1, h_{i+1})$ lies. Thus, there are eight types of paritied vertex.

For paths $h \in \mathcal{P}^{p,p'}_{a,b,c}(L)$, we define $h_{L+1} = c$, whereupon the shape and parity of the vertex at $i = L$ is well defined.

The weight function for the paths is best specified in terms of an (x, y)-coordinate system which is inclined at 45° to the original (i, h)-coordinate system and whose origin is at the path's initial point at $(i = 0, h = a)$. Specifically,

$$x = \frac{i-(h-a)}{2}, \qquad y = \frac{i+(h-a)}{2}.$$

Note that at each step in the path, either x or y is incremented and the other is constant. In this system, the path depicted in Figure 1 has its first few coordinates at $(0, 0)$, $(0, 1)$, $(0, 2)$, $(0, 3)$, $(1, 3)$, $(1, 4)$, $(1, 5)$, $(1, 6)$, $(2, 6)$, ...

Now, for $1 \leq i \leq L$, we define the weight $c_i = c(h_{i-1}, h_i, h_{i+1})$ of the ith vertex according to its shape, its parity and its (x, y)-coordinate, as specified in Table 1.

In Table 1, the lightly shaded bands can be either even or odd bands (or when $h_i = p' - 1$ or $h_i = 1$ in the lowermost four cases, not a band in the model at all). Note that for each vertex shape, only one parity case has nonzero weight in general. We shall refer to those four vertices, with assigned parity, for which the weight is in general nonzero, as *scoring* vertices. The other four vertices will be termed *nonscoring*.

We now define

$$wt(h) = \sum_{i=1}^{L} c_i. \tag{3}$$

To illustrate this procedure, consider again the path h depicted in Figure 1. The above table indicates that there are scoring vertices at $i = 3, 4, 5, 7, 8, 13$, and 14. This leads to

$$wt(h) = 0 + 3 + 1 + 1 + 6 + 7 + 6 = 24.$$

The generating function $\chi^{p,p'}_{a,b,c}(L)$ for the set of paths $\mathcal{P}^{p,p'}_{a,b,c}(L)$ is defined to be

$$\chi^{p,p'}_{a,b,c}(L;q) = \sum_{h\in\mathcal{P}^{p,p'}_{a,b,c}(L)} q^{wt(h)}. \tag{4}$$

Often, we drop the base q from the notation so that $\chi^{p,p'}_{a,b,c}(L) = \chi^{p,p'}_{a,b,c}(L;q)$. The same will be done for other functions without comment.

1.3 Bosonic generating function. By setting up recurrence relations for $\chi^{p,p'}_{a,b,c}(L)$, it may be readily verified that

$$\begin{aligned}\chi^{p,p'}_{a,b,c}(L) &= \sum_{\lambda=-\infty}^{\infty} q^{\lambda^2 pp'+\lambda(p'r-pa)} \begin{bmatrix} L \\ \frac{L+a-b}{2} - p'\lambda \end{bmatrix}_q \\ &\quad - \sum_{\lambda=-\infty}^{\infty} q^{(\lambda p+r)(\lambda p'+a)} \begin{bmatrix} L \\ \frac{L+a-b}{2} - p'\lambda - a \end{bmatrix}_q,\end{aligned} \tag{5}$$

where

$$r = \lfloor pc/p' \rfloor + (b-c+1)/2. \tag{6}$$

In the limit $L \to \infty$, we obtain

$$\lim_{L\to\infty} \chi^{p,p'}_{a,b,c}(L) = \chi^{p,p'}_{r,a}, \tag{7}$$

where r is defined in (6) and

$$\chi^{p,p'}_{r,s} = \frac{1}{(q)_\infty} \sum_{\lambda=-\infty}^{\infty} (q^{\lambda^2 pp'+\lambda(p'r-ps)} - q^{(\lambda p+r)(\lambda p'+s)}) \tag{8}$$

is, up to a normalization, the Rocha-Caridi expression [17] for the Virasoro character of central charge $c = 1 - 6(p'-p)^2/pp'$ and conformal dimension $\Delta^{p,p'}_{r,s} = \big((p'r-ps)^2 - (p'-p)^2\big)/4pp'$. Therefore, $\chi^{p,p'}_{a,b,c}(L)$ provides a finite analogue of the character $\chi^{p,p'}_{r,a}$.

2 Winged generating functions

For $h \in \mathcal{P}^{p,p'}_{a,b,c}(L)$, the values of b and c serve to specify a path *postsegment* that extends between (L,b) and $(L+1,c)$. We now define another set of paths which specifies both the direction of a postsegment and a *presegment*.

Let p and p' be positive coprime integers for which $0 < p < p'$. Then, given $a, b, L \in \mathbb{Z}_{\geq 0}$ such that $1 \leq a, b \leq p'-1$, $L+a-b \equiv 0 \pmod 2$, and $e, f \in \{0,1\}$, a path $h \in \mathcal{P}^{p,p'}_{a,b,e,f}(L)$ is a sequence $h_0, h_1, h_2, \ldots, h_L$, of integers such that

1. $1 \le h_i \le p' - 1$ for $0 \le i \le L$;
2. $h_{i+1} = h_i \pm 1$ for $0 \le i < L$;
3. $h_0 = a$, $h_L = b$.

If $f = 0$ (resp., $f = 1$), then the postsegment of each $h \in \mathcal{P}^{p,p'}_{a,b,e,f}(L)$ is defined to be in the NE (resp., SE) direction. If $e = 0$ (resp., $e = 1$), then the presegment of each $h \in \mathcal{P}^{p,p'}_{a,b,e,f}(L)$ is defined to be in the SE (resp., NE) direction. This enables a shape and a parity to be assigned to both the zeroth and the Lth vertices of h. For $h \in \mathcal{P}^{p,p'}_{a,b,e,f}(L)$, we define $e(h) = e$ and $f(h) = f$.

We now define a weight $\widetilde{wt}(h)$ for $h \in \mathcal{P}^{p,p'}_{a,b,e,f}(L)$. For $1 \le i < L$, set $\tilde{c}_i = c(h_{i-1}, h_i, h_{i+1})$ as above. Then set

$$\tilde{c}_L = \begin{cases} x & \text{if } h_L - h_{L-1} = 1 \text{ and } f(h) = 1, \\ y & \text{if } h_L - h_{L-1} = -1 \text{ and } f(h) = 0, \\ 0 & \text{otherwise,} \end{cases}$$

where (x, y) is the coordinate of the Lth vertex of h. We then designate this vertex as scoring if it is a peak vertex ($h_L = h_{L-1} - (-1)^{f(h)}$), and as nonscoring otherwise.

We define

$$\widetilde{wt}(h) = \sum_{i=1}^{L} \tilde{c}_i. \tag{9}$$

Consider the corresponding path $h' \in \mathcal{P}^{p,p'}_{a,b,c}(L)$ with $c = b + (-1)^f$, defined by $h'_i = h_i$ for $0 \le i \le L$. From Table 1, we see that $\widetilde{wt}(h) = wt(h')$ if the postsegment of h lies in an even band.

In what follows, we work entirely in terms of $\widetilde{wt}(h)$, and the generating functions that we derive from it. Only at the end of our work do we revert back to $wt(h)$ to obtain fermionic expressions for $\chi^{p,p'}_{a,b,c}(L)$.

Define the generating function

$$\tilde{\chi}^{p,p'}_{a,b,e,f}(L; q) = \sum_{h \in \mathcal{P}^{p,p'}_{a,b,e,f}(L)} q^{\widetilde{wt}(h)}, \tag{10}$$

where $\widetilde{wt}(h)$ is given by (9). Of course, $\tilde{\chi}^{p,p'}_{a,b,0,f}(L) = \tilde{\chi}^{p,p'}_{a,b,1,f}(L)$.

2.1 Striking sequence of a path. For each path h, define $\pi(h) \in \{0, 1\}$ to be the parity of the band between heights h_0 and h_1. (If $L(h) = 0$, we set $h_1 = h_0 + (-1)^{f(h)}$.) Thus for the path h shown in Figure 1, we have $\pi(h) = 1$. In addition, define $d(h) = 0$ when $h_1 - h_0 = 1$ and $d(h) = 1$ when $h_1 - h_0 = -1$. We then see that if $e(h)+d(h)+\pi(h)\equiv 0 \pmod 2$, then the zeroth vertex is a scoring vertex, and if $e(h)+d(h)+\pi(h)\equiv 1 \pmod 2$, then it is a nonscoring vertex.

Now consider each path $h\in\mathcal{P}^{p,p'}_{a,b,e,f}(L)$ as a sequence of straight lines, alternating in direction between NE and SE. Then, reading from the left, let the lines be of lengths $w_1, w_2, w_3, \ldots, w_l$ for some l with $w_i>0$ for $1\le i\le l$. Thence $w_1+w_2+\cdots+w_l=L(h)$, where $L(h)=L$ is the length of h.

For each of these lines, the last vertex will be considered to be part of the line but the first will not. Then the ith of these lines contains w_i vertices, the first w_i-1 of which are straight vertices. Then write $w_i=a_i+b_i$ so that b_i is the number of scoring vertices in the ith line. The striking sequence of h is then the array

$$\begin{pmatrix} a_1 & a_2 & a_3 & \cdots & a_l \\ b_1 & b_2 & b_3 & \cdots & b_l \end{pmatrix}^{(e(h),f(h),d(h))}.$$

With $\pi=\pi(h)$, $e=e(h)$, $f=f(h)$, and $d=d(h)$, we define

$$m(h)=\begin{cases} (e+d+\pi)\bmod 2+\sum_{i=1}^{l}a_i & \text{if } L>0,\\ |f-e| & \text{if } L=0,\end{cases}$$

whence $m(h)$ is the number of nonscoring vertices possessed by h (altogether, h has $L(h)+1$ vertices). We also define $\alpha(h)=(-1)^d((w_1+w_3+\cdots)-(w_2+w_4+\cdots))$ and for $L>0$,

$$\beta(h)=\begin{cases} (-1)^d((b_1+b_3+\cdots)-(b_2+b_4+\cdots)) \\ \qquad\qquad \text{if } e+d+\pi\equiv 0 \pmod 2,\\ (-1)^d((b_1+b_3+\cdots)-(b_2+b_4+\cdots))+(-1)^e \\ \qquad\qquad \text{otherwise.}\end{cases}$$

For $L=0$, we set $\beta(h)=f-e$.

For example, for the path shown in Figure 1 for which $d(h)=0$ and $\pi(h)=1$, the striking sequence is

$$\begin{pmatrix} 2&0&1&1&1&2&0\\ 1&1&2&1&0&1&1\end{pmatrix}^{(e,1,0)}.$$

In this case, $m(h)=8-e$, $\alpha(h)=2$, and $\beta(h)=2-e$.

We note that given the startpoint $h_0=a$ of the path, the path can be reconstructed from its striking sequence.[7] In particular, $h_L=b=a+\alpha(h)$. In addition, the nature of the final vertex may be deduced from a_l and b_l.[8]

Lemma 2.1. *Let the path h have the striking sequence* $\left(\begin{smallmatrix} a_1 & a_2 & a_3 & \cdots & a_l \\ b_1 & b_2 & b_3 & \cdots & b_l\end{smallmatrix}\right)^{(e,f,d)}$ *with* $w_i=a_i+b_i$ *for* $1\le i\le l$. *Then*

$$\widetilde{wt}(h)=\sum_{i=1}^{l} b_i(w_{i-1}+w_{i-3}+\cdots+w_{1+i \bmod 2}).$$

[7] We need only $w_1, w_2, \ldots, w_l$ together with d.

[8] Thus the value of f in the striking sequence is redundant—we retain it for convenience.

PROOF. For $L = 0$, both sides are clearly 0. So assume $L > 0$. First consider $d = 0$. For i odd, the ith line is in the NE direction and its x-coordinate is $w_2 + w_4 + \cdots + w_{i-1}$. By the prescription of the previous section and the definition of b_i, this line thus contributes $b_i(w_2 + w_4 + \cdots + w_{i-1})$ to the weight $\widetilde{wt}(h)$ of h. Similarly, for i even, the ith line is in the SE direction and contributes $b_i(w_1 + w_3 + \cdots + w_{i-1})$ to $\widetilde{wt}(h)$. The lemma then follows for $d = 0$. The case $d = 1$ is similar. □

2.2 Path parameters. We make the following definitions:

$$\begin{aligned}
\alpha^{p,p'}_{a,b} &= b - a,\\
\beta^{p,p'}_{a,b,e,f} &= \left\lfloor \tfrac{bp}{p'} \right\rfloor - \left\lfloor \tfrac{ap}{p'} \right\rfloor + f - e,\\
\delta^{p,p'}_{a,e} &= \begin{cases} 0 \text{ if } \left\lfloor \frac{(a+(-1)^e)p}{p'} \right\rfloor = \left\lfloor \frac{ap}{p'} \right\rfloor,\\ 1 \text{ if } \left\lfloor \frac{(a+(-1)^e)p}{p'} \right\rfloor \neq \left\lfloor \frac{ap}{p'} \right\rfloor. \end{cases}
\end{aligned}$$

(The superscripts of $\alpha^{p,p'}_{a,b}$ are superfluous, of course.) It may be seen that the value of $\delta^{p,p'}_{a,e}$ gives the parity of the band in which the path presegment resides.

Lemma 2.2. *Let $h \in \mathcal{P}^{p,p'}_{a,b,e,f}(L)$. Then $\alpha(h) = \alpha^{p,p'}_{a,b}$ and $\beta(h) = \beta^{p,p'}_{a,b,e,f}$.*

PROOF. That $\alpha(h) = \alpha^{p,p'}_{a,b}$ follows immediately from the definitions.

The second result is proved by induction on L. If $h \in \mathcal{P}^{p,p'}_{a,b,e,f}(0)$, then $a = b$, whence $\beta^{p,p'}_{a,b,e,f} = f - e = \beta(h)$ immediately from the definitions.

For $L > 0$, let $h \in \mathcal{P}^{p,p'}_{a,b,e,f}(L)$ and assume that the result holds for all $h' \in \mathcal{P}^{p,p'}_{a,b',e,f'}(L-1)$. We consider a particular h' by setting $h'_i = h_i$ for $0 \le i < L$, $b' = h_{L-1}$, and choosing $f' \in \{0, 1\}$ so that $f' = 0$ if either $b - b' = 1$ and the Lth segment of h lies in an even band or $b - b' = -1$ and the Lth segment of h lies in an odd band, and $f' = 1$ otherwise. It may easily be checked that the $(L-1)$th vertex of h' is scoring if and only if the $(L-1)$th vertex of h is scoring. Then, from the definition of $\beta(h)$, we see that

$$\beta(h) = \begin{cases} \beta(h') + 1 & \text{if } b - b' = 1 \text{ and } f = 1,\\ \beta(h') - 1 & \text{if } b - b' = -1 \text{ and } f = 0,\\ \beta(h') & \text{otherwise.} \end{cases}$$

The induction hypothesis gives $\beta(h') = \lfloor b'p/p' \rfloor - \lfloor ap/p' \rfloor + f' - e$. Then when the Lth segment of h lies in an even band so that $\lfloor bp/p' \rfloor = \lfloor b'p/p' \rfloor$, consideration of the four cases of $b - b' = \pm 1$ and $f \in \{0, 1\}$ shows that $\beta(h) = \lfloor bp/p' \rfloor - \lfloor ap/p' \rfloor + f - e$. When the Lth segment of h lies in an odd band so that $\lfloor bp/p' \rfloor = \lfloor b'p/p' \rfloor + b - b'$, consideration of the four cases of $b - b' = \pm 1$ and $f \in \{0, 1\}$ again shows that $\beta(h) = \lfloor bp/p' \rfloor - \lfloor ap/p' \rfloor + f - e$. The result follows by induction. □

2.3 Scoring generating functions. We now define a generating function for paths that have a particular number of nonscoring vertices. First define $\mathcal{P}^{p,p'}_{a,b,e,f}(L,m)$ to be the subset of $\mathcal{P}^{p,p'}_{a,b,e,f}(L)$ comprising those paths h for which $m(h)=m$. Then define

$$\chi^{p,p'}_{a,b,e,f}(L,m;q)=\sum_{h\in\mathcal{P}^{p,p'}_{a,b,e,f}(L,m)} q^{\widetilde{wt}(h)}. \tag{11}$$

Lemma 2.3. *Let $\beta=\beta^{p,p'}_{a,b,e,f}$. Then*

$$\chi^{p,p'}_{a,b,e,f}(L)=\sum_{\substack{m\equiv L+\beta\\ (\text{mod } 2)}} \chi^{p,p'}_{a,b,e,f}(L,m).$$

Proof. Let $h\in\mathcal{P}^{p,p'}_{a,b,e,f}(L)$. We claim that $m(h)+L(h)+\beta(h)\equiv 0 \pmod 2$. This will follow from showing that $L(h)-m(h)+(-1)^{d(h)}\beta(h)$ is even. If h has striking sequence $\left(\begin{smallmatrix} a_1 & a_2 & a_3 & \cdots & a_l\\ b_1 & b_2 & b_3 & \cdots & b_l\end{smallmatrix}\right)^{(e,f,d)}$, then $L(h)-m(h)=(b_1+b_2+\cdots+b_l)-(e+d+\pi)\bmod 2$, where $\pi=\pi(h)$. For $e+d+\pi\equiv 0 \pmod 2$, we immediately obtain $L(h)-m(h)+(-1)^d\beta(h)=2(b_1+b_3+\cdots)$. For $e+d+\pi\not\equiv 0 \pmod 2$, we obtain $L(h)-m(h)+(-1)^d\beta(h)=2(b_1+b_3+\ldots)-1+(-1)^{d+e}$, whence the claim is proved in all cases. The lemma follows once it is noted that $\beta(h)=\beta^{p,p'}_{a,b,e,f}$ via Lemma 2.2. □

Note 2.4. Since each element of $\mathcal{P}^{p,p'}_{a,b,e,f}(L,m)$ has $L+1$ vertices, it follows that $\chi^{p,p'}_{a,b,e,f}(L,m)$ is nonzero only if $0\le m\le L+1$. Therefore the sum in Lemma 2.3 may be further restricted to $0\le m\le L+1$.

2.4 A seed. The following result provides a seed on which the results of later sections will act.

Lemma 2.5. *If $L\ge 0$ is even, then*

$$\chi^{1,3}_{1,1,0,0}(L,m)=\chi^{1,3}_{2,2,1,1}(L,m)=\delta_{m,0}q^{\frac{1}{4}L^2}.$$

If $L>0$ is odd, then

$$\chi^{1,3}_{1,2,0,1}(L,m)=\chi^{1,3}_{2,1,1,0}(L,m)=\delta_{m,0}q^{\frac{1}{4}(L^2-1)}.$$

Proof. The $(1,3)$-model comprises one even band. Thus when L is even, there is precisely one $h\in\mathcal{P}^{1,3}_{1,1,0,0}(L)$. It has $h_i=1$ for i even and $h_i=2$ for i odd. We see that h has striking sequence $\left(\begin{smallmatrix} 0 & 0 & 0 & \cdots & 0\\ 1 & 1 & 1 & \cdots & 1\end{smallmatrix}\right)^{(0,0,0)}$ and $m(h)=0$. Lemma 2.1 then yields $\widetilde{wt}(h)=0+1+1+2+2+3+\cdots+(\frac{1}{2}L-1)+\frac{1}{2}L=(L/2)^2$, as required.

The other expressions follow in a similar way. □

2.5 Partitions. A partition $\lambda = (\lambda_1, \lambda_2, \ldots, \lambda_k)$ is a sequence of k integer parts $\lambda_1, \lambda_2, \ldots, \lambda_k$ satisfying $\lambda_1 \geq \lambda_2 \geq \cdots \geq \lambda_k > 0$. It is to be understood that $\lambda_i = 0$ for $i > k$. The weight wt (λ) of λ is given by wt $(\lambda) = \sum_{i=1}^{k} \lambda_i$.

We define $\mathcal{Y}(k, m)$ to be the set of all partitions λ with at most k parts, and for which $\lambda_1 \leq m$. A proof of the following well-known result may be found in [2].

Lemma 2.6. *The generating function*

$$\sum_{\lambda \in \mathcal{Y}(k,m)} q^{\mathrm{wt}\,(\lambda)} = \begin{bmatrix} m+k \\ m \end{bmatrix}_q .$$

3 The $\mathcal{B}$-transform

In this section, we introduce the $\mathcal{B}$-transform, which maps paths $\mathcal{P}^{p,p'}_{a,b,e,f}(L)$ into $\mathcal{P}^{p,p'+p}_{a',b',e,f}(L')$ for certain a', b', and various L'.

The band structure of the $(p, p'+p)$-model is easily obtained from that of the (p, p')-model. Indeed, according to Section 1.1, for $1 \leq r < p$, the rth odd band of the $(p, p'+p)$-model lies between heights $\lfloor r(p'+p)/p \rfloor = \lfloor rp'/p \rfloor + r$ and $\lfloor r(p'+p)/p \rfloor + 1 = \lfloor rp'/p \rfloor + r + 1$. Thus the height of the rth odd band in the $(p, p'+p)$-model is r greater than that in the (p, p')-model. Therefore, the $(p, p'+p)$-model may be obtained from the (p, p')-model by increasing the distance between neighboring odd bands by one unit and appending an extra even band to both the top and the bottom of the grid. For example, compare the (3, 8)-model of Figure 1 with the (3, 11)-model of Figure 3.

The $\mathcal{B}$-transform has three components, which we refer to as *path dilation*, *particle insertion*, and *particle motion*. These three components will also be known as the $\mathcal{B}_1$-, $\mathcal{B}_2$-, and $\mathcal{B}_3$-transforms, respectively. In fact, particle insertion is dependent on a parameter $k \in \mathbb{Z}_{\geq 0}$ and particle motion is dependent on a partition λ that has certain restrictions. Consequently, we sometimes refer to particle insertion and particle motion as $\mathcal{B}_2(k)$- and $\mathcal{B}_3(\lambda)$-transforms, respectively. Then combining the $\mathcal{B}_1$-, $\mathcal{B}_2(k)$-, and $\mathcal{B}_3(\lambda)$-transforms produces the $\mathcal{B}(k, \lambda)$-transform.

3.1 Path dilation. The $\mathcal{B}_1$-transform acts on a path $h \in \mathcal{P}^{p,p'}_{a,b,e,f}(L)$ to yield a path $h^{(0)} \in \mathcal{P}^{p,p'+p}_{a',b',e,f}(L^{(0)})$ for certain a', b', and $L^{(0)}$. First, the starting point a' of the new path $h^{(0)}$ is specified to be

$$a' = a + \left\lfloor \frac{ap}{p'} \right\rfloor + e.$$

If $r = \lfloor ap/p' \rfloor$, then r is the number of odd bands below $h = a$ in the (p, p')-model. Since the height of the rth odd band in the $(p, p'+p)$-model is r greater than that in the (p, p')-model, we thus see that under path dilation, the height of the startpoint above the next lowermost odd band (or if there isn't one, the bottom of the grid) has either increased by one or remained constant.

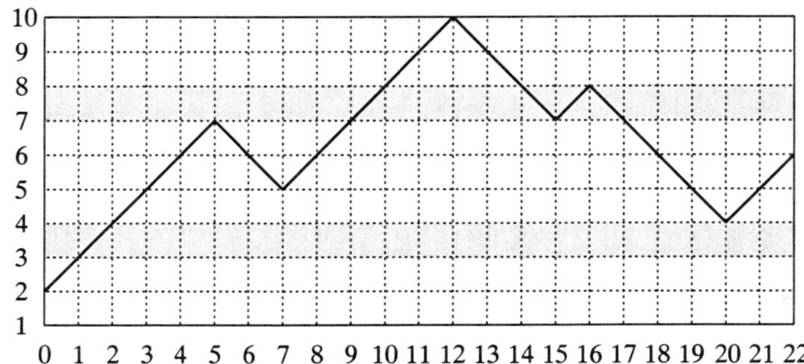

Figure 3.

We define $d(h^{(0)}) = d(h)$. The above definition specifies that $e(h^{(0)}) = e(h)$ and $f(h^{(0)}) = f(h)$.

In the case that $L = 0$ and $e = f$, we specify $h^{(0)}$ by setting $L^{(0)} = L(h^{(0)}) = 0$. When $L = 0$ and $e \neq f$, we leave the action of the $\mathcal{B}_1$-transform on h undefined. (It will not be used in this case.) Thus in Lemmas 3.3, 3.6, 3.7, 3.10, 3.13, 4.4, and 4.5 and Corollary 3.4, we implicitly exclude consideration of the case $L = 0$ and $e \neq f$. However, it must be considered in the proofs of Corollaries 3.14 and 4.6.

In the case $L > 0$ consider, as in Section 2.1, h to comprise l straight lines that alternate in direction, the ith of which is of length w_i and possesses b_i scoring vertices. $h^{(0)}$ is then defined to comprise l straight lines that alternate in direction (since $d(h^{(0)}) = d(h)$, the direction of the first line in $h^{(0)}$ is the same as that in h), the ith of which has length

$$w'_i = \begin{cases} w_i + b_i & \text{if } i \geq 2 \text{ or } e(h) + d(h) + \pi(h) \equiv 0 \pmod 2, \\ w_1 + b_1 + 2\pi(h) - 1 & \\ & \text{if } i = 1 \text{ and } e(h) + d(h) + \pi(h) \not\equiv 0 \pmod 2. \end{cases}$$

In particular, this determines $L^{(0)} = L(h^{(0)})$ and $b' = h^{(0)}_{L^{(0)}}$.

As an example, consider the path h shown in Figure 1 as an element of $\mathcal{P}^{3,8}_{2,4,e,1}(14)$. Here $d(h) = 0$, $\pi(h) = 1$, and $\lfloor ap/p' \rfloor = 0$. Thus when $e = 0$, the action of path dilation on h produces the path given in Figure 3. This path is an element of $\mathcal{P}^{3,11}_{2,6,e,1}(22)$. When $e = 1$, the action of path dilation on h produces the element of $\mathcal{P}^{3,11}_{3,6,e,1}(21)$ given in Figure 4.

The situation at the start point may be considered as falling into one of eight cases, corresponding to $e(h), d(h), \pi(h) \in \{0, 1\}$.[9] In Table 2, we illustrate the four cases that arise when $d(h) = 0$. (The four cases for $d(h) = 1$ may be obtained from these by an up–down reflection and changing the value of $e(h)$.)[10]

Lemma 3.1. *Let* $1 \leq p < p'$, $1 \leq a < p'$, $e \in \{0, 1\}$, *and* $a' = a + \lfloor ap/p' \rfloor + e$. *Then* $\lfloor a'p/(p' + p) \rfloor = \lfloor ap/p' \rfloor$ *and* $\delta^{p,p'+p}_{a',e} = 0$.

[9]Theses cases may be seen to correspond to the eight cases of vertex type as listed in Table 1.

[10]The examples here are such that $w_1 \geq 3$.

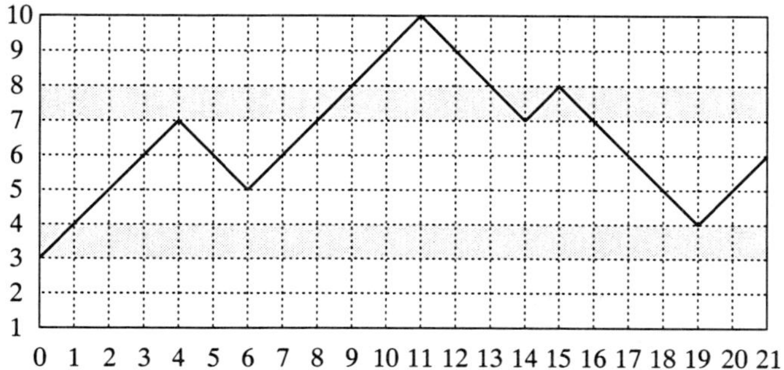

Figure 4.

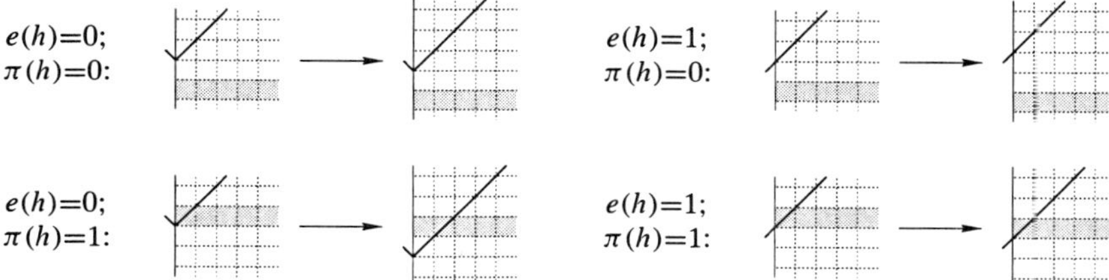

Table 2. $\mathcal{B}_1$-transforms at the startpoint.

PROOF. Let $r = \lfloor ap/p' \rfloor$, whence $p'r \leq pa < p'(r+1)$. Then for $x \in \{0, 1\}$, we have $(p'+p)r \leq p(a+r+x) < (p'+p)r+p'+xp$, so that $\lfloor (a+r+x)p/(p'+p) \rfloor = r$. In particular, $\lfloor a'p/(p'+p) \rfloor = r$ and $\lfloor (a+r+e+(-1)^e)p/(p'+p) \rfloor = r$. Thus $r = \lfloor a'p/(p'+p) \rfloor = \lfloor (a'+(-1)^e)p/(p'+p) \rfloor$, which gives the required results. □

This result asserts, among other things, that the presegment of $h^{(0)}$ always lies in an even band. This is also evident from Table 2.

Note 3.2. The action of path dilation on $h \in \mathcal{P}^{p,p'}_{a,b,e,f}(L)$ yields a path $h^{(0)} \in \mathcal{P}^{p,p'+p}_{a',b',e,f}(L^{(0)})$ that has, including the vertex at $i = 0$, no adjacent scoring vertices, except in the case where $\pi(h) = 1$ *and* $e(h) = d(h)$, when a single pair of scoring vertices occurs in $h^{(0)}$ at $i = 0$ and $i = 1$.

Also note that $\pi(h^{(0)}) = \pi(h)$, unless $\pi(h) = 1$ *and* $e(h) = d(h)$, in which case $\pi(h^{(0)}) = 0$.

Now compare the ith line of $h^{(0)}$ (which has length w'_i) with the ith line of h (which has length w_i). For the sake of the following argument, assume that there are odd bands immediately below (i.e., between heights 0 and 1), and immediately above (i.e., between heights $p'-1$ and p') the (p, p')-model and do likewise for the $(p, p'+p)$-model.

If the lines in question are in the NE direction, we claim that the height of the final vertex of that in $h^{(0)}$ above the next lower odd band is one greater than that in h. If the lines in question are in the SE direction, we claim that the height of the final vertex of that in $h^{(0)}$ below the next higher odd band is one greater than that in h. In particular, if either the first or last segment of the ith line is in an odd band, then the corresponding segment of $h^{(0)}$ lies in the same odd band.

We also claim that if that of h has a straight vertex that passes into the kth odd band in the (p, p')-model, then that of $h^{(0)}$ has a straight vertex that passes into the kth odd band in the $(p, p'+p)$-model.

These claims follow because in passing from the (p, p')-model to the $(p, p'+p)$-model, the distance between neighboring odd bands has increased by one and because the length of each line has increased by one for every scoring vertex and possibly a small adjustment made to the length of the first line. In effect, a new straight vertex has been inserted immediately prior to each scoring vertex and, if $e(h)+d(h)+\pi(h) \not\equiv 0 \pmod 2$, adjusting the length of the resulting first line by $2\pi(h)-1$.

Lemma 3.3. *Let* $h \in \mathcal{P}^{p,p'}_{a,b,e,f}(L)$ *have striking sequence* $\left(\begin{smallmatrix} a_1 & a_2 & a_3 & \cdots & a_l \\ b_1 & b_2 & b_3 & \cdots & b_l \end{smallmatrix}\right)^{(e,f,d)}$ *and let* $h^{(0)} \in \mathcal{P}^{p,p'+p}_{a',b',e,f}(L^{(0)})$ *be obtained from the action of the* $\mathcal{B}_1$*-transform on* h. *If* $e(h)+d(h)+\pi(h) \equiv 0 \pmod 2$, *then* $h^{(0)}$ *has striking sequence*

$$\begin{pmatrix} a_1+b_1 & a_2+b_2 & a_3+b_3 & \cdots & a_l+b_l \\ b_1 & b_2 & b_3 & \cdots & b_l \end{pmatrix}^{(e,f,d)},$$

and if $e(h)+d(h)+\pi(h) \not\equiv 0 \pmod 2$, *then* $h^{(0)}$ *has striking sequence*

$$\begin{pmatrix} a_1+b_1+\pi-1 & a_2+b_2 & a_3+b_3 & \cdots & a_l+b_l \\ b_1+\pi & b_2 & b_3 & \cdots & b_l \end{pmatrix}^{(e,f,d)}.$$

Moreover, if $m = m(h)$,

- $m(h^{(0)}) = L$;
- $L^{(0)} = \begin{cases} 2L-m+2 & \text{if } \pi = 1 \text{ and } e = d, \\ 2L-m & \text{otherwise;} \end{cases}$
- $\alpha(h^{(0)}) = \alpha(h) + \beta(h)$;
- $\beta(h^{(0)}) = \beta(h)$.

Proof. The form of the striking sequence for $h^{(0)}$ follows because for $i > 1$, every scoring vertex in the ith line of h accounts for an extra nonscoring vertex in that line. The same is true when $i = 1$, except in the case $(e(h)+d(h)+\pi(h)) \equiv 1$ (in proofs throughout this paper, we take all equivalences modulo 2), when the length of the new first line becomes $a_1 + 2b_1 + 2\pi - 1$. It follows from examining Table 2 that there are $b_1 + \pi$ scoring vertices in this case.

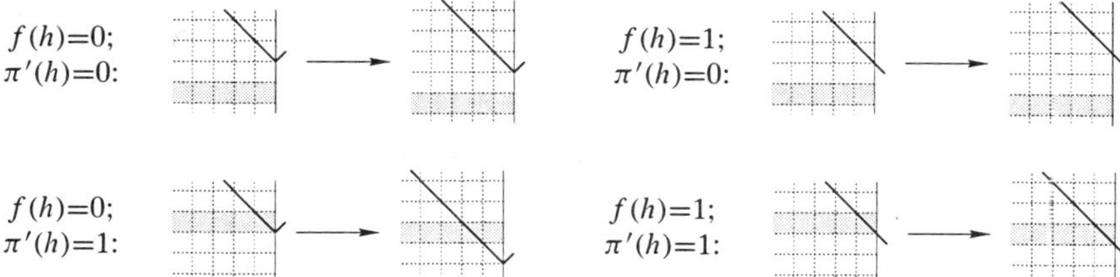

Table 3. $\mathcal{B}_1$-transforms at the endpoint.

Let $e = e(h)$, $d = d(h)$, $\pi = \pi(h)$, and $\pi' = \pi(h^{(0)})$. Then $e(h^{(0)}) = e$ and $d(h^{(0)}) = d$.

If $(e+d+\pi) \equiv 0$, then $(e+d+\pi') \equiv 0$ by Note 3.2. Thereupon $m^{(0)} = \sum_{i=1}^{l}(a_i + b_i) = L$. Additionally, $L^{(0)} = \sum_{i=1}^{l}(a_i+2b_i) = 2L - \sum_{i=1}^{l} a_i = 2L - m$. It follows immediately that $\beta(h^{(0)}) = \beta(h)$ and $\alpha(h^{(0)}) = \alpha(h) + \beta(h)$ in this case.

On the other hand, if $(e+d+\pi) \not\equiv 0$, then $\pi = 0 \Rightarrow e \neq d$ and $\pi = 1 \Rightarrow e = d$. In each instance, Note 3.2 implies that $\pi' = 0$. Thereupon, $m^{(0)} = (e + d + \pi') \bmod 2 + \pi - 1 + \sum_{i=1}^{l}(a_i+b_i) = \sum_{i=1}^{l}(a_i+b_i) = L$. Additionally, $L^{(0)} = 2\pi - 1 + \sum_{i=1}^{l}(a_i+2b_i) = 2L - (1+\sum_{i=1}^{l} a_i) + 2\pi = 2L - m + 2\pi$. This is the required value. Now in this case, $\beta(h) = (-1)^d((b_1+b_3+\cdots) - (b_2+b_4+\cdots)) + (-1)^e$. When $\pi = 0$ so that $(e + d + \pi') \equiv 1$, then $\beta(h^{(0)}) = \beta(h)$ follows immediately. When $\pi = 1$, we have $\beta(h^{(0)}) = (-1)^d((b_1 + 1 + b_3 + \cdots) - (b_2 + b_4 + \cdots))$. $\beta(h^{(0)}) = \beta(h)$ now follows in this case because $(e+d+\pi) \not\equiv 0$ implies that $e = d$. Finally, $\alpha(h^{(0)}) = \alpha(h) + (-1)^d((b_1+b_3+\cdots) - (b_2+b_4+\cdots)) + (-1)^d(2\pi - 1)$. Since $(-1)^d(2\pi - 1) = -(-1)^d(-1)^{\pi} = (-1)^e$, the lemma then follows. □

Corollary 3.4. *Let $h \in \mathcal{P}^{p,p'}_{a,b,e,f}(L)$ and $h^{(0)} \in \mathcal{P}^{p,p'+p}_{a',b',e,f}(L^{(0)})$ be the path obtained by the action of the $\mathcal{B}_1$-transform on h. Then $a' = a + \lfloor ap/p' \rfloor + e$ and $b' = b + \lfloor bp/p' \rfloor + f$.*

Proof. $a' = a + \lfloor ap/p' \rfloor + e$ by definition. Lemma 3.3 gives $\alpha(h^{(0)}) = \alpha(h) + \beta(h)$, whence Lemma 2.2 implies that $\alpha^{p,p'+p}_{a',b'} = \alpha^{p,p'}_{a,b} + \beta^{p,p'}_{a,b,e,f}$. Expanding this gives $b' - a' = b - a + \lfloor bp/p' \rfloor - \lfloor ap/p' \rfloor + f - e$, whence $b' = b + \lfloor bp/p' \rfloor + f$. □

The above result implies that the $\mathcal{B}_1$-transform maps $\mathcal{P}^{p,p'}_{a,b,e,f}(L)$ into a set of paths that have the same startpoint as one another and the same endpoint as one another. However, the lengths of these paths are not necessarily equal. We also see that the transformation of the endpoint is analogous to that which occurs at the startpoint. In particular, Lemma 3.1 implies that $\delta^{p,p'+p}_{b',f} = 0$ so that the path postsegment of $h^{(0)}$ always resides in an even band. For the four cases where $h_L = h_{L-1} - 1$, the $\mathcal{B}_1$-transform affects the endpoint as in Table 3 (the value $\pi'(h)$ is the parity of the band in which the Lth segment of h lies).

Lemma 3.5. *Let* $1 \le p < p'$, $1 \le a, b < p'$, $e, f \in \{0, 1\}$, $a' = a + \lfloor ap/p' \rfloor + e$, *and* $b' = b + \lfloor bp/p' \rfloor + f$. *Then* $\alpha^{p,p'+p}_{a',b'} = \alpha^{p,p'}_{a,b} + \beta^{p,p'}_{a,b,e,f}$ *and* $\beta^{p,p'+p}_{a',b',e,f} = \beta^{p,p'}_{a,b,e,f}$.

Proof. Lemma 3.1 implies that $\lfloor a'p/(p'+p) \rfloor = \lfloor ap/p' \rfloor$ and $\lfloor b'p/(p'+p) \rfloor = \lfloor bp/p' \rfloor$. The results then follow immediately from the definitions. □

Lemma 3.6. *Let* $h \in \mathcal{P}^{p,p'}_{a,b,e,f}(L)$ *and* $h^{(0)} \in \mathcal{P}^{p,p'+p}_{a',b',e,f}(L^{(0)})$ *be the path obtained by the action of the* $\mathcal{B}_1$*-transform on* h. *Then*

$$\widetilde{wt}(h^{(0)}) = \widetilde{wt}(h) + \frac{1}{4}\left((L^{(0)} - m^{(0)})^2 - \beta^2\right),$$

where $m^{(0)} = m(h^{(0)})$ *and* $\beta = \beta^{p,p'}_{a,b,e,f}$.

Proof. Let h have striking sequence $\left(\begin{smallmatrix} a_1 & a_2 & a_3 & \cdots & a_l \\ b_1 & b_2 & b_3 & \cdots & b_l \end{smallmatrix}\right)^{(e,f,d)}$ and let $\pi = \pi(h)$. If $(e + d + \pi) \equiv 0 \pmod 2$, then Lemmas 3.3 and 2.1 show that

$$\widetilde{wt}(h^{(0)}) - \widetilde{wt}(h) = (b_1 + b_3 + b_5 + \cdots)(b_2 + b_4 + b_6 + \cdots).$$

Via Lemma 3.3, we obtain $L^{(0)} - m^{(0)} = L - m(h) = b_1 + b_2 + \cdots + b_l$. Then since $\beta(h) = \pm((b_1 + b_3 + b_5 + \cdots) - (b_2 + b_4 + b_6 + \cdots))$, it follows that

$$\widetilde{wt}(h^{(0)}) - \widetilde{wt}(h) = \frac{1}{4}((L^{(0)} - m^{(0)})^2 - \beta(h)^2).$$

If $(e + d + \pi) \not\equiv 0 \pmod 2$, then Lemmas 3.3 and 2.1 show that

$$\begin{aligned}\widetilde{wt}(h^{(0)}) - \widetilde{wt}(h) &= (2\pi - 1 + b_1 + b_3 + b_5 + \cdots)(b_2 + b_4 + b_6 + \cdots) \\ &= \frac{1}{4}((L^{(0)} - m^{(0)})^2 - \beta(h)^2),\end{aligned}$$

the second equality resulting because $L^{(0)} - m^{(0)} = L - m(h) + 2\pi = b_1 + b_2 + \cdots + b_l + 2\pi - 1$ and

$$\begin{aligned}\beta(h) &= (-1)^d((b_1 + b_3 + b_5 + \cdots) - (b_2 + b_4 + b_6 + \cdots)) + (-1)^e \\ &= \pm((2\pi - 1 + b_1 + b_3 + b_5 + \cdots) - (b_2 + b_4 + b_6 + \cdots))\end{aligned}$$

on using $(-1)^{e+d} = -(-1)^\pi = 2\pi - 1$.

Finally, Lemma 2.2 gives $\beta(h) = \beta^{p,p'}_{a,b,e,f} = \beta$. □

3.2 Particle insertion. Let $p' > 2p$ so that the (p, p')-model has no two neighboring odd bands, and let $\delta^{p,p'}_{a',e} = 0$. Then if $h^{(0)} \in \mathcal{P}^{p,p'}_{a',b',e,f}(L^{(0)})$, the presegment of $h^{(0)}$ lies in an even band. By *inserting a particle* into $h^{(0)}$, we mean displacing $h^{(0)}$ two positions to the right and inserting two segments: the leftmost of these is in the NE (resp., SE) direction if $e = 0$ (resp., $e = 1$), and the rightmost is in the opposite direction, which is thus the direction of the presegment of $h^{(0)}$. In this way,

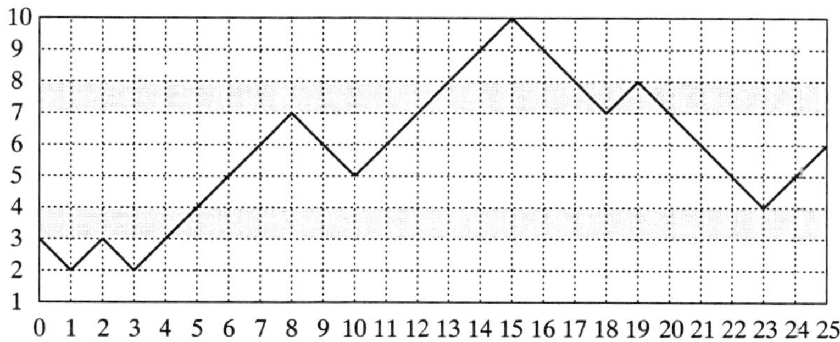

Figure 5.

we obtain a path $h^{(1)}$ of length $L^{(0)}+2$. We assign $e(h^{(1)}) = e$ and $f(h^{(1)}) = f$. Note also that $d(h^{(1)}) = e$ and $\pi(h^{(1)}) = 0$.

Thereupon, we may repeat this process of particle insertion. After inserting k particles into $h^{(0)}$, we obtain a path $h^{(k)} \in \mathcal{P}^{p,p'}_{a',b',e,f}(L^{(0)}+2k)$. We say that $h^{(k)}$ has been obtained by the action of a $\mathcal{B}_2(k)$-transform on $h^{(0)}$.

In the case of the element of $\mathcal{P}^{3,11}_{3,6,1,1}(21)$ shown in Figure 4, the insertion of two particles produces the element of $\mathcal{P}^{3,11}_{3,6,1,1}(25)$ shown in Figure 5.

Lemma 3.7. *Let $h \in \mathcal{P}^{p,p'}_{a,b,e,f}(L)$. Apply a $\mathcal{B}_1$-transform to h to obtain the path $h^{(0)} \in \mathcal{P}^{p,p'+p}_{a',b',e,f}(L^{(0)})$. Then obtain $h^{(k)} \in \mathcal{P}^{p,p'+p}_{a',b',e,f}(L^{(k)})$ by applying a $\mathcal{B}_2(k)$-transform to $h^{(0)}$. If $m^{(k)} = m(h^{(k)})$, then $L^{(k)} = L^{(0)}+2k$, $m^{(k)} = m^{(0)}$ and*

$$\widetilde{wt}(h^{(k)}) = \widetilde{wt}(h) + \frac{1}{4}((L^{(k)} - m^{(k)})^2 - \beta^2),$$

where $\beta = \beta^{p,p'}_{a,b,e,f}$.

Proof. It follows immediately from the definition of a $\mathcal{B}_2$-transform that $L^{(k)} = L^{(0)}+2k$. Lemma 3.6 yields

$$\widetilde{wt}(h^{(0)}) = \widetilde{wt}(h) + \frac{1}{4}\left((L^{(0)} - m(h^{(0)}))^2 - \beta^2\right). \tag{12}$$

Let the striking sequence of $h^{(0)}$ be $\left(\begin{smallmatrix} a_1 & a_2 & \cdots & a_l \\ b_1 & b_2 & \cdots & b_l \end{smallmatrix}\right)^{(e,f,d)}$ and let $\pi = \pi(h^{(0)})$.

If $e = d$, we are restricted to the case $\pi = 0$, since $\delta^{p,p'+p}_{a',e} = 0$ by Lemma 3.1. The striking sequence of $h^{(1)}$ is then $\left(\begin{smallmatrix} 0 & 0 & a_1 & a_2 & \cdots & a_l \\ 1 & 1 & b_1 & b_2 & \cdots & b_l \end{smallmatrix}\right)^{(e,f,e)}$. Thereupon $m(h^{(1)}) = \sum_{i=1}^{l} a_i = m(h^{(0)})$. In this case, Lemma 2.1 shows that $\widetilde{wt}(h^{(1)}) - \widetilde{wt}(h^{(0)}) = 1 + b_1 + b_2 + \cdots + b_l = L^{(0)} - m^{(0)} + 1$.

If $e \neq d$, the striking sequence of $h^{(1)}$ is $\left(\begin{smallmatrix} 0 & a_1+1-\pi & a_2 & \cdots & a_l \\ 1 & b_1+\pi & b_2 & \cdots & b_l \end{smallmatrix}\right)^{(e,f,e)}$. Then $m(h^{(1)}) = 1-\pi+\sum_{i=1}^{l} a_i$, which equals $m(h^{(0)}) = (e+d+\pi) \bmod 2 + \sum_{i=1}^{l} a_i$ for both $\pi = 0$ and $\pi = 1$. Here Lemma 2.1 shows that $\widetilde{wt}(h^{(1)}) - \widetilde{wt}(h^{(0)}) = \pi + b_1 + b_2 + \cdots + b_l$. Since $L^{(0)} - m^{(0)} = -(e+d+\pi) \bmod 2 + b_1 + b_2 + \cdots + b_l$, we once more have $\widetilde{wt}(h^{(1)}) - \widetilde{wt}(h^{(0)}) = L^{(0)} - m^{(0)} + 1$.

Repeated application of these results yields $m(h^{(k)}) = m(h^{(0)})$ and

$$\widetilde{wt}(h^{(k)}) = \widetilde{wt}(h^{(0)}) + k\left(L^{(0)} - m(h^{(0)})\right) + k^2.$$

Then on using (12) and $L^{(k)} = L^{(0)} + 2k$, the lemma follows. □

3.3 Particle moves. In this section, we once more restrict to the case $p' > 2p$ so that the (p, p')-model has no two neighboring odd bands, and consider only paths $h \in \mathcal{P}^{p,p'}_{a',b',e,f}(L')$, where $\delta^{p,p'}_{a',e} = \delta^{p,p'}_{b',f} = 0$.

We specify six types of local deformations of a path. These deformations will be known as *particle moves*. In each of the six cases, a particular sequence of four segments of a path is changed to a different sequence, the remainder of the path being unchanged. The moves are as follows—the path portion to the left of the arrow is changed to that on the right:

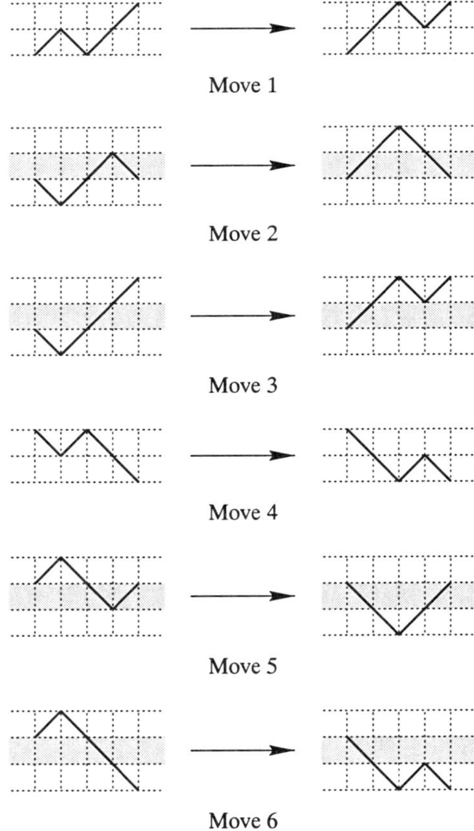

Since $p' > 2p$, each odd band is straddled by a pair of even bands. Thus there is no impediment to enacting moves 2 and 5 for paths in $\mathcal{P}^{p,p'}_{a',b',e,f}(L)$.

Note that moves 4–6 are inversions of moves 1–3. Also note that moves 2 and 3 (and likewise moves 5 and 6) may be considered to be the same move since in the two cases, the same sequence of three edges is changed.

In addition to the six moves described above, we permit certain deformations of a path close to its left and right extremities in certain circumstances. Each of these moves will be referred to as an *edge move*. They, together with their validities, are as follows:

If $e = 1$:

Edge move 1

If $e = 0$:

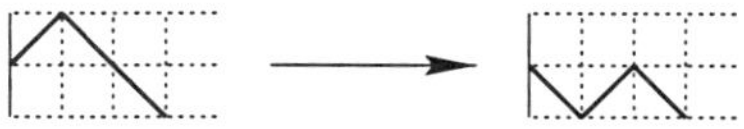

Edge move 2

If $f = 0$:

Edge move 3

If $f = 1$:

Edge move 4

In fact, the above four edge moves may be considered as instances of moves 1 and 4 described beforehand if we append the appropriate presegment to the path for edge moves 1 and 2 and we append the appropriate postsegment to the path for edge moves 3 and 4.

Lemma 3.8. *Let the path $\hat{h}$ differ from the path h in that four consecutive segments have changed according to one of the six moves described above or in that three consecutive segments have changed according to one of the four edge moves described above (subject to their restrictions). Then*

$$\widetilde{wt}(\hat{h}) = \widetilde{wt}(h) + 1.$$

Additionally, $L(\hat{h}) = L(h)$ *and* $m(\hat{h}) = m(h)$.

Proof. For each of the six moves and four edge moves, take the (x, y)-coordinate of the leftmost point of the depicted portion of h to be (x_0, y_0). Now consider the contribution to the weight of the three vertices in question before and after the move. (Although the vertex at (x_0, y_0) may change, its contribution does not.) In each of the ten cases, the contribution is $x_0 + y_0 + 1$ before the move and $x_0 + y_0 + 2$ afterwards. Thus $\widetilde{wt}(\hat{h}) = \widetilde{wt}(h) + 1$. The other statements are immediate on inspecting all ten moves. □

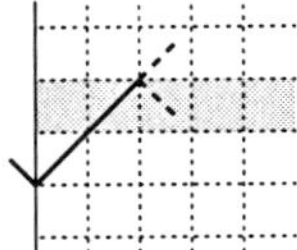

Figure 6. Not a particle.

Now observe that for each of the ten moves specified above, the sequence of path segments before the move consists of an adjacent pair of scoring vertices followed by a nonscoring vertex. The specified move replaces this combination with a nonscoring vertex followed by two scoring vertices. As anticipated above, the pair of adjacent scoring vertices is viewed as a particle. Thus each of the above ten moves describes a particle moving to the right by one step.

When $p' > 2p$ so that there are no two adjacent odd bands in the (p, p')-model, and noting that $\delta^{p,p'}_{b',f} = 0$, we see that each sequence comprising two scoring vertices followed by a nonscoring vertex is present among the ten configurations prior to a move, except for the case depicted in Figure 6 and its up–down reflection.

Only in these cases, where the zeroth and first segments are scoring and the first two segments are in the same direction, do we *not* refer to the adjacent pair of scoring vertices as a particle.

Also note that when $p' > 2p$ and $\delta^{p,p'}_{a',e} = \delta^{p,p'}_{b',f} = 0$, each sequence of a nonscoring vertex followed by two scoring vertices appears among the ten configurations that result from a move. In such cases, the move may thus be reversed.

3.4 The $\mathcal{B}_3$-transform. Since in each of the moves described in Section 3.3 a pair of scoring vertices shifts to the right by one step, we see that a succession of such moves is possible until the pair is followed by another scoring vertex. If this itself is followed by yet another scoring vertex, we forbid further movement. However, if it is followed by a nonscoring vertex, further movement is allowed after considering the latter two of the three consecutive scoring vertices to be the particle (instead of the first two).

As in Section 3.2, let $h^{(k)}$ be a path resulting from a $\mathcal{B}_2(k)$-transform acting on a path that itself is the image of a $\mathcal{B}_1$-transform. We now consider moving the k particles that have been inserted.

Lemma 3.9. *Let $\delta^{p,p'}_{b',f} = 0$. There is a bijection between the set of paths obtained by moving the particles in $h^{(k)}$ and $\mathcal{Y}(k, m)$, where $m = m(h^{(k)})$. This bijection is such that if $\lambda \in \mathcal{Y}(k, m)$ is the bijective image of a particular h, then*

$$\widetilde{wt}(h) = \widetilde{wt}(h^{(k)}) + \mathrm{wt}\,(\lambda).$$

Additionally, $L(h) = L(h^{(k)})$ and $m(h) = m(h^{(k)})$.

Proof. Since each particle moves by traversing a nonscoring vertex, since there are m of these to the right of the rightmost particle in $h^{(k)}$, and since there are no consecutive scoring vertices to its right, this particle can make λ_1 moves to the right with $0 \le \lambda_1 \le m$. Similarly, the next rightmost particle can make λ_2 moves to the right with $0 \le \lambda_2 \le \lambda_1$. Here the upper restriction arises because the two scoring vertices would then be adjacent to those of the first particle. Continuing in this way, we obtain that all possible final positions of the particles are indexed by $\lambda = (\lambda_1, \lambda_2, \ldots, \lambda_k)$ with $m \ge \lambda_1 \ge \lambda_2 \ge \cdots \ge \lambda_k \ge 0$, that is, by partitions of at most k parts with no part exceeding m. Moreover, since by Lemma 3.8 the weight increases by one for each move, the weight increase after the sequence of moves specified by a particular λ is equal to $\mathrm{wt}(\lambda)$. The final statement also follows from Lemma 3.8. □

We say that a path obtained by moving the particles in $h^{(k)}$ according to the partition λ has been obtained by the action of a $\mathcal{B}_3(\lambda)$-transform.

Having defined $\mathcal{B}_1, \mathcal{B}_2(k)$ for $k \ge 0$ and $\mathcal{B}_3(\lambda)$ for λ a partition with at most k parts, we now define a $\mathcal{B}(k,\lambda)$-transform as the composition $\mathcal{B}(k,\lambda) = \mathcal{B}_3(\lambda) \circ \mathcal{B}_2(k) \circ \mathcal{B}_1$.

Lemma 3.10. *Let $h' \in \mathcal{P}^{p,p'+p}_{a',b',e,f}(L')$ be obtained from $h \in \mathcal{P}^{p,p'}_{a,b,e,f}(L)$ by the action of the $\mathcal{B}(k,\lambda)$-transform. If $\pi = \pi(h)$ and $m = m(h)$, then*

- $L' = \begin{cases} 2L - m + 2k + 2 & \text{if } \pi = 1 \text{ and } e = d, \\ 2L - m + 2k & \text{otherwise}, \end{cases}$
- $m(h') = L$,
- $\widetilde{wt}(h') = \widetilde{wt}(h) + \frac{1}{4}\left((L'-L)^2 - \beta^2\right) + \mathrm{wt}(\lambda)$,

where $\beta = \beta^{p,p'}_{a,b,e,f}$.

Proof. These results follow immediately from Lemmas 3.3, 3.7, and 3.9. □

Note 3.11. Since particle insertion and the particle moves do not change the start-point, endpoint, or value $e(h)$ or $f(h)$ of a path h, then in view of Lemma 3.1 and Corollary 3.4, we see that the action of a $\mathcal{B}$-transform on $h \in \mathcal{P}^{p,p'}_{a,b,e,f}(L)$ yields a path $h' \in \mathcal{P}^{p,p'+p}_{a',b',e,f}(L')$, where $a' = a + \lfloor ap/p' \rfloor + e$, $b' = b + \lfloor bp/p' \rfloor + f$, and $\delta^{p,p'+p}_{a',e} = \delta^{p,p'+p}_{b',f} = 0$.

3.5 Particle content of a path. We again restrict to the case $p' > 2p$ so that the (p,p')-model has no two neighboring odd bands and let $h' \in \mathcal{P}^{p,p'}_{a',b',e,f}(L')$. In the following lemma, we once more restrict to the cases for which $\delta^{p,p'}_{a',e} = \delta^{p,p'}_{b',f} = 0$ and thus only consider the cases for which the presegment and the postsegment of h' lie in even bands.

Lemma 3.12. *For $1 \le p < p'$ with $p' > 2p$, let $1 \le a', b' < p'$ and $e, f \in \{0,1\}$ with $\delta^{p,p'}_{a',e} = \delta^{p,p'}_{b',f} = 0$. If $h' \in \mathcal{P}^{p,p'}_{a',b',e,f}(L')$, then there is a unique triple (h,k,λ), where $h \in \mathcal{P}^{p,p'-p}_{a,b,e,f}(L)$ for some a, b, L, such that the action of a $\mathcal{B}(k,\lambda)$-transform on h results in h'.*

PROOF. This is proved by reversing the constructions described in the previous sections. Locate the leftmost pair of consecutive scoring vertices in h' and move them leftward by reversing the particle moves until they occupy the zeroth and first positions. This is possible in all cases where $\delta^{p,p'}_{a',e} = \delta^{p,p'}_{b',f} = 0$. Now ignoring these two vertices, do the same with the next leftmost pair of consecutive scoring vertices, moving them leftward until they occupy the second and third positions. Continue in this way until all consecutive scoring vertices occupy the leftmost positions of the path. Denote this path by $h^{(\cdot)}$. At the leftmost end of $h^{(\cdot)}$, there will be a number of even segments (possibly zero) alternating in direction. Let this number be $2k$ or $2k+1$ according to whether is it even or odd. Clearly h' results from $h^{(\cdot)}$ by a $\mathcal{B}_3(\lambda)$-transform for a particular λ with at most k parts.

Removing the first $2k$ segments of $h^{(\cdot)}$ yields a path $h^{(0)} \in \mathcal{P}^{p,p'}_{a',b',e,f}$. This path thus has no two consecutive scoring vertices, except possibly at the zeroth and first positions, and then only if the first vertex is a straight vertex (as in Figure 6). Moreover, $h^{(k)}$ arises by the action of a $\mathcal{B}_2(k)$-transform on $h^{(0)}$.

Ignoring for the moment the case where there are scoring vertices at the zeroth and first positions, $h^{(0)}$ has by construction no pair of consecutive scoring vertices. Therefore, beyond the zeroth vertex, we may remove a nonscoring vertex before every scoring vertex to obtain a path $h \in \mathcal{P}^{p,p'-p}_{a,b,e,f}(L)$ for some a, b, L, from which $h^{(0)}$ arises by the action of a $\mathcal{B}_1$-transform.

On examining the third case depicted in Table 2, we see that the case where $h^{(0)}$ has a pair of scoring vertices at the zeroth and first positions arises similarly from a particular $h \in \mathcal{P}^{p,p'-p}_{a,b,e,f}(L)$ for some a, b, L. The lemma is then proved. □

The value of k obtained above will be referred to as the *particle content* of h'.

Lemma 3.13. *For $1 \le p < p'$, let $1 \le a, b < p'$ and $e, f \in \{0,1\}$ with $\delta^{p,p'}_{a,e} = 0$. Set $a' = a + e + \lfloor ap/p' \rfloor$ and $b' = b + f + \lfloor bp/p' \rfloor$. Fix $m_0, m_1 \ge 0$. Then the map $(h, k, \lambda) \mapsto h'$ effected by the action of a $\mathcal{B}(k,\lambda)$-transform on h, is a bijection between $\bigcup_k \mathcal{P}^{p,p'}_{a,b,e,f}(m_1, 2k+2m_1-m_0) \times \mathcal{Y}(k, m_1)$ and $\mathcal{P}^{p,p'+p}_{a',b',e,f}(m_0, m_1)$. Moreover,*

$$\widetilde{wt}(h') = \widetilde{wt}(h) + \frac{1}{4}\left((m_0 - m_1)^2 - \beta^2\right) + \mathrm{wt}\,(\lambda),$$

where $\beta = \beta^{p,p'}_{a,b,e,f}$.

PROOF. Given $h \in \mathcal{P}^{p,p'}_{a,b,e,f}(m_1, m)$, let h' be the result of a $\mathcal{B}(k,\lambda)$-transform on h. Since $\delta^{p,p'}_{a,e} = 0$ so that $\lfloor (a + (-1)^e)p/p') \rfloor = \lfloor ap/p' \rfloor$, it follows that if $\pi(h) = 1$, then $e(h) \ne d(h)$. Then with $m = 2m_1 + 2k - m_0$, we obtain $h' \in \mathcal{P}^{p,p'+p}_{a',b',e,f}(m_0, m_1)$ via Lemma 3.10.

Lemma 3.1 shows that $\delta^{p,p'+p}_{a',e} = \delta^{p,p'+p}_{b',f} = 0$. Thereupon, Lemma 3.12 shows that each $h' \in \mathcal{P}^{p,p'+p}_{a',b',e,f}(m_0, m_1)$ arises from a unique triple (h, k, λ), with $h \in \mathcal{P}^{p,p'}_{a,b,e,f}(m_1, m)$ for some m. The bijection follows.

The expression for $\widetilde{wt}(h')$ also results from Lemma 3.10. □

Note that the above lemma excludes consideration of the case for which $\delta^{p,p'}_{a,e} = 1$. In fact, similar results fail in that case. Nonetheless, it is necessary to tackle the $\delta^{p,p'}_{a,e} = 1$ case for a restricted set of paths in the more general analysis of [13].

Corollary 3.14. *For $1 \le p < p'$, let $1 \le a, b < p'$ and $e, f \in \{0, 1\}$ with $\delta^{p,p'}_{a,e} = 0$. Set $a' = a + e + \lfloor ap/p' \rfloor$ and $b' = b + f + \lfloor bp/p' \rfloor$. Fix $m_0, m_1 \ge 0$. Then*

$$\tilde{\chi}^{p,p'+p}_{a',b',e,f}(m_0, m_1) = q^{\frac{1}{4}((m_0-m_1)^2-\beta^2)} \sum_{\substack{m \equiv m_0 \\ (\mathrm{mod}\, 2)}} \begin{bmatrix} \frac{1}{2}(m_0 + m) \\ m_1 \end{bmatrix}_q \tilde{\chi}^{p,p'}_{a,b,e,f}(m_1, m),$$

where $\beta = \beta^{p,p'}_{a,b,e,f}$.

Proof. Apart from the case where $m_1 = 0$ and $e \ne f$, this follows immediately from Lemma 3.13 on setting $m = 2m_1 + 2k - m_0$, once it is noted, via Lemma 2.6, that $\left[{k+m_1 \atop m_1}\right]_q$ is the generating function for $\mathcal{Y}(k, m_1)$.

For the case $m_1 = 0$ and $e \ne f$, both sides are zero unless $a = b$ and m_0 is odd. In this case, $\mathcal{P}^{p,p'+p}_{a',b',e,f}(m_0, 0)$ has precisely one element h for which (via the same calculation as in the proof of 2.5) $\widetilde{wt}(h) = \frac{1}{4}(m_0^2 - 1)$. Thus the two sides are also equal in this case. □

4 The $\mathcal{D}$-transform

The *$\mathcal{D}$-transform* is defined to act on each $h \in \mathcal{P}^{p,p'}_{a,b,e,f}(L)$ to yield a path $\hat{h} \in \mathcal{P}^{p'-p,p'}_{a,b,1-e,1-f}(L)$ with exactly the same sequence of integer heights, i.e., $\hat{h}_i = h_i$ for $0 \le i \le L$. Note that by definition $e(\hat{h}) = 1 - e(h)$ and $f(\hat{h}) = 1 - f(h)$.

Since the band structure of the $(p' - p, p')$-model is obtained from that of the (p, p')-model simply by replacing odd bands by even bands and vice versa, then, ignoring the vertex at $i = 0$, each scoring vertex maps to a nonscoring vertex and vice versa. That $e(h)$ and $e(\hat{h})$ differ implies that the vertex at $i = 0$ is both scoring or both nonscoring in h and $\hat{h}$.

Lemma 4.1. *Let $\hat{h} \in \mathcal{P}^{p'-p,p'}_{a,b,1-e,1-f}(L)$ be obtained from $h \in \mathcal{P}^{p,p'}_{a,b,e,f}(L)$ by the action of the $\mathcal{D}$-transform. Then $\pi(\hat{h}) = 1 - \pi(h)$. Moreover, if $m = m(h)$, then*

- $L(\hat{h}) = L$,
- $m(\hat{h}) = \begin{cases} L - m & \text{if } e + d + \pi(h) \equiv 0 \pmod 2, \\ L - m + 2 & \text{if } e + d + \pi(h) \not\equiv 0 \pmod 2, \end{cases}$
- $\widetilde{wt}(\hat{h}) = \frac{1}{4}\left(L^2 - \alpha(h)^2\right) - \widetilde{wt}(h)$.

PROOF. Let h have striking sequence $\left(\begin{smallmatrix} a_1 & a_2 & a_3 & \cdots & a_l \\ b_1 & b_2 & b_3 & \cdots & b_l \end{smallmatrix}\right)^{(e,f,d)}$. Since beyond the zeroth vertex the $\mathcal{D}$-transform exchanges scoring vertices for nonscoring vertices and vice versa, it follows that the striking sequence for $\hat{h}$ is $\left(\begin{smallmatrix} b_1 & b_2 & b_3 & \cdots & b_l \\ a_1 & a_2 & a_3 & \cdots & a_l \end{smallmatrix}\right)^{(1-e,1-f,d)}$. It is immediate that $L(\hat{h}) = L$, $\pi(\hat{h}) = 1 - \pi(h)$, $e(\hat{h}) = 1 - e(h)$, and $d(\hat{h}) = d(h)$. Then $m(\hat{h}) = (e(\hat{h}) + d(\hat{h}) + \pi(\hat{h})) \bmod 2 + \sum_{i=1}^{l} b_i = (e + d + \pi(h)) \bmod 2 + L - \sum_{i=1}^{l} a_i = 2((e + d + \pi(h)) \bmod 2) + L - m(h)$.

Now let $w_i = a_i + b_i$ for $1 \le i \le l$. Then, using Lemma 2.1, we obtain

$$\begin{aligned} \widetilde{wt}(h) + \widetilde{wt}(\hat{h}) &= \sum_{i=1}^{l} b_i(w_{i-1} + w_{i-3} + \cdots + w_{1+i \bmod 2}) \\ &\quad + \sum_{i=1}^{l} a_i(w_{i-1} + w_{i-3} + \cdots + w_{1+i \bmod 2}) \\ &= \sum_{i=1}^{l} w_i(w_{i-1} + w_{i-3} + \cdots + w_{1+i \bmod 2}) \\ &= (w_1 + w_3 + w_5 + \cdots)(w_2 + w_4 + w_6 + \cdots). \end{aligned}$$

The lemma then follows because $(w_1 + w_3 + w_5 + \cdots) + (w_2 + w_4 + w_6 + \cdots) = L$ and $(w_1 + w_3 + w_5 + \cdots) - (w_2 + w_4 + w_6 + \cdots) = \pm\alpha(h)$. $\square$

Lemma 4.2. *Let* $1 \le p < p'$ *with* p *coprime to* p' *and* $1 \le a < p'$. *Then* $\lfloor a(p' - p)/p' \rfloor = a - 1 - \lfloor ap/p' \rfloor$.

If, in addition, a *is interfacial in the* (p, p')*-model and* $\delta_{a,e}^{p,p'} = 0$, *then* a *is interfacial in the* $(p' - p, p')$*-model and* $\delta_{a,1-e}^{p'-p,p'} = 0$.

PROOF. Since p and p' are coprime, $\lfloor ap/p' \rfloor < ap/p'$. Hence $\lfloor ap/p' \rfloor + \lfloor a(p' - p)/p' \rfloor = a - 1$.

Since the (p, p')-model differs from the $(p' - p, p')$-model only in that corresponding bands are of the opposite parity, a being interfacial in one model implies that it also is in the other. The final part follows immediately. $\square$

Corollary 4.3. *If* $1 \le p < p'$ *with* p *coprime to* p', $1 \le a, b < p'$, *and* $e, f \in \{0, 1\}$, *then* $\alpha_{a,b}^{p'-p,p'} = \alpha_{a,b}^{p,p'}$ *and* $\beta_{a,b,1-e,1-f}^{p'-p,p'} + \beta_{a,b,e,f}^{p,p'} = \alpha_{a,b}^{p,p'}$.

PROOF. Lemma 4.2 gives $\lfloor ap/p' \rfloor + \lfloor a(p' - p)/p' \rfloor = a - 1$ and likewise $\lfloor bp/p' \rfloor + \lfloor b(p' - p)/p' \rfloor = b - 1$. The required results follow immediately. $\square$

4.1 The $\mathcal{BD}$-pair. It will often be convenient to consider the combined action of a $\mathcal{D}$-transform followed immediately by a $\mathcal{B}$-transform. Such a pair will naturally be referred to as a $\mathcal{BD}$-transform and maps a path $h \in \mathcal{P}^{p'-p,p'}_{a,b,e,f}(L)$ to a path $h' \in \mathcal{P}^{p,p'+p}_{a',b',1-e,1-f}(L')$, where a', b', and L' are determined by our previous results.

In what follows, the $\mathcal{BD}$-transform will always follow a $\mathcal{B}$-transform. Thus we restrict consideration to where $2(p'-p) < p'$.

Lemma 4.4. *With $p' < 2p$, let $h \in \mathcal{P}^{p'-p,p'}_{a,b,e,f}(L)$. Let $h' \in \mathcal{P}^{p,p'+p}_{a',b',1-e,1-f}(L')$ result from the action of a $\mathcal{D}$-transform on h, followed by a $\mathcal{B}(k,\lambda)$-transform. Then*

- $L' = \begin{cases} L + m(h) + 2k - 2 & \text{if } \pi(h) = 1 \text{ and } e = d(h), \\ L + m(h) + 2k & \text{otherwise}, \end{cases}$
- $m(h') = L$,
- $\widetilde{wt}(h') = \frac{1}{4}\left(L^2 + (L'-L)^2 - \alpha^2 - \beta^2\right) + \mathrm{wt}\,(\lambda) - \widetilde{wt}(h)$,

where $\alpha = \alpha^{p,p'}_{a,b}$ and $\beta = \beta^{p,p'}_{a,b,1-e,1-f}$.

Proof. Let $\hat{h}$ result from the action of the $\mathcal{D}$-transform on h, and let $d = d(h)$, $\pi = \pi(h)$, $\hat{e} = e(\hat{h})$, $\hat{d} = d(\hat{h})$, and $\hat{\pi} = \pi(\hat{h})$. Then we immediately have $\hat{d} = d$, $\hat{e} = 1 - e$, and $\hat{\pi} = 1 - \pi$.

In the case where $\pi = 0$ and $e \neq d$, we have, using Lemmas 3.10 and 4.1, $L' = 2L(\hat{h}) - m(\hat{h}) + 2k + 2 = 2L - (L - m(h) + 2) + 2k + 2 = L + m(h) + 2k$.

In the case where $\pi = 1$ and $e = d$, we have, using Lemmas 3.10 and 4.1, $L' = 2L(\hat{h}) - m(\hat{h}) + 2k = 2L - (L - m(h) + 2) + 2k = L + m(h) + 2k - 2$.

In the other cases, $e + d + \pi \equiv 0 \pmod 2$ and so $\hat{e} + \hat{d} + \hat{\pi} \equiv 0 \pmod 2$. Lemmas 3.10 and 4.1 yield $L' = 2L(\hat{h}) - m(\hat{h}) + 2k = 2L - (L - m(h)) + 2k = L + m(h) + 2k$.

The expressions for $m(h')$ and $\widetilde{wt}(h')$ also follow immediately from Lemmas 3.10 and 4.1. □

We now obtain analogues of Lemma 3.13 and Corollary 3.14 that combine the $\mathcal{D}$-transform with the $\mathcal{B}$-transform. As above, we restrict to where $p' < 2p$.

Lemma 4.5. *For $1 \le p < p' < 2p$, let $1 \le a, b < p'$ and $e, f \in \{0,1\}$ with $\delta^{p'-p,p'}_{a,e} = 0$. Set $a' = a + 1 - e + \lfloor ap/p' \rfloor$ and $b' = b + 1 - f + \lfloor bp/p' \rfloor$. Fix $m_0, m_1 \ge 0$. Then the map $(h,k,\lambda) \mapsto h'$ effected by the action of a $\mathcal{D}$-transform on h followed by a $\mathcal{B}(k,\lambda)$-transform is a bijection between $\bigcup_k \mathcal{P}^{p'-p,p'}_{a,b,e,f}(m_1, m_0 - m_1 - 2k) \times \mathcal{Y}(k, m_1)$ and $\mathcal{P}^{p,p'+p}_{a',b',1-e,1-f}(m_0, m_1)$. Moreover,*

$$\widetilde{wt}(h') = \frac{1}{4}\left(m_1^2 + (m_0 - m_1)^2 - \alpha^2 - \beta^2\right) + \mathrm{wt}\,(\lambda) - \widetilde{wt}(h),$$

where $\alpha = \alpha^{p,p'}_{a,b}$ and $\beta = \beta^{p,p'}_{a,b,1-e,1-f}$.

Proof. Given $h \in \mathcal{P}^{p'-p,p'}_{a,b,e,f}(m_1,m)$, let $\hat{h}$ result from the action of a $\mathcal{D}$-transform on h and let h' be the result of a $\mathcal{B}(k,\lambda)$-transform on $\hat{h}$.

Since $\delta^{p'-p,p'}_{a,e} = 0$ so that $\lfloor (a+(-1)^e)(p'-p)/p')\rfloor = \lfloor a(p'-p)/p'\rfloor$, it follows that if $\pi(h)=1$, then $e(h)\neq d(h)$. Then for $m=m_0-m_1-2k$, we obtain $h' \in \mathcal{P}^{p,p'+p}_{a',b',1-e,1-f}(m_0,m_1)$ via Lemma 4.4.

Lemma 3.1 gives $\delta^{p,p'+p}_{a',1-e} = \delta^{p,p'+p}_{b',1-f} = 0$. Lemma 3.12 then shows that for arbitrary $h' \in \mathcal{P}^{p,p'+p}_{a',b',1-e,1-f}(m_0,m_1)$, there is a unique triple $(\hat{h},k,\lambda)$ with $\hat{h} \in \mathcal{P}^{p,p'}_{a,b,1-e,1-f}(m_1,m')$ for some m' such that the action of the $\mathcal{B}(k,\lambda)$-transform on $\hat{h}$ yields h'. Then, via the $\mathcal{D}$-transform, we obtain a unique $h \in \mathcal{P}^{p'-p,p'}_{a,b,e,f}(m_1,m'')$ for some m''. The bijection follows.

The expression for $\widetilde{wt}(h)$ also results from Lemma 4.4. □

Note that the above lemma excludes the case for which $\delta^{p'-p,p'}_{a,e}=1$. Once more, similar results fail in that case.

Corollary 4.6. *For $1\le p<p'<2p$, let $1\le a,b<p'$ and $e,f\in\{0,1\}$ with $\delta^{p'-p,p'}_{a,e}=0$. Set $a'=a+1-e+\lfloor ap/p'\rfloor$ and $b'=b+1-f+\lfloor bp/p'\rfloor$. Fix $m_0,m_1\ge 0$. Then*

$$\tilde{\chi}^{p,p'+p}_{a',b',1-e,1-f}(m_0,m_1;q) = q^{\frac14(m_1^2+(m_0-m_1)^2-\alpha^2-\beta^2)} \sum_{\substack{m\equiv m_0-m_1\\ (\mathrm{mod}\ 2)}} \begin{bmatrix} \frac12(m_0+m_1-m)\\ m_1\end{bmatrix}_q \tilde{\chi}^{p'-p,p'}_{a,b,e,f}(m_1,m;q^{-1}),$$

where $\alpha=\alpha^{p,p'}_{a,b}$ and $\beta=\beta^{p,p'}_{a,b,1-e,1-f}$.

Proof. Apart from the case where $m_1=0$ and $e\neq f$, this follows immediately from Lemma 4.5 on setting $m=m_0-m_1-2k$, once it is noted via Lemma 2.6 that $\left[{k+m_1 \atop m_1}\right]_q$ is the generating function for $\mathcal{Y}(k,m_1)$.

The case $m_1=0$ and $e\neq f$ is dealt with exactly as in the proof of Corollary 3.14. □

Lemma 4.7. *Let $1\le p<p'<2p$ with p coprime to p', $1\le a,b<p'$, and $e,f\in\{0,1\}$ and set $a'=a+1-e+\lfloor ap/p'\rfloor$ and $b'=b+1-f+\lfloor bp/p'\rfloor$. Then $\lfloor a'p/(p'+p)\rfloor = a-1-\lfloor a(p'-p)/p'\rfloor$ and $\lfloor b'p/(p'+p)\rfloor = b-1-\lfloor b(p'-p)/p'\rfloor$. In addition, $\alpha^{p,p'+p}_{a',b'} = 2\alpha^{p'-p,p'}_{a,b}-\beta^{p'-p,p'}_{a,b,e,f}$ and $\beta^{p,p'+p}_{a',b',1-e,1-f} = \alpha^{p'-p,p'}_{a,b}-\beta^{p'-p,p'}_{a,b,e,f}$.*

Proof. By Lemma 4.2 and Corollary 4.3, $\lfloor ap/p'\rfloor = a-1-\lfloor a(p'-p)/p'\rfloor$, $\lfloor bp/p'\rfloor = b-1-\lfloor b(p'-p)/p'\rfloor$, $\alpha^{p,p'}_{a,b}=\alpha^{p'-p,p'}_{a,b}$, and $\beta^{p,p'}_{a,b,1-e,1-f}=\alpha^{p'-p,p'}_{a,b}-\beta^{p'-p,p'}_{a,b,e,f}$. The current lemma then follows immediately from Lemmas 3.1 and 3.5. □

5 The structure of the (p, p')-model

5.1 Continued fractions. If p' and p are positive coprime integers and

$$\frac{p'}{p} = c_0 + \cfrac{1}{c_1 + \cfrac{1}{c_2 + \cfrac{1}{\ddots \, \cfrac{}{c_{n-1} + \cfrac{1}{c_n}}}}}$$

with $c_0 \geq 0$, $c_i \geq 1$ for $0 < i < n$, and $c_n \geq 2$, then $(c_0, c_1, c_2, \ldots, c_n)$ is said to be the *continued fraction* for p'/p.

We refer to n as the *height* of p'/p. We set $t = c_0 + c_1 + \cdots + c_n - 2$ and refer to it as the *rank* of p'/p. The height and rank of $\mathcal{P}^{p,p'}_{a,b,c}(L)$ are then defined to be equal to those of p'/p.

For $0 \leq k \leq n+1$, we also define

$$t_k = -1 + \sum_{i=0}^{k-1} c_i. \tag{13}$$

Then $t_{n+1} = t+1$ and $t_n \leq t-1$. We say that the index j with $0 \leq j \leq t_{n+1}$ is in zone k if $t_k < j \leq t_{k+1}$. We then write $k = \zeta(j)$. Note that there are $n+1$ zones and that for $0 \leq k \leq n$, zone k contains c_k indices.

5.2 The Takahashi and string lengths. Given positive coprime integers p and p' with p'/p having rank t, define the set $\{\kappa_i\}_{i=0}^t$ of *Takahashi lengths*, the set $\{\tilde{\kappa}_i\}_{i=0}^t$ of *truncated Takahashi lengths*, and the set $\{l_i\}_{i=0}^t$ of *string lengths* as follows. First, define y_k and z_k for $-1 \leq k \leq n+1$ by

$$\begin{aligned} y_{-1} &= 0; & z_{-1} &= 1, \\ y_0 &= 1; & z_0 &= 0; \\ y_k &= c_{k-1}y_{k-1} + y_{k-2}; & z_k &= c_{k-1}z_{k-1} + z_{k-2} \quad (1 \leq k \leq n+1). \end{aligned}$$

Now for $t_k < j \leq t_{k+1}$ and $0 \leq k \leq n$, set

$$\begin{aligned} \kappa_j &= y_{k-1} + (j - t_k)y_k, \\ \tilde{\kappa}_j &= z_{k-1} + (j - t_k)z_k, \\ l_j &= y_{k-1} + (j - t_k - 1)y_k. \end{aligned}$$

Note that $\kappa_j = l_{j+1}$ unless $j = t_k$ for some k, in which case $\kappa_{t_k} = y_k$ and $l_{t_k+1} = y_{k-1}$. We define $\mathcal{T} = \{\kappa_i\}_{i=0}^{t-1}$ and $\mathcal{T}' = \{p' - \kappa_i\}_{i=0}^{t-1}$. (We do not include κ_t in the former since it is present in the latter.) Then for $n > 0$, $\mathcal{T} \cap \mathcal{T}' = \emptyset$.[11]

[11]In fact, when $n = 0$, $\mathcal{T} \cap \mathcal{T}' = \{2, 3, \ldots, p'-2\}$. Then if $2 \leq a \leq p'-2$, different fermionic expressions for $\mathcal{P}^{p,p'}_{a,b,c}(L)$ arise by considering either $a \in \mathcal{T}$ or $a \in \mathcal{T}'$. The same holds for $2 \leq b \leq p'-2$. This $n = 0$ case was fully examined in [12].

For example, in the case $p' = 38$, $p = 11$, for which the continued fraction is $(3, 2, 5)$ so that $n = 2$, $(t_1, t_2, t_3) = (2, 4, 9)$ and $t = 8$. We then obtain

$$\begin{aligned}
(y_{-1}, y_0, y_1, y_2, y_3) &= (0, 1, 3, 7, 38),\\
(z_{-1}, z_0, z_1, z_2, z_3) &= (1, 0, 1, 2, 11),\\
(\kappa_0, \kappa_1, \kappa_2, \kappa_3, \kappa_4, \kappa_5, \kappa_6, \kappa_7) &= (1, 2, 3, 4, 7, 10, 17, 24),\\
(l_1, l_2, l_3, l_4, l_5, l_6, l_7, l_8) &= (1, 2, 1, 4, 3, 10, 17, 24),\\
(\tilde{\kappa}_0, \tilde{\kappa}_1, \tilde{\kappa}_2, \tilde{\kappa}_3, \tilde{\kappa}_4, \tilde{\kappa}_5, \tilde{\kappa}_6, \tilde{\kappa}_7) &= (1, 1, 1, 1, 2, 3, 5, 7).
\end{aligned}$$

An induction argument readily establishes that if $1 \le k \le n+1$, then $y_k z_{k-1} - y_{k-1} z_k = (-1)^k$, that y_k is coprime to z_k, and that y_k/z_k has continued fraction $(c_0, c_1, \ldots, c_{k-1})$. Thus, in particular, $y_{n+1} = p'$ and $z_{n+1} = p$.

6 Segmenting the model

6.1 Model comparisons. Here, we relate the parameters associated with the (p, p')-model for which the continued fraction is $(c_0, c_1, \ldots, c_n)$ to those associated with certain "simpler" models. In particular, if $c_0 > 1$, we compare them with those associated with the $(p, p' - p)$-model and if $c_0 = 1$, we compare them with those associated with the $(p' - p, p')$-model.

In the following two lemmas, the parameters associated with those simpler models will be primed to distinguish them from those associated with the (p, p')-model. In particular if $c_0 > 1$, $(p' - p)/p$ has continued fraction $(c_0 - 1, c_1, \ldots, c_n)$ so that in this case, $t' = t - 1$, $n' = n$, and $t'_k = t_k - 1$ for $1 \le k \le n$. If $c_0 = 1$, then $p'/(p' - p)$ has continued fraction $(c_1 + 1, c_2, \ldots, c_n)$ so that in this case, $t' = t$, $n' = n - 1$ and $t'_k = t_{k+1}$ for $1 \le k \le n'$.

Lemma 6.1. *Let $c_0 > 1$. For $1 \le k \le n$ and $0 \le j \le t$, let y_k, z_k, κ_j, and $\tilde{\kappa}_j$ be the parameters associated with the (p, p')-model as defined in Section 5.2. For $1 \le k \le n$ and $0 \le j \le t'$, let y'_k, z'_k, κ'_j, and $\tilde{\kappa}'_j$ be the corresponding parameters for the $(p, p' - p)$-model. Then*

- $y_k = y'_k + z'_k \quad (0 \le k \le n)$,
- $z_k = z'_k \quad (0 \le k \le n)$,
- $\kappa_j = \kappa'_{j-1} + \tilde{\kappa}'_{j-1} \quad (1 \le j \le t)$,
- $\tilde{\kappa}_j = \tilde{\kappa}'_{j-1} \quad (1 \le j \le t)$.

Proof. This result is a straightforward consequence of the definitions. □

Lemma 6.2. *Let $c_0 = 1$. For $1 \le k \le n$ and $0 \le j \le t$, let y_k, z_k, κ_j, and $\tilde{\kappa}_j$ be the parameters associated with the (p, p')-model as defined in Section 5.2. For $1 \le k \le n'$ and $0 \le j \le t$, let y'_k, z'_k, κ'_j, and $\tilde{\kappa}'_j$ be the corresponding parameters for the $(p' - p, p')$-model. Then*

- $y_k = y'_{k-1}$ $(1 \le k \le n)$,
- $z_k = y'_{k-1} - z'_{k-1}$ $(1 \le k \le n)$,
- $\kappa_j = \kappa'_j$ $(1 \le j \le t)$,
- $\tilde{\kappa}_j = \kappa'_j - \tilde{\kappa}'_j$ $(1 \le j \le t)$.

PROOF. Again, this result is a straightforward consequence of the definitions. □

Lemma 6.3. *If* $t_1 \le j \le t$, *then*[12]

$$\left\lfloor \frac{\tilde{\kappa}_j p'}{p} \right\rfloor = \kappa_j - \delta^{(2)}_{\zeta(j),1},$$

and if $0 \le j \le t$, *then*

$$\left\lfloor \frac{\kappa_j p}{p'} \right\rfloor = \tilde{\kappa}_j - \delta^{(2)}_{\zeta(j),0}.$$

PROOF. We prove the first of these two results by induction on the sum of the height and rank of p'/p. Since $\kappa_{t_1} = c_0$, $\tilde{\kappa}_{t_1} = 1$, $\zeta(t_1) = 0$, the required result always holds for the case $j = t_1$. In particular, it certainly holds in the case where the sum of the height and rank of p'/p is at most 1.

Now assume that the first part holds in the case that sum of height and rank is $n + t - 1$ and consider the case where p'/p has height n and rank t. First, assume that $p' > 2p$. For $j \ge t_1$, the induction hypothesis implies that $\kappa'_{j-1} - \delta^{(2)}_{\zeta'(j-1),1} < \tilde{\kappa}'_{j-1}(p'-p)/p < \kappa'_{j-1} - \delta^{(2)}_{\zeta'(j-1),1} + 1$, where the primed quantities pertain to the continued fraction of $(p'-p)/p$. Using Lemma 6.1 and noting that $\zeta'(j-1) = \zeta(j)$ readily yields $\kappa_j - \delta^{(2)}_{\zeta(j),1} < \tilde{\kappa}_j p'/p < \kappa_j - \delta^{(2)}_{\zeta(j),1} + 1$. This immediately gives the required result.

In the case where $p' < 2p$, first let $j \ge t_2$. The induction hypothesis implies that $\kappa'_j - \delta^{(2)}_{\zeta'(j),1} < \tilde{\kappa}'_j p'/(p'-p) < \kappa'_j - \delta^{(2)}_{\zeta'(j),1} + 1$, where the primed quantities pertain to the continued fraction of $p'/(p'-p)$. Using Lemma 6.2 and noting that $\zeta'(j) = \zeta(j) - 1$ readily yields $\kappa_j - \delta^{(2)}_{\zeta(j),1}(p'-p)/p < \tilde{\kappa}_j p'/p < \kappa_j + (1 - \delta^{(2)}_{\zeta(j),1})(p'-p)/p$. Since $(p'-p)/p < 1$, this implies the required result.

When $p' < 2p$, we have $c_0 = 1$ so that $t_1 = 0$ and $t_2 = c_1$. Then $\tilde{\kappa}_j = j$ for $t_1 < j \le t_2$, whereupon in view of the continued fraction expression for p'/p, we immediately obtain $\lfloor \tilde{\kappa}_j p'/p \rfloor = j = \kappa_j - 1$, as required.

The first part of the lemma then follows by induction. For $t_1 \le j \le t$, the second part readily follows from the first. For $0 \le j \le t_1 \le t$, both sides are clearly equal to 0. □

[12] We use the notation $\delta^{(2)}_{i,j} = 1$ if $i \equiv j \pmod 2$ and $\delta^{(2)}_{i,j} = 0$ if $i \not\equiv j \pmod 2$.

If $t_1 \le j \le t$, it follows from this result that with k such that $t_k < j \le t_{k+1}$, the $\tilde{\kappa}_j$th odd band in the (p, p')-model lies between heights $\kappa_j - 1$ and κ_j when k is odd and between heights κ_j and $\kappa_j + 1$ when k is even. Since there are no adjacent odd bands when $p' > 2p$, it follows that κ_j is interfacial when $j \ge t_1$. On switching the parity of each band, we obtain in the case where $p' < 2p$ that κ_j is interfacial when $j \ge t_2$.

Lemma 6.4. *If $1 \le p < p'$ and p is coprime to p', then for $1 \le s \le y_n - 2$, the sth band of the (p, p')-model is of the same parity as the sth band of the (z_n, y_n)-model.*

Proof. We must establish that $\lfloor s z_n / y_n \rfloor = \lfloor s z_{n+1} / y_{n+1} \rfloor$ for $1 \le s \le y_n - 1$.

With s such that $1 \le s < y_n$, let $r = \lfloor s z_{n+1} / y_{n+1} \rfloor$. Using $y_n z_{n+1} = y_{n+1} z_n + (-1)^n$ yields

$$r y_n - (-1)^n \frac{s}{y_{n+1}} \le s z_n < (r+1) y_n - (-1)^n \frac{s}{y_{n+1}}.$$

Since $1 \le s < y_n < y_{n+1}$, the first inequality here implies that $s z_n / y_n \ge r$. For the same reasons and noting that $s z_n / y_n$ is not integral, the second inequality here implies that $s z_n / y_n < r + 1$. The lemma then follows. □

This lemma shows that the (z_n, y_n)-model resides within the (p, p')-model between heights 1 and $y_n - 1$. The up–down symmetry of the (p, p')-model then also implies that the (z_n, y_n)-model also resides within the (p, p')-model between heights $p' - y_n + 1$ and $p' - 1$.

6.2 Interfacial retention. We now show that if h attains an interfacial height, then the path resulting from the action of a $\mathcal{B}$-transform on h attains the corresponding interfacial height.

Lemma 6.5. *Let $h \in \mathcal{P}^{p,p'}_{a,b,e,f}(L)$ and let $h \in \mathcal{P}^{p,p'+p}_{a',b',e,f}(L')$ result from the action of a $\mathcal{B}(k, \lambda)$-transform on h. Let s be interfacial in the (p, p')-model with $a \ne s \ne b$ and set $r = \lfloor (s+1)p/p' \rfloor$. Then $s + r$ is interfacial in the $(p, p' + p)$-model.*

If $h_i = s$ for $0 \le i \le L$ then $h'_j = s + r$ for some j with $0 \le j \le L'$. On the other hand, if $h'_j = s + r$ for $0 \le j \le L'$ then $h_i = s$ for some i with $0 \le i \le L$.

Proof. First, note that s borders the rth odd band in the (p, p')-model. If s is at the lower (resp., upper) edge of the rth odd band in the (p, p')-model, then $s + r$ is at the lower (resp., upper) edge of the rth odd band in the $(p, p' + p)$-model. In particular, this implies that $s + r$ is interfacial in the $(p, p' + p)$-model. Then note that in the (p, p')-model, there is at least one even band between the two odd bands on either side of s. (Assume that there is an odd band immediately above and immediately below the (p, p')-model grid if necessary.) Thus there are at least two even bands between the two odd bands on either side of $s + r$ in the $(p, p' + p)$-model.

Let $h^{(0)}$ result from the action of the $\mathcal{B}_1$-transform on h. The definition of this transform implies that if $h_i = s$ for some i, then $h^{(0)}_j = s + r$ for some j and vice versa. (When $\delta^{p,p'}_{a,e} = 1$ or $\delta^{p,p'}_{b,f} = 1$, this statement relies on $a \ne s \ne b$.)

If $h^{(k)}$ results from the action of the $\mathcal{B}_2(k)$-transform on $h^{(0)}$, then if $h_j^{(0)} = s+r$ for some j then $h_{j'}^{(k)} = s+r$ for some j' and vice versa. (This statement relies on the two odd bands either side of $s+r$ having at least two even bands between them.)

If h' results from the action of the $\mathcal{B}_3(\lambda)$-transform on $h^{(k)}$, then if $h_j^{(0)} = s+r$ for some j, examination of the ten particle moves and edge moves described in Section 3.3 shows that $h_{j'}^{(k)} = s+r$ for some j' and vice versa. (This statement also relies on the two odd bands either side of $s+r$ having at least two even bands between them.) Combining these results proves the lemma. □

We also need the analogue of this result for the $\mathcal{BD}$-transform.

Lemma 6.6. *Let $h \in \mathcal{P}^{p'-p,p'}_{a,b,e,f}(L)$ and let $h' \in \mathcal{P}^{p,p'+p}_{a',b',1-e,1-f}(L')$ result from the action of a $\mathcal{D}$-transform on h followed by a $\mathcal{B}(k,\lambda)$-transform. Let s be interfacial in the (p,p')-model with $a \neq s \neq b$ and set $r = \lfloor (s+1)p/p' \rfloor$. Then $s+r$ is interfacial in the $(p,p'+p)$-model.*

If $h_i = s$ for $0 \le i \le L$, then $h'_j = s+r$ for some j with $0 \le j \le L'$. On the other hand, if $h'_j = s+r$ for $0 \le j \le L'$, then $h_i = s$ for some i with $0 \le i \le L$.

Proof. This follows immediately from the above result after noting that if s is interfacial in the $(p'-p,p')$-model, then it also is in the (p,p')-model. □

A set $\mathcal{S}$ is said to be interfacial in the (p,p')-model if each $s \in \mathcal{S}$ is interfacial in the (p,p')-model. We now define $\mathcal{P}^{p,p'}_{a,b,e,f}(L,m)\{\mathcal{S}\}$ to be the subset of $\mathcal{P}^{p,p'}_{a,b,e,f}(L,m)$ comprising those paths h for which for each $s \in \mathcal{S}$ there exists i with $0 \le i \le L$ such that $h_i = s$. The generating function for this set is

$$\tilde{\chi}^{p,p'}_{a,b,e,f}(L;q)\{\mathcal{S}\} = \sum_{h \in \mathcal{P}^{p,p'}_{a,b,e,f}(L)\{\mathcal{S}\}} q^{\widetilde{wt}(h)}.$$

Of course, $\mathcal{P}^{p,p'}_{a,b,e,f}(L,m)\{\emptyset\} = \mathcal{P}^{p,p'}_{a,b,e,f}(L,m)$.

Given $\mathcal{S}$ as above, we now define $\mathcal{S}' = \{s + \lfloor (s+1)p/p' \rfloor : s \in \mathcal{S}\}$.

Corollary 6.7. *For $1 \le p < p'$, let $1 \le a, b < p'$ and $e, f \in \{0,1\}$ with $\delta^{p,p'}_{a,e} = 0$. Let $\mathcal{S}$ be interfacial in the (p,p')-model with $a \neq s \neq b$ for all $s \in \mathcal{S}$. Set $a' = a + e + \lfloor ap/p' \rfloor$ and $b' = b + f + \lfloor bp/p' \rfloor$. Fix $m_0, m_1 \ge 0$. Then*

$$\tilde{\chi}^{p,p'+p}_{a',b',e,f}(m_0,m_1)\{\mathcal{S}'\} = q^{\frac{1}{4}((m_0-m_1)^2-\beta^2)} \sum_{\substack{m \equiv m_0 \\ (\mathrm{mod}\ 2)}} \begin{bmatrix} \frac{1}{2}(m_0+m) \\ m_1 \end{bmatrix}_q \tilde{\chi}^{p,p'}_{a,b,e,f}(m_1,m)\{\mathcal{S}\},$$

where $\beta = \beta^{p,p'}_{a,b,e,f}$.

PROOF. Combining Lemmas 3.13 and 6.5 implies that the map $(h, k, \lambda) \mapsto h'$ effected by the action of a $\mathcal{B}(k, \lambda)$-transform on h is a bijection between $\bigcup_k \mathcal{P}^{p,p'}_{a,b,e,f}(m_1, 2k+2m_1-m_0)\{\mathcal{S}\} \times \mathcal{Y}(k, m_1)$ and $\mathcal{P}^{p,p'+p}_{a',b',e,f}(m_0, m_1)\{\mathcal{S}'\}$. The result follows as in the proof of Corollary 3.14. □

Corollary 6.8. *For $1 \le p < p' < 2p$, let $1 \le a, b < p'$ and $e, f \in \{0, 1\}$ with $\delta^{p'-p,p'}_{a,e} = 0$. Let $\mathcal{S}$ be interfacial in the (p, p')-model with $a \ne s \ne b$ for all $s \in \mathcal{S}$. Set $a' = a+1-e+\lfloor ap/p' \rfloor$ and $b' = b+1-f+\lfloor bp/p' \rfloor$. Fix $m_0, m_1 \ge 0$. Then*

$$\begin{aligned}
&\tilde{\chi}^{p,p'+p}_{a',b',1-e,1-f}(m_0, m_1; q)\{\mathcal{S}'\} \\
&\quad = q^{\frac{1}{4}(m_1^2+(m_0-m_1)^2-\alpha^2-\beta^2)} \\
&\qquad \times \sum_{\substack{m\equiv m_0-m_1\\(\mathrm{mod}\ 2)}} \begin{bmatrix} \frac{1}{2}(m_0+m_1-m) \\ m_1 \end{bmatrix}_q \tilde{\chi}^{p'-p,p'}_{a,b,e,f}(m_1, m; q^{-1})\{\mathcal{S}\},
\end{aligned}$$

where $\alpha = \alpha^{p,p'}_{a,b}$ and $\beta = \beta^{p,p'}_{a,b,1-e,1-f}$.

PROOF. Combining Lemmas 4.5 and 6.6 implies that the map $(h, k, \lambda) \mapsto h'$ effected by the action of a $\mathcal{D}$-transform on h immediately followed by a $\mathcal{B}(k, \lambda)$-transform is a bijection between $\bigcup_k \mathcal{P}^{p'-p,p'}_{a,b,e,f}(m_1, m_0-m_1-2k)\{\mathcal{S}\} \times \mathcal{Y}(k, m_1)$ and $\mathcal{P}^{p,p'+p}_{a',b',1-e,1-f}(m_0, m_1)\{\mathcal{S}'\}$. The result then follows as in the proof of Corollary 4.6. □

7 Extending and truncating paths

7.1 Extending paths. In this section, we specify a process by which a path $h \in \mathcal{P}^{p,p'}_{a,b,e,f}(L)$ may be extended by a single unit to its left or by a single unit to its right. One extension may follow the other to yield a path of length $L+2$.

Path extension on the left is restricted to where $\delta^{p,p'}_{a,e} = 0$ so that the presegment of h lies in the even band.

We obtain h' by defining $h'_0 = a' = a + (-1)^e$ and $h'_i = h_{i-1}$ for $1 \le i \le L+1$. In particular, $\pi(h') = 0$. We also define $e(h') = e' = 1 - e$, so that then $h' \in \mathcal{P}^{p,p'}_{a',b,e',f}(L+1)$.

This extending process is depicted in Figure 7.

Figure 7. Extending on the left.

Lemma 7.1. *Let $h \in \mathcal{P}^{p,p'}_{a,b,e,f}(L)$, where $\delta^{p,p'}_{a,e} = 0$. Let $h' \in \mathcal{P}^{p,p'}_{a',b,e',f}(L')$ be obtained from h by the above process of path extension. If $\Delta = a' - a$, then $\Delta = (-1)^e = -(-1)^{e'}$ and*

- $L' = L + 1$,
- $m(h') = m(h)$,
- $\widetilde{wt}(h') = \widetilde{wt}(h) + \frac{1}{2}(L - m(h) + \Delta\beta(h))$.

Furthermore, $\alpha^{p,p'}_{a',b} = \alpha^{p,p'}_{a,b} - \Delta$ and $\beta^{p,p'}_{a',b,e',f} = \beta^{p,p'}_{a,b,e,f} - \Delta$.

Proof. That $\Delta = (-1)^e = -(-1)^{e'}$ is immediate from the definition. Let h have striking sequence $\left(\begin{smallmatrix} a_1 & a_2 & a_3 & \cdots & a_l \\ b_1 & b_2 & b_3 & \cdots & b_l \end{smallmatrix}\right)^{(e,f,d)}$.

If $e = d$, we are restricted to the case $\pi(h) = 0$ since $\delta^{p,p'}_{a,e} = 0$. The striking sequence of h' is then $\left(\begin{smallmatrix} 0 & a_1 & a_2 & \cdots & a_l \\ 1 & b_1 & b_2 & \cdots & b_l \end{smallmatrix}\right)^{(e',f,e')}$. Thereupon, since $\pi(h') = 0$, we obtain $m(h') = m(h)$. In this case we immediately obtain, via Lemma 2.1, that $\widetilde{wt}(h') = \widetilde{wt}(h) + (b_1 + b_3 + \cdots)$. Thereupon, since $\Delta = (-1)^e = (-1)^d$, $\beta(h) = (-1)^d((b_1 + b_3 + \cdots) - (b_2 + b_4 + \cdots))$, and $m(h) = (a_1 + a_2 + a_3 + \cdots)$, we obtain $\widetilde{wt}(h') = \widetilde{wt}(h) + (L - m(h) + \Delta\beta(h))/2$.

If $e \neq d$, the striking sequence of h' is $\left(\begin{smallmatrix} a_1+1-\pi & a_2 & \cdots & a_l \\ b_1+\pi & b_2 & \cdots & b_l \end{smallmatrix}\right)^{(e',f,e')}$. Then $m(h') = 1 - \pi + \sum_{i=1}^{l} a_i$, which equals $m(h) = (e + d + \pi) \bmod 2 + \sum_{i=1}^{l} a_i$ for both $\pi = 0$ and $\pi = 1$. Here Lemma 2.1 implies that $\widetilde{wt}(h') = \widetilde{wt}(h) + (b_2 + b_4 + \cdots)$. Thereupon, since $\Delta = (-1)^e = -(-1)^d$, $\beta(h) = (-1)^d((b_1 + b_3 + \cdots) - (1 - \pi + b_2 + b_4 + \cdots))$, and $m(h) = (1 - \pi + a_1 + a_2 + a_3 + \cdots)$, we also obtain $\widetilde{wt}(h') = \widetilde{wt}(h) + (L - m(h) + \Delta\beta(h))/2$.

It is immediate that $\alpha^{p,p'}_{a',b} = \alpha^{p,p'}_{a,b} - \Delta$. Since $\pi(h') = 0$, $\lfloor a'p/p' \rfloor = \lfloor ap/p' \rfloor$. That $\beta^{p,p'}_{a',b,e',f} = \beta^{p,p'}_{a,b,e,f} - \Delta$ now follows. □

In the following lemma, we consider the special case when $2p < p' < 3p$ so that the first and second bands of the (p, p')-model are even and odd, respectively. We then only consider path extension into the first or the $(p' - 2)$th band of the (p, p')-model.

Lemma 7.2. *Let $2 < 2p < p' < 3p$ and either $a = 2$ and $e = 1$ or $a = p' - 2$ and $e = 0$. Then a is interfacial in the (p, p')-model. Let $\mathcal{S}$ be interfacial in the (p, p')-model and set $\Delta = (-1)^e$, $a' = a + \Delta$, and $e' = 1 - e$. Then*

$$\tilde{\chi}^{p,p'}_{a',b,e',f}(L, m)\{\mathcal{S} \cup \{a\}\} = q^{\frac{1}{2}(L-1-m+\Delta\beta)} \tilde{\chi}^{p,p'}_{a,b,e,f}(L - 1, m)\{\mathcal{S}\},$$

where $\beta = \beta^{p,p'}_{a,b,e,f}$.

In addition, $\alpha^{p,p'}_{a',b} = \alpha^{p,p'}_{a,b} - \Delta$ and $\beta^{p,p'}_{a',b,e',f} = \beta^{p,p'}_{a,b,e,f} - \Delta$.

Proof. Since $2p < p' < 3p$, it follows that $0 = \lfloor 2p/p' \rfloor \neq \lfloor 3p/p' \rfloor$, wherepon 2 and $p' - 2$ are both interfacial in the (p, p')-model and $\delta^{p,p'}_{a,e} = 0$.

Let $h \in \mathcal{P}^{p,p'}_{a,b,e,f}(L-1, m)\{\mathcal{S}\}$. Extend h on the left to obtain h' with $h'_0 = a' = a+\Delta$. Clearly, h' attains a. Then Lemma 7.1 implies that $h' \in \mathcal{P}^{p,p'}_{a',b,e',f}(L, m)\{\mathcal{S}\cup\{a\}\}$.

Conversely, any such h' arises from some $h \in \mathcal{P}^{p,p'}_{a,b,e,f}(L-1, m)\{\mathcal{S}\}$ in this way since either $h'_0 = 1$ and $e' = 0$ or $h'_0 = p' - 1$ and $e' = 1$. The result then follows from the expression for $\widetilde{wt}(h')$ given in Lemma 7.1 and $\beta(h) = \beta^{p,p'}_{a,b,e,f}$ from Lemma 2.2.

The final statement also follows from Lemma 7.1. □

For $h \in \mathcal{P}^{p,p'}_{a,b,e,f}(L)$, we now define path extension to the right in a similar way. Here we restrict path extension to the cases where $\delta^{p,p'}_{b,f} = 0$ so that the postsegment of h lies in the even band.

We obtain h' by defining $h'_i = h_i$ for $0 \le i \le L$ and $h'_{L+1} = b' = b + (-1)^f$. We also define $f(h') = f' = 1 - f$, so that then $h' \in \mathcal{P}^{p,p'}_{a,b',e,f'}(L+1)$.

This extending process is depicted in Figure 8.

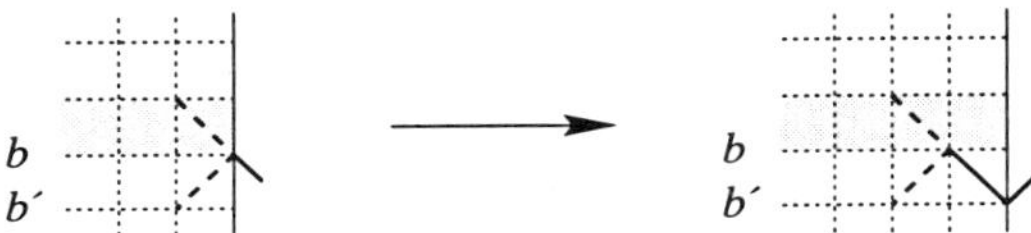

Figure 8. Extending on the right.

Lemma 7.3. *Let $h \in \mathcal{P}^{p,p'}_{a,b,e,f}(L)$, where $\delta^{p,p'}_{b,f} = 0$. Let $h' \in \mathcal{P}^{p,p'}_{a,b',e,f'}(L')$ be obtained from h by the above process of path extension. If $\Delta = b' - b$, then $\Delta = (-1)^f = -(-1)^{f'}$ and*

- $L' = L + 1$,
- $m(h') = m(h)$,
- $\widetilde{wt}(h') = \widetilde{wt}(h) + \frac{1}{2}(L - \Delta\alpha(h))$.

Furthermore, $\alpha^{p,p'}_{a,b'} = \alpha^{p,p'}_{a,b} + \Delta$ and $\beta^{p,p'}_{a,b',e,f'} = \beta^{p,p'}_{a,b,e,f} + \Delta$.

Proof. That $\Delta = (-1)^f = -(-1)^{f'}$ is immediate from the definition. Let h have striking sequence $\left(\begin{smallmatrix} a_1 & a_2 & a_3 & \cdots & a_l \\ b_1 & b_2 & b_3 & \cdots & b_l \end{smallmatrix}\right)^{(e,f,d)}$. It is easily checked that the Lth vertex of h' is scoring if and only if the Lth vertex of h is scoring.

Then if the extending segment is in the same direction as the Lth segment, h' has striking sequence $\left(\begin{smallmatrix} a_1 & a_2 & a_3 & \cdots & a_l \\ b_1 & b_2 & b_3 & \cdots & b_l+1 \end{smallmatrix}\right)^{(e,f',d)}$ and $\Delta = -(-1)^{d+l}$. It is immediate that $m(h') = m(h)$.

When the extending segment is in the direction opposite to that of the Lth segment, h' has striking sequence $\left(\begin{smallmatrix} a_1 & a_2 & \cdots & a_l & 0 \\ b_1 & b_2 & \cdots & b_l & 1 \end{smallmatrix}\right)^{(e,f',d)}$ and $\Delta = (-1)^{d+l}$. We immediately obtain $m(h') = m(h)$ in this case.

For $1 \le i \le l$, let $w_i = a_i + b_i$. We find $\alpha(h) = -(-1)^{d+l}((w_l + w_{l-2} \cdots) - (w_{l-1} + w_{l-3} + \cdots))$. In the first case above, Lemma 2.1 gives $\widetilde{wt}(h') = \widetilde{wt}(h) +$

$(w_{l-1}+w_{l-3}+w_{l-5}+\cdots)$, whereupon we obtain $\widetilde{wt}(h')=\widetilde{wt}(h)+\frac{1}{2}(L(h)-\Delta\alpha(h))$. In the second case above, Lemma 2.1 gives $\widetilde{wt}(h')=\widetilde{wt}(h)+(w_l+w_{l-2}+w_{l-4}+\cdots)$, and we again obtain $\widetilde{wt}(h')=\widetilde{wt}(h)+\frac{1}{2}(L(h)-\Delta\alpha(h))$.

It is immediate that $\alpha^{p,p'}_{a,b'}=\alpha^{p,p'}_{a,b}+\Delta$. That $\beta^{p,p'}_{a,b',e,f'}=\beta^{p,p'}_{a,b,e,f}+\Delta$ now follows because $\lfloor bp/p'\rfloor=\lfloor b'p/p'\rfloor$. □

Lemma 7.4. *Let $2<2p<p'<3p$ and either $b=2$ and $f=1$ or $b=p'-2$ and $f=0$. Then b is interfacial in the (p,p')-model. Let $\mathcal{S}$ be interfacial in the (p,p')-model and set $\Delta=(-1)^f$, $b'=b+\Delta$, and $f'=1-f$. Then*

$$\tilde{\chi}^{p,p'}_{a,b',e,f'}(L,m)\{\mathcal{S}\cup\{b\}\}=q^{\frac{1}{2}(L-1-\Delta\alpha)}\tilde{\chi}^{p,p'}_{a,b,e,f}(L-1,m)\{\mathcal{S}\},$$

where $\alpha=\alpha^{p,p'}_{a,b}$.

In addition, $\alpha^{p,p'}_{a,b'}=\alpha^{p,p'}_{a,b}+\Delta$ and $\beta^{p,p'}_{a,b',e,f'}=\beta^{p,p'}_{a,b,e,f}+\Delta$.

Proof. Since $2p<p'<3p$, it follows that $0=\lfloor 2p/p'\rfloor\neq\lfloor 3p/p'\rfloor$, wherepon 2 and $p'-2$ are both interfacial in the (p,p')-model, and $\delta^{p,p'}_{b,f}=0$.

Let $h\in\mathcal{P}^{p,p'}_{a,b,e,f}(L-1,m)\{\mathcal{S}\}$. Extend this path on the right to obtain h' with $h'_L=b'=b+\Delta$. Clearly, h' attains height b. Then, via Lemma 7.3, $h'\in\mathcal{P}^{p,p'}_{a,b',e,f'}(L,m)\{\mathcal{S}\cup\{b\}\}$. Conversely, any such h' arises in this way from some $h\in\mathcal{P}^{p,p'}_{a,b,e,f}(L-1,m)\{\mathcal{S}\}$ since either $h'_L=1$ and $f'=1$ or $h'_L=p'-1$ and $f'=0$. The required result then follows from the expression for $\widetilde{wt}(h')$ given in Lemma 7.3 and $\alpha(h)=\alpha^{p,p'}_{a,b}$ follows from Lemma 2.2.

The final statement follows from Lemma 7.3. □

7.2 Truncating paths. In this section, we specify a process by which a path $h\in\mathcal{P}^{p,p'}_{a,b,e,f}(L)$ for $L>0$ may be shortened by removing just the leftmost (first) segment, or by removing just the rightmost (Lth) segment. Consequently, the new path h' is of length $L'=L-1$. One shortening may follow the other to yield a path of length $L-2$.

In fact, we will only use these shortening processes when $p'>2p$ so that, in particular, the first and the $(p'-2)$th bands of the (p,p')-model are even.

Shortening on the left side will occur only when $a=1$ or $a=p'-1$ so that the removed segment is in an even band and will occur when the zeroth vertex is scoring.

Lemma 7.5. *Let $p'>2p$ and either $a=1$ and $e=0$ or $a=p'-1$ and $e=1$. Let $\mathcal{S}$ be interfacial in the (p,p')-model. Define $\Delta=-(-1)^e$, $e'=1-e$, and $a'=a-\Delta$. Then*

$$\tilde{\chi}^{p,p'}_{a',b,e',f}(L,m)\{\mathcal{S}\}=q^{-\frac{1}{2}(L+1-m+\Delta\beta)}\tilde{\chi}^{p,p'}_{a,b,e,f}(L+1,m)\{\mathcal{S}\},$$

where $\beta=\beta^{p,p'}_{a,b,e,f}$.

In addition, $\alpha^{p,p'}_{a',b} = \alpha^{p,p'}_{a,b} + \Delta$ *and* $\beta^{p,p'}_{a',b,e',f} = \beta^{p,p'}_{a,b,e,f} + \Delta$.

PROOF. Let $h \in \mathcal{P}^{p,p'}_{a,b,e,f}(L+1,m)\{\mathcal{S}\}$ and note that necessarily $h_1 = a'$. Let $h' \in \mathcal{P}^{p,p'}_{a',b,e',f}(L,m)\{\mathcal{S}\}$ be defined by $h'_i = h_{i+1}$ for $0 \le i \le L$. The lemma then follows on noting that $\delta^{p,p'}_{a',e'} = 0$ and using Lemma 7.1 after switching the roles of h and h' there. □

Shortening on the right side will occur only when $b = 1$ or $b = p' - 1$ so that the removed segment is in an even band and will occur when the Lth vertex is scoring.

Lemma 7.6. *Let* $p' > 2p$ *and either* $b = 1$ *and* $f = 0$ *or* $b = p' - 1$ *and* $f = 1$. *Let* $\mathcal{S}$ *be interfacial in the* (p,p')*-model. Define* $\Delta = -(-1)^f$, $f' = 1 - f$ *and* $b' = b - \Delta$. *Then*

$$\tilde{\chi}^{p,p'}_{a,b',e,f'}(L,m)\{\mathcal{S}\} = q^{-\frac{1}{2}(L+1-\Delta\alpha)}\tilde{\chi}^{p,p'}_{a,b,e,f}(L+1,m)\{\mathcal{S}\},$$

where $\alpha = \alpha^{p,p'}_{a,b}$.

In addition, $\alpha^{p,p'}_{a,b'} = \alpha^{p,p'}_{a,b} - \Delta$ *and* $\beta^{p,p'}_{a,b',e,f'} = \beta^{p,p'}_{a,b,e,f} - \Delta$.

PROOF. Let $h \in \mathcal{P}^{p,p'}_{a,b,e,f}(L+1,m)\{\mathcal{S}\}$ and note that necessarily $h_L = b'$. Let $h' \in \mathcal{P}^{p,p'}_{a,b',e,f'}(L,m)\{\mathcal{S}\}$ be defined by $h'_i = h_i$ for $0 \le i \le L$. The lemma then follows on noting that $\delta^{p,p'}_{b',f'} = 0$ and using Lemma 7.3 after switching the roles of h and h' there. □

8 Fermionic expressions

8.1 Results. In this section, we fix coprime p and p' and fix $a, b \in \mathcal{T} \cup \mathcal{T}'$ with $1 \le a, b < p'$. We make use of the definitions of Sections 5.1 and 5.2. For certain c, we present two fermionic expressions for $\mathcal{P}^{p,p'}_{a,b,c}(L)$. The value of c depends on b and for $p' > 2p$ is given by

$$c = \begin{cases} 2 & \text{if } b = 1, \\ b-1 & \text{if } 1 < b \le t_1, \\ p'-2 & \text{if } b = p'-1, \\ b+1 & \text{if } p' - t_1 \le b < p'-1, \\ b \pm 1 & \text{otherwise.} \end{cases} \tag{14}$$

For $p' < 2p$, change t_1 to t_2 in this definition.

The statement of these fermionic expressions requires the following notation. For convenience, set $a^L = a$ and $a^R = b$. Now for $A \in \{L, R\}$, define σ^A such that

$$\kappa_{\sigma^A} = \begin{cases} a^A & \text{if } a^A \in \mathcal{T}, \\ p' - a^A & \text{if } a^A \in \mathcal{T}'. \end{cases} \tag{15}$$

For $0 \le j \le t$, define[13] $\boldsymbol{e}_j = (e_1, e_2, \ldots, e_t)$ with $e_i = \delta_{ij}$. Then define

$$\boldsymbol{u}^A = \boldsymbol{e}_{\sigma^A} - \sum_{k:\sigma^A \le t_k < t} \boldsymbol{e}_{t_k} + \begin{cases} 0 & \text{if } a^A \in \mathcal{T}, \\ \boldsymbol{e}_t & \text{if } a^A \in \mathcal{T}', \end{cases} \tag{16}$$

and

$$\boldsymbol{\Delta}^A = \begin{cases} -\boldsymbol{e}_{\sigma^A} + \sum\limits_{k:\sigma^A \le t_k < t} \boldsymbol{e}_{t_k} & \text{if } a^A \in \mathcal{T}, \\ -\boldsymbol{e}_t + \boldsymbol{e}_{\sigma^A} - \sum\limits_{k:\sigma^A \le t_k < t} \boldsymbol{e}_{t_k} & \text{if } a^A \in \mathcal{T}'. \end{cases} \tag{17}$$

We define the matrix $\boldsymbol{C}$ to be the $t \times t$ tridiagonal matrix with entries $\boldsymbol{C}_{ij}$ for $0 \le i, j \le t-1$, where, when the indices are in this range,

$$\begin{array}{l} \boldsymbol{C}_{j,j-1} = -1,\ \boldsymbol{C}_{j,j} = 1,\ \boldsymbol{C}_{j,j+1} = 1 \ \text{ if } j = t_k, \quad k = 1, 2, \ldots, n, \\ \boldsymbol{C}_{j,j-1} = -1,\ \boldsymbol{C}_{j,j} = 2,\ \boldsymbol{C}_{j,j+1} = -1,\ 0 \le j < t \text{ otherwise.} \end{array} \tag{18}$$

It is also useful to define $\hat{\boldsymbol{C}}$ to be the $t \times t$ upper-triangular matrix with entries $\hat{\boldsymbol{C}}_{ij} = \boldsymbol{C}_{ij}$ as above with $1 \le i \le t$ and $0 \le j \le t-1$.

For example, in the case $p = 9$ and $p' = 31$, where the continued fraction of p'/p is $(3, 2, 4)$ and $t_1 = 2$, $t_2 = 4$, and $t_3 = 8$, we have

$$\boldsymbol{C} = \begin{pmatrix} 2 & -1 & . & . & . & . & . \\ -1 & 2 & -1 & . & . & . & . \\ . & -1 & 1 & 1 & . & . & . \\ . & . & -1 & 2 & -1 & . & . \\ . & . & . & -1 & 1 & 1 & . \\ . & . & . & . & -1 & 2 & -1 \\ . & . & . & . & . & -1 & 2 \end{pmatrix}, \qquad \hat{\boldsymbol{C}} = \begin{pmatrix} -1 & 2 & -1 & . & . & . & . \\ . & -1 & 1 & 1 & . & . & . \\ . & . & -1 & 2 & -1 & . & . \\ . & . & . & -1 & 1 & 1 & . \\ . & . & . & . & -1 & 2 & -1 \\ . & . & . & . & . & -1 & 2 \\ . & . & . & . & . & . & -1 \end{pmatrix}.$$

Since $\hat{\boldsymbol{C}}$ is upper-triangular, its inverse is readily obtained. Given a t-dimensional vector $\boldsymbol{u}$, we then define $Q_i \in \{0, 1\}$ for $0 \le i < t$, by[14]

$$(Q_0, Q_1, Q_2, \ldots, Q_{t-1})^T = \hat{\boldsymbol{C}}^{-1}\boldsymbol{u} \bmod 2. \tag{19}$$

We thus define the *parity vector* $\boldsymbol{Q}(\boldsymbol{u}) = (Q_1, Q_2, \ldots, Q_{t-1})$.

Now given a t-dimensional vector $\boldsymbol{u} = (u_1, u_2, \ldots, u_t)$, define the $(t-1)$-dimensional vector $\boldsymbol{u}^\flat = (u^\flat_1, u^\flat_2, \ldots, u^\flat_{t-1})$ by

$$u^\flat_j = \begin{cases} 0 & \text{if } t_k < j \le t_{k+1},\ k \equiv 0 \pmod 2, \\ u_j & \text{if } t_k < j \le t_{k+1},\ k \not\equiv 0 \pmod 2, \end{cases} \tag{20}$$

[13]In this paper, all vectors $\boldsymbol{Q}$, $\boldsymbol{m}$, $\hat{\boldsymbol{m}}$, $\boldsymbol{n}$, $\boldsymbol{u}$, $\boldsymbol{\Delta}$, and $\boldsymbol{e}$ should be considered as column vectors. However, for typographical convenience, we shall express their components in row vector form.

[14]For $\boldsymbol{v} = (v_1, v_2, \ldots, v_t)$, we define $\boldsymbol{v} \bmod 2 = (v_1 \bmod 2, v_2 \bmod 2, \ldots, v_t \bmod 2)$.

and the $(t-1)$-dimensional vector $\boldsymbol{u}^\sharp = (u_1^\sharp, u_2^\sharp, \dots, u_{t-1}^\sharp)$ by

$$u_j^\sharp = \begin{cases} u_j & \text{if } t_k < j \le t_{k+1},\ k \equiv 0 \,(\text{mod}\, 2), \\ 0 & \text{if } t_k < j \le t_{k+1},\ k \not\equiv 0 \,(\text{mod}\, 2). \end{cases} \tag{21}$$

Then, of course, $(\boldsymbol{u})_j = (\boldsymbol{u}^\flat + \boldsymbol{u}^\sharp)_j$ for $1 \le j < t$. For convenience, we sometimes write $\boldsymbol{u}_\flat$ and $\boldsymbol{u}_\sharp$ for $\boldsymbol{u}^\flat$ and $\boldsymbol{u}^\sharp$, respectively.

Finally, we define a value γ that depends on $\boldsymbol{\Delta}^L$ and $\boldsymbol{\Delta}^R$. This value is obtained by iteratively generating the sequences $(\beta_t, \beta_{t-1}, \dots, \beta_0)$, $(\alpha_t, \alpha_{t-1}, \dots, \alpha_0)$, and $(\gamma_t, \gamma_{t-1}, \dots, \gamma_0)$ as follows. Let $\alpha_t = \beta_t = \gamma_t = 0$. Now for $j = t, t-1, \dots, 1$, obtain α_{j-1}, β_{j-1}, and γ_{j-1} from α_j, β_j, and γ_j in the following three stages. First, obtain

$$(\beta'_{j-1}, \gamma'_{j-1}) = (\beta_j + (\boldsymbol{\Delta}^L)_j - (\boldsymbol{\Delta}^R)_j, \gamma_j + 2\alpha_j(\boldsymbol{\Delta}^R)_j). \tag{22}$$

Then obtain

$$(\alpha''_{j-1}, \gamma''_{j-1}) = (\alpha_j + \beta'_{j-1}, \gamma'_{j-1} - (\beta'_{j-1})^2). \tag{23}$$

Finally, set

$$(\alpha_{j-1}, \beta_{j-1}, \gamma_{j-1}) = \begin{cases} (\alpha''_{j-1}, \alpha''_{j-1} - \beta'_{j-1}, -(\alpha''_{j-1})^2 - \gamma''_{j-1}) & \text{if } j = t_k + 1,\ 1 \le k \le n, \\ (\alpha''_{j-1}, \beta'_{j-1}, \gamma''_{j-1}) & \text{otherwise.} \end{cases} \tag{24}$$

We then set $\gamma = \gamma_0$.

Theorem 8.1. *If $a, b \in \mathcal{T} \cup \mathcal{T}'$, define everything as above. Then*

$$\chi^{p,p'}_{a,b,c}(L) = \sum_{\boldsymbol{m} \equiv \boldsymbol{Q}(\boldsymbol{u}^L + \boldsymbol{u}^R)} q^{\frac14 \hat{\boldsymbol{m}}^T \boldsymbol{C} \hat{\boldsymbol{m}} - \frac14 L^2 - \frac12 (\boldsymbol{u}^L_\flat + \boldsymbol{u}^R_\sharp)\cdot \boldsymbol{m} + \frac14 \gamma} \prod_{j=1}^{t-1} \begin{bmatrix} m_j - \frac12 (\hat{\boldsymbol{C}} \hat{\boldsymbol{m}} - \boldsymbol{u}^L - \boldsymbol{u}^R)_j \\ m_j \end{bmatrix}_q$$
$$+ \begin{cases} \chi^{z_n, y_n}_{a,b,c}(L) & \text{if } a < y_n \text{ and } b < y_n, \\ \chi^{z_n, y_n}_{p'-a, p'-b, p'-c}(L) & \text{if } a > p' - y_n \text{ and } b > p' - y_n, \\ 0, & \text{otherwise.} \end{cases}$$

With $\boldsymbol{Q}(\boldsymbol{u}^L + \boldsymbol{u}^R) = (Q_1, Q_2, \dots, Q_{t-1})$, the summation here is over all vectors $\boldsymbol{m} = (m_1, m_2, \dots, m_{t-1})$ such that $m_j \in \mathbb{Z}_{\ge 0}$ and $m_j \equiv Q_j$ (mod 2) for $1 \le j < t$. Then $\hat{\boldsymbol{m}} = (L, m_1, m_2, \dots, m_{t-1})$.

The second fermionic expression for $\chi^{p,p'}_{a,b,c}(L)$ that we present, involves the modified form $\left[{A \atop B}\right]'_q$ of the Gaussian polynomial defined in (2).

Theorem 8.2. *If $a, b \in \mathcal{T} \cup \mathcal{T}'$, define everything as above. Then if $L \geq 0$,*

$$\chi^{p,p'}_{a,b,c}(L) = \sum_{\boldsymbol{m} \equiv \boldsymbol{Q}(\boldsymbol{u}^L+\boldsymbol{u}^R)} q^{\frac{1}{4}\hat{\boldsymbol{m}}^T C \hat{\boldsymbol{m}} - \frac{1}{4}L^2 - \frac{1}{2}(\boldsymbol{u}^L_\flat + \boldsymbol{u}^R_\sharp)\cdot \boldsymbol{m} + \frac{1}{4}\gamma} \prod_{j=1}^{t-1} \begin{bmatrix} m_j - \frac{1}{2}(\hat{\boldsymbol{C}}\hat{\boldsymbol{m}} - \boldsymbol{u}^L - \boldsymbol{u}^R)_j \\ m_j \end{bmatrix}'_q .$$

With $\boldsymbol{Q}(\boldsymbol{u}^L + \boldsymbol{u}^R) = (Q_1, Q_2, \ldots, Q_{t-1})$, the summation here is over all vectors $\boldsymbol{m} = (m_1, m_2, \ldots, m_{t-1})$ such that $m_j \in \mathbb{Z}_{\geq 0}$ and $m_j \equiv Q_j \pmod 2$ for $1 \leq j < t$. Then $\hat{\boldsymbol{m}} = (L, m_1, m_2, \ldots, m_{t-1})$.

8.2 Carrying out the induction. With p and p' fixed, employ the definitions of Section 5.1. Then for $0 \leq i \leq t$, let $k(i)$ be such that $t_{k(i)} \leq i < t_{k(i)+1}$ (i.e., $k(i) = \zeta(i+1)$) and define p_i and p'_i to be the positive coprime integers for which p'_i/p_i has continued fraction $(t_{k(i)+1} + 1 - i, c_{k(i)+1}, \ldots, c_n)$. Thus p'_i/p_i has rank $t - i$. As in Section 5.2, we obtain Takahashi lengths $\{\kappa^{(i)}_j\}_{j=0}^{t-i}$ and truncated Takahashi lengths $\{\tilde{\kappa}^{(i)}_j\}_{j=0}^{t-i}$ for p'_i/p_i.

Lemma 8.3. *Let $1 \leq i \leq t$. If $i \neq t_{k(i)}$, then*

$$\begin{aligned} p^{(i-1)\prime} &= p^{(i)\prime} + p^{(i)}, \\ p^{(i-1)} &= p^{(i)}, \\ \kappa^{(i-1)}_j &= \kappa^{(i)}_{j-1} + \tilde{\kappa}^{(i)}_{j-1} && (1 \leq j \leq t^{(i-1)}), \\ \tilde{\kappa}^{(i-1)}_j &= \tilde{\kappa}^{(i)}_{j-1} && (1 \leq j \leq t^{(i-1)}). \end{aligned}$$

If $i = t_{k(i)}$, then

$$\begin{aligned} p^{(i-1)\prime} &= 2p^{(i)\prime} - p^{(i)}, \\ p^{(i-1)} &= p^{(i)\prime} - p^{(i)}, \\ \kappa^{(i-1)}_j &= 2\kappa^{(i)}_{j-1} - \tilde{\kappa}^{(i)}_{j-1} && (2 \leq j \leq t^{(i-1)}), \\ \tilde{\kappa}^{(i-1)}_j &= \tilde{\kappa}^{(i)}_{j-1} - \tilde{\kappa}^{(i)}_{j-1} && (2 \leq j \leq t^{(i-1)}). \end{aligned}$$

Proof. If $i \neq t_{k(i)}$, then $k(i-1) = k(i)$. Then $p^{(i)\prime}/p^{(i)}$ and $p^{(i-1)\prime}/p^{(i-1)}$ have continued fractions $(t_{k(i)} + 1 - i, c_{k(i)+1}, \ldots, c_n)$ and $(t_{k(i)} + 2 - i, c_{k(i)+1}, \ldots, c_n)$, respectively. It follows immediately that $p^{(i-1)\prime} = p^{(i)\prime} + p^{(i)}$ and $p^{(i-1)} = p^{(i)}$. The expressions for $\kappa^{(i-1)}_j$ and $\tilde{\kappa}^{(i-1)}_j$ follow from Lemma 6.1.

If $i = t_{k(i)}$, then $k(i-1) = k(i) - 1$. Then $p^{(i)\prime}/p^{(i)}$ and $p^{(i-1)\prime}/p^{(i-1)}$ have continued fractions $(c_{k(i)}, c_{k(i)+1}, \ldots, c_n)$ and $(2, c_{k(i)}, c_{k(i)+1}, \ldots, c_n)$, respectively. It follows immediately that $p^{(i-1)\prime} = 2p^{(i)\prime} - p^{(i)}$ and $p^{(i-1)} = p^{(i)\prime} - p^{(i)}$. The expressions for $\kappa^{(i-1)}_j$ and $\tilde{\kappa}^{(i-1)}_j$ follow from combining Lemma 6.2 with Lemma 6.1. □

As above, take $A \in \{R, L\}$. If $a^A \in \mathcal{T}$, set

$$a_i^A = \begin{cases} 1 & \text{if } \sigma^A \le i < t, \\ \kappa^{(i)}_{\sigma^A - i} & \text{if } 0 \le i \le \sigma^A, \end{cases}$$

$$a_i^{A\prime} = \begin{cases} 1 + \delta_{i,t_{k(i)}} & \text{if } \sigma^A \le i < t, \\ \kappa^{(i-1)}_{\sigma^A - i + 1} & \text{if } 0 \le i < \sigma^A \end{cases}$$

and if $a^A \in \mathcal{T}'$, set

$$a_i^A = \begin{cases} p_i' - 1 & \text{if } \sigma^A \le i < t, \\ p_i' - \kappa^{(i)}_{\sigma^A - i} & \text{if } 0 \le i \le \sigma^A. \end{cases}$$

$$a_i^{A\prime} = \begin{cases} p_i' - 1 - \delta_{i,t_{k(i)}} & \text{if } \sigma^A \le i < t, \\ p_i' - \kappa^{(i-1)}_{\sigma^A - i + 1} & \text{if } 0 \le i < \sigma^A. \end{cases}$$

In addition, define k^A to be such that $t_{k^A} < \sigma^A \le t_{k^A+1}$. Then if $a^A \in \mathcal{T}$, set

$$e_i^A = \begin{cases} 0 & \text{if } \sigma^A \le i < t, \\ \delta^{(2)}_{k,k^A} & \text{if } 0 \le i < \sigma^A \end{cases}$$

and if $a^A \in \mathcal{T}'$, set

$$e_i^A = \begin{cases} 1 & \text{if } \sigma^A \le i < t, \\ 1 - \delta^{(2)}_{k,k^A} & \text{if } 0 \le i < \sigma^A. \end{cases}$$

Lemma 8.4. *Let* $1 \le i < t$. *Then for* $A \in \{L, R\}$,

$$a_i^{A\prime} = \begin{cases} a_i^A + \left\lfloor \dfrac{a_i^A p_i}{p_i'} \right\rfloor + e_i^A & \textit{if } i \ne t_{k(i)}, \\[2ex] 2a_i^A - \left\lfloor \dfrac{a_i^A p_i}{p_i'} \right\rfloor - e_i^A & \textit{if } i = t_{k(i)}. \end{cases}$$

PROOF. For p_i'/p_i, in view of the continued fraction specified above, the analogues of the quantities defined in (13) are $t_j' = t_{k(i)+j} - i$ for $1 \le j \le n - k(j) + 1$. For $i < \sigma^A$, the various cases are then readily proved using Lemmas 6.3 and 8.3. For $i \ge \sigma^A$, the results follow immediately. □

For each t-dimensional vector $\boldsymbol{u} = (u_1, u_2, \ldots, u_t)$, define the $(t-1)$-dimensional vector $\boldsymbol{u}^{(\flat,k)} = (u_1^{(\flat,k)}, u_2^{(\flat,k)}, \ldots, u_{t-1}^{(\flat,k)})$ by

$$u_j^{(\flat,k)} = \begin{cases} 0 & \text{if } t_{k'} < j \le t_{k'+1},\ k' \equiv k \pmod 2, \\ u_j & \text{if } t_{k'} < j \le t_{k'+1},\ k' \not\equiv k \pmod 2, \end{cases} \tag{25}$$

and the $(t-1)$-dimensional vector $\boldsymbol{u}^{(\sharp,k)} = (u_1^{(\sharp,k)}, u_2^{(\sharp,k)}, \ldots, u_{t-1}^{(\sharp,k)})$ by

$$u_j^{(\sharp,k)} = \begin{cases} u_j & \text{if } t_{k'} < j \le t_{k'+1},\ k' \equiv k \pmod 2, \\ 0 & \text{if } t_{k'} < j \le t_{k'+1},\ k' \not\equiv k \pmod 2. \end{cases} \tag{26}$$

For convenience, we sometimes write $\boldsymbol{u}_{(\flat,k)}$ instead of $\boldsymbol{u}^{(\flat,k)}$ and $\boldsymbol{u}_{(\sharp,k)}$ instead of $\boldsymbol{u}^{(\sharp,k)}$.

Now for $0 \le i \le t-2$, define

$$\begin{aligned} &F_{a,b}^{(i)}(\boldsymbol{u}^L, \boldsymbol{u}^R, m_i, m_{i+1}; q) \\ &= \sum \Biggl(q^{\frac{1}{4}\hat{\boldsymbol{m}}^{(i+1)T} \boldsymbol{C} \hat{\boldsymbol{m}}^{(i+1)} + \frac{1}{4}m_i^2 - \frac{1}{2}m_i m_{i+1} - \frac{1}{2}(\boldsymbol{u}^L_{(\flat,k(i))} + \boldsymbol{u}^R_{(\sharp,k(i))})\cdot \boldsymbol{m}^{(i)} + \frac{1}{4}\gamma_i''} \\ &\qquad \prod_{j=i+1}^{t-1} \begin{bmatrix} m_j - \frac{1}{2}(\hat{\boldsymbol{C}}\hat{\boldsymbol{m}}^{(i)} - \boldsymbol{u}^L - \boldsymbol{u}^R)_j \\ m_j \end{bmatrix}_q \Biggr), \end{aligned} \tag{27}$$

where the sum here is taken over all vectors $(m_{i+2}, m_{i+3}, \ldots, m_{t-1}) \equiv (Q_{i+2}, Q_{i+3}, \ldots, Q_{t-1})$, where $(Q_1, Q_2, \ldots, Q_{t-1}) = \boldsymbol{Q}(\boldsymbol{u}^L + \boldsymbol{u}^R)$. The $(t-1)$-dimensional vector $\boldsymbol{m}^{(i)} = (0, 0, \ldots, 0, m_{i+1}, m_{i+2}, m_{i+3}, \ldots, m_{t-1})$ has its first i components equal to zero. The t-dimensional vector $\hat{\boldsymbol{m}}^{(i)} = (0, 0, \ldots, 0, m_i, m_{i+1}, m_{i+2}, \ldots, m_{t-1})$ has its first i components equal to zero.

We also define

$$F_{a,b}^{(t-1)}(\boldsymbol{u}^L, \boldsymbol{u}^R, m_{t-1}, m_t; q) = q^{\frac{1}{4}m_{t-1}^2 + \frac{1}{4}\gamma_i''} \delta_{m_t,0}. \tag{28}$$

For convenience, we set $Q_t = 0$.

Since $\left[{m+n \atop m}\right]_{q^{-1}} = q^{-mn}\left[{m+n \atop m}\right]_q$, it follows that for $0 \le i \le t-2$,

$$\begin{aligned} &F_{a,b}^{(i)}(\boldsymbol{u}^L, \boldsymbol{u}^R, m_i, m_{i+1}; q^{-1}) \\ &= \sum \Biggl(q^{\frac{1}{4}\hat{\boldsymbol{m}}^{(i+1)T} \boldsymbol{C} \hat{\boldsymbol{m}}^{(i+1)} - \frac{1}{4}m_i^2 - \frac{1}{2}(\boldsymbol{u}^L_{(\flat,k(i)-1)} + \boldsymbol{u}^R_{(\sharp,k(i)-1)})\cdot \boldsymbol{m}^{(i)} - \frac{1}{4}\gamma_i''} \\ &\qquad \prod_{j=i+1}^{t-1} \begin{bmatrix} m_j - \frac{1}{2}(\hat{\boldsymbol{C}}\hat{\boldsymbol{m}}^{(i)} - \boldsymbol{u}^L - \boldsymbol{u}^R)_j \\ m_j \end{bmatrix}_q \Biggr), \end{aligned} \tag{29}$$

where the sum here is taken over all vectors $(m_{i+2}, m_{i+3}, \ldots, m_{t-1}) \equiv (Q_{i+2}, Q_{i+3}, \ldots, Q_{t-1})$ as above. Of course, we also have

$$F_{a,b}^{(t-1)}(\boldsymbol{u}^L, \boldsymbol{u}^R, m_{t-1}, m_t; q^{-1}) = q^{-\frac{1}{4}m_{t-1}^2 - \frac{1}{4}\gamma_i''} \delta_{m_t,0}. \tag{30}$$

Lemma 8.5. *Let* $0 \le i < t$, $m_i \equiv Q_i$, *and* $m_{i+1} \equiv Q_{i+1}$. *If*

$$\mathcal{S}^{(i)} = \begin{cases} \left\{\kappa^{(i)}_{t_n-i}\right\} & \text{if } i < t_n,\ \sigma^L < t_n,\ \sigma^R < t_n,\ \text{and } a, b \in \mathcal{T}, \\ \left\{p'_i - \kappa^{(i)}_{t_n-i}\right\} & \text{if } i < t_n,\ \sigma^L < t_n,\ \sigma^R < t_n,\ \text{and } a, b \in \mathcal{T}', \\ \emptyset & \text{otherwise}, \end{cases}$$

then

$$\tilde{\chi}^{p_i,p'_i}_{a^L_i,a^R_i,e^L_i,e^R_i}(m_i, m_{i+1})\left\{\mathcal{S}^{(i)}\right\} = F^{(i)}_{a,b}(\boldsymbol{u}^L, \boldsymbol{u}^R, m_i, m_{i+1}). \tag{31}$$

In addition, $\alpha^{p_i,p'_i}_{a^L_i,a^R_i} = \alpha''_i$ *and* $\beta^{p_i,p'_i}_{a^L_i,a^R_i,e^L_i,e^R_i} = \beta'_i$.

Proof. For $i = t-1$, we have $p'_i = 3$ and $p_i = 1$, and if $a^A \in \mathcal{T}$, then $a^A_i = 1$, $e^A_i = 0$, and $(\boldsymbol{\Delta}^A)_t = 0$; if $a^A \in \mathcal{T}'$, then $a^A_i = 2$, $e^A_i = 1$, and $(\boldsymbol{\Delta}^A)_t = -1$. Furthermore, we have $i \ge t_n$. Via (22) and (23), we obtain $\alpha''_{t-1} = \beta'_{t-1} = (\boldsymbol{\Delta}^L)_t - (\boldsymbol{\Delta}^R)_t$ and $\gamma''_{t-1} = -((\boldsymbol{\Delta}^L)_t - (\boldsymbol{\Delta}^R)_t)^2$. For $i = t-1$, the first statement of our induction proposition is now seen to hold via Lemma 2.5. The definitions of $\alpha^{p_i,p'_i}_{a^L_i,a^R_i}$ and $\beta^{p_i,p'_i}_{a^L_i,a^R_i,e^L_i,e^R_i}$ then yield the two final statements.

Now assume the result holds for a particular i with $1 \le i \le t-1$. As above, let $k(i)$ be such that $t_{k(i)} \le i < t_{k(i)+1}$. First, consider the case where $t_{k(i)} < i < t_{k(i)+1}$. Equation (24) gives $\alpha_i = \alpha''_i$, $\beta_i = \beta'_i$, and $\gamma_i = \gamma''_i$. Let $m_{i-1} \equiv Q_{i-1}$. On setting $M = m_{i-1} + u^L_i + u^R_i$, equation (19) implies that $M \equiv Q_{i+1}$. Then use of the induction hypothesis, Lemma 3.14 or Lemma 6.7 as appropriate, and Lemmas 8.3 and 8.4 yields

$$\begin{aligned} &\tilde{\chi}^{p_{i-1},p'_{i-1}}_{a^{L\prime}_i,a^{R\prime}_i,e^L_i,e^R_i}(M, m_i)\left\{\mathcal{S}^{(i-1)}\right\} \\ &\quad= \sum_{m_{i+1}\equiv Q_{i+1}} q^{\frac{1}{4}(M-m_i)^2-\frac{1}{4}\beta_i^2} \begin{bmatrix} \frac{1}{2}(M+m_{i+1}) \\ m_i \end{bmatrix}_q F^{(i)}_{a,b}(\boldsymbol{u}^L, \boldsymbol{u}^R, m_i, m_{i+1}). \end{aligned} \tag{32}$$

Here Lemma 3.5 also gives $\alpha^{p_{i-1},p'_{i-1}}_{a^{L\prime}_i,a^{R\prime}_i} = \alpha_i + \beta_i$ and $\beta^{p_{i-1},p'_{i-1}}_{a^{L\prime}_i,a^{R\prime}_i,e^L_i,e^R_i} = \beta_i$.

$\{\mathcal{S}^{(i-1)}\}$ appears on the left side here because, via Lemma 6.3, if $i < t_n$, then $\kappa^{(i)}_{t_n-i}$ is interfacial in the (p_i, p'_i)-model and borders the $\tilde{\kappa}^{(i)}_{t_n-i}$th odd band; and then $\kappa^{(i)}_{t_n-i} + \tilde{\kappa}^{(i)}_{t_n-i} = \kappa^{(i-1)}_{t_n-i+1}$ by Lemma 8.3; finally, noting that $i \ne t_n$, if $i \ge t_n$, then $i-1 \ge t_n$. (The up–down symmetry of the (p, p')-model implies that if $i < t_n$ then $p'_i - \kappa^{(i)}_{t_n-i}$ is interfacial in the (p_i, p'_i)-model and borders the $(p_i - \tilde{\kappa}^{(i)}_{t_n-i})$th odd band. Then we use $(p'_i - \kappa^{(i)}_{t_n-i}) + (p_i - \tilde{\kappa}^{(i)}_{t_n-i}) = p'_{i-1} - \kappa^{(i-1)}_{t_n-i+1}$ from Lemma 8.3.)

Since $M = m_{i-1} + u^L_i + u^R_i$, noting that $t_k < i < t_{k+1}$, we have

$$M + m_{i+1} = 2m_i - (\hat{\boldsymbol{C}}\hat{\boldsymbol{m}}^{(i-1)} - \boldsymbol{u}^L - \boldsymbol{u}^R)_i,$$

and

$$\begin{aligned}
&\hat{\boldsymbol{m}}^{(i+1)T}\boldsymbol{C}\hat{\boldsymbol{m}}^{(i+1)} + m_i^2 - 2m_i m_{i+1} + (M - m_i)^2\\
&\quad= \hat{\boldsymbol{m}}^{(i)T}\boldsymbol{C}\hat{\boldsymbol{m}}^{(i)} + M^2 - 2Mm_i\\
&\quad= \hat{\boldsymbol{m}}^{(i)T}\boldsymbol{C}\hat{\boldsymbol{m}}^{(i)} + m_{i-1}^2 - 2m_i m_{i-1}\\
&\qquad\qquad + (m_{i-1} - m_i)(u_i^L + u_i^R) + (u_i^L + u_i^R)^2.
\end{aligned}$$

(In the case where $i = t-1$, we require this expression after substituting $m_t = 0$.) Thence

$$\begin{aligned}
&\tilde{\chi}^{p_{i-1},p'_{i-1}}_{a_i^{L\prime},a_i^{R\prime},e_i^L,e_i^R}(m_{i-1} + u_i^L + u_i^R, m_i)\left\{\mathcal{S}^{(i-1)}\right\}\\
&\quad= \sum\Bigg(q^{\frac14\hat{\boldsymbol{m}}^{(i)T}\boldsymbol{C}\hat{\boldsymbol{m}}^{(i)} - \frac12 m_i m_{i-1} + \frac12(m_{i-1}-m_i)(u_i^L+u_i^R) - \frac12(\boldsymbol{u}^L_{(i,k(i))}+\boldsymbol{u}^R_{(i,k(i))})\cdot\boldsymbol{m}^{(i)}}\\
&\qquad\qquad q^{\frac14 m_{i-1}^2 + \frac14(u_i^L+u_i^R)^2 + \frac14\gamma_i - \frac14\beta_i^2}\prod_{j=i}^{t-1}\begin{bmatrix} m_j - \frac12(\hat{\boldsymbol{C}}\hat{\boldsymbol{m}}^{(i-1)} - \boldsymbol{u}^L - \boldsymbol{u}^R)_j \\ m_j \end{bmatrix}_q \Bigg),
\end{aligned}$$

where the sum is over all $(m_{i+1}, m_{i+2}, \ldots, m_{t-1}) \equiv (Q_{i+1}, Q_{i+2}, \ldots, Q_{t-1})$.

If $i = \sigma^R$, then $u_i^R = 1$. In this case, by definition, we have either $a_i^{R\prime} = 1$, $e_i^R = 0$, $a_{i-1}^R = 2$, and $e_{i-1}^R = 1$ or $a_i^{R\prime} = p'_{i-1} - 1$, $e_i^R = 1$, $a_{i-1}^R = p'_{i-1} - 2$, and $e_{i-1}^R = 0$. It is easily checked that $a_i^{R\prime} \notin \mathcal{S}^{(i-1)}$. Then Lemma 7.6 yields

$$\begin{aligned}
&\tilde{\chi}^{p_{i-1},p'_{i-1}}_{a_i^{L\prime},a_{i-1}^{R\prime},e_i^L,e_{i-1}^R}(m_{i-1} + u_i^L, m_i)\left\{\mathcal{S}^{(i-1)}\right\} \qquad (33)\\
&\quad= q^{-\frac12 u_i^R(m_{i-1}+u_i^L+u_i^R) + \frac12(\boldsymbol{\Delta}^R)_i(\alpha_i+\beta_i)}\,\tilde{\chi}^{p_{i-1},p'_{i-1}}_{a_i^{L\prime},a_i^{R\prime},e_i^L,e_i^R}(m_{i-1} + u_i^L + u_i^R, m_i)\left\{\mathcal{S}^{(i-1)}\right\}.
\end{aligned}$$

If $i \neq \sigma^R$, then (noting that $i \neq t_k$) $u_i^R = (\boldsymbol{\Delta}^R)_i = 0$, $e_{i-1}^R = e_i^R$, and $a_{i-1}^R = a_i^{R\prime}$. The preceding expression thus also holds in this case.

We also immediately obtain

$$\begin{aligned}
\alpha^{p_{i-1},p'_{i-1}}_{a_i^{L\prime},a_{i-1}^R} &= \alpha_i + \beta_i - (\boldsymbol{\Delta}^R)_i,\\
\beta^{p_{i-1},p'_{i-1}}_{a_i^{L\prime},a_{i-1}^R,e_i^L,e_{i-1}^R} &= \beta_i - (\boldsymbol{\Delta}^R)_i.
\end{aligned}$$

If $i = \sigma^L$, then $u_i^L = 1$. In this case, by definition we have either $a_i^{L\prime} = 1$, $e_i^L = 0$, $a_{i-1}^L = 2$, and $e_{i-1}^L = 1$ or $a_i^{L\prime} = p'_{i-1} - 1$, $e_i^L = 1$, $a_{i-1}^L = p'_{i-1} - 2$, and $e_{i-1}^L = 0$. It is easily checked that $a_i^{L\prime} \notin \mathcal{S}^{(i-1)}$. Then Lemma 7.5 yields

$$\begin{aligned}
&\tilde{\chi}^{p_{i-1},p'_{i-1}}_{a_{i-1}^L,a_{i-1}^R,e_{i-1}^L,e_{i-1}^R}(m_{i-1}, m_i)\left\{\mathcal{S}^{(i-1)}\right\} \qquad (34)\\
&\quad= q^{-\frac12 u_i^L(m_{i-1}-m_i+u_i^L) - \frac12(\boldsymbol{\Delta}^L)_i(\beta_i - (\boldsymbol{\Delta}^R)_i)}\,\tilde{\chi}^{p_{i-1},p'_{i-1}}_{a_i^{L\prime},a_{i-1}^R,e_i^L,e_{i-1}^R}(m_{i-1} + u_i^L, m_i)\left\{\mathcal{S}^{(i-1)}\right\}.
\end{aligned}$$

If $i \neq \sigma^R$, then (noting that $i \neq t_k$) $u_i^L = (\boldsymbol{\Delta}^L)_i = 0$, $e_{i-1}^L = e_i^L$, and $a_{i-1}^L = a_i^{L\prime}$. The preceding expression thus also holds in this case.

We also obtain

$$\begin{aligned} \alpha^{p_{i-1},p'_{i-1}}_{a^L_{i-1},a^R_{i-1}} &= \alpha_i + \beta_i - (\boldsymbol{\Delta}^R)_i + (\boldsymbol{\Delta}^L)_i, \\ \beta^{p_{i-1},p'_{i-1}}_{a^L_{i-1},a^R_{i-1},e^L_{i-1},e^R_{i-1}} &= \beta_i - (\boldsymbol{\Delta}^R)_i + (\boldsymbol{\Delta}^L)_i. \end{aligned}$$

Combining all of the above and using the expression for γ''_{i-1} given by (22) and (23) yields

$$\begin{aligned} &\tilde{\chi}^{p_{i-1},p'_{i-1}}_{a^L_{i-1},a^R_{i-1},e^L_{i-1},e^R_{i-1}}(m_{i-1},m_i)\left\{\mathcal{S}^{(i-1)}\right\} \\ &= \sum \left(q^{\frac{1}{4}\hat{\boldsymbol{m}}^{(i)T}\boldsymbol{C}\hat{\boldsymbol{m}}^{(i)}+\frac{1}{4}m^2_{i-1}-\frac{1}{2}m_{i-1}m_i-\frac{1}{2}(\boldsymbol{u}^L_{(\triangleright,k(i))}+\boldsymbol{u}^R_{(\triangleleft,k(i))})\cdot\boldsymbol{m}^{(i-1)}+\frac{1}{4}\gamma''_{i-1}} \right. \\ &\qquad \left. \prod_{j=i}^{t-1} \begin{bmatrix} m_j - \frac{1}{2}(\hat{\boldsymbol{C}}\hat{\boldsymbol{m}}^{(i-1)}-\boldsymbol{u}^L-\boldsymbol{u}^R)_j \\ m_j \end{bmatrix}_q \right) \\ &= F^{(i-1)}_{a,b}(\boldsymbol{u}^L,\boldsymbol{u}^R,m_{i-1},m_i), \end{aligned}$$

which is the required result when $i \neq t_k$ since $k(i) = k(i-1)$.

In this $i \neq t_k$ case, using (22) and (23), we also immediately obtain

$$\begin{aligned} \alpha^{p_{i-1},p'_{i-1}}_{a^L_{i-1},a^R_{i-1}} &= \alpha''_{i-1}, \\ \beta^{p_{i-1},p'_{i-1}}_{a^L_{i-1},a^R_{i-1},e^L_{i-1},e^R_{i-1}} &= \beta'_{i-1}. \end{aligned}$$

Now consider the case for which $i = t_k$. Equation (24) gives $\alpha_i = \alpha''_i$, $\beta_i = \alpha''_i - \beta'_i$, and $\gamma_i = -\alpha_i^2 - \gamma''_i$. Corollary 4.2 gives $\alpha^{p'_i-p_i,p'_i}_{a^L_i,a^R_i} = \alpha_i$ and $\beta^{p'_i-p_i,p'_i}_{a^L_i,a^R_i,1-e^L_i,1-e^R_i} = \beta_i$. Let $m_{i-1} \equiv Q_{i-1}$. On setting $M = m_{i-1} + u_i^L + u_i^R$, equation (19) implies that $M \equiv Q_{i+1}$. Then using the induction hypothesis, Lemma 4.6 or Lemma 6.8 as appropriate, and Lemmas 8.3 and 8.4 yields

$$\begin{aligned} &\tilde{\chi}^{p_i,p'_i}_{a^{L\prime}_i,a^{R\prime}_i,1-e^L_i,1-e^R_i}(M,m_i;q)\left\{\mathcal{S}^{(i)\prime}\right\} \\ &= \sum_{m_{i+1}\equiv Q_{i+1}} \left(q^{\frac{1}{4}(m_i^2+(M-m_i)^2-\alpha_i^2-\beta_i^2)} \begin{bmatrix} \frac{1}{2}(M+m_i-m_{i+1}) \\ m_i \end{bmatrix}_q \right. \\ &\qquad \left. F^{(i)}_{a,b}(\boldsymbol{u}^L,\boldsymbol{u}^R,m_i,m_{i+1};q^{-1}) \right), \end{aligned} \tag{35}$$

where

$$\mathcal{S}^{(i)\prime} = \begin{cases} \left\{\kappa_{t_n-i+1}^{(i-1)}\right\} & \text{if } i < t_n,\ \sigma^L < t_n,\ \sigma^R < t_n, \text{ and } a, b \in \mathcal{T}, \\ \left\{p_i' - \kappa_{t_n-i+1}^{(i-1)}\right\} & \text{if } i < t_n,\ \sigma^L < t_n,\ \sigma^R < t_n, \text{ and } a, b \in \mathcal{T}', \\ \emptyset & \text{otherwise,} \end{cases}$$

using an argument similar to that in the $i \neq t_{k(i)}$ case. Here Lemma 3.5 also gives $\alpha_{a_i^{L\prime},a_i^{R\prime}}^{p_{i-1},p_{i-1}'} = \alpha_i + \beta_i$ and $\beta_{a_i^{L\prime},a_i^{R\prime},1-e_i^L,1-e_i^R}^{p_{i-1},p_{i-1}'} = \beta_i$.

Now set $M = m_{i-1} + u_i^L + u_i^R$, whence, noting that $i = t_k$,

$$M + m_i - m_{i+1} = 2m_i - (\hat{\boldsymbol{C}}\hat{\boldsymbol{m}}^{(i-1)} - \boldsymbol{u}^L - \boldsymbol{u}^R)_i$$

(in the case where $i = t - 1$, we require this expression after substituting $m_t = 0$) and

$$\begin{aligned} \hat{\boldsymbol{m}}^{(i+1)T}\boldsymbol{C}\hat{\boldsymbol{m}}^{(i+1)} &- m_i^2 + m_i^2 + (M - m_i)^2 \\ &= \hat{\boldsymbol{m}}^{(i)T}\boldsymbol{C}\hat{\boldsymbol{m}}^{(i)} + M^2 - 2Mm_i \\ &= \hat{\boldsymbol{m}}^{(i)T}\boldsymbol{C}\hat{\boldsymbol{m}}^{(i)} + m_{i-1}^2 - 2m_i m_{i-1} \\ &\qquad + 2(m_{i-1} - m_i)(u_i^L + u_i^R) + (u_i^L + u_i^R)^2. \end{aligned}$$

Using (29) or (30) gives

$$\begin{aligned} &\tilde{\chi}_{a_i^{L\prime},a_i^{R\prime},1-e_i^L,1-e_i^R}^{p_{i-1},p_{i-1}'}(m_{i-1} + u_i^L + u_i^R, m_i)\left\{\mathcal{S}^{(i)\prime}\right\} \\ &= \sum \Bigg(q^{\frac{1}{4}\hat{\boldsymbol{m}}^{(i)T}\boldsymbol{C}\hat{\boldsymbol{m}}^{(i)} - \frac{1}{2}m_i m_{i-1} + \frac{1}{2}(m_{i-1}-m_i)(u_i^L+u_i^R) - \frac{1}{2}(\boldsymbol{u}^L_{(0.k(i)-1)} + \boldsymbol{u}^R_{(2.k(i-1))})\cdot\boldsymbol{m}^{(i)}} \\ &\qquad q^{\frac{1}{4}m_{i-1}^2 + \frac{1}{4}(u_i^L+u_i^R)^2 + \frac{1}{4}\gamma_i - \frac{1}{4}\beta_i^2} \prod_{j=i}^{t-1} \begin{bmatrix} m_j - \frac{1}{2}(\hat{\boldsymbol{C}}\hat{\boldsymbol{m}}^{(i-1)} - \boldsymbol{u}^L - \boldsymbol{u}^R)_j \\ m_j \end{bmatrix}_q \Bigg), \end{aligned}$$

where the sum is over all $(m_{i+1}, m_{i+2}, \ldots, m_{t-1}) \equiv (Q_{i+1}, Q_{i+2}, \ldots, Q_{t-1})$.

Now set $\mathcal{S}^{(i)R} = \mathcal{S}^{(i)\prime} \cup a_i^{R\prime}$ if $i > \sigma^R$ and $\mathcal{S}^{(i)R} = \mathcal{S}^{(i)\prime}$ otherwise. Since $i = t_k$, it follows that $u_i^R = -1$ if $i > \sigma^R$. In this case, by definition we have either $a_i^{R\prime} = 2$, $1 - e_i^R = 1$, $a_{i-1}^R = 1$ and $e_{i-1}^R = 0$ or $a_i^{R\prime} = p_{i-1}' - 2$, $1 - e_i^R = 0$, $a_{i-1}^R = p_{i-1}' - 1$, and $e_{i-1}^R = 1$. Then Lemma 7.4 yields

$$\begin{aligned} &\tilde{\chi}_{a_i^{L\prime},a_{i-1}^R,1-e_i^L,e_{i-1}^R}^{p_{i-1},p_{i-1}'}(m_{i-1} + u_i^L, m_i)\left\{\mathcal{S}^{(i)R}\right\} \\ &\quad = q^{-\frac{1}{2}u_i^R(m_{i-1}+u_i^L+u_i^R) + \frac{1}{2}(\boldsymbol{\Delta}^R)_i(\alpha_i+\beta_i)} \\ &\qquad \tilde{\chi}_{a_i^{L\prime},a_i^{R\prime},1-e_i^L,1-e_i^R}^{p_{i-1},p_{i-1}'}(m_{i-1} + u_i^L + u_i^R, m_i)\left\{\mathcal{S}^{(i)\prime}\right\}. \end{aligned} \tag{36}$$

In addition, the same expression clearly also holds in the case $i \leq \sigma^R$, for which $u_i^R = (\mathbf{\Delta}^R)_i = 0$, $e_{i-1}^R = 1 - e_i^R$, and $a_{i-1}^R = a_i^{R\prime}$. (In the $i = \sigma^R$ case, note that $k(i-1) = k(i) - 1 = k^R(i)$.)

Lemma 7.4 also implies that

$$\alpha^{p_{i-1},p'_{i-1}}_{a_i^{L\prime},a_{i-1}^R} = \alpha_i + \beta_i - (\mathbf{\Delta}^R)_i,$$
$$\beta^{p_{i-1},p'_{i-1}}_{a_i^{L\prime},a_{i-1}^R,e_i^L,e_{i-1}^R} = \beta_i - (\mathbf{\Delta}^R)_i.$$

Now set $\mathcal{S}^{(i)L} = \mathcal{S}^{(i)R} \cup a_i^{L\prime}$ if $i > \sigma^L$ and $\mathcal{S}^{(i)L} = \mathcal{S}^{(i)R}$ otherwise. Since $i = t_k$, it follows that $u_i^L = -1$ if $i > \sigma^L$. In this case, by definition we have either $a_i^{L\prime} = 2$, $1-e_i^L = 1$, $a_{i-1}^L = 1$, and $e_{i-1}^L = 0$ or $a_i^{L\prime} = p'_{i-1}-2$, $1-e_i^L = 0$, $a_{i-1}^L = p'_{i-1}-1$, and $e_{i-1}^R = 1$. Then Lemma 7.2 yields

$$\tilde{\chi}^{p_{i-1},p'_{i-1}}_{a_{i-1}^L,a_{i-1}^R,e_{i-1}^L,e_{i-1}^R}(m_{i-1},m_i)\left\{\mathcal{S}^{(i)L}\right\} \tag{37}$$
$$= q^{-\frac{1}{2}u_i^L(m_{i-1}-m_i+u_i^L)-\frac{1}{2}(\mathbf{\Delta}^L)_i(\beta_i-(\mathbf{\Delta}^R)_i)}\tilde{\chi}^{p_{i-1},p'_{i-1}}_{a_i^{L\prime},a_i^R,1-e_i^L,e_{i-1}^R}(m_{i-1}+u_i^L,m_i)\left\{\mathcal{S}^{(i)R}\right\}.$$

In addition, the same expression clearly also holds in the case $i \leq \sigma^L$, for which $u_i^L = (\mathbf{\Delta}^L)_i = 0$, $e_{i-1}^L = 1 - e_i^L$, and $a_{i-1}^L = a_i^{L\prime}$. (In the $i = \sigma^L$ case, note that $k(i-1) = k(i) - 1 = k^L(i)$.)

Lemma 7.2 also implies that

$$\alpha^{p_{i-1},p'_{i-1}}_{a_{i-1}^L,a_{i-1}^R} = \alpha_i + \beta_i - (\mathbf{\Delta}^R)_i + (\mathbf{\Delta}^L)_i,$$
$$\beta^{p_{i-1},p'_{i-1}}_{a_{i-1}^L,a_{i-1}^R,e_{i-1}^L,e_{i-1}^R} = \beta_i - (\mathbf{\Delta}^R)_i + (\mathbf{\Delta}^L)_i.$$

Combining all of the above cases for $i = t_k$ yields

$$\tilde{\chi}^{p_{i-1},p'_{i-1}}_{a_{i-1}^L,a_{i-1}^R,e_{i-1}^L,e_{i-1}^R}(m_{i-1},m_i)\left\{\mathcal{S}^{(i)L}\right\}.$$
$$= \sum\left(q^{\frac{1}{4}\hat{\boldsymbol{m}}^{(i)T}\boldsymbol{C}\hat{\boldsymbol{m}}^{(i)}+\frac{1}{4}m_{i-1}^2-\frac{1}{2}m_{i-1}m_i-\frac{1}{2}(\boldsymbol{u}^L_{(\supset,k(i)-1)}+\boldsymbol{u}^R_{(\subset,k(i)-1)})\cdot\boldsymbol{m}^{(i-1)}+\frac{1}{4}\gamma''_{i-1}}\right.$$
$$\left.\prod_{j=i}^{t-1}\begin{bmatrix} m_j - \frac{1}{2}(\hat{\boldsymbol{C}}\hat{\boldsymbol{m}}^{(i-1)}-\boldsymbol{u}^L-\boldsymbol{u}^R)_j \\ m_j \end{bmatrix}_q\right)$$
$$= F^{(i-1)}_{a,b}(\boldsymbol{u}^L,\boldsymbol{u}^R,m_{i-1},m_i).$$

Once it is established that

$$\mathcal{P}^{p_{i-1},p'_{i-1}}_{a_{i-1}^L,a_{i-1}^R,e_{i-1}^L,e_{i-1}^R}(m_{i-1},m_i)\left\{\mathcal{S}^{(i)L}\right\} = \mathcal{P}^{p_{i-1},p'_{i-1}}_{a_{i-1}^L,a_{i-1}^R,e_{i-1}^L,e_{i-1}^R}(m_{i-1},m_i)\left\{\mathcal{S}^{(i-1)}\right\},$$

we obtain the required result when $i = t_k$, since $k(i) = k(i-1) + 1$.

If $i = t_n$, then $\{\mathcal{S}^{(i)L}\} = \{\mathcal{S}^{(i-1)}\}$ immediately. Now let $i < t_n$. For $A \in \{L, R\}$, if $\sigma_i^A = -1$, then necessarily $\sigma_{t_n}^A = -1$. In the case where $a^A \in \mathcal{T}$, this implies that $\{2, \kappa_{t_n-i}^{(i)}\} \subset \mathcal{S}^{(i)L}$ and $\kappa_{t_n-i}^{(i)} \in \mathcal{S}^{(i-1)}$. Since $a_{i-1}^A = 1$, we may drop the element 2 from $\mathcal{S}^{(i)L}$ with no effect. Similar reasoning holds for $a^A \in \mathcal{T}'$, whereupon the claim is established.

In this $i = t_k$ case, making use of (22), (23), we also immediately obtain:

$$\alpha^{p_{i-1},p'_{i-1}}_{a^L_{i-1},a^R_{i-1}} = \alpha''_{i-1},$$
$$\beta^{p_{i-1},p'_{i-1}}_{a^L_{i-1},a^R_{i-1},e^L_{i-1},e^R_{i-1}} = \beta'_{i-1}.$$

The lemma then follows by induction. □

Before performing a sum over m_1, we require the following result.

Lemma 8.6. *For* $0 \le j \le t$,

$$\alpha''_j \equiv Q_j \pmod 2,$$
$$\beta'_j \equiv Q_j - Q_{j+1} \pmod 2.$$

Proof. Since $\alpha''_t = 0$, $\beta'_t = 0$, and $Q_t = Q_{t+1} = 0$, this result is manifest for $j = t$.

We now proceed by downward induction. Thus assume the result holds for a particular $j > 0$. When $j \ne t_{k(j)}$, equations (24) and (22) imply that $\beta'_{j-1} = \beta'_j + (\boldsymbol{u}^L)_j - (\boldsymbol{u}^R)_j$. Equation (19) implies that $Q_{j-1} \equiv Q_{j+1} - (\boldsymbol{u}^L)_j - (\boldsymbol{u}^R)_j$. Thus the induction hypothesis immediately gives $\beta'_{j-1} \equiv Q_{j-1} - Q_j$ in this case.

When $j = t_{k(j)}$, equations (24) and (22) imply that $\beta'_{j-1} = \alpha''_j - \beta'_j + (\boldsymbol{u}^L)_j - (\boldsymbol{u}^R)_j$. Equation (19) implies that $Q_{j-1} \equiv Q_j + Q_{j+1} - (\boldsymbol{u}^L)_j - (\boldsymbol{u}^R)_j$. Thus the induction hypothesis also gives $\beta'_{j-1} \equiv Q_{j-1} - Q_j$ in this case.

In both cases, equations (24), (22), and (23) give $\alpha''_{j-1} = \alpha''_j + \beta''_{j-1}$, whence the induction hypothesis immediately gives $\alpha''_j \equiv Q_{j-1}$ as required. □

Define

$$F_{a,b}(\boldsymbol{u}^L, \boldsymbol{u}^R, L; q) = \sum_{\boldsymbol{m}\equiv \boldsymbol{Q}(\boldsymbol{u}^L+\boldsymbol{u}^R)} q^{\frac14 \hat{\boldsymbol{m}}^T \boldsymbol{C}\hat{\boldsymbol{m}} - \frac14 L^2 - \frac12(\boldsymbol{u}^L_\flat+\boldsymbol{u}^R_\flat)\cdot\boldsymbol{m} + \frac14\gamma} \prod_{j=1}^{t-1} \begin{bmatrix} m_j - \frac12(\hat{\boldsymbol{C}}\hat{\boldsymbol{m}} - \boldsymbol{u}^L - \boldsymbol{u}^R)_j \\ m_j \end{bmatrix}_q.$$

The summation here is over all vectors $\boldsymbol{m} = (m_1, m_2, \ldots, m_{t-1})$ such that $m_j \in \mathbb{Z}_{\ge 0}$ and $m_j \equiv Q_j \pmod 2$ for $1 \le j < t$. Then $\hat{\boldsymbol{m}} = (m_0, m_1, m_2, \ldots, m_{t-1})$.

On defining

$$\mathcal{S} = \begin{cases} \{\kappa_i\} & \text{if } \sigma^L < t_n,\ \sigma^R < t_n, \text{ and } a, b \in \mathcal{T}, \\ \{p'_i - \kappa_i\} & \text{if } \sigma^L < t_n,\ \sigma^R < t_n, \text{ and } a, b \in \mathcal{T}', \\ \emptyset & \text{otherwise}, \end{cases}$$

we then obtain the following.

Lemma 8.7. *Let $p' > 2p$. If $L \equiv \alpha^{p,p'}_{a,b}$, then*

$$\tilde{\chi}^{p,p'}_{a,b,e^L_0,e^R_0}(L)\,\{\mathcal{S}\} = F_{a,b}(\boldsymbol{u}^L, \boldsymbol{u}^R, L).$$

In addition, $\delta^{p,p'}_{b,e^R_0} = 0$.

Proof. Lemma 8.6 implies that $L \equiv Q_0$. Lemma 2.3 requires the sum over all $m_1 \equiv L + \beta^{p,p'}_{a,b,e,f}$ of the $i = 0$ case of Lemma 8.5. This is applicable since for such m_1, Lemma 8.6 implies that $m_1 \equiv Q_1$.

The lemma follows after noting that in the $p' > 2p$ case, $\hat{\boldsymbol{m}}^{(1)T} \boldsymbol{C} \hat{\boldsymbol{m}}^{(1)} + L^2 - 2Lm_1 = \hat{\boldsymbol{m}}^T \boldsymbol{C} \hat{\boldsymbol{m}} - L^2$ and $\gamma''_0 = \gamma$. □

We now transfer this result to the original weighting function of (3). To do this we require the value of c given by (14). Then defining $\chi^{p,p'}_{a,b,c}(L)\,\{\mathcal{S}\}$ analogously to $\tilde{\chi}^{p,p'}_{a,b,e,f}(L)\,\{\mathcal{S}\}$, we obtain the following.

Lemma 8.8. *If $L \equiv \alpha^{p,p'}_{a,b}$ (mod 2), then*

$$\chi^{p,p'}_{a,b,c}(L)\,\{\mathcal{S}\} = F_{a,b}(\boldsymbol{u}^L, \boldsymbol{u}^R, L).$$

Proof. For the moment, assume that $p' > 2p$. Consider $h \in \mathcal{P}^{p,p'}_{a,b,e,f}(L)$ and $h' \in \mathcal{P}^{p,p'}_{a,b,c'}(L)$ given by $h'_i = h_i$ for $0 \le i \le L$. If $\delta^{p,p'}_{b,f} = 0$ and $c' = b + (-1)^f$, then, as noted in Section 2, $\widetilde{wt}(h) = wt(h')$. Consequently, $\tilde{\chi}^{p,p'}_{a,b,e,f}(L)\,\{\mathcal{S}\} = \chi^{p,p'}_{a,b,c'}(L)\,\{\mathcal{S}\}$. However, if b is interfacial, then the same is true for $c' = b \pm 1$. As noted at the end of Section 6.1, b is interfacial if $\sigma^R \ge t_1$. Otherwise, the current lemma follows from noting that for the c defined above, $c = b + (-1)^{e^R_0}$.

Now given $h \in \mathcal{P}^{p,p'}_{a,b,c}(L)$, define $\hat{h} \in \mathcal{P}^{p'-p,p'}_{a,b,c}(L)$ by $\hat{h}_i = h_i$ for $0 \le i \le L$. As in Lemma 4.1, $\mathrm{wt}\,(\hat{h}) = \frac{1}{4}(L^2 - \alpha^2) - \mathrm{wt}\,(h)$, where $\alpha = \alpha^{p,p'}_{a,b}$. Therefore, $\chi^{p,p'}_{a,b,c}(L)\,\{\mathcal{S}\} = q^{\frac{1}{4}(L^2-\alpha^2)}\chi^{p,p'}_{a,b,c}(L; q^{-1})\,\{\mathcal{S}\}$. Since $\alpha^{p,p'}_{a,b} = \alpha''_0$ by Lemma 8.5 and $\gamma_0 = -(\alpha''_0)^2 - \gamma''_0$ by (24), the $p' < 2p$ case follows from the $p' > 2p$ case obtained above after using $\left[{m+n \atop m}\right]_{q^{-1}} = q^{-mn}\left[{m+n \atop m}\right]_q$, and noting the change in the definition of $\boldsymbol{C}$. □

Proof of Theorem 8.1. First, consider the case where $a < y_n$ and $b < y_n$. Then necessarily $a, b \in \mathcal{T}$. Since $y_n = \kappa_{t_n}$, we have $\sigma^L < t_n$ and $\sigma^R < t_n$. Thereupon, $\mathcal{S} = \{y_n\}$. Let $h \in \mathcal{P}^{p,p'}_{a,b,c}(L) \backslash \mathcal{P}^{p,p'}_{a,b,c}(L)\{y_n\}$. Then $1 \le h_i < y_n$ for $0 \le i \le L$. Since by Lemma 6.4 the lowermost $y_n - 2$ bands of the (p, p')-model have exactly the same parities as the corresponding bands of the (z_n, y_n)-model, we see that if $h' \in \mathcal{P}^{z_n,y_n}_{a,b,c}(L)$ is defined by $h'_i = h_i$ for $0 \le i \le L$, then $wt(h') = wt(h)$. Since all of $\mathcal{P}^{z_n,y_n}_{a,b,c}(L)$ arises in this way, we have $\chi^{p,p'}_{a,b,c}(L) = \chi^{p,p'}_{a,b,c}(L)\{y_n\} + \chi^{z_n,y_n}_{a,b,c}(L)$. This proves the first case of Theorem 8.1.

The second case arises if $a > p' - y_n$ and $b > p' - y_n$. Here necessarily $a, b \in \mathcal{T}'$, whence again $\sigma^L < t_n$ and $\sigma^R < t_n$. The argument proceeds as above, noting that both the (p, p')- and (z_n, y_n)-models are up–down symmetric.

The other cases are immediate since $\mathcal{S} = \emptyset$. □

8.3 The *mn*-system. Each term in the fermionic expressions given by Theorem 8.1 or Theorem 8.2 corresponds to a vector $\boldsymbol{m} = (m_1, m_2, \ldots, m_{t-1})$, where $\boldsymbol{m} \equiv \boldsymbol{Q}(\boldsymbol{u}^L + \boldsymbol{u}^R)$. As usual, we set $\hat{\boldsymbol{m}} = (L, m_1, m_2, \ldots, m_{t-1})$. Now for each $\boldsymbol{m}$, define a vector $\boldsymbol{n} = (n_1, n_2, \ldots, n_t)$ by

$$\boldsymbol{n} = \frac{1}{2}(-\hat{\boldsymbol{C}}\hat{\boldsymbol{m}} + \boldsymbol{u}). \tag{38}$$

In view of (19), we see that $n_j \in \mathbb{Z}$ for $1 \leq j \leq t$. Then since

$$\frac{1}{2}(\boldsymbol{C}\hat{\boldsymbol{m}} - \boldsymbol{u}^L - \boldsymbol{u}^R)_j = -n_j, \tag{39}$$

in those terms that provide a nonzero contribution to the fermionic expression of Theorem 8.1, $n_j \geq 0$ for $1 \leq j \leq t$.

On examining the proof of Lemma 8.5, we see that n_i is the number of particles added at the ith induction step to pass from $\mathcal{P}^{p_i, p'_i}_{a^L_i, a^R_i, e^L_i, e^R_i}(m_i, m_{i+1})\{\mathcal{S}^{(i)}\}$ to $\mathcal{P}^{p_{i-1}, p'_{i-1}}_{a^L_{i-1}, a^R_{i-1}, e^L_{i-1}, e^R_{i-1}}(m_{i-1}, m_i)\{\mathcal{S}^{(i-1)}\}$.

The set of equations that link the two vectors $\hat{\boldsymbol{m}}$ and $\boldsymbol{n}$ is known as the ***mn***-system. On account of (18), the equations are more explicitly given for $1 \leq j \leq t$ by

$$m_{j-1} - m_{j+1} = m_j + 2n_j - u_j \qquad \text{if } j = t_k, \quad k = 1, 2, \ldots, n, \tag{40}$$
$$m_{j-1} + m_{j+1} = 2m_j + 2n_j - u_j \qquad \text{otherwise}, \tag{41}$$

where we set $m_t = m_{t+1} = 0$.

Using these two expressions and setting $m_0 = L$, it may be readily shown that

$$\sum_{i=1}^{t} l_i n_i = \frac{1}{2}\left(L + \sum_{i=1}^{t} l_i u_i\right). \tag{42}$$

Thereupon, the summands in the expression for $F_{a,b}(\boldsymbol{u}^L, \boldsymbol{u}^R, L)$ given in Theorem 8.1 correspond to solutions of (42) with each n_i a nonnegative integer.

8.4 The second fermionic form. The proof of Theorem 8.2 follows the same lines as that of Theorem 8.1. We will not give the full description but will indicate how the proof of Lemma 8.5 is affected by the use of the modified Gaussians. We first define $F^{(i)\prime}_{a,b}(\boldsymbol{u}^L, \boldsymbol{u}^R, m_i, m_{i+1}; q)$ for $0 \leq i < t$ in the same way as $F^{(i)}_{a,b}(\boldsymbol{u}^L, \boldsymbol{u}^R, m_i, m_{i+1}; q)$ in (27) and (28), except we will employ the modified Gaussians instead of the classical Gaussians. Note that this modified form of the

Gaussian differs from the form defined in (1) if and only if $A < 0$ and $B \geq 0$. In this case, $\left[{A \atop B}\right] = 0$. In addition, since $\left[{m+n \atop m}\right]'_{q^{-1}} = q^{-mn}\left[{m+n \atop m}\right]'_{q}$, it follows that the analogues of (29) and (30) hold.

Lemma 8.9. *Let* $0 \leq i < t$, $m_i \equiv Q_i$ *and* $m_{i+1} \equiv Q_{i+1}$. *If* $m_i \geq 0$, *then*

$$\tilde{\chi}^{p_i,p'_i}_{a^L_i,a^R_i,e^L_i,e^R_i}(m_i, m_{i+1}) = F^{(i)}_{a,b}(\boldsymbol{u}^L, \boldsymbol{u}^R, m_i, m_{i+1}). \tag{43}$$

In addition, $\alpha^{p_i,p'_i}_{a^L_i,a^R_i} = \alpha''_i$ *and* $\beta^{p_i,p'_i}_{a^L_i,a^R_i,e^L_i,e^R_i} = \beta'_i$.

PROOF. The proof proceeds much as in the proof of 8.5. However, we must certainly check that using the modified Gaussians does not introduce unwanted terms.

Consider the $i \neq t_{k(i)}$ case. Combining the analogues of (32), (33), and (34) yields

$$\begin{aligned}
&\tilde{\chi}^{p_{i-1},p'_{i-1}}_{a^L_{i-1},a^R_{i-1},e^L_{i-1},e^R_{i-1}}(m_{i-1}, m_i) \\
&\quad = \sum_{\substack{m_{i+1}\equiv Q_{i+1} \\ 0\leq m_{i+1}\leq m_i+1}} q^{\frac{1}{2}(m_i u^L_i - m_{i-1}(u^L_i+u^R_i) - u^L_i u^R_i - 2 + \beta_i((\boldsymbol{\Delta}^R)_i - (\boldsymbol{\Delta}^L)_i) + \alpha_i(\boldsymbol{\Delta}^R)_i + (\boldsymbol{\Delta}^L)_i(\boldsymbol{\Delta}^R)_i)} \\
&\qquad\qquad \times q^{\frac{1}{4}(M-m_i)^2 - \frac{1}{4}\beta_i^2} \left[{\frac{1}{2}(M+m_{i+1}) \atop m_i}\right]_q F^{(i)\prime}_{a,b}(\boldsymbol{u}^L, \boldsymbol{u}^R, m_i, m_{i+1}),
\end{aligned}$$

where $M = m_{i-1} + u^L_i + u^R_i$. Since $m_{i-1}, m_{i+1} \geq 0$ and $u^L_i, u^R_i \geq 0$ (because $i \neq t_{k(i)}$), we have

$$\left[{\frac{1}{2}(m_{i-1}+m_{i+1}+u^L_i+u^R_i) \atop m_i}\right]'_q = \left[{\frac{1}{2}(m_{i-1}+m_{i+1}+u^L_i+u^R_i) \atop m_i}\right]_q. \tag{44}$$

The induction step for $i \neq t_{k(i)}$ then proceeds exactly as in the proof of Lemma 8.5.

For the $i = t_{k(i)}$ case, combining the analogues of (35), (36), and (37) yields

$$\begin{aligned}
&\tilde{\chi}^{p_{i-1},p'_{i-1}}_{a^L_{i-1},a^R_{i-1},e^L_{i-1},e^R_{i-1}}(m_{i-1}, m_i)\left\{\tilde{S}\right\} \\
&\quad = \sum_{\substack{m_{i+1}\equiv Q_{i+1} \\ 0\leq m_{i+1}\leq m_i+1}} q^{\frac{1}{2}(m_i u^L_i - m_{i-1}(u^L_i+u^R_i) - u^L_i u^R_i + (\boldsymbol{\Delta}^L)_i(\boldsymbol{\Delta}^R)_i) - 1} \\
&\qquad\qquad \times q^{\frac{1}{2}(\beta_i((\boldsymbol{\Delta}^R)_i - (\boldsymbol{\Delta}^L)_i) + \alpha_i(\boldsymbol{\Delta}^R)_i) + \frac{1}{4}(m_i^2 + (M-m_i)^2 - \alpha_i^2 - \beta_i^2)} \\
&\qquad\qquad \times \left[{\frac{1}{2}(M+m_i-m_{i+1}) \atop m_i}\right]_q F^{(i)\prime}_{a,b}(\boldsymbol{u}^L, \boldsymbol{u}^R, m_i, m_{i+1}; q^{-1}),
\end{aligned} \tag{45}$$

where $M = m_{i-1} + u^L_i + u^R_i$ and $2 \in \tilde{S}$ if and only if either $a^L_i = 1$ or $a^R_i = 1$; $p' - 2 \in \tilde{S}$ if and only if either $a^L_i = p' - 1$ or $a^R_i = p' - 1$; and $\tilde{S}$ contains no other values.

We must check that (45) holds if the Gaussian is replaced by its modified form, and the "$\{\tilde{S}\}$" is removed.

If $u_i^L = u_i^R = 0$, then $\tilde{S} = \emptyset$. In addition, $m_{i+1} \le m_i + 1$ implies that

$$\begin{bmatrix} \frac{1}{2}(m_{i-1}+m_i-m_{i+1}+u_i^L+u_i^R) \\ m_i \end{bmatrix}_q' = \begin{bmatrix} \frac{1}{2}(m_{i-1}+m_i-m_{i+1}+u_i^L+u_i^R) \\ m_i \end{bmatrix}_q . \quad (46)$$

Thereupon, the induction step for this subcase of $i = t_{k(i)}$ follows as in the proof of Lemma 8.5.

Now consider $u_i^L \ne u_i^R$. We tackle the case $u_i^L = 0$ and $u_i^R = -1$. (The case $u_i^L = -1$ and $u_i^R = 0$ is similar.) This implies that $\sigma^L \ge t_{k(i)}$ and $\sigma^R < t_{k(i)}$. Then either $a_{i-1}^R = 1$ and $\tilde{S} = \{2\}$ or $a_{i-1}^R = p' - 1$ and $\tilde{S} = \{p' - 2\}$. In addition, $2 \le a_{i-1}^L \le p' - 2$. We immediately see that

$$\tilde{\chi}^{p_{i-1},p'_{i-1}}_{a_{i-1}^L,a_{i-1}^R,e_{i-1}^L,e_{i-1}^R}(m_{i-1}, m_i)\left\{\tilde{S}\right\} = \tilde{\chi}^{p_{i-1},p'_{i-1}}_{a_{i-1}^L,a_{i-1}^R,e_{i-1}^L,e_{i-1}^R}(m_{i-1}, m_i). \quad (47)$$

On the other hand, since $m_{i+1} \le m_i + 1$, (46) is valid here unless $m_{i-1} = m_i = 0$ and $m_{i+1} = 1$. Now $\sigma^L \ge t_{k(i)}$ implies that if $a_i^L = a_i^R$, then $\sigma^L = t_{k(i)}$ and $e_i^L = e_i^R$, whereupon $F_{a,b}^{(i)\prime}(\boldsymbol{u}^L, \boldsymbol{u}^R, 0, 1; q^{-1}) = 0$. In this case, since $a_{i-1}^L \ne a_{i-1}^R$, then $\tilde{\chi}^{p_{i-1},p'_{i-1}}_{a_{i-1}^L,a_{i-1}^R,e_{i-1}^L,e_{i-1}^R}(0, 0) = 0$. Therefore, the induction step holds in this $u_i^L \ne u_i^R$ case.

Now consider $u_i^L = u_i^R = -1$ so that $\sigma^L < t_{k(i)}$ and $\sigma^R < t_{k(i)}$. If $a^A \in \mathcal{T}$, then $a_{i-1}^A = 1$, and if $a^A \in \mathcal{T}'$, then $a_{i-1}^A = p' - 1$. Thereupon, (47) holds unless $m_{i-1} = m_i = 0$ and either both $a, b \in \mathcal{T}$ or both $a, b \in \mathcal{T}'$. In these cases,

$$\begin{aligned} &\tilde{\chi}^{p_{i-1},p'_{i-1}}_{a_{i-1}^L,a_{i-1}^R,e_{i-1}^L,e_{i-1}^R}(0, 0)\left\{\tilde{S}\right\} = 0, \\ &\tilde{\chi}^{p_{i-1},p'_{i-1}}_{a_{i-1}^L,a_{i-1}^R,e_{i-1}^L,e_{i-1}^R}(0, 0) = 1, \end{aligned} \quad (48)$$

by direct enumeration. On the other hand, (46) is valid here unless $m_{i-1} + m_i - m_{i+1} = 0$ and $m_i = 0$. If $m_{i-1} = m_i = 0$, then since $\left[{-1 \atop 0}\right]_q' = 1$ and $\alpha_i = \beta_i = 0$, the required analogue of (45) holds in this case. If $m_{i-1} = 1$ and $m_i = 0$, then both sides of the analogue of (45) are easily seen to be zero.

The induction step is now complete, whence the lemma follows. □

Note that at the ith step in the induction, an extra term arises due to the modified Gaussian only if $i = t_{k(i)}$, $\sigma^L < i$, $\sigma^R < i$, and either both $a, b \in \mathcal{T}$ or both $a, b \in \mathcal{T}'$. In this case, consider the term $F_{a,b}^{(i)\prime}(\boldsymbol{u}^L, \boldsymbol{u}^R, m_i, m_{i+1}; q^{-1})$ in (45), which enumerates the elements of $\mathcal{P}^{p_i,p'_i}_{a_i^L,a_i^R,e_i^L,e_i^R}(m_i, m_{i+1})$. In the case where the extra term arises, $m_i = m_{i+1} = 0$ and either both $a_i^L = a_i^R = 1$ and $e_i^L = e_i^R = 0$ or both $a_i^L = a_i^R = p' - 1$ and $e_i^L = e_i^R = 1$. Thus there is precisely one path $\tilde{h}$ of zero length.

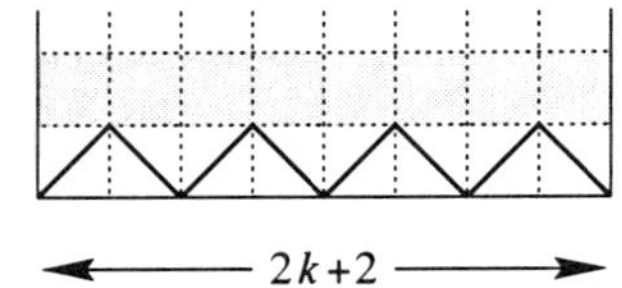

Figure 9.

Equation (45) encapsulates the action of a $\mathcal{D}$-transform, followed by a $\mathcal{B}(k,\lambda)$-transform on $\tilde{h}$, followed by extending the result on both sides (since $u_i^L = u_i^R = -1$). We thus obtain a path of length $m_{i-1} = 2k+2$ in the (p_{i-1}, p'_{i-1})-model. This path has the form given in Figure 9. The fact that this path contains $n_i = k$ particles is also encoded in (40).

When the classical Gaussians are employed, equation (45) thus fails to account for the case of a zero length path. Use of the modified Gaussian remedies this by permitting the case $n_i = -1$. This may be viewed as an annihilation of the $k = 0$ case of Figure 9, which, although appearing to be a particle (cf. Lemma 3.12), arises through solely the action of the $\mathcal{B}_1$-transform followed by path extension.

Acknowledgments. We would like to thank Professor Y. Pugai for collaboration on an earlier stage of this work and on related works and for many useful discussions. His contributions to this work are gratefully acknowledged. We also wish to thank Professors A. Berkovich, B. McCoy, and A. Schilling for many informative discussions on [7]. Finally, we wish to thank Professors M. Kashiwara and T. Miwa for the invitation to attend "Physical Combinatorics," where a preliminary version of this work was presented, and for their excellent hospitality.

This research was supported by the Australian Research Council (ARC).

References

[1] A. K. Agarwal and D. M. Bressoud, Lattice paths and multiple basic hypergeometric series, *Pacific J. Math.*, **136** (1989), 209–228.

[2] G. E. Andrews, *The Theory of Partitions*, Encyclopedia of Mathematics and Its Applications 2, Addison–Wesley, Reading, MA, 1976.

[3] G. E. Andrews, R. J. Baxter, D. M. Bressoud, W. H. Burge, P. J. Forrester, and G. X. Viennot, Partitions with prescribed hook differences, *European J. Combin.*, **8** (1987) 431–350.

[4] A. A. Belavin, A. M. Polyakov, and A. B. Zamolodchikov, Infinite conformal symmetry in two-dimensional quantum field theory, *Nuclear Phys. B*, **241** (1984), 333–380.

[5] A. Berkovich and B. M. McCoy, Continued fractions and fermionic representations for characters of $M(p, p')$ minimal models, *Lett. Math. Phys.*, **37** (1996), 49–66.

[6] A. Berkovich and B. M. McCoy, The perturbation $\phi_{2,1}$ of the $\mathcal{M}(p, p+1)$ models of conformal field theory and related polynomial character identities,

preprint ITP-SB-98-49, State University of New York at Stony Brook, Stony Brook, NY, 1998.

[7] A. Berkovich, B. M. McCoy and A. Schilling, Rogers–Schur–Ramanujan type identities for the $M(p, p')$ minimal models of conformal field theory, *Comm. Math. Phys.*, **191** (1998), 325–395.

[8] D. M. Bressoud, Lattice paths and the Rogers-Ramanujan identities, in K. Alladi, ed., *Proceedings of the International Ramanujan Centenary Conference, Madras,* 1987, Lecture Notes in Mathematics 1395, Springer-Verlag, Berlin, New York, Heidelberg, 1989.

[9] B. L. Feigen and D. B. Fuchs, Skew-symmetric differential operators on the line and Verma modules over the Virasoro algebra, *Functional Anal. Appl.*, **17** (1982), 114–126.

[10] O. Foda, K. S. M. Lee, Y. Pugai, and T. A. Welsh, Path generating transforms, preprint q-alg/9810043; *Contemp. Math.*, to appear.

[11] O. Foda, K. S. M. Lee, and T. A. Welsh, A Burge tree of Virasoro-type polynomial identities, *Internat. J. Modern Phys. A*, **13** (1998), 4967–5012.

[12] O. Foda and T. A. Welsh, Melzer's identities revisited, *Contemp. Math.*, **248** (1999), 207–234.

[13] O. Foda and T. A. Welsh, Polynomial fermionic characters of Forrester-Baxter models, in preparation.

[14] P. J. Forrester and R. J. Baxter, Further exact solutions of the eight-vertex SOS model and generalizations of the Rogers-Ramanujan identities, *J. Statist. Phys.*, **38** (1985), 435–472.

[15] G. Gasper and M. Rahman, *Basic Hypergeometric Series*, Encyclopedia of Mathematics and Its Applications 35, Cambridge University Press, London, Cambridge, 1990.

[16] M. Jimbo and T. Miwa, *Algebraic Analysis of Solvable Lattice Models*, CBMS Regional Conference Series in Mathematics 85, AMS, Providence 1995.

[17] A. Rocha-Caridi, Vacuum vector representations of the Virasoro algebra, in J. Lepowsky, S. Mandelstam, and I.M. Singer, eds., *Vertex Operators in Mathematics and Physics*, Springer-Verlag, Berlin, New York, Heidelberg, 1985.

Department of Mathematics and Statistics
University of Melbourne
Parkville, Victoria 3010
Australia
foda@maths.mu.oz.au
trevor@maths.mu.oz.au

Combinatorial R Matrices for a Family of Crystals: $C_n^{(1)}$ and $A_{2n-1}^{(2)}$ Cases

Goro Hatayama, Atsuo Kuniba, Masato Okado, and Taichiro Takagi

Abstract. The combinatorial R matrices are obtained for a family $\{B_l\}$ of crystals for $U'_q(C_n^{(1)})$ and $U'_q(A_{2n-1}^{(2)})$, where B_l is the crystal of the irreducible module corresponding to the one-row Young diagram of length l. The isomorphism $B_l \otimes B_k \simeq B_k \otimes B_l$ and the energy function are described explicitly in terms of a C_n-analogue of the Robinson–Schensted–Knuth-type insertion algorithm. As an application, a $C_n^{(1)}$-analogue of the Kostka polynomials is calculated for several cases.

1 Introduction

1.1 Background. *Physical combinatorics* might be defined naïvely as combinatorics guided by ideas or insights from physics. A distinguished example can be given by the Kostka polynomial. It is a polynomial $K_{\lambda\mu}(q)$ in q depending on two partitions λ, μ with the same number of nodes. Although there are a number of ways one can look at this polynomial, one can regard it as a q-analogue of the multiplicity of the irreducible $\mathfrak{sl}_n$-module V_λ in the m-fold tensor product $V_{(\mu_1)} \otimes V_{(\mu_2)} \otimes \cdots \otimes V_{(\mu_m)}$ ($\mu = (\mu_1, \ldots, \mu_m)$). Here for $\lambda = (\lambda_1, \ldots, \lambda_n)$ ($\lambda_1 \geq \cdots \geq \lambda_n \geq 0$) V_λ denotes the irreducible $\mathfrak{sl}_n$-module with highest weight $\sum_{i=1}^{n-1}(\lambda_i - \lambda_{i+1})\Lambda_i$, Λ_i being the fundamental weight of $\mathfrak{sl}_n$. In particular, $V_{(\mu_i)}$ is the symmetric tensor representation of degree μ_i.

In [KR], Kirillov and Reshetikhin presented the following expression for the Kostka polynomial:[1]

[1]This expression differs from the conventional definition of $K_{\lambda\mu}(q)$ by an overall power of q.

$$
\begin{aligned}
K_{\lambda\mu}(q) &= \sum_{\{m\}} q^{c(\{m\})} \prod_{\substack{1\le a\le n-1\\ i\ge 1}} \begin{bmatrix} p_i^{(a)} + m_i^{(a)} \\ m_i^{(a)} \end{bmatrix}, \qquad (1.1)\\
c(\{m\}) &= \frac{1}{2} \sum_{1\le a,b\le n-1} C_{ab} \sum_{i,j\ge 1} \min(i,j) m_i^{(a)} m_j^{(b)}\\
&\qquad - \sum_{i,j\ge 1} \min(i,\mu_j) m_i^{(1)},\\
p_i^{(a)} &= \delta_{a1} \sum_{j\ge 1} \min(i,\mu_j) - \sum_{1\le b\le n-1} C_{ab} \sum_{j\ge 1} \min(i,j) m_j^{(b)},
\end{aligned}
$$

where the sum $\sum_{\{m\}}$ is taken over $\{m_i^{(a)} \in \mathbb{Z}_{\ge 0} \mid 1 \le a \le n-1, i \ge 1\}$, satisfying $p_i^{(a)} \ge 0$ for $1 \le a \le n-1, i \ge 1$ and $\sum_{i\ge 1} i m_i^{(a)} = \lambda_{a+1} + \lambda_{a+2} + \cdots + \lambda_n$ for $1 \le a \le n-1$. $(C_{ab})_{1\le a,b\le n-1}$ is the Cartan matrix of $\mathfrak{sl}_n$, and $[{M \atop N}]$ is the q-binomial coefficient or Gaussian polynomial. An intriguing point is that this expression was obtained through the string hypothesis of the Bethe ansatz [Be] for the $\mathfrak{sl}_n$-invariant Heisenberg chain, which is certainly in the field of physics.

Another important idea comes from Baxter's corner transfer matrix (CTM) [Ba, ABF]. In the course of the study of CTM eigenvalues, the notion of one-dimensional sum (1dsum) has appeared [DJKMO], and it was recognized that 1dsums give affine Lie algebra characters. Such phenomena were clarified by the theory of perfect crystals [KMN1, KMN2]. As far as the Kostka polynomial is concerned, Nakayashiki and Yamada [NY] obtained the following expression:

$$
K_{\lambda\mu}(q) = \sum_{p} q^{E(p)}, \qquad (1.2)
$$

where p ranges over the elements $p = b_1 \otimes \cdots \otimes b_m$ of $B_{(\mu_1)} \otimes \cdots \otimes B_{(\mu_m)}$, satisfying $\tilde{e}_i p = 0$ $(i = 1, \dots, n-1)$ and $\mathrm{wt}\, p = \sum_{i=1}^{n-1} (\lambda_i - \lambda_{i+1}) \Lambda_i$. $B_{(\mu_i)}$ is the crystal base of the irreducible $U_q(\mathfrak{sl}_n)$-module with highest weight corresponding to (μ_i) and $\tilde{e}_i$ is the so-called Kashiwara operator. $E(p)$ is called the energy of p and is calculated by using the energy function H as

$$
E(p) = \sum_{1\le i<j\le m} H(b_i \otimes b_j^{(i+1)}),
$$

where $b_j^{(i)}$ is defined through the crystal isomorphism:

$$
\begin{aligned}
B_{(\mu_i)} \otimes B_{(\mu_{i+1})} \otimes \cdots \otimes B_{(\mu_j)} &\simeq B_{(\mu_j)} \otimes B_{(\mu_i)} \otimes \cdots \otimes B_{(\mu_{j-1})}\\
b_i \otimes b_{i+1} \otimes \cdots \otimes b_j &\mapsto b_j^{(i)} \otimes b_i' \otimes \cdots \otimes b_{j-1}'.
\end{aligned}
$$

In the two-fold tensor case, the crystal isomorphism $B_{(\mu_i)} \otimes B_{(\mu_j)} \simeq B_{(\mu_j)} \otimes B_{(\mu_i)}$: $b_i \otimes b_j \mapsto b_j' \otimes b_i'$ combined with the value $H(b_i \otimes b_j)$ is called the combinatorial R

matrix. The crystal base $B_{(l)}$ has a generalization to the rectangular shape $B_{(l^k)}$, and the corresponding generalization of the Kostka polynomial is considered in [SW, S].

In view of the equality (1.1) = (1.2), one is led to an application of the perfect crystal theory of $B_{(l)}$. Define a branching function $b_\lambda^V(q)$ for an $\widehat{\mathfrak{sl}}_n$-module V by

$$\begin{aligned} b_\lambda^V(q) &= \operatorname{tr}_{\mathcal{H}(V,\lambda)} q^{-d}, \\ \mathcal{H}(V,\lambda) &= \{v \in V \mid e_i v = 0\,(i = 1, \ldots, n-1), \operatorname{wt} v = \lambda\}. \end{aligned}$$

Here d is the degree operator. Let $V(l\Lambda_0)$ be the integrable $\widehat{\mathfrak{sl}}_n$-module with affine highest weight $l\Lambda_0$. Then (1.1) = (1.2) implies the spinon character formula:

$$b_\lambda^{V(l\Lambda_0)}(q) = \sum_\eta \frac{K_{\xi\eta}(q) F_\eta^{(l)'}(q)}{(q)_{\zeta_1} \cdots (q)_{\zeta_{n-1}}}. \tag{1.3}$$

For the definitions of ξ, $(\zeta_1, \ldots, \zeta_{n-1})$, $F_\eta^{(l)'}(q)$ along with the summing range of η; see Proposition 4.12 of [HKKOTY].

A key to the derivation of (1.3) is the fact that a suitable subset of the semiinfinite tensor product $\cdots \otimes B_{(l)} \otimes \cdots \otimes B_{(l)}$ can be identified with the crystal base $B(l\Lambda_0)$ of the integrable $U_q(\widehat{\mathfrak{sl}}_n)$-module with highest weight $l\Lambda_0$. Since all components are the same, such a case is called homogeneous. Recently, a generalization of such results to inhomogeneous cases was obtained [HKKOT]. For example, a suitable subset of

$$\cdots \otimes B_{(l_1+l_2)} \otimes B_{(l_2)} \otimes \cdots \otimes B_{(l_1+l_2)} \otimes B_{(l_2)} \otimes B_{(l_1+l_2)} \otimes B_{(l_2)}$$

can be identified with $B(l_1\Lambda_0) \otimes B(l_2\Lambda_0)$. Taking the corresponding limit of μ in the equality (1.1) = (1.2), one obtains an expression for the branching function $b_\lambda^{V(l_1\Lambda_0)\otimes V(l_2\Lambda_0)}(q)$.

Another important application of the inhomogeneous case is found in soliton cellular automata. Recently, several such automata have been related to known soliton equations through a limiting procedure called ultra-discretization [TS, TTMS]. Although they seem at first view to have nothing to do with the theory of crystals, recent studies revealed their underlying crystal structure [HKT, FOY, HHIKTT]. Namely, the combinatorial R matrix appears as the scattering rule of solitons as well as the time evolution operator for the automaton.

1.2 Present work. In the $\widehat{\mathfrak{sl}}_n$ case, a typical example of the isomorphism $B_{(3)} \otimes B_{(2)} \simeq B_{(2)} \otimes B_{(3)}$ is

$$112 \otimes 23 \mapsto 12 \otimes 123.$$

This case may be viewed as a scattering process of two composite particles 112 and 23. Through the collision, the constituent particles are reshuffled and then recombined into two other composite particles, 12 and 123.

In this paper, we study the combinatorial R matrices for a family of $U'_q(C_n^{(1)})$ and $U'_q(A_{2n-1}^{(2)})$ crystals. This includes a new type of example as

$$123 \otimes \bar{2}\bar{1} \mapsto 23 \otimes 0\bar{2}\bar{0} \quad \text{for } U'_q(C_n^{(1)}) \text{ case,}$$
$$\mapsto 13 \otimes 1\bar{1}\bar{1} \quad \text{for } U'_q(A_{2n-1}^{(2)}) \text{ case.}$$

Here we observe "antiparticles," which undergo a pair annihilation and a pair creation $(1) + (\bar{1}) \longrightarrow (0) + (\bar{0})$ or $(2) + (\bar{2}) \longrightarrow (1) + (\bar{1})$.

We shall consider a family $\{B_l \mid l \in \mathbb{Z}_{\geq 1}\}$ of crystals for $U'_q(C_n^{(1)})$ and $U'_q(A_{2n-1}^{(2)})$. The above example corresponds to $B_3 \otimes B_2 \simeq B_2 \otimes B_3$. Here B_l is the crystal of the irreducible U'_q-module corresponding to the l-fold symmetric "fusion" of the vector representation. For $U'_q(A_{2n-1}^{(2)})$, it was constructed in [KKM]. For $U'_q(C_n^{(1)})$, B_l in this paper denotes $B_{l/2}$ in [KKM] (B_l in [HKKOT]) when l is even (odd). Our main result is the explicit description of the isomorphism $B_l \otimes B_k \simeq B_k \otimes B_l$ and the associated energy function for any l and k. It will be done through a slight modification of the insertion algorithm for the C-tableaux introduced by T. H. Baker [B].[2] Since the two affine algebras $C_n^{(1)}$ and $A_{2n-1}^{(2)}$ share the common classical part C_n, they allow a parallel treatment and the results are similar in many respects.

Let us sketch them along the content of the paper.

In Section 2, we recall some basic facts about crystals. As a $U_q(C_n)$ crystal, it is known that $U'_q(C_n^{(1)})$ crystal B_l decomposes into the disjoint union of $B(l\Lambda_1)$, $B((l-2)\Lambda_1), \ldots,$ where $B(\lambda)$ denotes the crystal of the irreducible representation with highest weight λ. Within each $B(l'\Lambda_1)$ it is natural [KN, B] to parametrize the elements by length-l', one-row semistandard tableaux with letters $1 < \cdots < n < \bar{n} < \cdots < \bar{2} < \bar{1}$. Instead of doing so we will represent elements in B_l uniformly via length-l, one-row semistandard tableaux with letters $0 < 1 < \cdots < n < \bar{n} < \cdots < \bar{2} < \bar{1} < \bar{0}$. Here the number x_0 of 0 and $\overline{x}_0$ of $\bar{0}$ must be the same, according to which the elements belong to $B((l - 2x_0)\Lambda_1)$. Thus the number of letters in the tableaux has increased from $2n$ to $2(n + 1)$. In fact, under the insertion scheme in later sections, these tableaux will behave like those for $U_q(C_{n+1})$ [B] in some sense.

In Section 3, we first define an insertion algorithm for the tableaux introduced in Section 2. When there is no $(x, \bar{x})$ pair, it is the same as the well-known $\mathfrak{sl}_n$ case [F]. In general, our algorithm is essentially Baker's algorithm [B] for $U_q(C_{n+1})$ if $0 < \cdots < n < \bar{n} < \cdots < \bar{0}$ is regarded as $1 < \cdots < n+1 < \overline{n+1} < \cdots < \bar{1}$. (See Remark 3.2.) We describe it only for those tableaux with depth at most two, which suffices for our aim. We then state a main theorem, which describes the combinatorial R matrix of $U'_q(C_n^{(1)})$ explicitly in terms of the insertion scheme.

In Section 4, we prove the main theorem. As a $U_q(C_n)$ crystal, $B_l \otimes B_k$ decomposes into connected components that are isomorphic to the crystals of irreducible $U_q(C_n)$-modules. Within each component, the general elements are obtained by applying $\tilde{f}_i$s

[2] The tableaux and insertions employed in [B] and this paper are different from those in [Ber, KE, T].

$(1 \le i \le n)$ to the $U_q(C_n)$ highest elements. Our strategy is first to verify the theorem directly for the highest elements. For general elements, the theorem follows from the fact due to Baker that our insertion algorithm on letters $0, 1, \ldots, \overline{1}, \overline{0}$ can be regarded as the isomorphism of $U_q(C_{n+1})$ crystals. It turns out that $B_l \supset B_{l-2} \supset B_{l-4} \supset \cdots$ as the sets according to the number of $(0, \overline{0})$ pairs contained in the tableaux. We shall utilize this fact to remove the $(0, \overline{0})$ pairs before the insertion so as to avoid the pair annihilation of the boxes under the insertions and the resulting bumping–sliding transition in [B].

In Section 5, a parallel treatment is done for $U'_q(A^{(2)}_{2n-1})$. This case is simpler in that B_l coincides with $B(l\Lambda_1)$ as a $U_q(C_n)$ crystal. Consequently, we do not have letters 0 and $\overline{0}$ in the tableaux. The main difference from the $U'_q(C^{(1)}_n)$ case is to remove 1 and $\overline{1}$ appropriately before the insertion.

In Appendix A, we detail the calculation for the proof of Proposition 4.1.

In Appendix B, another rule for finding the image under $B_l \otimes B_k \simeq B_k \otimes B_l$ is given for the $U'_q(C_n)$ case. In practical calculations, it is often more efficient than the one based on the insertion scheme in the main text.

In Appendix C, the $C^{(1)}_n$-analogue $X_{\lambda,\mu}(t)$ of the Kostka polynomials in the sense of Section 1.1 is listed up to $|\mu| = 6$. They coincide with the Kostka polynomial if $|\lambda| = |\mu|$. We note that our generalization of the Kostka polynomial is a q-analogue of the tensor product multiplicities. Except for the $U'_q(A^{(1)}_n)$ case, it is different from the q-analogue of weight multiplicities by Lusztig [L] in general.

Finally, we remark that the isomorphism $B_l \otimes B_k \simeq B_k \otimes B_l$ for $U'_q(C^{(1)}_n)$ in this paper has been identified with the two-body scattering rule in the soliton cellular automaton [HKT].

2 Definitions

2.1 Brief summary of crystals. Let I be an index set. A crystal B is a set B with the maps

$$\tilde{e}_i, \tilde{f}_i : B \sqcup \{0\} \longrightarrow B \sqcup \{0\} \quad (i \in I)$$

satisfying the following properties:

- $\tilde{e}_i 0 = \tilde{f}_i 0 = 0$;
- for any b and i, there exists $n > 0$ such that $\tilde{e}_i^n b = \tilde{f}_i^n b = 0$;
- for $b, b' \in B$ and $i \in I$, $\tilde{f}_i b = b'$ if and only if $b = \tilde{e}_i b'$.

For an element b of B, we set

$$\varepsilon_i(b) = \max\{n \in \mathbb{Z}_{\ge 0} \mid \tilde{e}_i^n b \ne 0\}, \qquad \varphi_i(b) = \max\{n \in \mathbb{Z}_{\ge 0} \mid \tilde{f}_i^n b \ne 0\}.$$

For two crystals B and B', the tensor product $B \otimes B'$ is defined:

$$B \otimes B' = \{b \otimes b' \mid b \in B, b' \in B'\}.$$

The actions of $\tilde{e}_i$ and $\tilde{f}_i$ are defined by

$$\tilde{e}_i(b \otimes b') = \begin{cases} \tilde{e}_i b \otimes b' & \text{if } \varphi_i(b) \geq \varepsilon_i(b'), \\ b \otimes \tilde{e}_i b' & \text{if } \varphi_i(b) < \varepsilon_i(b'), \end{cases} \tag{2.1}$$

$$\tilde{f}_i(b \otimes b') = \begin{cases} \tilde{f}_i b \otimes b' & \text{if } \varphi_i(b) > \varepsilon_i(b'), \\ b \otimes \tilde{f}_i b' & \text{if } \varphi_i(b) \leq \varepsilon_i(b'). \end{cases} \tag{2.2}$$

Here $0 \otimes b$ and $b \otimes 0$ are understood to be 0.

2.2 The energy function and the combinatorial R matrix. Let $\mathfrak{g}$ be an affine Lie algebra and let B and B' be two $U_q'(\mathfrak{g})$ crystals. We assume that B and B' are finite sets and that $B \otimes B'$ is connected. The algebra $U_q'(\mathfrak{g})$ is a subalgebra of $U_q(\mathfrak{g})$. Their definitions are given in Section 2.1 (resp., Section 3.2) of [KMN1] for $U_q(\mathfrak{g})$ (resp., $U_q'(\mathfrak{g})$).

Suppose $b \otimes b' \in B \otimes B'$ is mapped to $\tilde{b}' \otimes \tilde{b} \in B' \otimes B$ under the isomorphism $B \otimes B' \simeq B' \otimes B$ of $U_q'(\mathfrak{g})$ crystals. A $\mathbb{Z}$-valued function H on $B \otimes B'$ is called an *energy function* if for any i and $b \otimes b' \in B \otimes B'$ such that $\tilde{e}_i(b \otimes b') \neq 0$, it satisfies

$$H(\tilde{e}_i(b \otimes b')) = \begin{cases} H(b \otimes b') + 1 & \text{if } i = 0, \varphi_0(b) \geq \varepsilon_0(b'), \varphi_0(\tilde{b}') \geq \varepsilon_0(\tilde{b}), \\ H(b \otimes b') - 1 & \text{if } i = 0, \varphi_0(b) < \varepsilon_0(b'), \varphi_0(\tilde{b}') < \varepsilon_0(\tilde{b}), \\ H(b \otimes b') & \text{otherwise.} \end{cases} \tag{2.3}$$

When we want to emphasize $B \otimes B'$, we write $H_{BB'}$ for H. This definition of the energy function is due to (3.4.e) of [NY], which is a generalization of the definition for the $B = B'$ case in [KMN1]. The energy function is unique up to an additive constant since $B \otimes B'$ is connected. By definition, $H_{BB'}(b \otimes b') - H_{B'B}(\tilde{b}' \otimes \tilde{b})$ is a constant independent of $b \otimes b'$. In this paper, we choose the constant to be 0. We call the isomorphism $B \otimes B' \simeq B' \otimes B$ endowed with the energy function $H_{BB'}$ the *combinatorial R-matrix*.

2.3 $C_n^{(1)}$ crystals. Given a nonnegative integer l, we consider a $U_q'(C_n^{(1)})$ crystal denoted by B_l.

If l is even, B_l is the same as that defined in [KKM]. (Their B_l is identical to our B_{2l}.) If l is odd, B_l is defined in [HKKOT]. B_ls are the crystals associated with the crystal bases of the irreducible finite-dimensional representation of the quantum affine algebra $U_q'(C_n^{(1)})$. As a set, B_l reads

$$B_l = \left\{ (x_1, \ldots, x_n, \overline{x}_n, \ldots, \overline{x}_1) \;\middle|\; x_i, \overline{x}_i \in \mathbb{Z}_{\geq 0}, \sum_{i=1}^{n} (x_i + \overline{x}_i) \in \{l, l-2, \ldots\} \right\}.$$

The crystal structure is given by (2.7).

B_l is isomorphic to $\bigoplus_{0 \leq j \leq l,\, j \equiv l \pmod 2} B(j\Lambda_1)$ as crystals for $U_q(C_n)$, where $B(j\Lambda_1)$ is the one associated with the irreducible representation of with highest-weight $j\Lambda_1$. As a special case of the more general family of $U_q(C_n)$ crystals [KN], the crystal $B(j\Lambda_1)$ has a description with the semistandard C-tableaux. The entries are $1, \dots, n$ and $\overline{1}, \dots, \overline{n}$ with total order

$$1 < 2 < \cdots < n < \overline{n} < \cdots < \overline{2} < \overline{1}.$$

In this description, $b = (x_1, \dots, x_n, \overline{x}_n, \dots, \overline{x}_1) \in B(j\Lambda_1)$ is depicted by

$$b = \boxed{\overbrace{1 \cdots 1}^{x_1}} \cdots \boxed{\overbrace{n \cdots n}^{x_n}}\boxed{\overbrace{\overline{n} \cdots \overline{n}}^{\overline{x}_n}} \cdots \boxed{\overbrace{\overline{1} \cdots \overline{1}}^{\overline{x}_1}}. \tag{2.4}$$

The length of this one-row tableau is equal to j, namely, $\sum_{i=1}^{n}(x_i + \overline{x}_i) = j$. Here and in the remaining part of this paper, we denote

$$\overbrace{\boxed{i}\boxed{i}\boxed{\cdots}\boxed{i}}^{x}$$

by

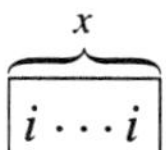

or, more simply, by

$$\overset{x}{\boxed{\quad i \quad}}.$$

We shall depict the elements of B_l by one-row tableaux with length l by supplying pairs of $\boxed{0}$ and $\boxed{\overline{0}}$. Adding 0 and $\overline{0}$ into the set of the entries of the tableaux, we assume the total order $0 < 1 < \cdots < \overline{1} < \overline{0}$. Thus we depict $b = (x_1, \dots, x_n, \overline{x}_n, \dots, \overline{x}_1) \in B_l$ by

$$\mathbb{T}(b) = \boxed{\overbrace{0 \cdots 0}^{x_0}}\boxed{\overbrace{1 \cdots 1}^{x_1}} \cdots \boxed{\overbrace{n \cdots n}^{x_n}}\boxed{\overbrace{\overline{n} \cdots \overline{n}}^{\overline{x}_n}} \cdots \boxed{\overbrace{\overline{1} \cdots \overline{1}}^{\overline{x}_1}}\boxed{\overbrace{\overline{0} \cdots \overline{0}}^{\overline{x}_0}}, \tag{2.5}$$

where $x_0 = \overline{x}_0 = (l - \sum_{i=1}^{n}(x_i + \overline{x}_i))/2$. If $x_0 = 0$ we say that $\mathbb{T}(b)$ has no $\boxed{0}$. Sometimes we identify $\mathbb{T}(b)$ with b and omit the frame of $\mathbb{T}(b)$, e.g., $0\overline{2}\overline{0} = (0, \dots, 0, 1, 0) \in B_3$.

This description means that we have embedded B_l as a set into $U_q(C_{n+1})$ crystal $B(l\Lambda_1)$. Let us denote this embedding by ς:

$$\varsigma : U_q(C_n) \text{ crystal } B_l \text{ as a set} \hookrightarrow U_q(C_{n+1}) \text{ crystal } B(l\Lambda_1). \tag{2.6}$$

It shifts the entries of the tableaux as $\varsigma(i) = i + 1$ and $\varsigma(\overline{i}) = \overline{i+1}$ for $i = 0, 1, \dots, n$. For example, $\varsigma(0\overline{2}\overline{0}) = 1\overline{3}\overline{1}$. For $b \in B_l$ and $i = 1, \dots, n$ one has $\varsigma(\tilde{e}_i b) = \tilde{e}_{i+1}\varsigma(b)$ and $\varsigma(\tilde{f}_i b) = \tilde{f}_{i+1}\varsigma(b)$.

The crystal structure of B_l is given by

$$\begin{aligned}
\tilde{e}_0 b &= \begin{cases} (x_1-2, x_2, \ldots, \overline{x}_2, \overline{x}_1) & \text{if } x_1 \geq \overline{x}_1 + 2, \\ (x_1-1, x_2, \ldots, \overline{x}_2, \overline{x}_1+1) & \text{if } x_1 = \overline{x}_1 + 1, \\ (x_1, x_2, \ldots, \overline{x}_2, \overline{x}_1+2) & \text{if } x_1 \leq \overline{x}_1, \end{cases} \\
\tilde{e}_n b &= (x_1, \ldots, x_n+1, \overline{x}_n-1, \ldots, \overline{x}_1), \\
\tilde{e}_i b &= \begin{cases} (x_1, \ldots, x_i+1, x_{i+1}-1, \ldots, \overline{x}_1) & \text{if } x_{i+1} > \overline{x}_{i+1}, \\ (x_1, \ldots, \overline{x}_{i+1}+1, \overline{x}_i-1, \ldots, \overline{x}_1) & \text{if } x_{i+1} \leq \overline{x}_{i+1}, \end{cases} \\
\tilde{f}_0 b &= \begin{cases} (x_1+2, x_2, \ldots, \overline{x}_2, \overline{x}_1) & \text{if } x_1 \geq \overline{x}_1, \\ (x_1+1, x_2, \ldots, \overline{x}_2, \overline{x}_1-1) & \text{if } x_1 = \overline{x}_1 - 1, \\ (x_1, x_2, \ldots, \overline{x}_2, \overline{x}_1-2) & \text{if } x_1 \leq \overline{x}_1 - 2, \end{cases} \\
\tilde{f}_n b &= (x_1, \ldots, x_n-1, \overline{x}_n+1, \ldots, \overline{x}_1), \\
\tilde{f}_i b &= \begin{cases} (x_1, \ldots, x_i-1, x_{i+1}+1, \ldots, \overline{x}_1) & \text{if } x_{i+1} \geq \overline{x}_{i+1}, \\ (x_1, \ldots, \overline{x}_{i+1}-1, \overline{x}_i+1, \ldots, \overline{x}_1) & \text{if } x_{i+1} < \overline{x}_{i+1}, \end{cases}
\end{aligned} \tag{2.7}$$

where $b = (x_1, \ldots, x_n, \overline{x}_n, \ldots, \overline{x}_1)$ and $i = 1, \ldots, n-1$. For this b, we have

$$\begin{aligned}
\varphi_i(b) &= x_i + (\overline{x}_{i+1} - x_{i+1})_+ \quad \text{for } i = 0, 1, \ldots, n-1, \\
\varepsilon_i(b) &= \overline{x}_i + (x_{i+1} - \overline{x}_{i+1})_+ \quad \text{for } i = 0, 1, \ldots, n-1, \\
\varphi_n(b) &= x_n, \quad \varepsilon_n(b) = \overline{x}_n.
\end{aligned} \tag{2.8}$$

Here $(x)_+ := \max(x, 0)$.

Except for Section 5 concerning $A^{(2)}_{2n-1}$, we normalize the energy function for the $C^{(1)}_n$ case as

$$H_{B_l B_k}((l, 0, \ldots, 0) \otimes (0, k, 0, \ldots, 0)) = 0,$$

regardless of $l < k$ or $l \geq k$. (For $A^{(2)}_{2n-1}$ we will employ a different normalization. See (5.4).)

3 Explicit description of isomorphism and energy function

3.1 The algorithm of column insertions. Set an alphabet $\mathcal{X} = \mathcal{A} \sqcup \bar{\mathcal{A}}$, $\mathcal{A} = \{0, 1, \ldots, n\}$ and $\bar{\mathcal{A}} = \{\overline{0}, \overline{1}, \ldots, \overline{n}\}$ with total order $0 < 1 < \cdots < n < \overline{n} < \cdots < \overline{1} < \overline{0}$. Unless otherwise stated, a tableau means a (column-strict) semistandard one with entries taken from $\mathcal{X}$ in Sections 3 and 4. For the alphabet $\mathcal{X}$, we follow the convention that Greek letters $\alpha, \beta, \ldots$ belong to $\mathcal{A} \sqcup \bar{\mathcal{A}}$ while Latin letters $x, y, \ldots$ (resp., $\overline{x}, \overline{y}, \ldots$) belong to $\mathcal{A}$ (resp., $\bar{\mathcal{A}}$).

Given a letter $\alpha \in \mathcal{X}$ and the tableau T that have at most two rows, we define a tableau denoted by $(\boxed{\alpha} \to T)$, and call such an algorithm a "column insertion of a letter α into a tableau T." (We sometimes identify a letter α with a box $\boxed{\alpha}$.) Let us begin with such Ts that have at most one column. The procedure of the column insertion ($\boxed{\alpha} \to T$) can be summarized as follows:

Case 1a. $\left(\boxed{\alpha} \to \emptyset\right) = \boxed{\alpha}$.

Case 2a. $\left(\boxed{\beta} \to \boxed{\alpha}\right) = \begin{array}{|c|}\hline \alpha \\ \hline \beta \\ \hline \end{array}$ if $\alpha < \beta$.

Case 1b. $\left(\boxed{\alpha} \to \boxed{\beta}\right) = \begin{array}{|c|c|}\hline \alpha & \beta \\ \hline \end{array}$ if $\alpha \leq \beta$.

Case 2b. $\left(\boxed{\beta} \to \begin{array}{|c|}\hline \alpha \\ \hline \gamma \\ \hline \end{array}\right) = \begin{array}{|c|c|}\hline \alpha & \gamma \\ \hline \beta & \multicolumn{1}{c}{} \\ \cline{1-1} \end{array}$ if $\alpha < \beta \leq \gamma$ and $(\alpha, \gamma) \neq (x, \overline{x})$.

Case 3b. $\left(\boxed{\alpha} \to \begin{array}{|c|}\hline \beta \\ \hline \gamma \\ \hline \end{array}\right) = \begin{array}{|c|c|}\hline \alpha & \beta \\ \hline \gamma & \multicolumn{1}{c}{} \\ \cline{1-1} \end{array}$ if $\alpha \leq \beta < \gamma$ and $(\alpha, \gamma) \neq (x, \overline{x})$.

Case 4b. $\left(\boxed{\beta} \to \begin{array}{|c|}\hline x \\ \hline \overline{x} \\ \hline \end{array}\right) = \begin{array}{|c|c|}\hline x-1 & \overline{x-1} \\ \hline \beta & \multicolumn{1}{c}{} \\ \cline{1-1} \end{array}$ if $x \leq \beta \leq \overline{x}$ and $x \neq 0$.

Case 5b. $\left(\boxed{x} \to \begin{array}{|c|}\hline \beta \\ \hline \overline{x} \\ \hline \end{array}\right) = \begin{array}{|c|c|}\hline x+1 & \beta \\ \hline \overline{x+1} & \multicolumn{1}{c}{} \\ \cline{1-1} \end{array}$ if $x < \beta < \overline{x}$ and $x \neq n$.

Cases 2b–5b do not cover all the tableaux with two rows, but in this paper we deal only with these situations. In particular, the tableaux generated by these insertions have at most two rows. Note that the algorithm, except for Cases 4b and 5b, agrees with the Knuth-type column insertion. We call Cases 1b–5b the "bumping cases."

When T is a general tableau with at most two rows, we repeat the above procedure: We insert a box into the leftmost column of T according to the above formula. If it is not a bumping case, replace the column by the right-hand side of the formula. Otherwise, replace the column by the right-hand side of the formula without the right box. We consider that this right box is bumped. We insert it into the second column of T from the left and repeat the procedure above until we come to a nonbumping Case 1a or 2a.

Example 3.1. $n = 4$.

$$\left(\boxed{2} \to \begin{array}{|c|c|c|c|}\hline 0 & 3 & 4 & \bar{0} \\ \hline 4 & \bar{3} & \bar{1} & \multicolumn{1}{c}{} \\ \cline{1-3} \end{array}\right) = \underset{\substack{\uparrow \\ \boxed{4}}}{\begin{array}{|c|c|c|c|}\hline 0 & 3 & 4 & \bar{0} \\ \hline 2 & \bar{3} & \bar{1} & \multicolumn{1}{c}{} \\ \cline{1-3} \end{array}} = \underset{\substack{\uparrow \\ \boxed{\bar{2}}}}{\begin{array}{|c|c|c|c|}\hline 0 & 2 & 4 & \bar{0} \\ \hline 2 & 4 & \bar{1} & \multicolumn{1}{c}{} \\ \cline{1-3} \end{array}} = \underset{\substack{\uparrow \\ \boxed{\bar{1}}}}{\begin{array}{|c|c|c|c|}\hline 0 & 2 & 4 & \bar{0} \\ \hline 2 & 4 & \bar{2} & \multicolumn{1}{c}{} \\ \cline{1-3} \end{array}} = \begin{array}{|c|c|c|c|c|}\hline 0 & 2 & 4 & \bar{1} & \bar{0} \\ \hline 2 & 4 & \bar{2} & \multicolumn{2}{c}{} \\ \cline{1-3} \end{array}$$

For a tableau T, we denote by $w(T)$ the Japanese order of word reading of T. Thus $w(T)$ is a sequence of letters that is created by reading all letters on T from the rightmost column to the leftmost column, and in each column from the top to the bottom. For instance,

$$w\left(\begin{array}{|c|c|c|c|}\hline \alpha_1 & \alpha_2 & \cdots & \alpha_j \\ \hline\end{array}\right) = \alpha_j \cdots \alpha_2\alpha_1,$$

and

$$w\left(\begin{array}{|c|c|c|c|c|c|c|}\hline \alpha_1 & \alpha_2 & \cdots & \alpha_i & \alpha_{i+1} & \cdots & \alpha_j \\ \hline \beta_1 & \beta_2 & \cdots & \beta_i \\ \cline{1-4}\end{array}\right) = \alpha_j \cdots \alpha_{i+1}\alpha_i\beta_i \cdots \alpha_2\beta_2\alpha_1\beta_1,$$

and so on. Let T and T' be one-row tableaux. By abuse of notation, we denote by $T' \to T$ the tableau constructed by successive column insertions of the letters of the word $w(T')$ into T. Namely, if

$$w(T') = \tau_j \tau_{j-1} \cdots \tau_1,$$

then we write

$$(T' \to T) = (\tau_1 \to \cdots (\tau_{j-1} \to (\tau_j \to T)) \cdots).$$

(Following a usual convention in type A [F], it might be written as a *product tableau*, $T' \cdot T$.) In particular $(T \to \emptyset) = T$ for any T.

Throughout this paper we let T_1 be the length of the first row of a tableau T.

Remark 3.2. Our algorithm is a specialization of the column insertion for the C_n case [B]. Let b_i be an element of $U_q(C_{n+1})$ crystal $B(l_i\Lambda_1)$ $(i = 1, 2)$. Denote by $b_2 \stackrel{*}{\longrightarrow} b_1$ the tableau obtained by successive column insertions with the original definition [B] of $w(\mathbb{T}(b_2))$ into b_1. (In [B], $b_2 \stackrel{*}{\longrightarrow} b_1$ is denoted by $b_1 * b_2$.) Then for any element b_i of $U'_q(C_n^{(1)})$ crystal B_{l_i} such that $\mathbb{T}(b_1)$ or $\mathbb{T}(b_2)$ has no $\boxed{0}$, our $(\mathbb{T}(b_2) \to \mathbb{T}(b_1))$ has been determined so that

$$\varsigma\Big(\big(\mathbb{T}(b_2) \to \mathbb{T}(b_1)\big)\Big) = \varsigma\big(\mathbb{T}(b_2)\big) \stackrel{*}{\longrightarrow} \varsigma\big(\mathbb{T}(b_1)\big).$$

We will calculate $(\mathbb{T}(b_2) \to \mathbb{T}(b_1))$ only when $\mathbb{T}(b_1)$ or $\mathbb{T}(b_2)$ has no $\boxed{0}$ in this paper. Under such a situation no pair annihilation takes place during the insertion in the right-hand side. See also Remark 4.9.

We shall also use the reverse bumping algorithm [B]. In our case, where the tableau has at most two rows, the algorithm is rather simple. We use it only in the following five cases.

Case 1c.
$$\begin{array}{c}\boxed{\beta}\\ \downarrow \\ \boxed{\alpha}\end{array} \quad = \quad \begin{array}{cc}\boxed{\alpha} & \\ \downarrow & \\ & \boxed{\beta}\end{array} \qquad \text{if } \alpha \le \beta.$$

Case 2c.
$$\begin{array}{c}\boxed{\gamma}\\ \downarrow \\ \begin{array}{|c|}\hline \alpha \\ \hline \beta \\ \hline\end{array}\end{array} \quad = \quad \begin{array}{cc}\boxed{\beta} & \\ \downarrow & \\ & \begin{array}{|c|}\hline \alpha \\ \hline \gamma \\ \hline\end{array}\end{array} \qquad \text{if } \alpha < \beta \le \gamma \text{ and } (\alpha, \gamma) \ne (x, \overline{x}).$$

Case 3c. $$\begin{array}{c}\boxed{\beta}\\ \downarrow \\ \begin{array}{|c|}\hline \alpha\\ \hline \gamma\\ \hline\end{array}\end{array} \quad = \quad \begin{array}{c}\boxed{\alpha}\\ \downarrow \\ \begin{array}{|c|}\hline \beta\\ \hline \gamma\\ \hline\end{array}\end{array} \qquad \text{if } \alpha \leq \beta < \gamma \text{ and } (\alpha, \gamma) \neq (x, \overline{x}).$$

Case 4c. $$\begin{array}{c}\boxed{\overline{x}}\\ \downarrow \\ \begin{array}{|c|}\hline x\\ \hline \beta\\ \hline\end{array}\end{array} \quad = \quad \begin{array}{c}\boxed{\beta}\\ \downarrow \\ \begin{array}{|c|}\hline x+1\\ \hline \overline{x+1}\\ \hline\end{array}\end{array} \qquad \text{if } x < \beta < \overline{x} \text{ and } x \neq n.$$

Case 5c. $$\begin{array}{c}\boxed{\beta}\\ \downarrow \\ \begin{array}{|c|}\hline x\\ \hline \overline{x}\\ \hline\end{array}\end{array} \quad = \quad \begin{array}{c}\boxed{x-1}\\ \downarrow \\ \begin{array}{|c|}\hline \beta\\ \hline \overline{x-1}\\ \hline\end{array}\end{array} \qquad \text{if } x \leq \beta \leq \overline{x} \text{ and } x \neq 0.$$

Here

$$\begin{array}{c}\boxed{\alpha}\\ \downarrow \\ C\end{array} \quad = \quad \begin{array}{c}\boxed{\beta}\\ \downarrow \\ C'\end{array}$$

means that if a letter β is column inserted into a column C', then the column is changed to C and a letter α is bumped out.

3.2 Main theorem: $C_n^{(1)}$ case. Fix $l, k \in \mathbb{Z}_{\geq 1}$. Given $b_1 \otimes b_2 \in B_l \otimes B_k$, we define the element $b_2' \otimes b_1' \in B_k \otimes B_l$ and $l', k', m \in \mathbb{Z}_{\geq 0}$ by the following rule.

Rule 3.3. Set $z = \min(\sharp\,\boxed{0} \text{ in } \mathbb{T}(b_1), \sharp\,\boxed{0} \text{ in } \mathbb{T}(b_2)) = \min(\sharp\,\boxed{\overline{0}} \text{ in } \mathbb{T}(b_1), \text{and } \sharp\,\boxed{\overline{0}}$ in $\mathbb{T}(b_2))$. Remove $(\boxed{0}, \boxed{\overline{0}})$ pairs simultaneously from $\mathbb{T}(b_1)$ and $\mathbb{T}(b_2)$ z times. Denote the resulting tableaux by $\hat{\mathbb{T}}(b_1)$ and $\hat{\mathbb{T}}(b_2)$, and set $l' = \hat{\mathbb{T}}(b_1)_1 = l - 2z$ and $k' = \hat{\mathbb{T}}(b_2)_1 = k - 2z$. ($T_1$ is the length of the first row of a tableau T.) Operate the column insertion and set $\hat{\mathbb{P}}(b_2 \to b_1) = (\hat{\mathbb{T}}(b_2) \longrightarrow \hat{\mathbb{T}}(b_1))$. It has the form

$$\begin{array}{|c|c|}\hline j_1 \cdots\cdots j_{k'} & i_{m+1} \ \cdots \ i_{l'} \\ \hline i_1 \cdots i_m & \multicolumn{1}{c}{} \\ \cline{1-1}\end{array},$$

where m is the length of the second row; hence that of the first row is $l' + k' - m$ $(0 \leq m \leq k')$.

Next, we bump out l' letters from the tableau $T^{(0)} = \hat{\mathbb{P}}(b_2 \to b_1)$ by the reverse bumping algorithm. For the boxes containing $i_{l'}, i_{l'-1}, \ldots, i_1$ in the above tableau, we do it first for $i_{l'}$, then $i_{l'-1}$, and so on. Correspondingly, let w_1 be the first letter that is bumped out from the leftmost column, w_2 be the second, and so on. Denote by $T^{(i)}$ the resulting tableau when w_i is bumped out ($1 \le i \le l'$). Note that $w_1 \le w_2 \le \cdots \le w_{l'}$. Now $b'_1 \in B_l$ and $b'_2 \in B_k$ are uniquely specified by

$$\mathbb{T}(b'_2) = \boxed{\overbrace{0\cdots0}^{z} \mid T^{(l')} \mid \overbrace{\bar{0}\cdots\bar{0}}^{z}},$$

$$\mathbb{T}(b'_1) = \boxed{\overbrace{0\cdots0}^{z} \mid w_1 \mid \cdots \mid w_{l'} \mid \overbrace{\bar{0}\cdots\bar{0}}^{z}}.$$

Our main result for $U'_q(C_n^{(1)})$ is the following.

Theorem 3.4. *Given $b_1 \otimes b_2 \in B_l \otimes B_k$, specify $b'_2 \otimes b'_1 \in B_k \otimes B_l$ and l', k', m by Rule 3.3. Let $\iota : B_l \otimes B_k \xrightarrow{\sim} B_k \otimes B_l$ be the isomorphism of a $U'_q(C_n^{(1)})$ crystal. Then we have*

$$\iota(b_1 \otimes b_2) = b'_2 \otimes b'_1,$$
$$H_{B_l B_k}(b_1 \otimes b_2) = \min(l', k') - m.$$

In Appendix B, we give an alternative algorithm equivalent to Rule 3.3, which is analogous to the type-A case (Rule 3.11 of [NY]). In practical calculations, it is often more efficient than Rule 3.3 based on the insertion algorithm.

Remark 3.5. Associated with the tableau $\hat{\mathbb{P}}(b_2 \to b_1) = (\hat{\mathbb{T}}(b_2) \longrightarrow \hat{\mathbb{T}}(b_1))$, we have the recording tableau $\hat{\mathbb{Q}}(b_2 \to b_1)$, as in the Robinson–Schensted–Knuth correspondence [F]. $\hat{\mathbb{Q}}(b_2 \to b_1)$ has a common shape with $\hat{\mathbb{P}}(b_2 \to b_1)$, and its entries are the consecutive integers from 1 to $l'+k'$. By considering a C-analogue of the column bumping lemma [F, p. 187], it is easy to see that the integers from $l'+1$ to $l'+m$ are in the second row. With $\hat{\mathbb{Q}}(b_2 \to b_1)$, we can reverse the column insertion procedure and recover $\hat{\mathbb{T}}(b_1)$ and $\hat{\mathbb{T}}(b_2)$ from $\hat{\mathbb{P}}(b_2 \to b_1)$. The recording tableau $\hat{\mathbb{Q}}(b'_1 \to b'_2)$ for the column insertion $(\hat{\mathbb{T}}(b'_1) \longrightarrow \hat{\mathbb{T}}(b'_2))$ is defined similarly. It has a common shape with $\hat{\mathbb{P}}(b_2 \to b_1)$, and its entries are the consecutive integers from 1 to $l' + k'$. (Integers from $k' + 1$ to $k' + m$ are in the second row.) In Rule 3.3, we have constructed $T^{(l')}$ and $\boxed{w_1 \cdots w_{l'}}$ from $\hat{\mathbb{P}}(b_2 \to b_1)$ with the help of the recording tableau $\hat{\mathbb{Q}}(b'_1 \to b'_2)$.

Example 3.6. Let us assume $n \ge 4$ and take $b_1 = (2, 0, 1, 1, 0, \ldots, 0, 1, 1, 1) \in B_9$ and $b_2 = (0, \ldots, 0, 3, 0, 0, 2) \in B_7$. Then we have $z = 1$, $l' = 7$, $k' = 5$, and

$$\hat{\mathbb{T}}(b_1) = 1134\bar{3}\bar{2}\bar{1}, \qquad \hat{\mathbb{T}}(b_2) = \bar{4}\bar{4}\bar{4}\bar{1}\bar{1}.$$

In this examplem we have

$$\hat{\mathbb{P}}(b_2 \to b_1) = \begin{matrix} 0034\bar{3}\bar{2}\bar{1} \\ \bar{4}\bar{4}\bar{4}0\bar{0} \end{matrix}, \; \hat{\mathbb{Q}}(b_2 \to b_1) = \begin{matrix} 1234567 \\ 890\acute{1}\acute{2} \end{matrix}, \; \hat{\mathbb{Q}}(b'_1 \to b'_2) = \begin{matrix} 12345\acute{1}\acute{2} \\ 6789\acute{0} \end{matrix}.$$

Here we have written 10, 11, 12 as $\acute{0}, \acute{1}, \acute{2}$. The column insertion $(\hat{\mathbb{T}}(b_2) \longrightarrow \hat{\mathbb{T}}(b_1))$ appears as

$$\begin{array}{l} 1134\bar{3}\bar{2}\bar{1} \\ \bar{1} \end{array}, \quad \begin{array}{l} 0134\bar{3}\bar{2}\bar{1} \\ \bar{1}\bar{0} \end{array}, \quad \begin{array}{l} 0134\bar{3}\bar{2}\bar{1} \\ \bar{4}\bar{1}\bar{0} \end{array}, \quad \begin{array}{l} 0034\bar{3}\bar{2}\bar{1} \\ \bar{4}\bar{4}\bar{0}\bar{0} \end{array}, \quad \begin{array}{l} 0034\bar{3}\bar{2}\bar{1} \\ \bar{4}\bar{4}\bar{4}\bar{0}\bar{0} \end{array}.$$

The reverse bumping according to the recording tableau $\hat{\mathbb{Q}}(b_1' \to b_2')$ appears as

$$\begin{array}{l} 0034\bar{3}\bar{2}\bar{1} \\ \bar{4}\bar{4}\bar{4}\bar{0}\bar{0} \end{array}, \quad \begin{array}{l} 034\bar{3}\bar{2}\bar{1} \\ \bar{4}\bar{4}\bar{4}\bar{0}\bar{0} \end{array}, \quad \begin{array}{l} 034\bar{2}\bar{1} \\ \bar{4}\bar{4}\bar{3}\bar{0}\bar{0} \end{array}, \quad \begin{array}{l} 044\bar{2}\bar{1} \\ \bar{4}\bar{4}\bar{0}\bar{0} \end{array}, \quad \begin{array}{l} 044\bar{2}\bar{1} \\ \bar{4}\bar{0}\bar{0} \end{array}, \quad \begin{array}{l} 144\bar{2}\bar{1} \\ \bar{1}\bar{0} \end{array}, \quad \begin{array}{l} 144\bar{2}\bar{1} \\ \bar{0} \end{array}.$$

Adding the $(\boxed{0}, \boxed{\bar{0}})$ pair $z = 1$ time, we get

$$\mathbb{T}(b_2') = 0144\bar{2}\bar{1}\bar{0}, \qquad \mathbb{T}(b_1') = 00\bar{4}\bar{4}\bar{4}\bar{1}\bar{0}\bar{0}.$$

Therefore, we obtain

$$b_1' = (0, \ldots, 0, 4, 0, 0, 1) \in B_9, \quad b_2' = (1, 0, 0, 2, 0, \ldots, 0, 1, 1) \in B_7.$$

4 Proof: $C_n^{(1)}$ case

We call an element b of a $U_q'(C_n^{(1)})$ crystal a $U_q(C_n)$ *highest element* if it satisfies $\tilde{e}_i b = 0$ for $i = 1, 2, \ldots, n$. Let $b_2' \otimes b_1' = \iota(b_1 \otimes b_2)$ under the isomorphism of $U_q'(C_n^{(1)})$ crystals $\iota : B_l \otimes B_k \stackrel{\sim}{\to} B_k \otimes B_l$. By definition, if $b_1 \otimes b_2$ is a $U_q(C_n)$ highest element, so is $b_2' \otimes b_1'$. In Section 4.1, we prove Proposition 4.1. This verifies Theorem 3.4 when $b_1 \otimes b_2$ is a $U_q(C_n)$ highest element and either $\mathbb{T}(b_1)$ or $\mathbb{T}(b_2)$ is free of $\boxed{0}$. In Section 4.2, we prove that if both $\mathbb{T}(b_1)$ and $\mathbb{T}(b_2)$ have at least one $\boxed{0}$, then the combinatorial R on $B_l \otimes B_k$ is reduced to the combinatorial R on $B_{l-2} \otimes B_{k-2}$ by removing a $(\boxed{0}, \boxed{\bar{0}})$ pair. In Section 4.3, we quote a proposition [B] that assures the compatibility of the column insertion algorithm with a $U_q(C_n)$ crystal isomorphism. Based on these preparations, we complete the proof of Theorem 3.4 for general elements in Section 4.4.

4.1 Combinatorial R for a class of highest elements.

Proposition 4.1. *Given $b_1 \otimes b_2 \in B_l \otimes B_k$, let $b_2' \otimes b_1' = \iota(b_1 \otimes b_2) \in B_k \otimes B_l$ be the image under the isomorphism. Suppose that $b_1 \otimes b_2$ is a $U_q(C_n)$ highest element and $\mathbb{T}(b_1)$ or $\mathbb{T}(b_2)$ has no $\boxed{0}$. Then $\mathbb{T}(b_2')$ or $\mathbb{T}(b_1')$ also has no $\boxed{0}$, and their column insertions give a common tableau:*

$$(\mathbb{T}(b_2) \longrightarrow \mathbb{T}(b_1)) = \left(\mathbb{T}(b_1') \longrightarrow \mathbb{T}(b_2')\right). \tag{4.1}$$

The value of the energy function is given by

$$H(b_1 \otimes b_2) = (\mathbb{T}(b_2) \longrightarrow \mathbb{T}(b_1))_1 - \max(l, k).$$

We give a proof of Proposition 4.1 by a case checking in Appendix A. In this subsection, we only list all the $U_q(C_n)$ highest elements of the above type. We also list the values of their energy functions. Let $(x_1, x_2, —, \overline{x}_1)$ stand for $(x_1, x_2, 0, \ldots, 0, \overline{x}_1)$.

Lemma 4.2. *We have*

$$\iota : (l, 0, —, 0) \otimes (k, 0, —, 0) \mapsto (k, 0, —, 0) \otimes (l, 0, —, 0)$$

under the isomorphism $\iota : B_l \otimes B_k \xrightarrow{\sim} B_k \otimes B_l$.

Proof. They are the unique elements in $B_l \otimes B_k$ and $B_k \otimes B_l$, respectively, that do not vanish when $(\tilde{e}_0)^{l+k}$ is applied. (They are also the unique elements such that $\varphi_i - \varepsilon_i = (l+k)\delta_{i,1}$.) □

Lemma 4.3. *Let* $b_1 \otimes b_2 \in B_l \otimes B_k$ $(l \geq k)$. *Suppose that* $b_1 \otimes b_2$ *is a* $U_q(C_n)$ *highest element and that* $\mathbb{T}(b_1)$ *or* $\mathbb{T}(b_2)$ *has no* $\boxed{0}$. *Then it has either the form*

$$(l, 0, —, 0) \otimes (x_1, x_2, —, \overline{x}_1)$$

with $x_1, x_2, \overline{x}_1 \in \mathbb{Z}_{\geq 0}$ *and* $x_1 + x_2 + \overline{x}_1 \leq k$ *or the form*

$$(l - 2y_0, 0, —, 0) \otimes (x_1, x_2, —, k - x_1 - x_2)$$

with $y_0 (\neq 0), x_1, x_2 \in \mathbb{Z}_{\geq 0}$, $l - k \geq 2y_0 - x_1$, *and* $x_1 + x_2 \leq k$.

We call the former a *type* I and the latter a *type* II $U_q(C_n)$ highest element. They are exclusive.

Proof. Let $b_1 = (y_1, \ldots, y_n, \overline{y}_n, \ldots, \overline{y}_1)$ and $b_2 = (x_1, \ldots, x_n, \overline{x}_n, \ldots, \overline{x}_1)$. Since $b_1 \otimes b_2$ is a $U_q(C_n)$ highest element, $\varepsilon_i(b_1 \otimes b_2) = \max(\varepsilon_i(b_1), \varepsilon_i(b_1) + \varepsilon_i(b_2) - \varphi_i(b_1)) = 0$ for $i = 1, \ldots, n$. This means that $\varepsilon_i(b_1) = 0$ and $\varphi_i(b_1) \geq \varepsilon_i(b_2)$ for $i = 1, \ldots, n$. Thus we have $\varepsilon_n(b_1) = \overline{y}_n = 0$, and then $\varepsilon_{n-1}(b_1) = \overline{y}_{n-1} + (y_n - \overline{y}_n)_+ = 0$, i.e., $\overline{y}_{n-1} = y_n = 0$. Repeating the same process, we come to $\varepsilon_1(b_1) = \overline{y}_1 + (y_2 - \overline{y}_2)_+ = 0$, i.e., $\overline{y}_1 = y_2 = 0$. Thus $b_1 = (l - 2y_0, 0, —, 0)$ and $\varphi_i(b_1) = (l - 2y_0)\delta_{i,1}$. Thus we have $\varepsilon_n(b_2) = \overline{x}_n = 0$, and then $\varepsilon_{n-1}(b_2) = \overline{x}_{n-1} + (x_n - \overline{x}_n)_+ = 0$, i.e., $\overline{x}_{n-1} = x_n = 0$. Repeating the same process, we come to $\varepsilon_2(b_2) = \overline{x}_2 + (x_3 - \overline{x}_3)_+ = 0$, i.e., $\overline{x}_2 = x_3 = 0$. Thus $b_2 = (x_1, x_2, —, \overline{x}_1)$ and $\varepsilon_1(b_2) = x_2 + \overline{x}_1$. Therefore, we have a condition $\varphi_1(b_1) = l - 2y_0 \geq x_2 + \overline{x}_1$. If $\mathbb{T}(b_1)$ has no $\boxed{0}$, then $y_0 = 0$ and this condition certainly holds. If $\mathbb{T}(b_2)$ has no $\boxed{0}$, then $x_2 + \overline{x}_1 = k - x_1$; thus we impose the condition $l - k \geq 2y_0 - x_1$. □

Lemma 4.4. *Under the isomorphism of* $U'_q(C_n^{(1)})$ *crystals*

$$\iota : B_l \otimes B_k \xrightarrow{\sim} B_k \otimes B_l \quad (l \geq k),$$

the type I $U_q(C_n)$ *highest element is mapped as*

$$\begin{aligned}&(l, 0, -, 0) \otimes (x_1, x_2, -, \overline{x}_1)\\ &\quad \mapsto (k, 0, -, 0) \otimes (x_1 + l - k - y, x_2, -, \overline{x}_1 - y),\end{aligned}$$

where $y = \min[l - k, (\overline{x}_1 - x_1)_+]$. *The value of the energy function for this element is* $x_0 + (x_1 - \overline{x}_1)_+$ *with* $x_0 = (k - x_1 - x_2 - \overline{x}_1)/2$.

PROOF. For a set of operators $\mathcal{O}_1, \mathcal{O}_2, \dots$ on the $U_q'(C_n^{(1)})$ crystals, we define $\prod_i^{2\swarrow\nwarrow 1} \mathcal{O}_i$ by $\mathcal{O}_2\mathcal{O}_1$ for $n = 2$, $\mathcal{O}_2\mathcal{O}_3\mathcal{O}_2\mathcal{O}_1$ for $n = 3$, $\mathcal{O}_2\mathcal{O}_3\mathcal{O}_4\mathcal{O}_3\mathcal{O}_2\mathcal{O}_1$ for $n = 4$, etc. Let $l = 2m$ or $l = 2m - 1$. The lemma can be proved by applying the sequence of operators

$$\tilde{f}_0^{m+x_0+(x_1-\overline{x}_1)_+} \tilde{e}_1^{\min(x_1,\overline{x}_1)} \left(\prod_i^{2\swarrow\nwarrow 1} \tilde{e}_i^{x_2+\min(x_1,\overline{x}_1)} \right) \tilde{e}_0^{k+m} \tag{4.2}$$

to both sides of Lemma 4.2.

In what follows, we will show

$$\begin{aligned}&H((l, 0, -, 0) \otimes (x_1, x_2, -, \overline{x}_1))\\ &\quad = H((l, 0, -, 0) \otimes (k, 0, -, 0)) - k + x_0 + (x_1 - \overline{x}_1)_+.\end{aligned}$$

In the case where $l = 2m$ and $m \geq k$, the value of the energy function was lowered by k when the first to the kth $\tilde{e}_0$s were applied and raised by $x_0 + (x_1 - \overline{x}_1)_+$ when the $(m + 1)$st to the last $\tilde{f}_0$s were applied. In the case where $l = 2m$ and $m < k$, in addition to the same change as in the previous case, the value of the energy function was raised by $k - m$ when the $(2m + 1)$st to the last $\tilde{e}_0$s were applied and lowered by the same amount when the first to the $(k - m)$th $\tilde{f}_0$s were applied.

In the case where $l = 2m - 1$ and $m - 1 > k$, the value of the energy function was lowered by k when the first to the kth $\tilde{e}_0$s were applied and raised by $x_0 + (x_1 - \overline{x}_1)_+$ when the $(m + 1)$st to the last $\tilde{f}_0$s were applied. In the case where $l = 2m - 1$ and $m - 1 \leq k$, in addition to the same change as the previous case, the value of the energy function was raised by $k - m + 1$ when the $2m$th to the last $\tilde{e}_0$s were applied and lowered by the same amount when the first to the $(k - m + 1)$st $\tilde{f}_0$s were applied.

Recall that we have normalized the energy function as $H_{B_l B_k}((l, 0, -, 0) \otimes (0, k, -, 0)) = 0$. Thus we have $H((l, 0, -, 0) \otimes (x_1, x_2, -, \overline{x}_1)) = x_0 + (x_1 - \overline{x}_1)_+$. □

Corollary 4.5. *For any* $l, k \in \mathbb{Z}_{\geq 0}$, *we have*

$$H_{B_l B_k}((l, 0, -, 0) \otimes (k, 0, -, 0)) = \min(l, k).$$

Lemma 4.6. *Under the isomorphism of* $U_q'(C_n^{(1)})$ *crystals*

$$\iota : B_l \otimes B_k \xrightarrow{\sim} B_k \otimes B_l \quad (l \geq k),$$

the type II $U_q(C_n)$ *highest element is mapped as*

$$b_1 \otimes b_2 := (l - 2y_0, 0, —, 0) \otimes (x_1, x_2, —, k - x_1 - x_2) \mapsto b_2' \otimes b_1',$$

where $b_2' \otimes b_1'$ *is given by the following two condidions. Let* $\overline{x}_1 = k - x_1 - x_2$.

1. *If* $l - k > y_0 \geq (x_1 - \overline{x}_1)_+$,

$$b_2' \otimes b_1' = (k, 0, —, 0) \otimes (x_1 + l - k - y_0 - z, x_2, —, \overline{x}_1 + y_0 - z),$$

where $z = \min[y_0 + \overline{x}_1 - x_1, l - k - y_0]$. $H(b_1 \otimes b_2) = 0$.

2. *If* $l - k \leq y_0$ *or* $y_0 < (x_1 - \overline{x}_1)_+$,

$$b_2' \otimes b_1' = (k - 2y_0 + 2w, 0, —, 0) \otimes (x_1 + l - k - w, x_2, —, \overline{x}_1 + w),$$

where $w = \min[l - k, (2y_0 - x_1 + \overline{x}_1)_+]$. $H(b_1 \otimes b_2) = \max[y_0 - l + k, x_1 - \overline{x}_1 - y_0]$.

PROOF. If $y_0 \geq (x_1 - \overline{x}_1)_+$, let $l = 2m$ or $l = 2m - 1$. The lemma can be proved by applying the following sequence of operators

$$\tilde{f}_0^{m-y_0} \tilde{e}_1^{x_1} \left(\prod_i^{2 \swarrow \nwarrow 1} \tilde{e}_i^{x_2 + x_1} \right) \tilde{e}_0^{k+m} \tag{4.3}$$

to both sides of Lemma 4.2. In the case where $l = 2m$ and $m \geq k$, the value of the energy function was lowered by k when the first to the kth $\tilde{e}_0$s were applied. In the where case $l = 2m$, $m < k$, and $2m - k > y_0$ (resp., $2m - k \leq y_0$), in addition to the same change in the previous case, the value of the energy function was raised by $k - m$ when the $(2m + 1)$st to the last $\tilde{e}_0$s were applied and lowered by $k - m$ (resp., $m - y_0$) when the first to the $(k - m)$th (resp., the last) $\tilde{f}_0$s were applied. In the case where $l = 2m - 1$ and $m - 1 > k$, the value of the energy function was lowered by k when the first to the kth $\tilde{e}_0$s were applied. In the case where $l = 2m - 1$, $m - 1 \leq k$, and $2m - 1 - k > y_0$ (resp., $2m - 1 - k \leq y_0$), in addition to the same change in the previous case, the value of the energy function was raised by $k - m + 1$ when the $2m$th to the last $\tilde{e}_0$s were applied and lowered by $k - m + 1$ (resp., $m - y_0$) when the first to the $(k - m + 1)$st (resp., the last) $\tilde{f}_0$s were applied.

If $y_0 < (x_1 - \overline{x}_1)_+ \leq 2y_0$, one can check that $\tilde{e}_0^{x_1 - \overline{x}_1 - y_0}(b_1 \otimes b_2) = b_1 \otimes \tilde{e}_0^{x_1 - \overline{x}_1 - y_0} b_2$. Lemma 4.7 and the previous case of the present lemma enable us to obtain its image under the map ι. They also tell us that now the value of the energy function is equal to $(2y_0 - x_1 + \overline{x}_1 + k - l)_+$. Then apply $\tilde{f}_0^{x_1 - \overline{x}_1 - y_0}$. Since it again turns out to hit the right component of the tensor product, the value of the energy function is raised by $x_1 - \overline{x}_1 - y_0$.

If $2y_0 < (x_1 - \overline{x}_1)_+$, one can check that $\tilde{e}_0^{x_1 - \overline{x}_1 - y_0}(b_1 \otimes b_2) = b_1 \otimes \tilde{e}_0^{x_1 - \overline{x}_1 - y_0} b_2$. Lemma 4.7 and 4.4 enable us to obtain its image under the map ι. They also tell us that now the value of the energy function is equal to 0. Then apply $\tilde{f}_0^{x_1 - \overline{x}_1 - y_0}$. Since it again hits the right component of the tensor product, the value of the energy function is raised by $x_1 - \overline{x}_1 - y_0$. □

4.2 Relation of R on $B_l \otimes B_k$ and $B_{l-2} \otimes B_{k-2}$. Let $l \geq 3$. For any $b = (x_1, \dots, \bar{x}_1) \in B_{l-2}$, we define $\tau^l_{l-2}(b) \in B_l$ to be the unique element such that the tableau $\mathbb{T}(\tau^l_{l-2}(b))$ is made from the tableau $\mathbb{T}(b)$ by adding a $(\boxed{0}, \boxed{\bar{0}})$ pair. Note that $\tau^l_{l-2}(b)$ also has the same presentation $(x_1, \dots, \bar{x}_1)$ in B_l. The map $\tau^l_{l-2} : B_{l-2} \to B_l$ is injective and has the following property:

$$\begin{aligned} \tilde{f}_i \tau^l_{l-2}(b) &= \tau^l_{l-2}(\tilde{f}_i b) \quad (0 \leq i \leq n) \qquad \text{if } \tilde{f}_i b \neq 0, \\ \tilde{e}_i \tau^l_{l-2}(b) &= \tau^l_{l-2}(\tilde{e}_i b) \quad (0 \leq i \leq n) \qquad \text{if } \tilde{e}_i b \neq 0. \end{aligned} \tag{4.4}$$

Lemma 4.7. *We have $\tau^l_{l-2}(b_1) \otimes \tau^k_{k-2}(b_2) \simeq \tau^k_{k-2}(b'_2) \otimes \tau^l_{l-2}(b'_1)$ under the isomorphism $B_l \otimes B_k \simeq B_k \otimes B_l$ if and only if $b_1 \otimes b_2 \simeq b'_2 \otimes b'_1$ under $B_{l-2} \otimes B_{k-2} \simeq B_{k-2} \otimes B_{l-2}$. We also have $H_{B_l B_k}(\tau^l_{l-2}(b_1) \otimes \tau^k_{k-2}(b_2)) = H_{B_{l-2} B_{k-2}}(b_1 \otimes b_2)$.*

Proof. Since τ^l_{l-2} and τ^k_{k-2} are injective, the *only if* part of the statement follows immediately after the *if* part is proved. Without loss of generality, we assume $l \geq k$. Set

$$b^{(l)} = (l, 0, \text{—}, 0) \in B_l.$$

First, consider the case where $b_1 = b^{(l-2)} \in B_{l-2}$ and $b_2 = b^{(k-2)} \in B_{k-2}$. Then $b'_2 = b^{(k-2)}$, $b'_1 = b^{(l-2)}$, and $H_{B_{l-2} B_{k-2}}(b^{(l-2)} \otimes b^{(k-2)}) = k - 2$ by Corollary 4.5. On the other hand, we have

$$\begin{aligned} \psi\left(b^{(l)} \otimes b^{(k)}\right) &= \tau^l_{l-2}(b^{(l-2)}) \otimes \tau^k_{k-2}(b^{(k-2)}) \in B_l \otimes B_k, \\ \psi\left(b^{(k)} \otimes b^{(l)}\right) &= \tau^k_{k-2}(b^{(k-2)}) \otimes \tau^l_{l-2}(b^{(l-2)}) \in B_k \otimes B_l, \end{aligned}$$

where

$$\begin{aligned} \psi = {} & \tilde{e}_0 (\tilde{e}_1)^{l+k-2} (\tilde{e}_2)^{l+k-2} \cdots (\tilde{e}_{n-1})^{l+k-2} (\tilde{e}_n)^{l+k-2} \\ & \times (\tilde{e}_{n-1})^{l+k-2} \cdots (\tilde{e}_2)^{l+k-2} (\tilde{e}_1)^{l+k-2} (\tilde{e}_0)^{l+k-1}. \end{aligned}$$

By Lemma 4.2, one has

$$\tau^l_{l-2}(b^{(l-2)}) \otimes \tau^k_{k-2}(b^{(k-2)}) \simeq \tau^k_{k-2}(b^{(k-2)}) \otimes \tau^l_{l-2}(b^{(l-2)}) \tag{4.5}$$

under the isomorphism $B_l \otimes B_k \simeq B_k \otimes B_l$. The energy was lowered by k when the first to the kth $\tilde{e}_0$s were applied and raised by $k - 1$ when the $(l + 1)$st to the $(l + k - 1)$st $\tilde{e}_0$s were applied. Then it was lowered by 1 when the leftmost $\tilde{e}_0$ was applied. Thus we have

$$\begin{aligned} H_{B_l B_k}(\tau^l_{l-2}(b^{(l-2)}) \otimes \tau^k_{k-2}(b^{(k-2)})) &= H_{B_l B_k}(b^{(l)} \otimes b^{(k)}) - 2 = k - 2 \\ &= H_{B_{l-2} B_{k-2}}(b^{(l-2)} \otimes b^{(k-2)}). \end{aligned} \tag{4.6}$$

The proof is finished in this special case from Corollary 4.5.

Now we consider the general elements $b_1 \otimes b_2 \in B_{l-2} \otimes B_{k-2}$ and $b_2' \otimes b_1' \in B_{k-2} \otimes B_{l-2}$ that are mapped to each other under the isomorphism. Take any finite sequence ψ' made of $\tilde{e}_i$s and $\tilde{f}_i$s $(i = 0, 1, \ldots, n)$ such that

$$b_1 \otimes b_2 = \psi'(b^{(l-2)} \otimes b^{(k-2)}), \tag{4.7}$$

which is equivalent to

$$b_2' \otimes b_1' = \psi'(b^{(k-2)} \otimes b^{(l-2)}). \tag{4.8}$$

For any operator in ψ', the rules (2.1)–(2.2) determine whether it should hit the left or the right component of the tensor product. For any $c_1 \otimes c_2 \in B_{l-2} \otimes B_{k-2}$, we have $\varphi_i(\tau^l_{l-2}(c_1)) = \varphi_i(c_1) + \delta_{i,0}$ and $\varepsilon_i(\tau^k_{k-2}(c_2)) = \varepsilon_i(c_2) + \delta_{i,0}$ from (2.8). Thus the alternatives in (2.1)–(2.2) are not changed by $\tau^l_{l-2} \otimes \tau^k_{k-2}$. From (4.4), it follows that $(\tau^l_{l-2} \otimes \tau^k_{k-2})(\psi'(c_1 \otimes c_2)) = \psi'(\tau^l_{l-2}(c_1) \otimes \tau^k_{k-2}(c_2))$. Applying $\tau^l_{l-2} \otimes \tau^k_{k-2}$ (resp., $\tau^k_{k-2} \otimes \tau^l_{l-2}$) to (4.7) (resp., (4.8)), we thus get

$$\tau^l_{l-2}(b_1) \otimes \tau^k_{k-2}(b_2) = \psi'(\tau^l_{l-2}(b^{(l-2)}) \otimes \tau^k_{k-2}(b^{(k-2)})), \tag{4.9}$$

$$\tau^k_{k-2}(b_2') \otimes \tau^l_{l-2}(b_1') = \psi'(\tau^k_{k-2}(b^{(k-2)}) \otimes \tau^l_{l-2}(b^{(l-2)})). \tag{4.10}$$

From (4.5) it follows that

$$\tau^l_{l-2}(b_1) \otimes \tau^k_{k-2}(b_2) \simeq \tau^k_{k-2}(b_2') \otimes \tau^l_{l-2}(b_1')$$

under the isomorphism $B_l \otimes B_k \simeq B_k \otimes B_l$. When comparing (4.7) and (4.9), change of the value of the energy function caused by ψ' is not affected by $\tau^l_{l-2} \otimes \tau^k_{k-2}$. Therefore, from (4.6), we have $H_{B_l B_k}(\tau^l_{l-2}(b_1) \otimes \tau^k_{k-2}(b_2)) = H_{B_{l-2} B_{k-2}}(b_1 \otimes b_2)$. □

4.3 Column insertion and $U_q(C_n)$ crystal morphism. The next proposition is due to Baker [B, Proposition 7.1]. For a dominant integral weight λ of the C_n root system, let $B(\lambda)$ be the $U_q(C_n)$ crystal associated with the irreducible highest-weight representation $V(\lambda)$ [KN]. The elements of $B(\lambda)$ can be represented by the semistandard C-tableaux of shape λ [KN].

Proposition 4.8. *Let $B(\mu) \otimes B(\nu) \simeq \bigoplus_j B(\lambda_j)^{\oplus m_j}$ be the tensor product decomposition of crystals. Here λ_js are distinct highest weights and $m_j (\geq 1)$ is the multiplicity of $B(\lambda_j)$. Forgetting the multiplicities, we have the canonical morphism from $B(\mu) \otimes B(\nu)$ to $\bigoplus_j B(\lambda_j)$. Define ψ_C by*

$$\psi_C(b_1 \otimes b_2) = \left(b_2 \stackrel{*}{\longrightarrow} b_1\right).$$

Then ψ_C gives the unique crystal morphism from $B(\mu) \otimes B(\nu)$ to $\bigoplus_j B(\lambda_j)$.

Here $b_2 \stackrel{*}{\longrightarrow} b_1$ is the tableau obtained from successive column insertions of letters of the Japanese order of word reading of b_2 into b_1 by the original definition in [B]. (In [B], $b_2 \stackrel{*}{\longrightarrow} b_1$ is denoted by $b_1 * b_2$.)

Remark 4.9. The insertion $b_2 \stackrel{*}{\longrightarrow} b_1$ may include such a process that $\boxed{x}$ and $\boxed{\overline{x}}$ annihilate pairwise and an empty box thereby produced slides out. In [B], this process was called a *bumping–sliding transition*. Consider the case where both b_1 and b_2 are one-row tableaux. (We shall omit the symbol $\mathbb{T}$ here.) In this case the bumping–sliding transition can occur only when $\boxed{\overline{1}}$ is inserted into $\boxed{1}$. We defined our column insertion ($\longrightarrow$) so that $\varsigma((b_2 \longrightarrow b_1))$ is equivalent to $(\varsigma(b_2) \stackrel{*}{\longrightarrow} \varsigma(b_1))$. In Rule 3.3, for a $U_q'(C_n^{(1)})$ combinatorial R matrix, we have removed $(\boxed{0}, \boxed{\overline{0}})$ pairs beforehand, which become $(\boxed{1}, \boxed{\overline{1}})$ under ς. Thus we have avoided the occurrence of a bumping–sliding transition.

4.4 Proof of Theorem 3.4. With no loss of generality, we assume $l \geq k$. Let $b_2' \otimes b_1'$ be the image of $b_1 \otimes b_2$ under the isomorphism of $U_q'(C_n^{(1)})$ crystals ι : $B_l \otimes B_k \stackrel{\sim}{\rightarrow} B_k \otimes B_l$. In order to prove Theorem 3.4 we are to show the claims:

1. Let
$$\begin{aligned} z_0 &= \min(\sharp\boxed{0} \text{ in } \mathbb{T}(b_1), \sharp\boxed{0} \text{ in } \mathbb{T}(b_2)), \\ z_0' &= \min(\sharp\boxed{0} \text{ in } \mathbb{T}(b_1'), \sharp\boxed{0} \text{ in } \mathbb{T}(b_2')). \end{aligned}$$
Then $z_0' = z_0$.

2. Remove $(\boxed{0}, \boxed{\overline{0}})$ pairs z_0 times from $\mathbb{T}(b_1)$, $\mathbb{T}(b_2)$, $\mathbb{T}(b_1')$ and $\mathbb{T}(b_2')$. Call the resulting tableaux $\hat{\mathbb{T}}(b_1)$, $\hat{\mathbb{T}}(b_2)$, $\hat{\mathbb{T}}(b_1')$ and $\hat{\mathbb{T}}(b_2')$, respectively. Then we have
$$\left(\hat{\mathbb{T}}(b_2) \longrightarrow \hat{\mathbb{T}}(b_1)\right) = \left(\hat{\mathbb{T}}(b_1') \longrightarrow \hat{\mathbb{T}}(b_2')\right). \tag{4.11}$$

3. $H_{B_l B_k}(b_1 \otimes b_2) = (\hat{\mathbb{T}}(b_2) \longrightarrow \hat{\mathbb{T}}(b_1))_1 - \hat{\mathbb{T}}(b_1)_1$.

Proof. Thanks to Lemma 4.7, it suffices to verify the above claims only when $\mathbb{T}(b_1)$ or $\mathbb{T}(b_2)$ has no $\boxed{0}$. Such a case can be reduced to Proposition 4.1 by the argument as follows.

Let $b_1 \otimes b_2$ be an element of $B_l \otimes B_k$ that is not necessarily a $U_q(C_n)$ highest element and let either $\mathbb{T}(b_1)$ or $\mathbb{T}(b_2)$ be free of $\boxed{0}$. Let $b_2' \otimes b_1' = \iota(b_1 \otimes b_2)$ under the isomorphism $\iota : B_l \otimes B_k \rightarrow B_k \otimes B_l$. There exists a sequence $i_1, i_2, \dots, i_s$ $(1 \leq i_\alpha \leq n,\ \alpha = 1, 2, \dots, s)$ such that $\dot{b}_1 \otimes \dot{b}_2 := \tilde{e}_{i_s} \cdots \tilde{e}_{i_1}(b_1 \otimes b_2)$ is a $U_q(C_n)$ highest element. Then we have $b_2' \otimes b_1' = \tilde{f}_{i_1} \cdots \tilde{f}_{i_s} \circ \iota \circ \tilde{e}_{i_s} \cdots \tilde{e}_{i_1}(b_1 \otimes b_2)$. Let $\dot{b}_2' \otimes \dot{b}_1' = \iota(\dot{b}_1 \otimes \dot{b}_2)$. Since $\tilde{e}_i$ $(1 \leq i \leq n)$ does not change $\sharp\boxed{0}$, $\mathbb{T}(\dot{b}_1)$ or $\mathbb{T}(\dot{b}_2)$ also has no $\boxed{0}$. Therefore from Proposition 4.1 we have
$$\left(\mathbb{T}(\dot{b}_2) \longrightarrow \mathbb{T}(\dot{b}_1)\right) = \left(\mathbb{T}(\dot{b}_1') \longrightarrow \mathbb{T}(\dot{b}_2')\right), \tag{4.12}$$

where $\mathbb{T}(\dot{b}'_1)$ or $\mathbb{T}(\dot{b}'_2)$ has no $\boxed{0}$. Since $b'_2 \otimes b'_1 = \tilde{f}_{i_1} \cdots \tilde{f}_{i_s}(\dot{b}'_2 \otimes \dot{b}'_1)$ and $1 \le i_\alpha \le n$, we conclude that $\mathbb{T}(b'_1)$ or $\mathbb{T}(b'_2)$ has no $\boxed{0}$. Thus claim 1 is indeed valid as $z_0 = z'_0 = 0$. By Remark 3.2, (4.12) is equivalent to

$$\left(\varsigma(\mathbb{T}(\dot{b}_2)) \xrightarrow{*} \varsigma(\mathbb{T}(\dot{b}_1))\right) = \left(\varsigma(\mathbb{T}(\dot{b}'_1)) \xrightarrow{*} \varsigma(\mathbb{T}(\dot{b}'_2))\right) =: \mathbb{P}. \tag{4.13}$$

Here ς is defined in (2.6). Regarding $\mathbb{P}$ as an element of a $U_q(C_{n+1})$ crystal, we apply Proposition 4.8 to get

$$\left(\varsigma(\mathbb{T}(b_2)) \xrightarrow{*} \varsigma(\mathbb{T}(b_1))\right) = \tilde{f}_{i_1+1} \cdots \tilde{f}_{i_s+1}(\mathbb{P}) = \left(\varsigma(\mathbb{T}(b'_1)) \xrightarrow{*} \varsigma(\mathbb{T}(b'_2))\right), \tag{4.14}$$

which is equivalent to

$$(\mathbb{T}(b_2) \longrightarrow \mathbb{T}(b_1)) = \left(\mathbb{T}(b'_1) \longrightarrow \mathbb{T}(b'_2)\right), \tag{4.15}$$

proving claim 2. Since $\tilde{f}_i$ $(1 \le i \le n)$ does not change the shape of the tableaux [B] and $H(b_1 \otimes b_2) = H(\dot{b}_1 \otimes \dot{b}_2)$, claim 3 follows from Proposition 4.1. □

5 $U'_q(A^{(2)}_{2n-1})$ crystal case

5.1 Definitions. Given a nonnegative integer l, let us denote by B_l the $U'_q(A^{(2)}_{2n-1})$ crystal defined in [KKM]. (Their B_l is identical to our B_l.) As a set, B_l reads

$$B_l = \left\{ (x_1, \ldots, x_n, \overline{x}_n, \ldots, \overline{x}_1) \;\middle|\; x_i, \overline{x}_i \in \mathbb{Z}_{\ge 0}, \sum_{i=1}^{n} (x_i + \overline{x}_i) = l \right\}.$$

B_l is isomorphic to $B(l\Lambda_1)$ as a crystal for $U_q(C_n)$. The crystal structure is given by

$$\begin{aligned}
\tilde{e}_0 b &= \begin{cases} (x_1, x_2 - 1, \ldots, \overline{x}_2, \overline{x}_1 + 1) & \text{if } x_2 > \overline{x}_2, \\ (x_1 - 1, x_2, \ldots, \overline{x}_2 + 1, \overline{x}_1) & \text{if } x_2 \le \overline{x}_2, \end{cases} \\
\tilde{e}_n b &= (x_1, \ldots, x_n + 1, \overline{x}_n - 1, \ldots, \overline{x}_1), \\
\tilde{e}_i b &= \begin{cases} (x_1, \ldots, x_i + 1, x_{i+1} - 1, \ldots, \overline{x}_1) & \text{if } x_{i+1} > \overline{x}_{i+1}, \\ (x_1, \ldots, \overline{x}_{i+1} + 1, \overline{x}_i - 1, \ldots, \overline{x}_1) & \text{if } x_{i+1} \le \overline{x}_{i+1}, \end{cases} \\
\tilde{f}_0 b &= \begin{cases} (x_1, x_2 + 1, \ldots, \overline{x}_2, \overline{x}_1 - 1) & \text{if } x_2 \ge \overline{x}_2, \\ (x_1 + 1, x_2, \ldots, \overline{x}_2 - 1, \overline{x}_1) & \text{if } x_2 < \overline{x}_2, \end{cases} \\
\tilde{f}_n b &= (x_1, \ldots, x_n - 1, \overline{x}_n + 1, \ldots, \overline{x}_1), \\
\tilde{f}_i b &= \begin{cases} (x_1, \ldots, x_i - 1, x_{i+1} + 1, \ldots, \overline{x}_1) & \text{if } x_{i+1} \ge \overline{x}_{i+1}, \\ (x_1, \ldots, \overline{x}_{i+1} - 1, \overline{x}_i + 1, \ldots, \overline{x}_1) & \text{if } x_{i+1} < \overline{x}_{i+1}, \end{cases}
\end{aligned} \tag{5.1}$$

where $b = (x_1, \ldots, x_n, \overline{x}_n, \ldots, \overline{x}_1)$ and $i = 1, \ldots, n-1$. For this b, we have

$$\begin{aligned} \varphi_0(b) &= \overline{x}_1 + (\overline{x}_2 - x_2)_+, \qquad \varepsilon_0(b) = x_1 + (x_2 - \overline{x}_2)_+, \\ \varphi_i(b) &= x_i + (\overline{x}_{i+1} - x_{i+1})_+ \quad \text{for } i = 1, \ldots, n-1, \\ \varepsilon_i(b) &= \overline{x}_i + (x_{i+1} - \overline{x}_{i+1})_+ \quad \text{for } i = 1, \ldots, n-1, \\ \varphi_n(b) &= x_n, \quad \varepsilon_n(b) = \overline{x}_n. \end{aligned} \tag{5.2}$$

We shall depict the element $b = (x_1, \ldots, x_n, \overline{x}_n, \ldots, \overline{x}_1) \in B_l$ with the tableau

$$\mathcal{T}(b) = \boxed{\overbrace{1 \cdots 1}^{x_1}} \cdots \boxed{\overbrace{n \cdots n}^{x_n}}\boxed{\overbrace{\overline{n} \cdots \overline{n}}^{\overline{x}_n}} \cdots \boxed{\overbrace{\overline{1} \cdots \overline{1}}^{\overline{x}_1}}. \tag{5.3}$$

The length of this one-row tableau is equal to l, namely, $\sum_{i=1}^{n}(x_i + \overline{x}_i) = l$.

In this section, we normalize the energy function as

$$H_{B_l B_k}((l, 0, \text{—}, 0) \otimes (0, 0, \text{—}, k)) = 0, \tag{5.4}$$

irrespective of $l < k$ or $l \geq k$.

5.2 Main theorem: $A^{(2)}_{2n-1}$ case. The insertion symbol $\longrightarrow$ and the reverse bumping in this section are the same ones as in Section 3.

Given $b_1 \otimes b_2 \in B_l \otimes B_k$, we define the element $b'_2 \otimes b'_1 \in B_k \otimes B_l$ and $l', k', m \in \mathbb{Z}_{\geq 0}$ by the following rule.

Rule 5.1. Set $z = \min(\sharp\boxed{1} \text{ in } \mathcal{T}(b_1), \sharp\boxed{\overline{1}} \text{ in } \mathcal{T}(b_2))$. Remove $\boxed{1}$s (resp., $\boxed{\overline{1}}$s) from $\mathcal{T}(b_1)$ (resp., $\mathcal{T}(b_2)$) z times and call the resulting tableaux $\acute{\mathcal{T}}(b_1)$ (resp., $\grave{\mathcal{T}}(b_2)$). Let $l' = \acute{\mathcal{T}}(b_1)_1 = l - z$ and $k' = \grave{\mathcal{T}}(b_2)_1 = k - z$. Operate the column insertion and set $\mathbb{P}(b_2 \stackrel{*}{\to} b_1) = (\grave{\mathcal{T}}(b_2) \longrightarrow \acute{\mathcal{T}}(b_1))$. (This $\mathbb{P}(b_2 \stackrel{*}{\to} b_1)$ coincides with the column insertion $\mathcal{T}(b_2) \stackrel{*}{\longrightarrow} \mathcal{T}(b_1)$, because of $(\boxed{\overline{1}} \stackrel{*}{\longrightarrow} \boxed{1}) = \emptyset$.) $\mathbb{P}(b_2 \stackrel{*}{\to} b_1)$ has the form

$$\begin{array}{|c|c|} \hline j_1 \cdots\cdots j_{k'} & i_{m+1} \cdots i_{l'} \\ \hline i_1 \cdots i_m & \\ \cline{1-1} \end{array},$$

where m is the length of the second row; hence that of the first row is $l' + k' - m$ $(0 \leq m \leq k')$.

Next, we bump out l' letters from the tableau $T^{(0)} = \mathbb{P}(b_2 \stackrel{*}{\to} b_1)$ by the reverse bumping algorithm. For the boxes containing $i_{l'}, i_{l'-1}, \ldots, i_1$ in the above tableau, we do this first for $i_{l'}$, then $i_{l'-1}$, and so on. Correspondingly, let w_1 be the first letter that is bumped out from the leftmost column, w_2 be the second, and so on. Denote by $T^{(i)}$ the resulting tableau when w_i is bumped out $(1 \leq i \leq l')$. Now $b'_1 \in B_l$ and $b'_2 \in B_k$ are uniquely specified by

$$\mathcal{T}(b'_2) = \boxed{\overbrace{1 \cdots 1}^{z}}\boxed{T^{(l')}}, \qquad \mathcal{T}(b'_1) = \boxed{w_1}\cdots\boxed{w_{l'}}\boxed{\overbrace{\overline{1} \cdots \overline{1}}^{z}}.$$

Our main result for $U'_q(A^{(2)}_{2n-1})$ is the following.

Theorem 5.2. *Given $b_1 \otimes b_2 \in B_l \otimes B_k$, specify $b'_2 \otimes b'_1 \in B_k \otimes B_l$ and l', k', m by Rule* 5.1. *Let $\iota : B_l \otimes B_k \xrightarrow{\sim} B_k \otimes B_l$ be the isomorphism of a $U'_q(A^{(2)}_{2n-1})$ crystal. Then we have*

$$\iota(b_1 \otimes b_2) = b'_2 \otimes b'_1,$$
$$H_{B_l B_k}(b_1 \otimes b_2) = 2\min(l', k') - m.$$

Example 5.3. If $1123 \otimes 1\bar{1}\bar{1}$ is regarded as an element of the $U'_q(A^{(2)}_{2n-1})$ crystal $B_4 \otimes B_3$, it is mapped to $113 \otimes 12\bar{1}\bar{1} \in B_3 \otimes B_4$ under the isomorphism. Here $\mathbb{P}(1\bar{1}\bar{1} \overset{*}{\rightarrow} 1123) = 123$ and $H(1123 \otimes 1\bar{1}\bar{1}) = 2$. If $1123 \otimes 1\bar{1}\bar{1}$ is regarded as an element of the $U'_q(C^{(1)}_n)$ crystal $B_4 \otimes B_3$, it is mapped to $123 \otimes 01\bar{1}\bar{0} \in B_3 \otimes B_4$ under the isomorphism. Here $\hat{\mathbb{P}}(1\bar{1}\bar{1} \rightarrow 1123) = \begin{smallmatrix} 0 & 1 & 2 & 3 \\ 1 & \bar{1} & \bar{0} & \end{smallmatrix}$ and $H(1123 \otimes 1\bar{1}\bar{1}) = 0$.

5.3 Proof: $A^{(2)}_{2n-1}$ case. Given $b_1 \otimes b_2 \in B_k \otimes B_k$, determine $b'_2 \otimes b'_1 \in B_k \otimes B_l$ by Rule 5.1. To prove Theorem 5.2, we are to show the following claims:

$$\left(\grave{T}(b_2) \longrightarrow \acute{T}(b_1)\right) = \left(\grave{T}(b'_1) \longrightarrow \acute{T}(b'_2)\right). \tag{5.5}$$

$$H_{B_l B_k}(b_1 \otimes b_2) = \left(\grave{T}(b_2) \longrightarrow \acute{T}(b_1)\right)_1 - |l - k|. \tag{5.6}$$

Lemma 5.4. *We have*

$$\iota : (l, 0, —, 0) \otimes (k, 0, —, 0) \mapsto (k, 0, —, 0) \otimes (l, 0, —, 0)$$

under the isomorphism $B_l \otimes B_k \xrightarrow{\sim} B_k \otimes B_l$.

Proof. They are the unique elements in $B_l \otimes B_k$ and $B_k \otimes B_l$, respectively, that do not vanish when $(\tilde{e}_0)^{l+k}$ is applied and do not vanish when $(\tilde{f}_1)^{l+k}$ is applied. (They are also the unique elements such that $\varphi_i - \varepsilon_i = (l + k)\delta_{i,1}$.) □

Proof of Theorem 5.2. Claim (5.5) is due to Proposition 4.8 and the fact that the irreducible decomposition of the $U_q(C_n)$ module $V(l\Lambda_1) \otimes V(k\Lambda_1)$ is multiplicity free (for generic q).

We call an element b of a $U'_q(A^{(2)}_{2n-1})$ crystal a $U_q(C_n)$ *highest element* if it satisfies $\tilde{e}_i b = 0$ for $i = 1, 2, \dots, n$. To show claim (5.6), it suffices to check it for $U_q(C_n)$ highest elements. Then the general case follows from Proposition 4.8 because $H(\tilde{f}_{i_1} \cdots \tilde{f}_{i_j}(b_1 \otimes b_2)) = H(b_1 \otimes b_2)$ for $i_1, \dots, i_j \in \{1, \dots, n\}$ if $\tilde{f}_{i_1} \cdots \tilde{f}_{i_j}(b_1 \otimes b_2) \neq 0$. We assume $l \geq k$ with no loss of generality. Suppose that $b_1 \otimes b_2 \simeq b'_2 \otimes b'_1$ is a $U_q(C_n)$ highest element. In general, it has the form

$$b_1 \otimes b_2 = (l, 0, —, 0) \otimes (x_1, x_2, —, \overline{x}_1),$$

where x_1, x_2 and $\overline{x}_1$ are arbitrary as long as $k = x_1 + x_2 + \overline{x}_1$. Applying

$$\tilde{e}_0^{\overline{x}_1} \tilde{e}_2^{x_2+\overline{x}_1} \cdots \tilde{e}_{n-1}^{x_2+\overline{x}_1} \tilde{e}_n^{x_2+\overline{x}_1} \tilde{e}_{n-1}^{x_2+\overline{x}_1} \cdots \tilde{e}_2^{x_2+\overline{x}_1} \tilde{e}_0^{x_2+\overline{x}_1}$$

to both sides of Lemma 5.4, we find

$$(l, 0, —, 0) \otimes (x_1, x_2, —, \overline{x}_1) \simeq (k, 0, —, 0) \otimes (x_1', x_2, —, \overline{x}_1).$$

Here $x_1' = l - x_2 - \overline{x}_1$. In the course of the application of $\tilde{e}_i$s, the value of the energy function has changed as

$$H((l, 0, —, 0) \otimes (x_1, x_2, —, \overline{x}_1)) = H((l, 0, —, 0) \otimes (k, 0, —, 0)) - x_2 - 2\overline{x}_1.$$

Thus according to our normalization (5.4), we have $H(b_1 \otimes b_2) = 2k - x_2 - 2\overline{x}_1$. On the other hand, for this highest element, the column insertions (5.5) lead to a tableau

1 · · · · · · 1
2 · · · 2

,

whose first row has the length $l + k - x_2 - 2\overline{x}_1$. This completes the proof of claim (5.6). □

Remark 5.5. For $b = (x_1, \ldots, \overline{x}_1) \in B_{l-1}$ $(l \geq 2)$ define

$$\acute{\tau}_{l-1}^{l}(b) = (x_1 + 1, x_2, \ldots, \overline{x}_1) \in B_l,$$
$$\grave{\tau}_{l-1}^{l}(b) = (x_1, \ldots, \overline{x}_2, \overline{x}_1 + 1) \in B_l.$$

Then we have $\acute{\tau}_{l-1}^{l}(c_1) \otimes \grave{\tau}_{k-1}^{k}(c_2) \simeq \acute{\tau}_{k-1}^{k}(c_2') \otimes \grave{\tau}_{l-1}^{l}(c_1')$ under the isomorphism $B_l \otimes B_k \simeq B_k \otimes B_l$ if and only if $c_1 \otimes c_2 \simeq c_2' \otimes c_1'$ under $B_{l-1} \otimes B_{k-1} \simeq B_{k-1} \otimes B_{l-1}$. We also have $H_{B_l B_k}(\acute{\tau}_{l-1}^{l}(c_1) \otimes \grave{\tau}_{k-1}^{k}(c_2)) = H_{B_{l-1} B_{k-1}}(c_1 \otimes c_2)$.

Appendix A Proof of Proposition 4.1

In this appendix, we assume that $l \geq k$.

A.1 Column insertions of type I $U_q(C_n)$ highest elements. Let us consider an element in $B_l \otimes B_k$ depicted by

$$b_1 \otimes b_2 = \overbrace{\boxed{\qquad\qquad 1 \qquad\qquad}}^{l} \otimes \underset{x_0}{\boxed{0}}\underset{x_1}{\boxed{1}}\underset{x_2}{\boxed{2}}\underset{\overline{x}_1}{\boxed{\overline{1}}}\underset{x_0}{\boxed{\overline{0}}}\,.$$

(In this appendix, we denote $\mathbb{T}(b)$ simply by b.) It is a $U_q(C_n)$ highest element. We denote by $b_2' \otimes b_1'$ the image of this element under the isomorphism $\iota : B_l \otimes B_k \to B_k \otimes B_l$.

A.1.1. Let $\bar{x}_1 \le x_1$. Then $b'_2 \otimes b'_1$ is depicted by

$$b'_2 \otimes b'_1 = \overset{k}{\boxed{\quad 1 \quad}} \otimes \underset{x_0}{\boxed{0}}\overset{x_1+l-k}{\boxed{\quad 1 \quad}}\underset{x_2}{\boxed{2}}\underset{\bar{x}_1}{\boxed{\bar{1}}}\underset{x_0}{\boxed{\bar{0}}}\,.$$

The column insertions $(b_2 \longrightarrow b_1)$ and $(b'_1 \longrightarrow b'_2)$ lead to the same intermediate result:

$$\underset{x_0}{\boxed{0}}\underset{x_1-\bar{x}_1}{\boxed{1}} \longrightarrow \begin{array}{|c|c|c|l} \hline 0 & \multicolumn{3}{c|}{\overset{l-\bar{x}_1}{1}} \\ \hline 1 & 2 & \bar{0} & \\ \hline \bar{x}_1 & x_2 & \bar{x}_1+x_0 & \end{array}\,.$$

The value of the energy function is $x_0 + x_1 - \bar{x}_1$.

A.1.2. Let $\bar{x}_1 > x_1$. Then $b'_2 \otimes b'_1$ is depicted by

$$b'_2 \otimes b'_1 = \overset{k}{\boxed{\quad 1 \quad}} \otimes \underset{x_0+y}{\boxed{0}}\overset{x_1+l-k-y}{\boxed{\quad 1 \quad}}\underset{x_2}{\boxed{2}}\underset{\bar{x}_1-y}{\boxed{\bar{1}}}\underset{x_0+y}{\boxed{\bar{0}}}\,,$$

where

$$y = \min[l-k, \bar{x}_1 - x_1]. \tag{A.1}$$

The column insertions $(b_2 \longrightarrow b_1)$ and $(b'_1 \longrightarrow b'_2)$ lead to the same intermediate result.

For $x_1 + x_2 > \bar{x}_1$,

$$\underset{x_0}{\boxed{\quad 0 \quad}} \longrightarrow \begin{array}{|c|c|c|l} \hline \multicolumn{2}{|c|}{\overset{\bar{x}_1}{0}} & \multicolumn{2}{c|}{\overset{l-\bar{x}_1}{1}} \\ \hline 1 & 2 & \bar{0} & \\ \hline x_1 & x_2 & \bar{x}_1+x_0 & \end{array}\,.$$

For $x_1 + x_2 \le \bar{x}_1$,

$$\underset{x_0}{\boxed{\quad 0 \quad}} \longrightarrow \begin{array}{|c|c|c|c|c|l} \hline \multicolumn{2}{|c|}{0} & \overset{\bar{x}_1-x_1-x_2}{0} & 1 & \multicolumn{2}{c|}{\overset{l-\bar{x}_1}{1}} \\ \hline 1 & 2 & \bar{1} & \bar{0} & \bar{0} & \\ \hline x_1 & x_2 & & & x_0+x_1+x_2 & \end{array}\,.$$

The value of the energy function is x_0. Here and in the following we use the notation

$$\overset{m}{\begin{array}{|c||c|} \hline 0 & 1 \\ \hline \bar{1} & \bar{0} \\ \hline \end{array}} = \overset{\frac{m}{2}\ \frac{m}{2}}{\begin{array}{|c|c|} \hline 0 & 1 \\ \hline \bar{1} & \bar{0} \\ \hline \end{array}} \ (m\text{: even}), \quad \underset{\frac{m+1}{2}\ \frac{m-1}{2}}{\overset{\frac{m-1}{2}\ \frac{m+1}{2}}{\begin{array}{|c|c|} \hline 0 & 1 \\ \hline \bar{1} & \bar{0} \\ \hline \end{array}}} \ (m\text{: odd}).$$

A.2 Column insertions of type II $U_q(C_n)$ highest elements.

Let

$$b_1 \otimes b_2 = \begin{array}{|c|c|c|}\hline \underset{y_0}{0} & \overset{l-2y_0}{1} & \underset{y_0}{\bar{0}} \\ \hline \end{array} \otimes \begin{array}{|c|c|c|}\hline \underset{x_1}{1} & \underset{x_2}{2} & \underset{\bar{x}_1}{\bar{1}} \\ \hline \end{array}$$

be a $U_q(C_n)$ highest element in $B_l \otimes B_k$. Thus we assume $l - 2y_0 \geq x_2 + \bar{x}_1$.

A.2.1. Let $l - k > y_0 \geq x_1 - \bar{x}_1$. Then $b_2' \otimes b_1'$ is depicted by

$$b_2' \otimes b_1' = \begin{array}{|c|}\hline \overset{k}{1} \\ \hline \end{array} \otimes \begin{array}{|c|c|c|c|c|}\hline \underset{z}{0} & \overset{x_1+l-k-y_0-z}{1} & \underset{x_2}{2} & \overset{\bar{x}_1+y_0-z}{\bar{1}} & \underset{z}{\bar{0}} \\ \hline \end{array},$$

where

$$z = \min[y_0 + \bar{x}_1 - x_1, l - k - y_0]. \tag{A.2}$$

The column insertions $(b_2 \longrightarrow b_1)$ and $(b_1' \longrightarrow b_2')$ give the same result.

For $y_0 \geq k$,

$$\begin{array}{|c|c|c|c|}\hline \overset{y_0}{0} & & \overset{l-2y_0}{1} & \overset{y_0}{\bar{0}} \\ \hline \underset{x_1}{1}\ \ \underset{x_2}{2}\ \ \underset{\bar{x}_1}{\bar{1}} & \multicolumn{3}{c}{} \\ \cline{1-1} \end{array}.$$

For $k > y_0 \geq x_1 + x_2$,

$$\begin{array}{|c|c|c|c|c|}\hline \overset{y_0}{0} & 0 & 1 & \overset{l-k-y_0}{1} & \overset{y_0}{\bar{0}} \\ \hline \underset{x_1}{1}\ \underset{x_2}{2}\ \ \bar{1} & \bar{1} & \bar{0} & \multicolumn{2}{c}{} \\ \cline{1-3} \multicolumn{1}{c}{} & \multicolumn{2}{c}{k-y_0} & \multicolumn{2}{c}{} \end{array}.$$

For $x_1 + x_2 > y_0 \geq x_1 + x_2 - \bar{x}_1$,

$$\begin{array}{|c|c|c|c|c|}\hline 0 & 0 & 1 & \overset{l-2y_0-\bar{x}_1}{1} & \overset{y_0}{\bar{0}} \\ \hline \underset{x_1}{1}\ \ \underset{x_2}{2} & \bar{1} & \bar{0} & \underset{x_1+x_2-y_0}{\bar{0}} & \multicolumn{1}{c}{} \\ \cline{1-4} \end{array}$$

(with $\bar{x}_1+y_0-x_1-x_2$ above the columns $\begin{smallmatrix}0 & 1\\ \bar{1} & \bar{0}\end{smallmatrix}$).

For $x_1 + x_2 - \bar{x}_1 > y_0$,

$$\begin{array}{|c|c|c|}\hline \overset{y_0+\bar{x}_1}{0} & \overset{l-2y_0-\bar{x}_1}{1} & \overset{y_0}{\bar{0}} \\ \hline \underset{x_1}{1}\ \ \ \underset{x_2}{2} & \underset{\bar{x}_1}{\bar{0}} & \multicolumn{1}{c}{} \\ \cline{1-2} \end{array}.$$

The value of the energy function is 0.

A.2.2. Let $l-k \le y_0$ and $2y_0+k-l-x_1+\bar{x}_1 > 0$. Then $b_2' \otimes b_1'$ is depicted by

$$b_2' \otimes b_1' = \overbrace{\boxed{\underbrace{0}_{y_0-l+k}\ \Big|\ 1\ \Big|\ \underbrace{\bar{0}}_{y_0-l+k}}}^{2l-k-2y_0} \otimes \boxed{\underbrace{1}_{x_1}\ \Big|\ \underbrace{2}_{x_2}\ \Big|\ \underbrace{\bar{1}}_{\bar{x}_1+l-k}}.$$

The column insertions $(b_2 \longrightarrow b_1)$ and $(b_1' \longrightarrow b_2')$ lead to the same intermediate result.

For $l-2y_0 \ge 2\bar{x}_1$,

$$\underbrace{\boxed{1}}_{y_0-l+k} \longrightarrow \begin{array}{|c|c|c|c|}\hline \multicolumn{2}{|c|}{\overset{y_0+\bar{x}_1}{0}} & \overset{l-2y_0-\bar{x}_1}{1} & \overset{y_0}{\bar{0}} \\ \hline \underset{l-y_0-\bar{x}_1-x_2}{1} & \underset{x_2}{2} & \underset{\bar{x}_1}{\bar{0}} & \\ \cline{1-3} \end{array}.$$

For $l-2y_0 < 2\bar{x}_1$,

$$\underbrace{\boxed{1}}_{y_0-l+k} \longrightarrow \begin{array}{|c|c|c|c|c|}\hline \multicolumn{2}{|c|}{0} & \overset{2\bar{x}_1-l+2y_0}{0\,\|\,1} & 1 & \overset{y_0}{\bar{0}} \\ \hline \underset{l-y_0-\bar{x}_1-x_2}{1} & \underset{x_2}{2} & \bar{1}\,\|\,\bar{0} & \underset{l-2y_0-\bar{x}_1}{\bar{0}} & \\ \cline{1-4} \end{array}.$$

The value of the energy function is y_0-l+k.

A.2.3. Let $y_0 < x_1-\bar{x}_1$ and $2y_0+k-l-x_1+\bar{x}_1 \le 0$. Then $b_2' \otimes b_1'$ is depicted by

$$b_2' \otimes b_1' = \overbrace{\boxed{\underbrace{0}_{y_0-w}\ \Big|\ 1\ \Big|\ \underbrace{\bar{0}}_{y_0-w}}}^{k-2y_0+2w} \otimes \boxed{\overbrace{1}^{x_1+l-k-w}\ \Big|\ \underbrace{2}_{x_2}\ \Big|\ \underbrace{\bar{1}}_{\bar{x}_1+w}},$$

where $w=(2y_0-x_1+\bar{x}_1)_+$. The column insertions $(b_2 \longrightarrow b_1)$ and $(b_1' \longrightarrow b_2')$ lead to the same intermediate result:

$$\underbrace{\boxed{1}}_{x_1-\bar{x}_1-y_0} \longrightarrow \begin{array}{|c|c|c|c|}\hline 0 & \multicolumn{2}{c|}{\overset{l-2y_0-\bar{x}_1}{1}} & \overset{y_0}{\bar{0}} \\ \hline \underset{y_0+\bar{x}_1}{1} & \underset{x_2}{2} & \underset{\bar{x}_1}{\bar{0}} & \\ \cline{1-3} \end{array}.$$

The value of the energy function is $x_1-\bar{x}_1-y_0$.

Appendix B Alternative rule for $C_n^{(1)}$

We assume $l \ge k$.

B.1 Algorithm for the isomorphism. Let $b_1 = (x_1, \ldots, \overline{x}_1) \in B_l$, $b_2 = (y_1, \ldots, \overline{y}_1) \in B_k$. We are going to show the rule for finding the image of $b_1 \otimes b_2$ under the isomorphism

$$\begin{aligned} \iota : B_l \otimes B_k &\stackrel{\sim}{\to} B_k \otimes B_l \\ b_1 \otimes b_2 &\mapsto b_2' \otimes b_1'. \end{aligned}$$

Let $x_0 = \overline{x}_0 = (l - \sum_{i=1}^{n}(x_i + \overline{x}_i))/2$ and $y_0 = \overline{y}_0 = (k - \sum_{i=1}^{n}(y_i + \overline{y}_i))/2$. We start with the following initial diagram.

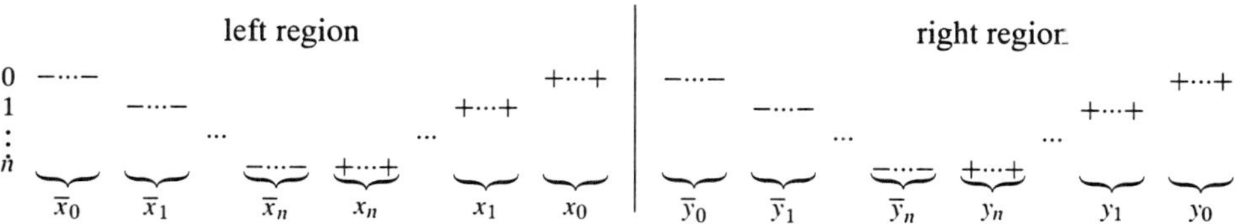

By using Lemma 4.7, one can remove ($\boxed{0}$, $\boxed{\overline{0}}$) pairs from b_1 and b_2 simultaneously as many times as possible. Thus we assume in the following that either x_0 or y_0 is equal to 0 throughout. Then the general procedure to obtain the isomorphism and energy function is as follows.

0. Each symbol $+$ or $-$ is marked or unmarked. In the initial diagram all the symbols are unmarked.

1. There are three regions (left, right, and middle—the latter is empty in the initial diagram). Pick the leftmost symbol a in the right region. Find a's partner b in the left region according to rule 2–3. Apply (a), repeat this procedure as many times as possible, and then apply (b). During the procedure if a symbol named a is a $+$ (resp., $-$) symbol we call it $+_a$ (resp., $-_a$).

   (a) If a exists and there is a partner b, mark b according to rule 4. Put a new line on the right of a, which forms the new boundary between the middle region and the right region. (In the second turn or later, delete the old line on the left of a.)

   (b) If a does not exist or there is no partner of a, then stop. Enumerate the cardinality of the symbols in the right region and denote it by h. This h is equal to the value of the energy function, which is so normalized as the minimal value is equal to 0. Proceed to (c) or (d) according to the value of h.

   (c) If $h = 0$, the procedure is finished. See (e).

   (d) If $h > 0$, give up the diagram. Go back to the initial diagram and mark the leftmost h symbols in the left region. Then start the procedure again from rule 1 in this new setting and stop it, leaving the rightmost h symbols in the right region untouched. Then see (e).

(e) The isomorphism ι is obtained as follows. At the end of the procedure, the marked symbols signify the contents of b_2', and the unmarked symbols signify the contents of b_1'.

2. If a is a $-$ symbol $(-_a)$ in the ith row, look at the ith row in the left region.

   (a) If there are unmarked $+$ symbols in the ith row in the left region, pick one of them and call it $+_c$.

      i. If there are no unmarked symbols (besides $+_c$), neither in the ith row nor in the lower rows in the left region, then $+_c$ itself is identified with the partner $b(= +_b)$.

      ii. If there are unmarked symbols (besides $+_c$), either in the ith row or in the lower rows in the left region, move $-_a$ and $+_c$ to the $(i-1)$st row. Then apply procedure (b).

   (b) If there are no unmarked $+$ symbols in the ith row in the left region or if one has already done procedure (a)ii,

      i. If there are unmarked $-$ symbols in the left region whose positions are lower than that of $-_a$, then the partner $b(= -_b)$ is chosen from one of those $-$ symbols that has the highest position.

      ii. If there are no unmarked $-$ symbols in the left region whose positions are lower than that of $-_a$, then the partner $b(= +_b)$ is chosen from one of the unmarked $+$ symbols in the left region that has the lowest position.

3. If a is a $+$ symbol $(+_a)$, then the partner $b(= +_b)$ is chosen from one of the unmarked $+$ symbols whose positions are higher than that of $+_a$ but the lowest among them.

4. If the partner b is a $-$ symbol $(-_b)$, mark it. If b is a $+$ symbol $(+_b)$ in the jth row, look at the jth row in the left and the middle region.

   (a) If there are unmarked $-$ symbols in the jth row, either in the left region or in the middle region, pick the leftmost one of them and call it $-_d$. Move $+_b$ and $-_d$ to the $(j+1)$st row and then mark the $+_b$.

   (b) If there are no unmarked $-$ symbols in the jth row, neither in the left region nor in the middle region, then mark the $+_b$.

This description of the rule is derived from the column insertion rule in Section 3.2 accompanied with the *reverse row insertion procedure* for the C-tableaux. We do not describe the latter procedure in this paper.

In rule 4(a), we have chosen $-_d$ to be the leftmost one. However, the final result of the procedure is actually the same for any choice of the $-$ symbols in the jth row of the middle and left regions.

B.2 Examples. Let us present two examples. We signify the marked symbols with circles.

B.2.1 *Example* 1. Let us derive

$$\iota : 1134\bar{3}\bar{2}\bar{1} \otimes \bar{4}\bar{4}\bar{4}\bar{1}\bar{1} \mapsto 144\bar{2}\bar{1} \otimes 0\bar{4}\bar{4}\bar{4}\bar{4}\bar{1}\bar{0} \tag{B.1}$$

under the isomorphism $B_7 \otimes B_5 \simeq B_5 \otimes B_7$ of the $U_q'(C_4^{(1)})$ crystals. The value of the energy function is 0 for this element. The initial diagram is as follows.

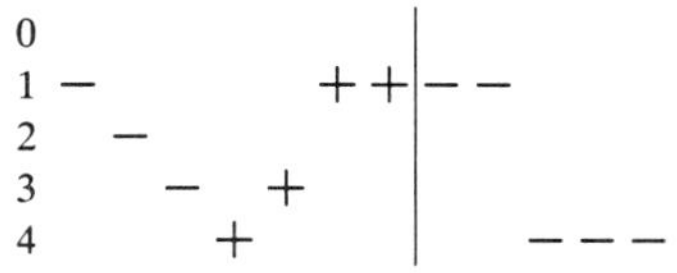

```
0                         |
1  −                 +  + | −  −
2     −                   |
3        −     +          |
4           +             |       −  −  −
```

We apply 2(a)ii and then 2(b)i.

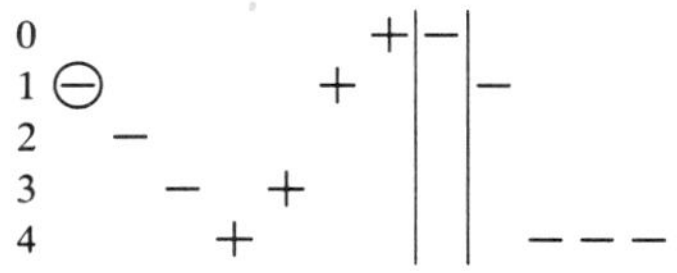

```
0                       + | − |
1  ⊖                 +    |   | −
2     −                   |   |
3        −     +          |   |
4           +             |   |   −  −  −
```

We again apply 2(a)ii and then 2(b)i.

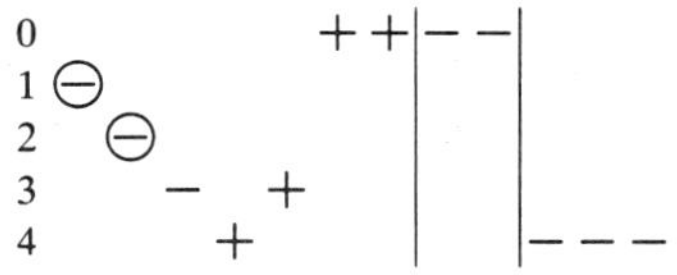

```
0                    +  + | −  − |
1  ⊖                      |      |
2     ⊖                   |      |
3        −     +          |      |
4           +             |      | −  −  −
```

We apply 2(a)i.

```
0                    +  + | −  −    |
1  ⊖                      |         |
2     ⊖                   |         |
3        −     +          |         |
4           ⊕             |       − | −  −
```

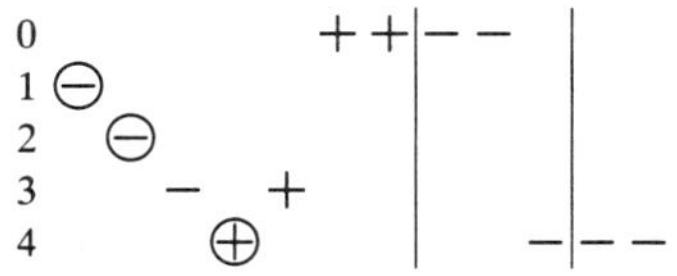

We apply 2(b)ii to find the partner and then apply 4(a) to mark the partner.

```
0                    +  + | −  −       |
1  ⊖                      |            |
2     ⊖                   |            |
3                         |            |
4        −  ⊕  ⊕          |       −  − | −
```

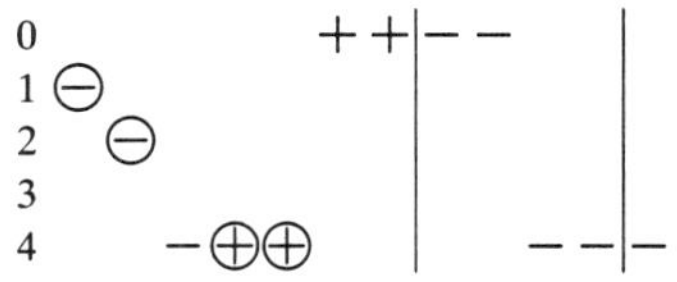

We again apply 2(b)ii to find the partner and then apply 4(a) to mark the partner.

```
0                       + | −             |
1  ⊖                 ⊕    |    −          |
2     ⊖                   |               |
3                         |               |
4        −  ⊕  ⊕          |       −  −  − |
```

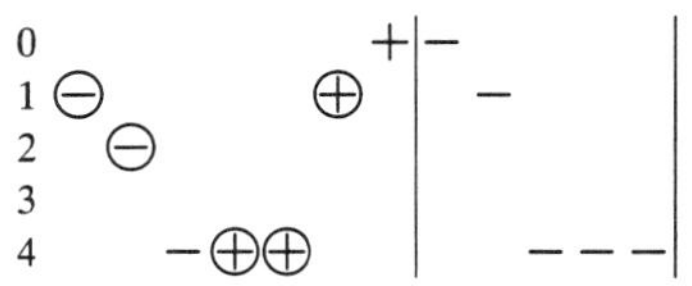

The procedure is finished. Here the set of marked symbols stands for $144\bar{2}\bar{1} \in B_5$ and the set of unmarked symbols stands for $0\bar{4}\bar{4}\bar{4}\bar{4}\bar{1}\bar{0} \in B_7$.

Let us check the isomorphism by definition:

$$\begin{array}{ccc} 1134\bar{3}\bar{2}\bar{1} \otimes \bar{4}\bar{4}\bar{4}\bar{1}\bar{1} & \xrightarrow{\iota} & 144\bar{2}\bar{1} \otimes 0\bar{4}\bar{4}\bar{4}\bar{4}\bar{1}\bar{0} \\ \psi_1 \downarrow & & \psi_1 \downarrow \\ 1111111 \otimes 122\bar{1}\bar{1} & \xrightarrow{\iota} & 11111 \otimes 01122\bar{1}\bar{0}, \end{array} \tag{B.2}$$

where

$$\psi_1 = (\tilde{e}_2)^2(\tilde{e}_3)^2(\tilde{e}_1)^6(\tilde{e}_2)^6(\tilde{e}_3)^4(\tilde{e}_4)^6(\tilde{e}_3)^4(\tilde{e}_2)^2\tilde{e}_1. \tag{B.3}$$

We arrive at a $U_q(C_4)$ highest element:

$$\begin{array}{ccc} 1111111 \otimes 122\bar{1}\bar{1} & \xrightarrow{\iota} & 11111 \otimes 01122\bar{1}\bar{0} \\ \psi_2 \downarrow & & \psi_2 \downarrow \\ 1111111 \otimes 11111 & \xrightarrow{\iota} & 11111 \otimes 1111111, \end{array} \tag{B.4}$$

where

$$\psi_2 = (\tilde{f}_0)^8(\tilde{f}_1)^3(\tilde{f}_2)^3(\tilde{f}_3)^3(\tilde{f}_4)^3(\tilde{f}_3)^3(\tilde{f}_2)^3\tilde{f}_1(\tilde{e}_0)^3. \tag{B.5}$$

The energy was raised by 1 when the third $\tilde{e}_0$ was applied and lowered by 1 when the first $\tilde{f}_0$ was applied. Then it was raised by 5 when the fourth to the eighth $\tilde{f}_0$s were applied.

B.2.2 *Example* 2. Let us derive

$$\iota : 0\bar{2}\bar{2}\bar{1}\bar{1}\bar{1}\bar{0} \otimes 11122\bar{2}\bar{1} \mapsto 0\bar{2}\bar{1}\bar{1}\bar{1}\bar{0} \otimes 1112\bar{2}\bar{2}\bar{1} \tag{B.6}$$

under the isomorphism $B_7 \otimes B_6 \simeq B_6 \otimes B_7$ of the $U'_q(C_2^{(1)})$ crystals. The value of the energy function is 4 for this element. The initial diagram is as follows.

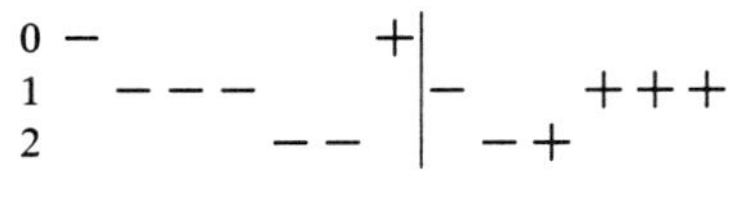

```
0  −                  + |
1     − − −             | −         + + +
2               − −     |      − +
```

We apply 2(b)i.

```
0  −                  + | |
1     − − −             | − |       + + +
2               − ⊖     |   | − +
```

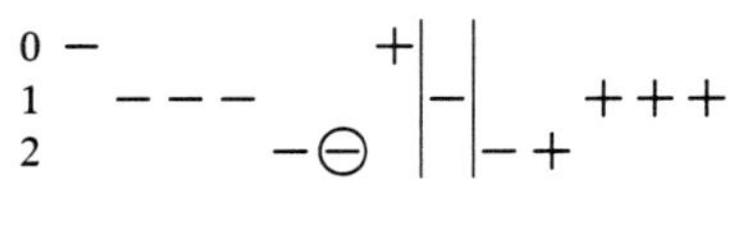

We apply 2(b)ii to find the partner and then apply 4(a) to mark the partner.

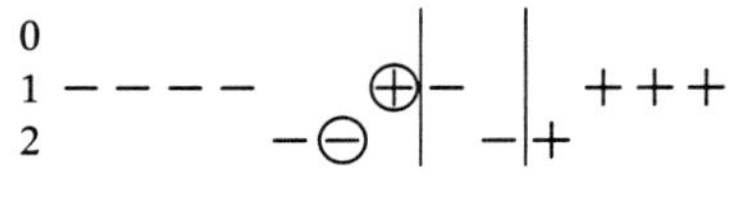

```
0                       |       |
1  − − − −            ⊕ | −     |   + + +
2               − ⊖     |     − | +
```

This time, we find that there is no partner in the left region for the leftmost $+$ symbol in the right region. We interrupt the procedure here according to 1(b). Since there are four symbols in the right region, we find that the value of the energy function is equal to 4. Following 1(d), we give up this diagram and go to the initial diagram with four marked symbols.

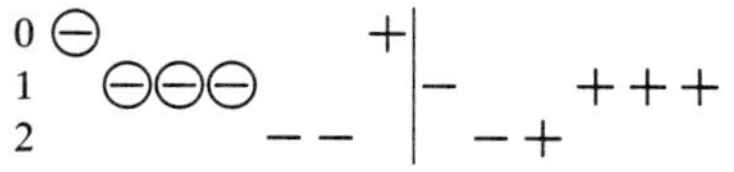

$$
\begin{array}{l|l}
0\ \ominus \qquad\qquad\qquad\quad + & \\
1\ \quad \ominus\ominus\ominus & - \qquad\quad +++ \\
2\ \qquad\qquad\qquad - - & \quad - +
\end{array}
$$

We apply 2(b)i.

$$
\begin{array}{l|l|l}
0\ \ominus \qquad\qquad\qquad\quad + & & \\
1\ \quad \ominus\ominus\ominus & - & \qquad +++ \\
2\ \qquad\qquad\qquad - \ominus & & - +
\end{array}
$$

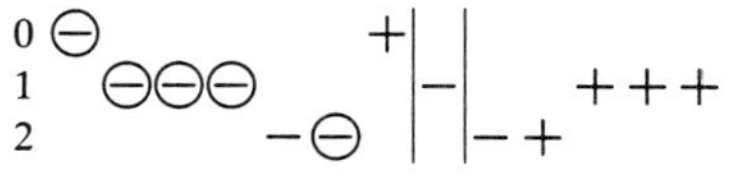

We apply 2(b)ii to find the partner and then apply 4(b) to mark the partner.

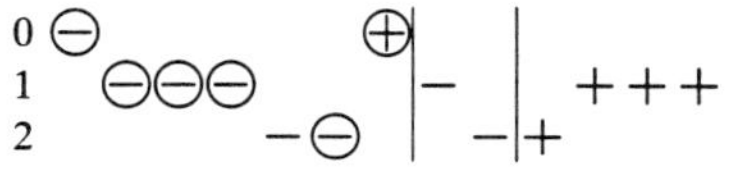

$$
\begin{array}{l|l|l}
0\ \ominus \qquad\qquad\qquad\quad \oplus & & \\
1\ \quad \ominus\ominus\ominus & - & \quad +++ \\
2\ \qquad\qquad\qquad - \ominus & \quad - & +
\end{array}
$$

The procedure is finished. Here the set of marked symbols stands for $0\bar{2}\bar{1}\bar{1}\bar{1}\bar{0} \in B_6$ and the set of unmarked symbols stands for $1112\bar{2}\bar{2}\bar{1} \in B_7$.

Appendix C $C_n^{(1)}$ Kostka polynomials

Let $\mu_1 \geq \cdots \geq \mu_L$ (≥ 1) be a set of integers. We set $\mu = (\mu_1, \mu_2, \ldots, \mu_L)$. Consider the tensor product of $U'_q(C_n^{(1)})$ crystals $B_{\mu_1} \otimes \cdots \otimes B_{\mu_L}$. Let $\lambda = (\lambda_1, \ldots, \lambda_n)$ be another partition satisfying $|\lambda| \leq |\mu|$ and $|\lambda| \equiv |\mu| \pmod 2$. We define a classically restricted 1dsum (cf. [HKOTY]):

$$X_{\lambda,\mu}(t) = \sum{}^{*}\, t^{\sum_{0 \leq i < j \leq L} H(b_i \otimes b_j^{(i+1)})}, \tag{C.1}$$

where the sum $\sum^*$ is taken over all $b_1 \otimes \cdots \otimes b_L \in B_{\mu_1} \otimes \cdots \otimes B_{\mu_L}$ satisfying

$$\tilde{e}_i(b_1 \otimes \cdots \otimes b_L) = 0, \;\; \varphi_i(b_1 \otimes \cdots \otimes b_L) = \lambda_i - \lambda_{i+1}, \quad 1 \leq i \leq n \;\; (\lambda_{n+1} = 0).$$

This condition is equivalent to

$$(b_L \xrightarrow{*} \cdots \xrightarrow{*} (b_3 \xrightarrow{*} (b_2 \xrightarrow{*} b_1)) \cdots) = T(\lambda). \tag{C.2}$$

$T(\lambda)$ is the unique tableau of both shape and weight λ. Namely, all the letters in the first row are 1, those in the second row are 2, and so on. (We distinguished b_j from $\mathbb{T}(b_j)$.) In the summation we set $b_i^{(i)} = b_i$, and $b_j^{(i)}$ $(i < j)$ are defined by successive use of the crystal isomorphism,

$$\begin{aligned} B_{\mu_i} \otimes \cdots \otimes B_{\mu_{j-1}} \otimes B_{\mu_j} &\simeq B_{\mu_i} \otimes \cdots \otimes B_{\mu_j} \otimes B_{\mu_{j-1}} \\ b_i \otimes \cdots \otimes b_{j-1} \otimes b_j &\mapsto b_i \otimes \cdots \otimes b_j^{(j-1)} \otimes b'_{j-1} \\ \cdots &\simeq B_{\mu_j} \otimes B_{\mu_i} \otimes \cdots \otimes B_{\mu_{j-1}} \\ \cdots &\mapsto b_j^{(i)} \otimes b'_i \otimes \cdots \otimes b'_{j-1}. \end{aligned}$$

Condition (C.2) implies that $b_j^{(1)}$ in the tableau presentation should have the form $\mathbb{T}(b_j^{(1)}) = \boxed{0\cdots 0}\boxed{1\cdots 1}\boxed{\bar{0}\cdots\bar{0}}$. b_0 is chosen so that $H(b_0 \otimes b_j^{(1)}) = -\sharp(\boxed{0}$ in $\mathbb{T}(b_j^{(1)}))$. Up to an additive constant, this agrees with the choice of b_0 in [HKOTY]. Up to an overall power of t, this is a polynomial which may be viewed as a $C_n^{(1)}$-analogue of the Kostka polynomial. In fact, if $|\lambda| = |\mu|$, $X_{\lambda,\mu}(t)$ coincides with the ordinary Kostka polynomial $K_{\lambda,\mu}(t)$. To see this, note that $|\lambda| = |\mu|$ implies that there are no bar letters in $b_1 \otimes \cdots \otimes b_L$ contributing to the sum (C.1). In such a case, the function H and the isomorphism $B_l \otimes B_k \simeq B_k \otimes B_l$ are the same as the $A_n^{(1)}$ case.

Following the tables of [Ma, pp. 239-240], we give a list of $X_{\lambda,\mu}(t)$ or the matrices $X(t) := \{X_{\lambda,\mu}(t)\}$ for $|\mu| \le 6$ and $|\lambda| = |\mu| - 2, |\mu| - 4, \dots$ with $n \ge L$. $X_{\lambda,\mu}(t)$ is independent of n if $n \ge L$. If $n < L$, it is n-dependent in general. For instance, let $\lambda = (1^3)$ and $\mu = (1^5)$. The element $\boxed{1}\otimes\boxed{2}\otimes\boxed{3}\otimes\boxed{4}\otimes\boxed{\bar{4}} \in (B_1)^{\otimes 5}$ contributes to $X_{(1^3),(1^5)}(t)$ for $n \ge 4$ but does not for $n = 2, 3$. We have checked that all the data in the table agrees with the fermionic formula in [HKOTY]. In the tables below, a row (resp., column) specifies λ (resp., μ) in $X_{\lambda,\mu}(t)$.

$$X_{\emptyset,(2)}(t) = t^{-1}, \qquad X_{\emptyset,(1^2)}(t) = 1.$$

$$X_{(1),(3)}(t) = t^{-1}, \qquad X_{(1),(21)}(t) = t^{-1} + 1, \qquad X_{(1),(1^3)}(t) = 1 + t + t^2.$$

	(4)	(31)	(2^2)	(21^2)	(1^4)
$\emptyset$	t^{-2}	t^{-1}	$t^{-2}+1$	$t^{-1}+t$	$1+t^2+t^4$
(2)	t^{-1}	$t^{-1}+1$	$t^{-1}+1+t$	$2+t+t^2$	$t+t^2+2t^3+t^4+t^5$
(1^2)		t^{-1}	1	$t^{-1}+1+t$	$1+t+2t^2+t^3+t^4$

	(5)	(41)	(32)	(31^2)	(2^21)
(1)	t^{-2}	$t^{-2}+t^{-1}$	$t^{-2}+t^{-1}+1$	$2t^{-1}+1+t$	$t^{-2}+t^{-1}+2+t+t^2$
(3)	t^{-1}	$t^{-1}+1$	$t^{-1}+1+t$	$2+t+t^2$	$1+2t+t^2+t^3$
(21)		t^{-1}	$t^{-1}+1$	$t^{-1}+2+t$	$t^{-1}+2+2t+t^2$
(1^3)				t^{-1}	1

	(21^3)	(1^5)
(1)	$t^{-1}+2+2t+2t^2+t^3+t^4$	$1+t+2t^2+2t^3+3t^4+2t^5+2t^6+t^7+t^8$
(3)	$t+2t^2+2t^3+t^4+t^5$	$t^3+t^4+2t^5+2t^6+2t^7+t^8+t^9$
(21)	$2+3t+3t^2+2t^3+t^4$	$t+2t^2+3t^3+4t^4+4t^5+3t^6+2t^7+t^8$
(1^3)	$t^{-1}+1+t+t^2$	$1+t+2t^2+2t^3+2t^4+t^5+t^6$

	(6)	(51)	(42)	(41^2)	(3^2)	(321)
∅	t^{-3}	t^{-2}	$t^{-3}+t^{-1}$	$t^{-2}+1$	$t^{-2}+1$	$t^{-2}+t^{-1}+t$
(2)	t^{-2}	$t^{-2}+t^{-1}$	$2t^{-2}+t^{-1}+1$	$3t^{-1}+1+t$	$2t^{-1}+1+t$	$t^{-2}+2t^{-1}+3+t+t^2$
(1^2)		t^{-2}	t^{-1}	$t^{-2}+t^{-1}+1$	$t^{-2}+1$	$t^{-2}+2t^{-1}+1+t$
(4)	t^{-1}	$t^{-1}+1$	$t^{-1}+1+t$	$2+t+t^2$	$1+t+t^2$	$1+2t+t^2+t^3$
(31)		t^{-1}	$t^{-1}+1$	$t^{-1}+2+t$	$t^{-1}+1+t$	$t^{-1}+3+2t+t^2$
(21^2)				t^{-1}		$t^{-1}+1$
(2^2)			t^{-1}	1	1	$t^{-1}+1+t$
(1^4)						

	(31^3)	(2^3)
∅	$t^{-1}+1+t+t^3$	$t^{-3}+t^{-1}+1+t+t^3$
(2)	$t^{-1}+3+3t+3t^2+t^3+t^4$	$t^{-2}+t^{-1}+3+2t+3t^2+t^3+t^4$
(1^2)	$2t^{-1}+2+3t+t^2+t^3$	$t^{-1}+1+2t+t^2+t^3$
(4)	$t+2t^2+2t^3+t^4+t^5$	$t+t^2+2t^3+t^4+t^5$
(31)	$2+3t+4t^2+2t^3+t^4$	$1+2t+3t^2+2t^3+t^4$
(21^2)	$t^{-1}+2+2t+t^2$	$1+t+t^2$
(2^2)	$1+2t+t^2+t^3$	$t^{-1}+1+2t+t^2+t^3$
(1^4)	t^{-1}	

	(2^21^2)	(21^4)
∅	$t^{-2}+2+t+t^2+t^4$	$t^{-1}+2t+t^2+2t^3+t^4+t^5+t^7$
(2)	$2t^{-1}+2+5t+3t^2+3t^3+t^4+t^5$	$2+2t+5t^2+4t^3+6t^4+3t^5+3t^6+t^7+t^8$
(1^2)	$t^{-2}+t^{-1}+4+2t+3t^2+t^3+t^4$	$t^{-1}+2+4t+4t^2+5t^3+3t^4+3t^5+t^6+t^7$
(4)	$2t^2+2t^3+2t^4+t^5+t^6$	$t^3+t^4+3t^5+2t^6+2t^7+t^8+t^9$
(31)	$1+4t+4t^2+4t^3+2t^4+t^5$	$t+3t^2+5t^3+6t^4+5t^5+4t^6+2t^7+t^8$
(21^2)	$t^{-1}+2+3t+2t^2+t^3$	$2+3t+5t^2+4t^3+4t^4+2t^5+t^6$
(2^2)	$2+2t+3t^2+t^3+t^4$	$2t+2t^2+4t^3+3t^4+3t^5+t^6+t^7$
(1^4)	1	$t^{-1}+1+t+t^2+t^3$

	(1^6)
∅	$1+t^2+t^3+2t^4+t^5+3t^6+t^7+2t^8+t^9+t^{10}+t^{12}$
(2)	$t+t^2+3t^3+3t^4+6t^5+5t^6+7t^7+5t^8+6t^9+3t^{10}+3t^{11}+t^{12}+t^{13}$
(1^2)	$1+t+3t^2+3t^3+6t^4+5t^5+7t^6+5t^7+6t^8+3t^9+3t^{10}+t^{11}+t^{12}$
(4)	$t^6+t^7+2t^8+2t^9+3t^{10}+2t^{11}+2t^{12}+t^{13}+t^{14}$
(31)	$t^3+2t^4+4t^5+5t^6+7t^7+7t^8+7t^9+5t^{10}+4t^{11}+2t^{12}+t^{13}$
(21^2)	$t+2t^2+4t^3+5t^4+7t^5+7t^6+7t^7+5t^8+4t^9+2t^{10}+t^{11}$
(2^2)	$t^2+t^3+3t^4+3t^5+5t^6+4t^7+5t^8+3t^9+3t^{10}+t^{11}+t^{12}$
(1^4)	$1+t+2t^2+2t^3+3t^4+2t^5+2t^6+t^7+t^8$

Acknowledgments. The authors thank T. H. Baker for useful discussions. They also thank M. Kashiwara and T. Miwa for organizing the conference "Physical Combinatorics" in Kyoto during January 29–February 2, 1999.

References

[ABF] G. E. Andrews, R. J. Baxter, and P. J. Forrester, Eight vertex SOS model and generalized Rogers-Ramanujan-type identities, *J. Statist. Phys.*, **35** (1984), 193–266.

[B] T. H. Baker, An insertion scheme for C_n crystals, in M. Kashiwara and T. Miwa, eds., *Physical Combinatorics*, Birkhäuser, Boston, 2000, 1–48 (this volume).

[Ba] R. J. Baxter, *Exactly Solved Models in Statistical Mechanics*, Academic Press, London, 1982.

[Be] H. A. Bethe, Zur Theorie der Metalle I: Eigenwerte und Eigenfunktionen der linearen Atomkette, *Z. Phys.*, **71** (1931), 205–231.

[Ber] A. Berele, A Schensted-type correspondence for the symplectic group, *J. Combin. Theory Ser. A*, **43** (1986), 320–328.

[DJKMO] E. Date, M. Jimbo, A. Kuniba, T. Miwa, and M. Okado, One dimensional configuration sums in vertex models and affine Lie algebra characters, *Lett. Math. Phys.*, **17** (1989), 69–77.

[F] W. Fulton, *Young Tableaux: With Applications to Representation Theory and Geometry*, London Mathematical Society Student Texts 35, Cambridge University Press, Cambridge, 1997.

[FOY] K. Fukuda, M. Okado, and Y. Yamada, Energy functions in box ball systems, preprint math.QA/9908116.

[HHIKTT] G. Hatayama, K. Hikami, R. Inoue, A. Kuniba, T. Takagi, and T. Tokihiro, The $A^{(1)}_M$ automata related to crystals of symmetric tensors, preprint math.QA/9912209.

[HKKOT] G. Hatayama, Y. Koga, A. Kuniba, M. Okado, and T. Takagi, Finite crystals and paths, preprint math.QA/9901082.

[HKKOTY] G. Hatayama, A. N. Kirillov, A. Kuniba, M. Okado, T. Takagi, and Y. Yamada, Character formulae of $\widehat{sl}_n$-modules and inhomogeneous paths, *Nuclear Phys. B*, **536** (1999), 575–616.

[HKOTY] G. Hatayama, A. Kuniba, M. Okado, T. Takagi, and Y. Yamada, Remarks on fermionic formula, in N. Jing and K. C. Misra, eds., *Recent Developments in Quantum Affine Algebras and Related Topics*, Contemporary Mathematics 248, AMS, Providence, 1999, 243–291.

[HKT] G. Hatayama, A. Kuniba, and T. Takagi, Soliton cellular automata associated with finite crystals, preprint solv-int/9907020.

[KE] R. C. King and N. G. I. El-Sharkaway, Standard Young tableaux and weight multiplicities of the classical Lie groups, *J. Phys. A*, **16** (1983), 3153–3177.

[KKM] S-J. Kang, M. Kashiwara, and K. C. Misra, Crystal bases of Verma modules for quantum affine Lie algebras, *Compositio Math.*, **92** (1994), 299–325.

[KMN1] S.-J. Kang, M. Kashiwara, K. C. Misra, T. Miwa, T. Nakashima, and A. Nakayashiki, Affine crystals and vertex models, *Internat. J. Modern Phys. A*, **7**-1A (1992), 449–484.

[KMN2] S-J. Kang, M. Kashiwara, K. C. Misra, T. Miwa, T. Nakashima, and A. Nakayashiki, Perfect crystals of quantum affine Lie algebras, *Duke Math. J.*, **68** (1992), 499–607.

[KN] M. Kashiwara and T. Nakashima, Crystal graph for representations of the q-analogue of classical Lie algebras, *J. Algebra*, **165** (1994), 295–345.

[KR] A. N. Kirillov and N. Yu. Reshetikhin, The Bethe ansatz and the combinatorics of Young tableaux, *J. Soviet Math.*, **41** (1988), 925–955.

[L] G. Lusztig, Singularities, character formulas, and a q-analogue of weight multiplicities, *Astérisque*, **101–102** (1983), 208–227.

[Ma] I. Macdonald, *Symmetric Functions and Hall Polynomials*, 2nd ed., Oxford University Press, New York, 1995.

[NY] A. Nakayashiki and Y. Yamada, Kostka polynomials and energy functions in solvable lattice models, *Sel. Math.*, **3** (1997), 547–599.

[S] M. Shimozono, Affine type A crystal structure on tensor product of rectangles, Demazure characters, and nilpotent varieties, preprint math.QA/9804039.

[SW] A. Schilling and S. O. Warnaar, Inhomogeneous lattice paths, generalized Kostka polynomials and A_{n-1} supernomials, *Comm. Math. Phys.*, **202** (1999), 359–401.

[T] I. Terada, A Robinson-Schensted-type correspondence for a dual pair on spinors, *J. Combin. Theory Ser. A*, **63** (1993), 90–109.

[TS] D. Takahashi and J. Satsuma, A soliton cellular automaton, *J. Phys. Soc. Japan*, **59** (1990), 3514–3519.

[TTMS] T. Tokihiro, D. Takahashi, J. Matsukidaira, and J. Satsuma, From soliton equations to integrable cellular automata through a limiting procedure, *Phys. Rev. Lett.*, **76** (1996), 3247–3250.

Goro Hatayama and Atsuo Kuniba
Institute of Physics
University of Tokyo
Komaba, Tokyo 153-8902
Japan
hata@gokutan.c.u-tokyo.ac.jp
atsuo@gokutan.c.u-tokyo.ac.jp

Masato Okado
Department of Informatics
and Mathematical Science
Graduate School of Engineering Science
Osaka University
Toyonaka, Osaka 560-8531
Japan
okado@sigmath.es.osaka-u.ac.jp

Taichiro Takagi
Department of Mathematics and Physics
National Defense Academy
Yokosuka 239-8686
Japan
takagi@cc.nda.ac.jp

Theta Functions Associated with Affine Root Systems and the Elliptic Ruijsenaars Operators

Yasushi Komori

Abstract. We study a family of mutually commutative difference operators associated with affine root systems. These operators act on the space of meromorphic functions on the Cartan subalgebra of the affine Lie algebra. We show that the space spanned by characters of a fixed positive level is invariant under the action of these operators.

1 Introduction

In [27], a family of mutually commutative operators, whose coefficients consist of theta functions, were introduced as a relativistic quantum many-body system, i.e., an elliptic difference analogue of the Calogero–Sutherland model. Since then, these operators have been studied extensively from various points of view, especially by analogy with the Macdonald operators. The eigenvectors of Macdonald operators are a two-parameter extension of Schur functions or the characters of finite-dimensional simple Lie algebras. Then it is natural to expect this structure in the elliptic case. In fact, it was clarified in [11, 12] that the elliptic analogues of type $A_l^{(1)}$ and $C_2^{(1)}$ have an invariant subspace in the meromorphic functions and that this space is actually spanned by the characters of the corresponding affine Lie algebra. These facts are found through studies of the intertwining vectors between face models and vertex models. Independently, in [9], the Boltzmann weight of the matrix elements of Belavin's elliptic R-matrix was calculated using this fact implicitly.

In a series of Cherednik's papers [3, 6, 7], it was proved that the double affine Hecke algebra plays an essential role in Macdonald theory. There are some algebras that are considered as describing the structure of elliptic analogues [5, 8, 10, 31]. In this paper, we employ yet another approach—the root algebra—to these operators. Although this algebra was introduced by Cherednik, our construction is novel even in the trigonometric case due to the existence of spectral parameters. To be more precise, by setting these parameters appropriately, the difference operators automatically become invariant under the action of the Weyl group, while in the theory of the Hecke algebra, they are obtained through symmetrization. Following [21], where we studied nontwisted cases, we construct a family of mutually commuting difference operators associated with arbitrary affine root systems. These operators are shown to act on the vector space of the Weyl group invariant meromorphic functions and, furthermore, on the space spanned by the characters of a fixed positive level.

This paper is organized as follows: In Section 2, we present the notation and definitions used in this paper. In Section 3, we define the root algebras that were introduced by Cherednik in the development of the theory of affine Hecke algebras. In Section 4, we demonstrate some examples of the generators of a commutative subalgebra in the root algebras. In Section 5, we give some representations of root algebras with a spectral parameter, which consist of Jacobi's theta functions and act on the meromorphic functions on the Cartan subalgebra. We show that when we assign a special value to the spectral parameter, the difference operators preserve the Weyl group invariant subspace. By construction, they form a commutative family. In section 6, we calculate the explicit forms of these operators at this spectral parameter and observe that they can be regarded as an elliptic analogue of the Macdonald operators. In twisted cases, we have the difference and quantum versions of the systems that was recently proposed and is dealt with in terms of the Lax formalism [1]. We also prove that the generators are algebraically independent, and thus the commutative subalgebra is isomorphic to a polynomial ring. In section 7, we show in our main theorem (Theorem 7.5) that they have an infinite-dimensional invariant subspace (finite-rank submodule) of the theta functions of positive level, where the key of the proof is due to [9, 19]. The last section is devoted to concluding remarks.

To end this section, we present two elliptic difference operators which take the simplest form among the generators, respectively, in the root systems of type $A_{l-1}^{(1)}$ and $A_{2l}^{(2)}$:

$$\widehat{Y}^{-\lambda_1}_{A_{l-1}^{(1)}} = \sum_{j=1}^{l} \prod_{j\neq k}^{l} \frac{\vartheta_1(x_j - x_k - \mu)}{\vartheta_1(x_j - x_k)} t_j(\kappa) \prod_{j=1}^{l} t_j(-\kappa/l), \tag{1.1}$$

$$\begin{aligned}
\widehat{Y}^{-\lambda_1}_{A_{2l}^{(2)}} = \sum_{j=1}^{l} & \left(\prod_{\substack{k=1\\k\neq j}}^{l} \frac{\vartheta_1(x_j - x_k - \mu)}{\vartheta_1(x_j - x_k)} \frac{\vartheta_1(x_j + x_k - \mu)}{\vartheta_1(x_j + x_k)} \right) \\
& \times \left(\prod_{r=0}^{3} \frac{\vartheta_{r+1}(x_j - \nu_r)}{\vartheta_{r+1}(x_j)} \frac{\vartheta_{r+1}(x_j + \kappa/2 - \bar{\nu}_r)}{\vartheta_{r+1}(x_j + \kappa/2)} \right) t_j(\kappa) \\
& + \sum_{j=1}^{l} \left(\prod_{\substack{k=1\\k\neq j}}^{l} \frac{\vartheta_1(-x_j - x_k - \mu)}{\vartheta_1(-x_j - x_k)} \frac{\vartheta_1(-x_j + x_k - \mu)}{\vartheta_1(-x_j + x_k)} \right)
\end{aligned}$$

$$\times \left(\prod_{r=0}^{3} \frac{\vartheta_{r+1}(-x_j - \nu_r)}{\vartheta_{r+1}(-x_j)} \frac{\vartheta_{r+1}(-x_j + \kappa/2 - \bar{\nu}_r)}{\vartheta_{r+1}(-x_j + \kappa/2)} \right) t_j(-\kappa)$$

$$- \sum_{p=0}^{3} \left(\frac{\pi}{\vartheta_1'(0)} \right)^2 \frac{2}{\vartheta_1(-\mu)\vartheta_1(-\kappa - \mu)}$$

$$\times \left(\prod_{r=0}^{3} \vartheta_{r+1}(-\kappa - \nu_{\pi_p r})\vartheta_{r+1}(-\bar{\nu}_{\pi_p r}) \right)$$

$$\times \left(\prod_{j=1}^{N} \frac{\vartheta_{p+1}(x_j - \kappa/2 - \mu)}{\vartheta_{p+1}(x_j - \kappa/2)} \frac{\vartheta_{p+1}(-x_j - \kappa/2 - \mu)}{\vartheta_{p+1}(-x_j - \kappa/2)} \right). \tag{1.2}$$

Here we have realized the root systems in $\mathbb{C}^l$ in the standard way; $\vartheta_j(x) = \vartheta_j(x;\tau)$ is the Jacobi theta function and $t_i(\kappa)$ is a translation of the variable x_i by κ. π_r $(r = 0, 1, 2, 3)$ denotes the permutation $\pi_0 = id$, $\pi_1 = (01)(23)$, $\pi_2 = (02)(13)$, and $\pi_3 = (03)(12)$. The parameters κ, μ, ν_r, and $\bar{\nu}_r$ $(r = 0, 1, 2, 3)$ are arbitrary constants. The operator (1.1) was introduced in [27] together with the whole family of commuting difference operators, while the operator (1.2) was conjectured to be a member of a commutative family in [32, 33]. The $A_{2l}^{(2)}$-type model was referred to as D-type or BC-type in previous papers.

If we set $\kappa = l\mu/k$ in the $A_{l-1}^{(1)}$ case and $\kappa = (\nu + 2\bar{\nu} + 2(l-1)\mu)/k$ in the $A_{2l}^{(2)}$ case, where $\nu = \sum \nu_r$ and $\bar{\nu} = (\sum \bar{\nu}_r)/2$, then these operators have an invariant subspace which consists of characters of level k corresponding to each affine Lie algebra. When the parameters μ, ν, $\bar{\nu}$ are set to unity, we see that κ reduces to $h^\vee/k$, where $h^\vee$ is the dual Coxeter number. For the derivation of these facts in each case, see [11, 13, 14, 20, 22].

2 Affine root systems

We give some well-known facts about affine root systems and affine Weyl groups [2, 4, 15], which are standard tools in the theory of affine Hecke algebras. Some of the definitions are slightly changed and extended so that they include twisted affine root systems. The notation is mainly due to [16].

Let $\mathfrak{g} = \mathfrak{g}(A)$ be the affine Lie algebra associated with the generalized Cartan matrix A of type $X_N^{(r)}$, $\mathfrak{h}$ be its Cartan subalgebra, $\dim \mathfrak{h} = l + 2$ be the rank of $\mathfrak{g}$, $I = \{0, \dots, l\}$ be a set of indices, $\Pi = \{\alpha_i | i \in I\} \subset \mathfrak{h}^*$ be the set of simple roots, $\Pi^\vee = \{\alpha_i^\vee | i \in I\} \subset \mathfrak{h}$ be the set of simple coroots, Δ be the root system, Q and $Q^\vee$ be the root and coroot lattices, and P and $P^\vee$ be the weight and coweight lattices:

$$Q = \bigoplus_{i\in I} \mathbb{Z}\alpha_i \subset P = \bigoplus_{i\in I} \mathbb{Z}\Lambda_i \oplus \mathbb{C}\delta \subset \mathfrak{h}^*, \tag{2.1}$$

$$Q^\vee = \bigoplus_{i\in I} \mathbb{Z}\alpha_i^\vee \subset P^\vee = \bigoplus_{i\in I} \mathbb{Z}\Lambda_i^\vee \oplus \mathbb{C}K \subset \mathfrak{h}, \tag{2.2}$$

where $\langle \alpha_i, \Lambda_j^\vee \rangle = \delta_{ij}$, $\langle \Lambda_i, \alpha_j^\vee \rangle = \delta_{ij}$, and $d = \Lambda_0^\vee$. Since the normalized invariant form is nondegenerate on $\mathfrak{h}$, we have an isomorphism $\nu : \mathfrak{h} \to \mathfrak{h}^*$ defined by

$$\langle \nu(h), h_1 \rangle = (h|h_1), \quad h, h_1 \in \mathfrak{h} \tag{2.3}$$

and the induced bilinear form $(.|.)$ on $\mathfrak{h}^*$. Let $\overset{\circ}{I} = \{1, \ldots, l\}$, $\overset{\circ}{\Pi} = \{\alpha_i | i \in \overset{\circ}{I}\}$, and $\overset{\circ}{\Pi}{}^\vee = \{\alpha_i^\vee | i \in \overset{\circ}{I}\}$. Let $\overset{\circ}{\mathfrak{h}}{}^*$ be the subspace of $\mathfrak{h}^*$ spanned by $\overset{\circ}{\Pi}$ over $\mathbb{C}$. For $\lambda \in \mathfrak{h}^*$, denote by $\overline{\lambda}$ the orthogonal projection of λ on $\overset{\circ}{\mathfrak{h}}{}^*$. Let $\overset{\circ}{Q}$ be the sublattice of Q generated by $\overset{\circ}{\Pi}$ and $\overset{\circ}{P}$ be the projection of P on $\overset{\circ}{\mathfrak{h}}{}^*$. The dual notions $\overset{\circ}{\mathfrak{h}}$, $\overline{h}$, $\overset{\circ}{Q}{}^\vee$, and $\overset{\circ}{P}{}^\vee$ are defined similarly:

$$\overset{\circ}{Q} = \bigoplus_{i \in \overset{\circ}{I}} \mathbb{Z}\alpha_i \subset \overset{\circ}{P} = \bigoplus_{i \in \overset{\circ}{I}} \mathbb{Z}\overline{\Lambda_i} \subset \overset{\circ}{\mathfrak{h}}{}^*, \tag{2.4}$$

$$\overset{\circ}{Q}{}^\vee = \bigoplus_{i \in \overset{\circ}{I}} \mathbb{Z}\alpha_i^\vee \subset \overset{\circ}{P}{}^\vee = \bigoplus_{i \in \overset{\circ}{I}} \mathbb{Z}\overline{\Lambda_i^\vee} \subset \overset{\circ}{\mathfrak{h}}. \tag{2.5}$$

For $\alpha \in \Delta^{re}$, let r_α be a reflection defined by

$$r_\alpha(\lambda) := \lambda - \langle \lambda, \alpha^\vee \rangle \alpha, \quad \lambda \in \mathfrak{h}^*. \tag{2.6}$$

The Weyl group $\overset{\circ}{W}$ is generated by the fundamental reflections $\{r_i := r_{\alpha_i} | i \in \overset{\circ}{I}\}$ on $\mathfrak{h}^*$ and the affine Weyl group W is generated by $\{r_i | i \in I\}$. The defining relations are given by $r_i^2 = id$ and the Coxeter relations

$$(r_i r_j)^{m_{ij}} = id \quad \text{for } i \neq j \in I, \tag{2.7}$$

where $m_{ij} = 2$ if α_i and α_j are disconnected in the Dynkin diagram $S(A)$ and $m_{ij} = 3, 4, 6$ if $1, 2, 3$ lines, respectively, connect α_i and α_j in $S(A)$. We note that there is no Coxeter relation in the affine root systems of rank 2. For $\alpha \in \overset{\circ}{\mathfrak{h}}{}^*$, we define endomorphisms t_α and t_α^ι of the vector space $\mathfrak{h}^*$ for $\kappa \in \mathbb{C}$ by (cf. [5])

$$t_\alpha(\lambda) := \lambda + \langle \lambda, K \rangle \alpha - \left((\lambda|\alpha) + \frac{1}{2}|\alpha|^2 \langle \lambda, K \rangle \right) \delta, \tag{2.8}$$

$$t_\alpha^\iota(\lambda) := \lambda - \kappa(\lambda|\alpha)\delta. \tag{2.9}$$

Here t_α^ι is associated with an endomorphism of $\mathfrak{h}^*$, $\iota(\lambda) := \lambda + (\kappa - 1)\langle \lambda, d \rangle \delta$ as follows:

$$\iota \circ t_\alpha(\overset{\circ}{\lambda}) = t_\alpha^\iota(\overset{\circ}{\lambda}). \tag{2.10}$$

Let a_i and $a_i^\vee$ be the labels of the Dynkin diagram from [16, Table Aff]. Note that $a_0 = 2$ if A is of type $A_{2l}^{(2)}$ and $a_0 = 1$ otherwise. Let $\theta := \delta - a_0\alpha_0 \in \overset{\circ}{\Delta}_+$, $M := \nu(\mathbb{Z}(\overset{\circ}{W} \cdot \theta^\vee)) \subset \overset{\circ}{\mathfrak{h}}{}^*$, and T_M be the corresponding group of translations of M. Then we have the following.

Proposition 2.1. *The group W is the semidirect product $W = \overset{\circ}{W} \ltimes T_M$.*

For $\alpha \in \Delta^{re}$, let $\gamma_\alpha := r$ if $\alpha \in \Delta_l$ and $\gamma_\alpha := 1$ otherwise. Then the real roots are written as

$$\Delta^{re} = \begin{cases} \{\alpha + n\gamma_\alpha\delta | \alpha \in \overset{\circ}{\Delta}, n \in \mathbb{Z}\} & \text{if } A \text{ is not of type } A_{2l}^{(2)}, \\ \{\alpha + n\gamma_\alpha\delta | \alpha \in \overset{\circ}{\Delta}, n \in \mathbb{Z}\} & \\ \quad \cup \left\{\frac{1}{2}(\alpha + (2n-1)\delta) | \alpha \in \overset{\circ}{\Delta}_l, n \in \mathbb{Z}\right\} & \text{if } A \text{ is of type } A_{2l}^{(2)}. \end{cases} \tag{2.11}$$

Let $\widehat{M} := \{\lambda \in \overset{\circ}{\mathfrak{h}}{}^* | \alpha \in \Delta^{re}, (\alpha|\lambda) \in \gamma_\alpha\mathbb{Z}\}$. Then we see that $\widehat{M} \subset \overset{c}{P}$ and $T_{\widehat{M}}$ is normalized by $\overset{\circ}{W}$.

Definition 2.2. The extended affine Weyl group $\widehat{W}$ is the semidirect product $\widehat{W} := \overset{\circ}{W} \ltimes T_{\widehat{M}}$.

The lattice $\widehat{M}$ is defined so that the extended affine Weyl group acts on Δ. Here are the explicit description of $\widehat{M}$ and its canonical basis $\{\lambda_i | i \in \overset{\circ}{I}\}$:

$$\widehat{M} = \begin{cases} \nu(\overset{\circ}{P}{}^\vee) & \text{if } r = 1, \\ \overset{\circ}{P} & \text{otherwise}, \end{cases} \qquad \lambda_i = \begin{cases} \nu(\overline{\Lambda_i^\vee}) & \text{if } r = 1, \\ \overline{\Lambda_i} & \text{otherwise}. \end{cases} \tag{2.12}$$

We also use $\widehat{M}_- := \oplus_{i \in \overset{\circ}{I}} \mathbb{Z}_{\leq 0}\lambda_i$. The action of $\widehat{W}$ is naturally induced on $\mathfrak{h}$ via the form $\langle \cdot, \cdot \rangle$.

Let Ω be the subgroup of $\widehat{W}$ which stabilizes the affine Weyl chamber C.

Proposition 2.3. *The subgroup Ω is isomorphic to $\widehat{W}/W \simeq T_{\widehat{M}}/T_M$ and thus Abelian. The extended affine Weyl group $\widehat{W}$ is isomorphic to the semidirect product $W \rtimes \Omega$.*

Definition 2.4.

1. The length $\ell(w)$ of $w \in W$ is the length ℓ of the reduced decomposition:

$$w = r_{i_1} \cdots r_{i_\ell} \quad \text{for } i_k \in I, \tag{2.13}$$

$$\ell(id) = 0. \tag{2.14}$$

2. The length $\ell(\hat{w})$ of $\hat{w} \in \widehat{W}$ is the number of the positive roots made negative by $\hat{w}^{-1}$:

$$\ell(\hat{w}) := |\Delta_{\hat{w}}|, \tag{2.15}$$

$$\Delta_{\hat{w}} := \{\alpha \in \Delta_+ \cap -\hat{w}\Delta_+\}, \tag{2.16}$$

which is equivalent to the definition $\ell(w)$ for $w \in W$. The reduced decomposition of $\hat{w} \in \widehat{W}$ is $\hat{w} = w\omega = r_{i_1} \cdots r_{i_\ell}\omega$, where $\omega \in \Omega$ and $\ell = \ell(\hat{w}) = \ell(w)$.

The set $\Delta_{\hat{w}}$ is explicitly described as $\Delta_{\hat{w}} = \{\alpha^1 = \alpha_{i_1}, \alpha^2 = r_{i_1}(\alpha_{i_2}), \dots, \alpha^\ell = w r_{i_\ell}(\alpha_{i_\ell})\}$. By definition, $\Delta_{\hat{w}}$ is independent of reduced expressions. One sees that $\Omega = \{\omega \in \widehat{W}, \ell(\omega) = 0\}$.

Definition 2.5. A weight $\lambda \in \widehat{M}$ is said to be minuscule if $\Delta_{t_{-\lambda}} \subset \mathring{\Delta}_+$.

We use the following useful formulas, which can easily be derived from the definitions above:

$$\Delta_{t_{\lambda_-}} = \begin{cases} \left\{\alpha - n\gamma_\alpha\delta \middle| \alpha \in \mathring{\Delta}_+, 0 \geq n > \dfrac{1}{\gamma_\alpha}(\lambda_-|\alpha)\right\} \\ \qquad \text{if } A \text{ is not of type } A_{2l}^{(2)}, \\ \left\{\alpha - n\gamma_\alpha\delta \middle| \alpha \in \mathring{\Delta}_+, 0 \geq n > \dfrac{1}{\gamma_\alpha}(\lambda_-|\alpha)\right\} \\ \quad \cup \left\{\dfrac{1}{2}(\alpha - (2n-1)\delta) \middle| \alpha \in (\mathring{\Delta}_+)_l, 0 \geq n > \dfrac{1}{2}(\lambda_-|\alpha)\right\} \\ \qquad \text{if } A \text{ is of type } A_{2l}^{(2)}, \end{cases} \tag{2.17a}$$

$$\ell(t_{\lambda_-}) = \begin{cases} \displaystyle\sum_{\alpha \in \mathring{\Delta}_+} \left|\frac{1}{\gamma_\alpha}(\alpha|\lambda_-)\right| & \text{if } A \text{ is not of type } A_{2l}^{(2)}, \\ \displaystyle\sum_{\alpha \in \mathring{\Delta}_+} |(\alpha|\lambda_-)| & \text{if } A \text{ is of type } A_{2l}^{(2)}, \end{cases} \tag{2.17b}$$

$$\ell(r_j t_{-\lambda_i}) = \ell(t_{-\lambda_i}) + 1, \tag{2.17c}$$
$$\ell(r_i t_{-\lambda_i}) = \ell(t_{-\lambda_i}) - 1, \tag{2.17d}$$
$$\ell(t_{\lambda_-} w) = \ell(t_{\lambda_-}) + \ell(w), \tag{2.17e}$$
$$\ell(t_{\lambda_- + \lambda'_-}) = \ell(t_{\lambda_-}) + \ell(t_{\lambda'_-}), \tag{2.17f}$$

where $i \neq j \in \mathring{I}$, $\lambda_-, \lambda'_- \in \widehat{M}_-$, and $w \in \mathring{W}$.

3 Root algebras

We shall define root algebras after Cherednik [4]. Let $\mathcal{T}$ be the tensor algebra over $\mathbb{C}$ generated by independent variables $\{R_\alpha | \alpha \in \Delta^{re}\}$. Then the action of $\hat{w} \in \widehat{W}$ on Δ^{re} induces an action on $\mathcal{T}$ by ${}^{\hat{w}} : R_\alpha \mapsto R_{\hat{w}(\alpha)}$.

Definition 3.1. Let $\mathcal{I}$ be the ideal in $\mathcal{T}$, which is generated by all the elements of the form for $i \neq j \in I$, and let $\hat{w} \in \widehat{W}$:

$${}^{\hat{w}}(\underbrace{R_{\alpha_i} \otimes R_{r_i\alpha_j} \otimes R_{r_ir_j\alpha_i} \otimes \cdots}_{m_{ij} \text{ factors}}) - {}^{\hat{w}}(\underbrace{R_{\alpha_j} \otimes R_{r_j\alpha_i} \otimes R_{r_jr_i\alpha_j} \otimes \cdots}_{m_{ij} \text{ factors}}). \tag{3.1}$$

The root algebra $\widetilde{\mathcal{R}}$ is $\mathcal{T}/\mathcal{I}$. $\{R_\alpha | \alpha \in \Delta^{re}\}$ are called the R-matrices.

Because of the $\widehat{W}$-invariance of $\mathcal{I}$, the action of $\widehat{W}$ is induced on $\widetilde{\mathcal{R}}$. For simplicity, we write products in $\widetilde{\mathcal{R}}$ in the usual way for associative algebras.

Theorem 3.2.

1. *There exists a unique set* $\{R_{\hat{w}} | \hat{w} \in \widehat{W}\} \subset \widetilde{\mathcal{R}}$ *satisfying the relations*

$$R_{v\,w} = R_v\, {}^{v}R_w, \qquad R_{r_i} = R_{\alpha_i} \quad (i \in I), \qquad R_\omega = 1, \tag{3.2}$$

where $\omega \in \Omega$, $v, w \in \widehat{W}$, *and* $\ell(v\,w) = \ell(v) + \ell(w)$.

2. *We have the* R*-matrix for* $\hat{w} \in \widehat{W}$ *and its arbitrary reduced decomposition* $\hat{w} = w\omega = r_{i_1} \cdots r_{i_\ell}\omega$ *as*

$$\begin{aligned} R_{\hat{w}} &= R_{\alpha^1} \cdots R_{\alpha^\ell}, \\ \alpha^1 &= \alpha_{i_1}, \quad \alpha^2 = r_{i_1}(\alpha_{i_2}), \ldots, \alpha^\ell = w r_{i_\ell}(\alpha_{i_\ell}) \in \Delta_{\hat{w}}. \end{aligned} \tag{3.3}$$

Instead of the original root algebra, we use the following extension, where $\widetilde{\mathcal{R}}$ is combined with the translation group $T_{\widehat{M}}$.

Definition 3.3. $\mathcal{R} := \widetilde{\mathcal{R}} \rtimes T_{\widehat{M}}$:

$$(R\,t_\lambda)(R'\,t_\mu) = R\,({}^{t_\lambda}R')\,t_{\lambda+\mu}, \tag{3.4}$$

where $R, R' \in \widetilde{\mathcal{R}}$ and $\lambda, \mu \in \widehat{M}$.

We see that $\mathcal{R}$ is generated by $\{t_{\lambda_i}, R_\alpha | i \in \mathring{I}, \alpha \in \mathring{\Delta}\}$ if A is not of type $A_{2l}^{(2)}$ and $\{t_{\lambda_i}, R_\alpha | i \in \mathring{I}, \alpha \in \mathring{\Delta}, 2\alpha - \delta \in \mathring{\Delta}\}$ if A is of type $A_{2l}^{(2)}$.

Theorem 3.4. *The subalgebra* $\mathcal{S} \subset \mathcal{R}$ *generated by* $\{Y^\lambda := R_{t_\lambda} t_\lambda | \lambda \in \widehat{M}_-\}$ *forms a commutative algebra and is generated by* $\{Y^{-\lambda_i} | i \in \mathring{I}\}$.

Proof. The proof is straightforward by (2.17) and Definition 3.3. □

4 Affine root systems of rank 3

We present some examples of the above construction. The rank of the lowest nontrivial affine root system is 3, and there are six types of affine root systems of rank 3. We denote $\alpha = \alpha_1$ and $\beta = \alpha_2$, where $|\alpha_1| \geq |\alpha_2|$, and $\lambda = \lambda_1$ and $\mu = \lambda_2$, respectively. We have the following systems that are mutually commutative by construction:

$A_2^{(1)}$

$$Y^{-\lambda} = R_\alpha\, R_{\alpha+\beta}\, t_{-\lambda}, \tag{4.1a}$$

$$Y^{-\mu} = R_\beta\, R_{\alpha+\beta}\, t_{-\mu}, \tag{4.1b}$$

$C_2^{(1)}$

$$Y^{-\lambda} = R_\alpha\, R_{\alpha+\beta}\, R_{\alpha+2\beta}\, t_{-\lambda}, \tag{4.1c}$$

$$Y^{-\mu} = R_\beta\, R_{\alpha+2\beta}\, R_{\alpha+\beta}\, R_{\alpha+2\beta+\delta}\, t_{-\mu}, \tag{4.1d}$$

$G_2^{(1)}$
$$Y^{-\lambda} = R_\alpha\, R_{\alpha+\beta}\, R_{2\alpha+3\beta}\, R_{\alpha+2\beta}\, R_{\alpha+3\beta}\, R_{2\alpha+3\beta+\delta}\, t_{-\lambda}, \tag{4.1e}$$
$$Y^{-\mu} = R_\beta\, R_{\alpha+3\beta}\, R_{\alpha+2\beta}\, R_{2\alpha+3\beta}\, R_{\alpha+\beta}\, R_{\alpha+3\beta+\delta}\, R_{2\alpha+3\beta+\delta}\, R_{\alpha+2\beta+\delta}\, R_{\alpha+3\beta+2\delta}\, R_{2\alpha+3\beta+2\delta}\, t_{-\mu}, \tag{4.1f}$$

$A_4^{(2)}$
$$Y^{-\lambda} = R_\alpha\, R_{\alpha+\beta}\, R_{\alpha+2\beta}\, R_{\frac{1}{2}\alpha+\frac{1}{2}\delta}\, R_{\alpha+\beta+\delta}\, R_{\frac{1}{2}\alpha+\beta+\frac{1}{2}\delta}\, t_{-\lambda}, \tag{4.1g}$$
$$Y^{-\mu} = R_\beta\, R_{\alpha+2\beta}\, R_{\alpha+\beta}\, R_{\frac{1}{2}\alpha+\beta+\frac{1}{2}\delta}\, t_{-\mu}, \tag{4.1h}$$

$D_3^{(2)}$
$$Y^{-\lambda} = R_\alpha\, R_{\alpha+\beta}\, R_{\alpha+2\beta}\, R_{\alpha+\beta+\delta}\, t_{-\lambda}, \tag{4.1i}$$
$$Y^{-\mu} = R_\beta\, R_{\alpha+2\beta}\, R_{\alpha+\beta}\, t_{-\mu}, \tag{4.1j}$$

$D_4^{(3)}$
$$Y^{-\lambda} = R_\alpha\, R_{\alpha+\beta}\, R_{2\alpha+3\beta}\, R_{\alpha+2\beta}\, R_{\alpha+3\beta}\, R_{\alpha+\beta+\delta}\, R_{\alpha+2\beta+\delta}\, R_{2\alpha+3\beta+3\delta}\, R_{\alpha+\beta+2\delta}\, R_{\alpha+2\beta+2\delta}\, t_{-\lambda}, \tag{4.1k}$$
$$Y^{-\mu} = R_\beta\, R_{\alpha+3\beta}\, R_{\alpha+2\beta}\, R_{2\alpha+3\beta}\, R_{\alpha+\beta}\, R_{\alpha+2\beta+\delta}\, t_{-\mu}. \tag{4.1l}$$

5 Elliptic R-matrices

For $\alpha \in \Delta^{re}$, let $\mu_\alpha \in \mathbb{C}$ be $\widehat{W}$-invariant constants: $\mu_{\hat{w}(\alpha)} = \mu_\alpha$ for $\hat{w} \in \widehat{W}$. Let $Y := \{h \in \mathfrak{h} \mid \operatorname{Re}\langle \delta, h\rangle > 0\}$ and let $\mathcal{M}$ be the set of meromorphic functions on Y. We define an action of $w = \mathring{w} t_\lambda \in \widehat{W}$ on $\mathcal{M}$ as $(w\, f)(h) := f(t^\iota_{-\lambda} \mathring{w}^{-1}(h))$.

Fix $\kappa \in \mathbb{C}$ and $\xi \in \mathring{\mathfrak{h}}^*$. We define $\widehat{R}_\alpha \in \operatorname{End}_{\mathbb{C}}\mathcal{M}$ for $\alpha \in \Delta^{re}$ by

$$\widehat{R}_\alpha := H_\alpha(\mu_\alpha) - H_\alpha(\langle \xi, \alpha^\vee\rangle)\, r_\alpha, \tag{5.1}$$

with the following function (see the definitions in the appendix):

$$H_\alpha(\nu) := \frac{\vartheta^1(-\gamma_\alpha\mu_\alpha\delta;\gamma_\alpha)}{\vartheta^{1\prime}(0;\gamma_\alpha)}\sigma_{\gamma_\alpha\nu}(\iota(\alpha);\gamma_\alpha). \tag{5.2}$$

Theorem 5.1. *The map $\pi : R_\alpha \mapsto \widehat{R}_\alpha, t_\lambda \mapsto t_\lambda$ induces a homomorphism from $\mathcal{R}$ to* $\operatorname{End}_{\mathbb{C}}\mathcal{M}$. *These R-matrices satisfy the unitarity*

$$\widehat{R}_\alpha\, \widehat{R}_{-\alpha} = \left(\frac{\vartheta^1(-\gamma_\alpha\mu_\alpha\delta;\gamma_\alpha)}{\vartheta^{1\prime}(0;\gamma_\alpha)}\right)^2 \left(\wp^0(\gamma_\alpha\mu_\alpha\delta;\gamma_\alpha) - \wp^0(\gamma_\alpha\langle\xi,\alpha^\vee\rangle\delta;\gamma_\alpha)\right)\mathrm{Id}_{\mathcal{M}}. \tag{5.3}$$

Besides the above representation, we have more general forms that depend on the relation among Q, $Q^\vee$, M. For $\alpha \in \Delta^{re}$, let

$$N_\alpha := \left\{ \phi^j_\alpha := (m_\alpha, n_\alpha) \in \mathbb{R}^2_{>0} \;\middle|\; \begin{array}{ll} \overline{\alpha^\vee} \in m_\alpha \mathring{Q}^\vee, & m_\alpha\langle \alpha, \mathring{Q}^\vee\rangle \subset \mathbb{Z}, \\ n_\alpha\gamma_\alpha\overline{\alpha^\vee} \in m_\alpha M, & m_\alpha(M|\alpha) \subset n_\alpha\gamma_\alpha\mathbb{Z} \end{array} \right\}. \tag{5.4}$$

This condition is required when the root algebra acts on the vector space spanned by theta functions (Proposition 7.1) and is an elliptic analogue in the representation of the Hecke algebras [24].

We enumerate the set N_α as follows.

	ϕ_α^1	ϕ_α^2	ϕ_α^3	ϕ_α^4
A is of type $C_l^{(1)}$ and α is long	(1,1)	(1,2)	(1/2,1)	(1/2,1/2)
A is of type $A_{2l-1}^{(2)}$ and α is long	(1,1)			(1/2,1/2)
A is of type $D_{l+1}^{(2)}$ and α is short	(1,1)	(1,2)		
A is of type $A_{2l}^{(2)}$ and α is short	(2,1)	(2,2)	(1,1)	(1,1/2)
A is of type $A_{2l}^{(2)}$ and α is long	(1,1/2)	(1,1)	(1/2,1/2)	(1/2,1/4)
otherwise	(1,1)			

Here we have numbered the elements of N_α for later convenience. Let $\zeta_\alpha^j \in \mathbb{C}$ for $1 \leq j \leq 4$ $\widehat{W}$-invariant constants. If $\phi_\alpha^j \notin N_\alpha$, set $\zeta_\alpha^j = 0$. In place of (5.2), we define

$$H_\alpha(\nu) := \sum_{\phi_\alpha^j=(m_\alpha,n_\alpha)\in N_\alpha} \zeta_\alpha^j \frac{\vartheta^1(-n_\alpha\gamma_\alpha\mu_\alpha\delta/m_\alpha;n_\alpha\gamma_\alpha)}{\vartheta^{1\prime}(0;n_\alpha\gamma_\alpha)}\sigma_{n_\alpha\gamma_\alpha\nu/m_\alpha}(\iota(m_\alpha\alpha);n_\alpha\gamma_\alpha). \tag{5.5}$$

Then we have a more general representation of $\mathcal{R}$ including Theorem 5.1.

Theorem 5.2. *The map π in Theorem* 5.1 *with* (5.5) *induces a homomorphism from $\mathcal{R}$ to* $\mathrm{End}_{\mathbb{C}}\mathcal{M}$. *These R-matrices satisfy the unitarity*

$$\widehat{R}_\alpha\,\widehat{R}_{-\alpha} = u_\alpha(\delta)\mathrm{Id}_{\mathcal{M}}, \tag{5.6}$$

where $u_\alpha(\delta)$ depends only on δ and vanishes if $\langle\xi,\alpha^\vee\rangle = \pm\mu_\alpha$.

More precisely, we have $u_\alpha(\delta) = ((p_1.\zeta_a)^2,(p_2.\zeta_a)^2,(p_3.\zeta_a)^2,(p_4.\zeta_a)^2).S.d_\alpha$,

$$S = \frac{1}{4}\begin{pmatrix} 1 & 0 & 0 & 0 \\ -1 & 0 & 0 & 1 \\ -1 & 4 & 0 & 0 \\ 1 & -4 & 4 & -1 \end{pmatrix}, \qquad \zeta_\alpha = \begin{pmatrix} \tilde{\zeta}_\alpha^1 \\ \tilde{\zeta}_\alpha^2 \\ \tilde{\zeta}_\alpha^3 \\ \tilde{\zeta}_\alpha^4 \end{pmatrix}, \qquad d_\alpha = \begin{pmatrix} d_\alpha^1 \\ d_\alpha^2 \\ d_\alpha^3 \\ d_\alpha^4 \end{pmatrix},$$

$$p_1 = (2,1,1,2), \quad p_2 = (0,0,1,2), \quad p_3 = (0,1,1,0), \quad p_4 = (0,0,1,0),$$

$$\tilde{\zeta}_\alpha^j = \zeta_\alpha^j \frac{\vartheta^1(-n_\alpha\gamma_\alpha\mu_\alpha\delta/m_\alpha;n_\alpha\gamma_\alpha)}{\vartheta^{1\prime}(0;n_\alpha\gamma_\alpha)},$$

$$d_\alpha^j = \wp^0(n_\alpha\gamma_\alpha\mu_\alpha\delta/m_\alpha;n_\alpha\gamma_\alpha) - \wp^0(n_\alpha\gamma_\alpha\langle\xi,\alpha^\vee\rangle\delta/m_\alpha;n_\alpha\gamma_\alpha).$$

Proof. We can verify the relations (3.1) case by case by a direct substitution of (5.5); for details, see [18, 19, 30]. □

We employ these operators even for the affine root systems of rank 2, though they do not have any Coxeter relations.

We shall clarify some properties of the operators $\widehat{Y}^\lambda = \pi(Y^\lambda)$.

Lemma 5.3. *The R-matrices $\widehat{R}$ satisfy the following relations:*

$$\widehat{R}_{-\alpha_j}\widehat{R}_{t_{-\lambda_i}} = {}^{r_j}\widehat{R}_{t_{-\lambda_i}}\widehat{R}_{-\alpha_j} \quad \textit{for } j \neq i, \tag{5.7}$$

$$\widehat{R}_{t_{-\lambda_i}} = \widehat{R}_{\alpha_i} R, \tag{5.8}$$

where R is a product of some R-matrices.

PROOF. Combining (2.17), the unitarity (5.6), and an equality $r_j\, t_{-\lambda_i} = t_{-\lambda_i}\, r_j$, we obtain (5.7) for generic ξ and thus for all $\xi \in \overset{\circ}{\mathfrak{h}}{}^*$. The form (5.8) is due to the fact that $\ell(r_i\, t_{-\lambda_i}) = \ell(t_{-\lambda_i}) - 1$ implies the exchange condition [2] $t_{-\lambda_i} = r_{i_1}\cdots r_{i_\ell}\omega = r_i r_{i_1}\cdots r_{i_{m-1}} r_{i_{m+1}}\cdots r_{i_\ell}\omega$ for some m. □

If the parameter ξ satisfies $\langle \xi, \alpha_i^\vee\rangle = -\mu_{-\alpha_i}$, then the R-matrix $\widehat{R}_{-\alpha_i}$ reduces to the form $\widehat{R}_{-\alpha_i} = 2\, H_{-\alpha}(\mu_{-\alpha}) P_i^{(-)}$, where $P_i^{(-)}$ is the antisymmetric projection $\frac{1}{2}(1 - r_i)$. Let

$$\overset{\circ}{\rho}_\mu := \sum_{i\in \overset{\circ}{I}} \mu_{\alpha_i}\overline{\Lambda_i} = \frac{1}{2}\sum_{\alpha\in\overset{\circ}{\Delta}_+} \mu_\alpha \alpha. \tag{5.9}$$

From these properties, we have the following theorem.

Theorem 5.4. *Let $\mathcal{V} := \mathcal{M}^{\overset{\circ}{W}}$, the $\overset{\circ}{W}$-invariant subspace of $\mathcal{M}$, and let $\xi = -\overset{\circ}{\rho}_\mu$. Then $\widehat{Y}^\lambda \in \mathrm{End}_{\mathbb{C}}\mathcal{V}$.*

PROOF. It is sufficient to check it for the generators $\widehat{Y}^{-\lambda_i}$. By Lemma 5.3, we see that $\widehat{R}_{-\alpha_j}\widehat{Y}^{-\lambda_i}|_{\mathcal{V}} = 0$ for $j \neq i$ by (5.7) and for $j = i$ by (5.8), noting that the unitarity (5.6) vanishes. Hence $\widehat{Y}^{-\lambda_i}|_{\mathcal{V}} = r_j\,\widehat{Y}^{-\lambda_i}|_{\mathcal{V}}$ for all $j \in \overset{\circ}{I}$. □

The symbol Y^λ is adopted since in a certain limit, it reduces to the same one, up to a constant factor, as in the affine Hecke algebras, where Y^λ is defined for all $\lambda \in \widehat{M}$. We remark that $\hat{Y}^\lambda$ has its inverse in $\mathrm{End}_{\mathbb{C}}\mathcal{M}$ for generic $\xi \in \overset{\circ}{\mathfrak{h}}{}^*$ but loses its inverse when $\xi = -\overset{\circ}{\rho}_\mu$.

6 Elliptic difference operators

In this section, we calculate the explicit forms of the operators $\widehat{Y}^\lambda$ for some λ on the space $\mathcal{V}$. Throughout this section, we fix $\xi = -\overset{\circ}{\rho}_\mu$.

Theorem 6.1. *Let $(-\lambda)$ be minuscule. Then we have*

$$\widehat{Y}^\lambda|_{\mathcal{V}} = \frac{1}{\left|\overset{\circ}{W}_\lambda\right|}\sum_{w\in\overset{\circ}{W}} w\left(\prod_{\substack{\alpha\in\overset{\circ}{\Delta}_+\\ (\lambda|\alpha)=-\gamma_\alpha}} H_\alpha(\mu_\alpha)\, t_\lambda\right)\Bigg|_{\mathcal{V}}, \tag{6.1}$$

where $\overset{\circ}{W}_\lambda$ is the stabilizer of λ in $\overset{\circ}{W}$.

PROOF. First, notice that R_{t_λ} consists of nonaffine R-matrices, R_α for $\alpha \in \overset{\circ}{\Delta}_+$, because $(-\lambda)$ is minuscule. Substituting the R-matrices (5.1) into $\widehat{Y}^\lambda$ and expanding them, we see that every term includes a translation operator of the form $t_{w(\lambda)}\, w$, where

$$w = r_{\alpha\{p\}} \cdots r_{\alpha\{1\}} \in \overset{\circ}{W}, \tag{6.2}$$

$$\alpha\{q\} := \alpha^{m_q}, \quad 1 \le m_p < m_{p-1} < \cdots < m_2 < m_1 \le \ell(t_\lambda), \tag{6.3}$$

and $\Delta_{t_\lambda} = \{\alpha^1 = \alpha_{i_1}, \alpha^2 = r_{i_1}(\alpha_{i_2}), \ldots, \alpha^\ell = r_{i_1} r_{i_2} \cdots r_{i_{\ell-1}}(\alpha_{i_\ell})\}$ for a reduced express $t_\lambda = r_{i_1} r_{i_2} \cdots r_{i_\ell} \omega$. Let us show that $w(\lambda) = \lambda$ implies $w = id$. Suppose $w(\lambda) = \lambda$ and $w \neq id$; then we have $\ell(w\, t_\lambda) = \ell(t_\lambda\, w)$. From (2.17), $\ell(t_\lambda\, w) = \ell(t_\lambda) + \ell(w) > \ell(t_\lambda)$, while $\ell(w\, t_\lambda) < \ell(t_\lambda)$ by a direct calculation, which leads to a contradiction. This implies that the term including $t_\lambda\, w$, $w \in \overset{\circ}{W}$ appears if and only if $w = id$. The coefficient of this term can be easily calculated:

$$\prod_{\substack{\alpha \in \overset{\circ}{\Delta}_+ \\ (\lambda|\alpha) = -\gamma_\alpha}} H_\alpha(\mu_\alpha). \tag{6.4}$$

The $\overset{\circ}{W}$-invariance of the operator $\widehat{Y}^\lambda$ yields the form (6.1). □

It is worth noting that, as in the trigonometric case [26], we can rewrite $\widehat{Y}^\lambda$ in a simply laced root system as follows:

$$\widehat{Y}^\lambda|_{\mathcal{V}} = \frac{1}{\left|\overset{\circ}{W}_\lambda\right|} \sum_{w \in \overset{\circ}{W}} \frac{(t_{-\mu w\lambda/\kappa} A_\rho)}{A_\rho}\, t_{w\lambda} \Bigg|_{\mathcal{V}}, \tag{6.5}$$

where we have set $\mu = \mu_\alpha$ and $\zeta_\alpha^1 = 1$.

In general, it is complicated and difficult to compute explicit forms of the operators when λ is not minuscule. This is the case even in the framework of the affine Hecke algebras. There is no minuscule weight available in the root systems of type $E_8^{(1)}$, $F_4^{(1)}$, $G_2^{(1)}$, $A_{2l}^{(2)}$, $E_6^{(2)}$, and $D_4^{(3)}$. However, every root system possesses a "quasi-minuscule" weight $\nu(\theta^\vee)$ in the sense of the following properties.

Lemma 6.2.

1. $\Delta_{t_{-\nu(\theta^\vee)}} = \Delta_{r_\theta} \cup \{a_0^{-1}(\delta + \theta)\}$.
2. $(\nu(\theta^\vee)|\alpha) = 0$ *or* γ_α *for* $\alpha \in \overset{\circ}{\Delta}_+$, $\alpha \neq \theta$, *and* $(\nu(\theta^\vee)|\theta) = 2$.
3. $\nu(\theta^\vee) = \lambda_i$, *where* α_i *is the unique vertex connected to* α_0 *if* A *is not of type* $A_l^{(1)}$.

Proof. We see that $r_\theta \alpha_0 = r_\theta(a_0^{-1}(\delta - \theta)) = a_0^{-1}(\delta + \theta) \in \Delta_+^{re}$, which implies the first statement due to the expression $r_\theta\, r_0 = t_{-\nu(\theta^\vee)}$. The second statement is immediate from the first and (2.17). Since $\langle \nu(\theta^\vee), \alpha_i^\vee \rangle = \langle a_0^{-1}\theta, \alpha_i^\vee \rangle = \langle a_0^{-1}\delta - \alpha_0, \alpha_i^\vee \rangle = -\langle \alpha_0, \alpha_i^\vee \rangle$, we have $\nu(\theta^\vee) = -\sum_{i \in \mathring{I}} \langle \alpha_0, \alpha_i^\vee \rangle \overline{\Lambda_i}$. Then the last statement follows from the tables in [2, 16]. □

Since every λ_i is minuscule in the root system of type $A_l^{(1)}$, we have the explicit form of $\widehat{Y}^{-\nu(\theta^\vee)}$ by Theorem 6.1. Thus we concentrate on the other root systems. Fix i as in Lemma 6.2.

By the expression $t_{-\nu(\theta^\vee)} = r_\theta\, r_0$, we have $Y^{-\nu(\theta^\vee)} = R_{r_\theta}\, R_{a_0^{-1}(\theta+\delta)}\, t_{-\nu(\theta^\vee)} = R_{r_\theta}\, t_{-\nu(\theta^\vee)}\, R_{-\alpha_0}$. For the operator $\widehat{Y}^{-\nu(\theta^\vee)}$, an analogous statement to Lemma 5.3 holds.

Lemma 6.3. *The R-matrices $\widehat{R}$ satisfy the following relations:*

$$\widehat{R}_{-\alpha_j}\widehat{R}_{r_\theta} = {}^{r_j}\widehat{R}_{r_\theta}\widehat{R}_{-\alpha_j} \quad \textit{for } j \neq i, \tag{6.6}$$

$$\widehat{R}_{r_\theta} = \widehat{R}_{\alpha_i} R, \tag{6.7}$$

where R is a product of some R-matrices.

Proof. We have $r_j\, t_{-\nu(\theta^\vee)} = t_{-\nu(\theta^\vee)}\, r_j$ and $r_j\, r_0 = r_0\, r_j$ for $j \neq i$ since α_i is the unique vertex connected to α_0. Then r_j and $r_\theta = t_{-\nu(\theta^\vee)}\, r_0$ commute, which implies $\ell(r_j\, r_\theta) = \ell(r_\theta) + 1$ and thus (6.6). The form (6.7) follows from the fact that $\ell(r_i\, t_{-\nu(\theta^\vee)}) = \ell(t_{-\nu(\theta^\vee)}) - 1$ implies $\ell(r_i\, r_\theta) = \ell(r_\theta) - 1$ and the exchange condition. □

Let $\mathring{W}_i$ be the parabolic subgroup generated by $\{r_j | j \in \mathring{I}, j \neq i\}$ and $\mathcal{V}_i$ be the $\mathring{W}_i$-invariant subspace of $\mathcal{M}$.

Lemma 6.4. *The operator $\widehat{R}_{r_\theta}\, t_{-\nu(\theta^\vee)}$ maps $\mathcal{V}_i$ to $\mathcal{V}$ and the operator $\widehat{R}_{-\alpha_0}$ maps $\mathcal{V}$ to $\mathcal{V}_i$.*

Proof. The former statement can be shown in the same way as Theorem 5.4 and the latter can be shown directly. □

Theorem 6.5.

$$\widehat{Y}^{-\nu(\theta^\vee)}|_{\mathcal{V}} = \frac{1}{\left|\mathring{W}_{\nu(\theta^\vee)}\right|} \tag{6.8}$$

$$\times \sum_{w \in \mathring{W}} w\left(\left(\prod_{\substack{\alpha \in \mathring{\Delta}_+ \\ \langle \alpha, \theta^\vee \rangle > 0}} H_\alpha(\mu_\alpha)\right)\left(H_{a_0^{-1}(\theta+\delta)}(\mu_{\alpha_0})\, t_{-\nu(\theta^\vee)} - H_{a_0^{-1}(\theta+\delta)}((\mathring{\rho}_\mu|\theta))\right)\right)\Bigg|_{\mathcal{V}}.$$

Proof. The explicit form of $\widehat{R}_{r_\theta}\, t_{-\nu(\theta^\vee)}$ on $\mathcal{V}_i$ can be computed in a similar way to Theorem 6.1. Since $Y^{-\nu(\theta^\vee)} = R_{r_\theta}\, t_{-\nu(\theta^\vee)}\, R_{-\alpha_0}$, we obtain the form (6.8). □

The operator (1.2) is actually (6.8) of type $A_{2l}^{(2)}$, where the terms without translations are gathered by use of identities of the theta functions. In [20, 22], we calculated the explicit forms of $\widehat{Y}^{-\lambda_j}|_{\mathcal{V}}$ for all $j \in \overset{\circ}{I}$ in this root system. The operator (6.8) in the affine root systems of type $E_8^{(1)}$, $F_4^{(1)}$, $G_2^{(1)}$, and $A_{2l}^{(2)}$ should be compared to the Macdonald(–Koornwinder) operator $D_{\theta^\vee}$ of type E_8, F_4, G_2, and BC_l, respectively, while the operator (6.1) in the remaining root systems of type $X_l^{(1)}$ should be compared to $E_{\nu^{-1}(\lambda_i)}$ of type X_l [23, 25].

In order to investigate a general $\widehat{Y}^\lambda$, let us define a partial order in $\widehat{M}_-$. We remark that this partial order is different from that in the affine Hecke algebras.

Definition 6.6. Let $\lambda, \lambda' \in \widehat{M}_-$. We write $\lambda \succeq \lambda'$ if $\ell(t_\lambda) > \ell(t_{\lambda'})$ or $\lambda = \lambda'$.

For an arbitrary weight $\lambda \in \widehat{M}_-$, we have the "leading term" of $\widehat{Y}^\lambda$ with respect to the order $\succ$.

Theorem 6.7. *Let $\lambda \in \widehat{M}_-$. Then we have*

$$\widehat{Y}^\lambda|_{\mathcal{V}} = \frac{1}{|\overset{\circ}{W}_\lambda|} \sum_{w \in \overset{\circ}{W}} w \left(g_\lambda^\lambda t_\lambda + \sum_{\lambda \succ \lambda'} g_{\lambda'}^\lambda t_{\lambda'} \right) \Bigg|_{\mathcal{V}}, \tag{6.9}$$

where $g_{\lambda'}^\lambda \in \mathcal{M}$. In particular, we have $g_\lambda^\lambda = \prod_{\alpha \in \Delta_{t_\lambda}} H_\alpha(\mu_\alpha)$.

Proof. Because $\widehat{Y}^\lambda$ is $\overset{\circ}{W}$-invariant, it is sufficient to calculate the coefficients of the translations of antidominant weights. A translation $t_{\lambda'}$, $\lambda' \in \widehat{M}_-$ in the expansion of $\widehat{Y}^\lambda$ appears as $wt_\lambda = t_{\lambda'}\overset{\circ}{w}$ where $\overset{\circ}{w} \in \overset{\circ}{W}$ and

$$\begin{aligned} w &= r_{\alpha\{p\}} \cdots r_{\alpha\{1\}} \in W, & (6.10)\\ \alpha\{q\} &= \alpha^{m_q}, \quad 1 \le m_p < m_{p-1} < \cdots < m_2 < m_1 \le \ell(t_\lambda). & (6.11) \end{aligned}$$

Then $\ell(t_\lambda) \ge \ell(wt_\lambda) = \ell(t_{\lambda'}\overset{\circ}{w}) = \ell(t_{\lambda'}) + \ell(\overset{\circ}{w})$, which implies $\ell(t_\lambda) > \ell(t_{\lambda'})$ if $w \ne id$. Hence we have the expression (6.9). □

Theorem 6.8 (cf. [4]). $\{\widehat{Y}^{-\lambda_i}|i \in \overset{\circ}{I}\}$ *are algebraically independent on $\mathcal{V}$.*

Proof. Consider $Y = \sum_\lambda a_\lambda Y^\lambda \in \mathcal{S}$ with $a_\lambda \in \mathbb{C}$. Let M_Y be the set of all the maximal antidominant weights in the expansion of $\widehat{Y}$ on $\mathcal{V}$. Let $M'_Y := \cup_{\lambda \in M_Y} \{\lambda' \in \widehat{M}_- | \lambda' \preceq \lambda\}$. Then we have

$$\widehat{Y}|_{\mathcal{V}} = \sum_{\lambda \in M_Y} \sum_{w \in \overset{\circ}{W}} w \left(a_\lambda g_\lambda^\lambda t_\lambda + \text{lower terms } (\lambda' \prec \lambda) \right) \Bigg|_{\mathcal{V}}. \tag{6.12}$$

Fix $\lambda \in M_Y$. There exists $h_0 \in \mathfrak{h}$ such that

$$\begin{aligned} &\{wh_0 - h_0 | w \in \overset{\circ}{W}\} & (6.13)\\ &\quad \cap \{\kappa^{-1}\langle \delta, h_0\rangle(\nu^{-1}(w\lambda') - \nu^{-1}(w'\lambda'))| w, w' \in \overset{\circ}{W}, \lambda' \in M'_Y\} = \emptyset \end{aligned}$$

and $g_\lambda^\lambda(h_0) \neq 0$. Suppose $\widehat{Y} f = 0$ for all $f \in \mathcal{V}$. Then $(\widehat{Y} f)(h_0) = 0$. Since $\{(t_{w\lambda'} f)(h_0) | \lambda' \in M'_Y, w \in \mathring{W}\}$ can be made arbitrary for suitable $f \in \mathcal{V}$, it follows that $a_\lambda = 0$, and hence we have the result. □

Corollary 6.9. $\mathcal{S} \simeq \mathbb{C}[T_{\widehat{M}_-}]$.

7 Action on theta functions of level k

The aim of this section is to show that the operators $\widehat{Y}^\lambda$ in the previous sections act on $(\widetilde{Th}^k)^{\mathring{W}}$, the $\mathring{W}$-invariant space of the theta functions of level k, or the space of the characters. To be more precise, we identify $\widehat{Y}^\lambda$ with an operator on $(\widetilde{Th}^k)^{\mathring{W}}$ by restricting the domain. We regard this space both as a $\mathbb{C}$-vector space and as an $\mathcal{O}$-module. The basic idea is from [9, 19], where the matrix elements of Belavin's $\mathbb{Z}_k$-symmetric elliptic R-matrix and associated K-matrices are calculated. Now it turns out that they are the elliptic difference operators of type $A_1^{(1)}$ or $A_2^{(2)}$.

First, let us outline our strategy. Since the representation π in Theorem 5.2 does not preserve $\widetilde{Th}^k$ for general $\xi \in \mathring{\mathfrak{h}}^*$, we introduce another representation $\bar{\pi}$ that always preserves this space. The images of $\mathcal{S}$ by π and $\bar{\pi}$ coincide when we set $\xi = -\mathring{\rho}_\mu$. As was shown, $\pi(\mathcal{S})$ at this value preserves the $\mathring{W}$-invariant subspace, so does $\bar{\pi}(\mathcal{S})$. On the other hand, $\bar{\pi}(\mathcal{S})$ preserves $\widetilde{Th}^k$ by construction; so does $\pi(\mathcal{S})$. Therefore, we can deduce that $\pi(\mathcal{S}) = \bar{\pi}(\mathcal{S})$ acts on $(\widetilde{Th}^k)^{\mathring{W}}$.

Let $h_\mu^\vee := (\mathring{\rho}_\mu|\theta) + \mu_{\alpha_0} = \sum_{i \in I} \mu_{\alpha_i} a_i^\vee$ and $\Xi := \frac{\xi + \mathring{\rho}_\mu}{h_\mu^\vee}$. Throughout this section, we fix $\kappa = \frac{h_\mu^\vee}{k}$, though some of the following statements do not require this condition.

We extend the action of t_λ on $\mathcal{M}$ for arbitrary $\lambda \in \mathring{\mathfrak{h}}^*$ by $(t_\lambda f)(h) := f(t^\iota_{-\lambda} h)$. Let $\bar{R}_\alpha \in \mathrm{End}_{\mathbb{C}} \mathcal{M}$ be defined by

$$\bar{R}_\alpha := t_{\epsilon_\alpha^1} \widehat{R_{\overline{\alpha}}}\, t_{\epsilon_\alpha^2}, \tag{7.1}$$

where $\epsilon_\alpha^1 := \frac{1}{h_\mu^\vee}(-\frac{1}{2}\mu_\alpha \overline{\alpha} - \xi + \eta_\alpha)$, $\epsilon_\alpha^2 := \frac{1}{h_\mu^\vee}(-\frac{1}{2}\mu_\alpha \overline{\alpha} + \xi - \eta_\alpha)$, and $\eta_\alpha \in \mathring{\mathfrak{h}}^*$ is taken arbitrary such that $\langle \eta_\alpha, \overline{\alpha^\vee} \rangle = 0$. Then $\bar{R}_\alpha$ does not depend on the choice of η_α and is thus well defined. According to our plan, we show that this operator acts on $\widetilde{Th}^k$.

Proposition 7.1. *For arbitrary* $\xi \in \mathring{\mathfrak{h}}^*$, $\bar{R}_\alpha \in \mathrm{End}_{\mathcal{O}}(\widetilde{Th}^k)$.

Proof. We note that

$$t^\iota_\alpha t_\beta = t_\beta t^\iota_\alpha f^{(\alpha|\beta)}, \tag{7.2}$$

where $\alpha \in \widehat{M}$, $\beta \in M$, and $f(\lambda) := \lambda - \kappa \langle \lambda, K \rangle \delta$ [5]. By using this relation and condition (5.4), we can check behavior under the action of the Heisenberg group (see the appendix) and holomorphy on the domain Y. Then we see that $\bar{R}_\alpha \in \mathrm{End}_{\mathbb{C}}(\widetilde{Th}^k)$. Since $\widehat{W}$ fixes δ, we have the proof. □

Here we shall make crucial steps to the main statement.

Lemma 7.2. *Let $\hat{w} = r_{i_1} \cdots r_{i_\ell}\omega \in \widehat{W}$ be a reduced expression. Let*

$$\eta_n := -\overset{\circ}{\rho}_\mu + \sum_{m=1}^{n-1} \nu_m \overline{\alpha^m} + \frac{1}{2}\nu_n \overline{\alpha^n}, \tag{7.3}$$

where $\Delta_{\hat{w}} = \{\alpha^1 = \alpha_{i_1}, \alpha^2 = r_{i_1}(\alpha_{i_2}), \ldots, \alpha^\ell = wr_{i_\ell}(\alpha_{i_\ell})\}$,

$$\nu_n := \begin{cases} \mu_n & \text{if } \alpha_{i_n} \neq \alpha_0, \\ -(\overset{\circ}{\rho}_\mu|\theta) & \text{if } \alpha_{i_n} = \alpha_0, \end{cases} \tag{7.4}$$

and $\mu_n := \mu_{\alpha^n}$. Then $\langle \eta_n, \overline{(\alpha^n)^\vee} \rangle = 0$.

Proof. First, observe that if $\alpha^n = r_{i_1} \cdots r_{i_{n-1}}\alpha_{i_n}$, then $\overline{\alpha^n} = \bar{r}_{i_1} \cdots \bar{r}_{i_{n-1}}\overline{\alpha_{i_n}}$ and $\overline{(\alpha^n)^\vee} = \bar{r}_{i_1} \cdots \bar{r}_{i_{n-1}}\overline{\alpha_{i_n}^\vee}$, where $\bar{r}_i := r_i$ for $i \neq 0$ and $\bar{r}_0 := r_\theta$.

$$\begin{aligned} \langle -\overset{\circ}{\rho}_\mu, \overline{(\alpha^n)^\vee} \rangle &= \langle -\overset{\circ}{\rho}_\mu, \bar{r}_{i_1} \cdots \bar{r}_{i_{n-1}}\overline{\alpha_{i_n}^\vee} \rangle \\ &= \langle -\overset{\circ}{\rho}_\mu + \nu_1\overline{\alpha^1}, \bar{r}_{i_2} \cdots \bar{r}_{i_{n-1}}\overline{\alpha_{i_n}^\vee} \rangle \\ &= \langle -\overset{\circ}{\rho}_\mu, \bar{r}_{i_2} \cdots \bar{r}_{i_{n-1}}\overline{\alpha_{i_n}^\vee} \rangle - \nu_1\langle \overline{\alpha^1}, \overline{(\alpha^n)^\vee} \rangle \\ &\ \ \vdots \\ &= -\sum_{m=1}^{n-1} \nu_m \langle \overline{\alpha^m}, \overline{(\alpha^n)^\vee} \rangle - \nu_n. \end{aligned} \tag{7.5}$$

Then we have

$$\left\langle -\overset{\circ}{\rho}_\mu + \sum_{m=1}^{n-1} \nu_m \overline{\alpha^m} + \frac{1}{2}\nu_n \overline{\alpha^n}, \overline{(\alpha^n)^\vee} \right\rangle = 0, \tag{7.6}$$

and the proof is complete. □

Proposition 7.3. *Let $\hat{w} = r_{i_1} \cdots r_{i_\ell}\omega \in \widehat{W}$ be a reduced expression. Then*

$$\bar{R}_{\alpha^1}\bar{R}_{\alpha^2} \cdots \bar{R}_{\alpha^\ell} = t_{-\Xi}\widehat{R}_{\alpha^1}\widehat{R}_{\alpha^2} \cdots \widehat{R}_{\alpha^\ell}t_\lambda t_\Xi, \tag{7.7}$$

where $\lambda = -\frac{1}{h_\mu^\vee}\sum_{n=1}^{\ell} \mu_n \overline{\alpha^n} = -\frac{1}{h_\mu^\vee}\sum_{\alpha \in \Delta_{\hat{w}}} \mu_\alpha \overline{\alpha}$.

Proof. We set $\eta_{\alpha^n} = \eta_n$, obtained in Lemma 7.2, and set

$$\epsilon_0 := \epsilon^1_{\alpha^1} = -\Xi + \frac{1}{2h^\vee_\mu}(\nu_1 - \mu_1)\overline{\alpha^1}, \tag{7.8}$$

$$\epsilon_n := \epsilon^2_{\alpha^n} + \epsilon^1_{\alpha^{n+1}} = \frac{1}{h^\vee_\mu}\left(\frac{1}{2}(\nu_n - \mu_n)\overline{\alpha^n} + \frac{1}{2}(\nu_{n+1} - \mu_{n+1})\overline{\alpha^{n+1}}\right), \quad 1 \le n \le \ell - 1, \tag{7.9}$$

$$\epsilon_\ell := \epsilon^2_{\alpha^\ell} = \Xi + \frac{1}{h^\vee_\mu}\left(-\sum_{m=1}^{\ell-1} \nu_m \overline{\alpha^m} - \frac{1}{2}(\nu_\ell + \mu_\ell)\overline{\alpha^\ell}\right). \tag{7.10}$$

Because $r_{i_1} \cdots r_{i_\ell}\omega$ is a reduced expression, we have for $1 \le n \le \ell - 1$ that

$$\epsilon_n = \begin{cases} -\frac{1}{2}\overline{\alpha^n} & \text{if } \alpha_{i_n} = \alpha_0, \\ -\frac{1}{2}\overline{\alpha^{n+1}} & \text{if } \alpha_{i_{n+1}} = \alpha_0, \\ 0 & \text{otherwise.} \end{cases} \tag{7.11}$$

Let $\bar{w}_n = \bar{r}_{i_1} \cdots \bar{r}_{i_n}$. If $\alpha_{i_n} = \alpha_0$, then $t_{(-\overline{\alpha^n}/2)}\widehat{R}_{(\overline{\alpha^n})}t_{(-\overline{\alpha^n}/2)} = \widehat{R}_{(\bar{w}_{n-1}\alpha_0)}t_{(\bar{w}_{n-1}\nu(\theta^\vee))}$, and if $\alpha_{i_n} \neq \alpha_0$, then $\widehat{R}_{(\overline{\alpha^n})} = \widehat{R}_{(\bar{w}_{n-1}\alpha_{i_n})}$. By using the identity

$$\alpha^n = r_{i_1} \cdots r_{i_{n-1}}\alpha_{i_n} = \left(\prod_{\substack{m<n \\ \alpha_{i_m}=\alpha_0}} t_{(\bar{w}_{m-1}\nu(\theta^\vee))}\right)\bar{w}_{n-1}\alpha_{i_n}, \tag{7.12}$$

we arrive at (7.7). □

Apply this proposition to an element that has two reduced expressions of the form

$$\hat{w} = \underbrace{r_i r_j r_i \cdots}_{m_{ij} \text{ factors}} = \underbrace{r_j r_i r_j \cdots}_{m_{ij} \text{ factors}} \tag{7.13}$$

for $i \neq j \in I$. Then the relation

$$\bar{R}_{\alpha_i}\bar{R}_{r_i\alpha_j}\bar{R}_{r_ir_j\alpha_i} \cdots = \bar{R}_{\alpha_j}\bar{R}_{r_j\alpha_i}\bar{R}_{r_jr_i\alpha_j} \cdots \tag{7.14}$$

immediately follows. Regarding $w\Pi$ for $w \in \overset{\circ}{W}$ as a set of fundamental roots in Lemma 7.2 and Proposition 7.3, we have proved the following theorem.

Theorem 7.4. *The map* $\bar{\pi} : R_\alpha \mapsto \bar{R}_\alpha, t_\lambda \mapsto \mathrm{Id}_{\mathcal{M}}$ *induces a homomorphism from* $\mathcal{R}$ *to* $\mathrm{End}_{\mathbb{C}}\mathcal{M}$ *and* $\mathrm{End}_{\mathcal{O}}(\widetilde{Th}^k)$.

For $\lambda \in \widehat{M}_-$, we set $\bar{Y}^\lambda := \bar{\pi}(Y^\lambda) = \bar{R}_{\alpha^1}\bar{R}_{\alpha^2} \cdots \bar{R}_{\alpha^\ell} \in \mathrm{End}_{\mathcal{O}}(\widetilde{Th}^k)$. Now we are in position to prove the main theorem.

Theorem 7.5. *Let* $\kappa = \frac{h^\vee_\mu}{k}$ *and* $\xi = -\overset{\circ}{\rho}_\mu$. *Then* $\widehat{Y}^\lambda = \bar{Y}^\lambda \in \mathrm{End}_{\mathcal{O}}((\widetilde{Th}^k)^{\overset{\circ}{W}})$.

By Proposition 7.3, we have already shown that

$$\tilde{Y}^{\lambda} = t_{-\Xi}\widehat{R}_{\alpha^1}\cdots\widehat{R}_{\alpha^\ell}t_{\lambda'}t_{\Xi} = t_{-\Xi}\widehat{R}_{t_\lambda}t_{\lambda'}t_{\Xi}, \tag{7.15}$$

where $\lambda' = -\frac{1}{h_\mu^\vee}\sum_{n=1}^{\ell}\mu_n\overline{\alpha^n} = -\frac{1}{h_\mu^\vee}\sum_{\alpha\in\Delta_{t_\lambda}}\mu_\alpha\overline{\alpha}$. Since $\Xi = 0$, if we set $\xi = -\mathring{\rho}_\mu$, we have only to show that $\lambda' = \lambda$.

Due to (2.17), we have another description of λ', which can be regarded as an image of λ by some linear map:

$$-\sum_{\alpha\in\Delta_{t_\lambda}}\mu_\alpha\overline{\alpha} = \begin{cases} \displaystyle\sum_{\alpha\in\mathring{\Delta}_+}\frac{1}{\gamma_\alpha}\mu_\alpha(\alpha|\lambda)\alpha & \text{if } A \text{ is not of type } A_{2l}^{(2)}, \\ \displaystyle\sum_{\alpha\in\mathring{\Delta}_+}\frac{1}{\gamma_\alpha}\mu_\alpha(\alpha|\lambda)\alpha + \frac{1}{4}\mu_{\alpha_0}\sum_{\alpha\in(\mathring{\Delta}_+)_l}(\alpha|\lambda)\alpha & \text{if } A \text{ is of type } A_{2l}^{(2)}. \end{cases} \tag{7.16}$$

Lemma 7.6. *Let $L : \mathring{\mathfrak{h}}^* \to \mathring{\mathfrak{h}}^*$ be a linear map defined by $L : \lambda \mapsto \frac{1}{2}\sum_{\alpha\in\mathring{\Delta}}\nu_\alpha(\alpha|\lambda)\alpha$, where ν_α is $\mathring{W}$-invariant constant. Then $L = a\mathrm{Id}_{\mathring{\mathfrak{h}}^*}$, where $a \in \mathbb{C}$.*

Proof. We see $L \in \mathrm{End}_{\mathbb{C}[\mathring{W}]}(\mathring{\mathfrak{h}}^*)$. Since $\mathbb{C}[\mathring{W}]$ acts on $\mathring{\mathfrak{h}}^*$ irreducibly, the statement follows from Schur's lemma. □

By Lemma 7.6, we see that $-\sum_{\alpha\in\Delta_{t_\lambda}}\mu_\alpha\overline{\alpha} = a\,\lambda$ for some $a \in \mathbb{C}$. The following proposition completes the proof of Theorem 7.5.

Proposition 7.7. $-\sum_{\alpha\in\Delta_{t_\lambda}}\mu_\alpha\overline{\alpha} = h_\mu^\vee\,\lambda.$

Proof. Let L be a linear map defined in the right-hand side of (7.16). Due to Lemma 7.6, we can evaluate the factor a at any element of $\mathring{\mathfrak{h}}^*$. Recall that every root system has a quasi-minuscule weight $\nu(\theta^\vee)$, whose properties we have already investigated.

- A is not of type $A_{2l}^{(2)}$:

$$L(\nu(\theta^\vee)) = \sum_{\alpha\in\mathring{\Delta}_+}\frac{1}{\gamma_\alpha}\mu_\alpha\langle\alpha,\theta^\vee\rangle\alpha = \sum_{\substack{\alpha\in\mathring{\Delta}_+\\ \langle\alpha,\theta^\vee\rangle\neq 0}}\mu_\alpha\alpha + \mu_\theta\theta = a\,\nu(\theta^\vee), \tag{7.17}$$

where we have used Lemma 6.2. By applying $(.|\theta)$ in the last equality, we obtain

$$a = \frac{1}{2}\sum_{\alpha\in\mathring{\Delta}_+}\mu_\alpha(\alpha|\theta) + \mu_\theta = (\mathring{\rho}_\mu|\theta) + \mu_{\alpha_0}. \tag{7.18}$$

- A is of type $A_{2l}^{(2)}$: In a similar manner, we have

$$L(\nu(\theta^\vee)) = \sum_{\substack{\alpha\in\mathring{\Delta}_+ \\ \langle\alpha,\theta^\vee\rangle\neq 0}} \mu_\alpha\alpha + \sum_{\substack{\alpha\in(\mathring{\Delta}_+)_l \\ \langle\alpha,\theta^\vee\rangle\neq 0}} \frac{1}{2}\mu_{\alpha_0}\alpha = a\,\nu(\theta^\vee) \tag{7.19}$$

and, consequently,

$$a = (\mathring{\rho}_\mu|\theta) + \frac{1}{4}\mu_{\alpha_0}(\rho_l|\theta) = (\mathring{\rho}_\mu|\theta) + \mu_{\alpha_0}, \tag{7.20}$$

where $\rho_l = \sum_{\alpha\in(\mathring{\Delta}_+)_l} \alpha = 2\overline{\Lambda_l}$.

In any case, $L(\nu(\theta^\vee)) = h_\mu^\vee\,\nu(\theta^\vee)$ and we have $-\sum_{\alpha\in\Delta_{t_\lambda}} \mu_\alpha\overline{\alpha} = L(\lambda) = h_\mu^\vee\,\lambda$, as required. □

Note that we also showed that

$$\sum_{\alpha\in\mathring{\Delta}} (\lambda|\alpha)(\mu|\alpha) = 2h^\vee(\lambda|\mu) \quad \text{for } \lambda,\mu\in\mathring{\mathfrak{h}}^*, \tag{7.21}$$

in the nontwisted root systems. See [16, Corollary 8.7].

8 Concluding remarks

We constructed mutually commuting difference operators by means of root algebras. Since the operator is represented in a single product of affine R-matrices, we had only to pursue the image of each R-matrix, and therefore we suceeded in proving that they act on the characters of the irreducible representations of affine Lie algebras. However, the procedure of diagonalization has yet to be solved. Prior to tackling this difficult problem, we may need to show self-adjointness on the space of the characters with respect to some inner product since there is no certainty that they can be diagonalized. In the $A_2^{(1)}$ case, self-adjointness was established in [31] for an arbitrary level of positive integer.

Since this operator was originally introduced as a quantum many-body system, self-adjointness should be also an important problem in this sense. In the trigonometric case, we readily see that Macdonald operators are essentially self-adjoint on the polynomials of exponentials since the operators are diagonalized in terms of the Macdonald polynomials. In the elliptic case, however, this problem is less investigated. See, for example, [28, 29], where the two-body system is extensively studied by constructing the explicit eigenvectors, or [17], where the extensibility to positive self-adjoint operators is shown by introducing a certain measure on a torus. These systems correspond to negative levels in terms of the affine Lie algebras, and if we treat positive level cases, the measure includes discrete parts.

We hope the construction developed in this paper sheds light on these problems.

Appendix Fundamental functions and identities

We define an action of $n = (v, \lambda, u) \in \overset{\circ}{\mathfrak{h}} \times \overset{\circ}{\mathfrak{h}}{}^* \times \mathbb{C}$ on a holomorphic function F on Y by

$$(nF)(h) := F(t_{-\lambda}(h) - 2\pi i v - (u + \pi i \langle \lambda, v \rangle)K). \tag{A.1}$$

Definition A.1. The Heisenberg group is $N_{\mathbb{Z}} = \{(v, \lambda, u) \in \overset{\circ}{\mathfrak{h}} \times \overset{\circ}{\mathfrak{h}}{}^* \times i\mathbb{R} | v \in \overset{\circ}{Q}{}^{\vee}, \lambda \in M, u + \pi i \langle \lambda, v \rangle \in 2\pi i \mathbb{Z}\}$ with multiplication:

$$(v, \lambda, u)(v', \lambda', u') := (v + v', \lambda + \lambda', u + u' + \pi i(\langle \lambda', v \rangle - \langle \lambda, v' \rangle)). \tag{A.2}$$

Definition A.2. Fix a nonnegative integer k. A theta function of level k is a holomorphic function F on the domain Y such that the following two conditions hold:

$$n(F) = F \quad \text{for all } n \in N_{\mathbb{Z}}, \tag{A.3}$$
$$n(F) = e^{-ka} F \quad \text{for all } n = (0, 0, a) \in (0, 0, \mathbb{C}). \tag{A.4}$$

Let $\widetilde{Th}^k$ denote the vector space over $\mathbb{C}$ of the theta functions of level k. It is known that $\mathcal{O} := \widetilde{T}\langle'$ is the set of holomorphic functions of $\langle \delta, h \rangle$.

For $\lambda \in \mathfrak{h}^*$ such that level$(\lambda) = k > 0$, we set

$$\Theta_\lambda := e^{-\frac{|\lambda|^2}{2k}\delta} \sum_{t \in T_M} e^{t(\lambda)}. \tag{A.5}$$

It is known that $\{\Theta_\lambda |\, \text{level}(\lambda) = k\}$ is an $\mathcal{O}$-basis of $\widetilde{Th}^k$.

Consider the root system of type $A_1^{(1)}$. Then $\Pi = \{\alpha_0, \alpha_1\}, M = \mathbb{Z}\alpha_1, (\alpha_1|\alpha_1) = 2$. We have four theta functions of level 2 for $k \in \mathbb{Z}/4\mathbb{Z}$;

$$\Theta_{2\Lambda_0 + k\overline{\Lambda_1}} = e(2\Lambda_0) \sum_{n \in \mathbb{Z}} e\left(-\frac{1}{2}\left(2n + \frac{k}{2}\right)^2 \delta + \left(2n + \frac{k}{2}\right)\alpha_1\right), \tag{A.6}$$

where $e(\lambda)(h) := \exp(\langle \lambda, h \rangle)$ for $\lambda \in \mathfrak{h}^*$. We see that level$(\rho) = h^\vee = 2$.

$$A_\rho = \sum_{w \in \overset{\circ}{W}} \varepsilon(w)\Theta_{w(\rho)} = \Theta_{2\Lambda_0 + \frac{1}{2}\alpha_1} - \Theta_{2\Lambda_0 - \frac{1}{2}\alpha_1} = e\left(\rho - \frac{1}{8}\delta\right) \prod_{\alpha \in \Delta_+} (1 - e(-\alpha)). \tag{A.7}$$

Motivated by these equations, we define theta functions for $\lambda \in \mathfrak{h}^*$ and $\gamma > 0$ by

$$\vartheta^1(\lambda;\gamma) := \sum_{n\in\mathbb{Z}} (-1)^n e\left(-\frac{1}{2}\left(n+\frac{1}{2}\right)^2 \gamma\delta + \left(n+\frac{1}{2}\right)\lambda\right), \tag{A.8a}$$

$$\vartheta^2(\lambda;\gamma) := \sum_{n\in\mathbb{Z}} e\left(-\frac{1}{2}\left(n+\frac{1}{2}\right)^2 \gamma\delta + \left(n+\frac{1}{2}\right)\lambda\right), \tag{A.8b}$$

$$\vartheta^3(\lambda;\gamma) := \sum_{n\in\mathbb{Z}} e\left(-\frac{1}{2}n^2\gamma\delta + n\lambda\right), \tag{A.8c}$$

$$\vartheta^0(\lambda;\gamma) := \sum_{n\in\mathbb{Z}} (-1)^n e\left(-\frac{1}{2}n^2\gamma\delta + n\lambda\right), \tag{A.8d}$$

and eta function

$$\eta(\delta) := e\left(-\frac{1}{24}\delta\right) \prod_{n\in\mathbb{Z}_{\geq 1}} (1 - e(-n\delta)). \tag{A.9}$$

Then by (A.6) and (A.7), we have

$$\vartheta^1(\lambda;\gamma) = e\left(\frac{\lambda}{2} - \frac{1}{8}\gamma\delta\right)(1-e(-\lambda)) \prod_{\substack{\lambda'\in\{-\lambda,0,\lambda\} \\ n\in\mathbb{Z}_{\geq 1}}} (1 - e(-\lambda' - n\gamma\delta)). \tag{A.10}$$

It is well known that

$$A_\rho = e(h^\vee \Lambda_0) f(\delta) \prod_{\alpha\in\mathring{\Delta}_+} \vartheta^1(\alpha;\gamma_\alpha) \tag{A.11}$$

for some function $f(\delta)$ which depends only on δ. We symbolically set

$$\vartheta^{1\prime}(0;\gamma) := (\eta(\gamma\delta))^3. \tag{A.12}$$

For $\lambda \in \mathfrak{h}^*$ and $\nu \in \mathbb{C}$, we define

$$\sigma_\nu(\lambda;\gamma) := \frac{\vartheta^1(\lambda-\nu\delta;\gamma)\vartheta^{1\prime}(0;\gamma)}{\vartheta^1(\lambda;\gamma)\vartheta^1(-\nu\delta;\gamma)}, \qquad \wp^0(\lambda;\gamma) := \left(\frac{\vartheta^0(\lambda;\gamma)\vartheta^{1\prime}(0;\gamma)}{\vartheta^1(\lambda;\gamma)\,\vartheta^0(0;\gamma)}\right)^2. \tag{A.13}$$

We see that these theta functions are related to the classical Jacobi theta functions $\vartheta_j(z;\tau)$, Weierstrass function $\wp(z;1,\tau)$, and Dedekind eta function $\eta(\tau)$ as

$$\vartheta^1(\lambda;\gamma)(h) = -i\vartheta_1(\langle\lambda,\mathring{h}\rangle;\gamma\tau), \tag{A.14}$$

$$\vartheta^j(\lambda;\gamma)(h) = \vartheta_j(\langle\lambda,\mathring{h}\rangle;\gamma\tau), \tag{A.15}$$

$$\wp^0(\lambda;\gamma)(h) = -\frac{1}{4\pi^2}(\wp(\langle\lambda,\mathring{h}\rangle;1,\gamma\tau) - \wp(\gamma\tau/2;1,\gamma\tau)), \tag{A.16}$$

$$\eta(\delta)(h) = \eta(\gamma\tau), \tag{A.17}$$

where we have set $h = 2\pi i(\mathring{h} - \tau d + uK)$, $\mathring{h} \in \mathfrak{h}$, and $\tau, u \in \mathbb{C}$.

Acknowledgments. The author expresses his sincere gratitude to Professors Masaki Kashiwara and Tetsuji Miwa, who kindly allowed him to speak about his research. Thanks are due also to Professors Atsuo Kuniba, Jun'ichi Shiraishi, Junji Suzuki, Yuji Yamada, and Koji Hasegawa and Drs. T. H. Baker, Goro Hatayama, Takeshi Ikeda, Tetsuya Kikuchi, Kazuhiro Hikami, and Akinori Nishino for fruitful discussions and helpful comments. In addition, the author would like to thank Professor Miki Wadati for his kind interest in this work.

The author is a Research Fellow of the Japan Society for the Promotion of Science.

References

[1] A. J. Bordner and R. Sasaki, Calogero-Moser models III: Elliptic potentials and twisting, preprint hep-th/9812232.

[2] N. Bourbaki, *Groupes et algèbre de Lie*, Hermann, Paris, 1969.

[3] I. Cherednik, Double affine Hecke algebras, Knizhnik-Zamolodchikov equations, and Macdonald's operators, *Internat. Math. Res. Notices*, **9** (1992), 171–180.

[4] I. Cherednik, Quantum Knizhnik-Zamolodchikov equations and affine root systems, *Comm. Math. Phys.*, **150** (1992), 109–136.

[5] I. Cherednik, Difference-elliptic operators and root systems, *Internat. Math. Res. Notices*, **1** (1995), 43–59.

[6] I. Cherednik, Double affine Hecke algebras and Macdonald's conjectures, *Ann. Math.*, **141** (1995), 191–216.

[7] I. Cherednik, Intertwining operators of double affine Hecke algebras, *Sel. Math.*, **3** (1997).

[8] P. I. Etingof, Central elements for quantum affine algebras and affine Macdonald's operators, *Math. Res. Lett.*, **2** (1995), 611–628.

[9] G. Felder and V. Pasquier, A simple construction of elliptic R-matrices, *Lett. Math. Phys.*, **32** (1994), 167.

[10] G. Felder and A. Varchenko, Elliptic quantum groups and Ruijsenaars models, preprint q-alg/9704005.

[11] K. Hasegawa, Ruijsenaars' commuting difference operators as commuting transfer matrices, *Comm. Math. Phys.*, **187** (1997), 289–325.

[12] K. Hasegawa, T. Ikeda, and T. Kikuchi, Commuting difference operators arising from the elliptic $C_2^{(1)}$-face model, preprint math.QA/9810062.

[13] K. Hikami and Y. Komori, Diagonalization of the elliptic Ruijsenaars model: Correspondence with the Belavin model, *European Phys. J.*, **B5** (1998), 583–588.

[14] K. Hikami and Y. Komori, Diagonalization of the elliptic Ruijsenaars model of type-BC, *J. Phys. Soc. Japan*, **67** (1998), 4037–4044.

[15] J. E. Humphreys, *Reflection Groups and Coxeter Groups*, Cambridge University Press, Cambridge, 1990.

[16] V. G. Kac, *Infinite Dimensional Lie Algebras*, Cambridge University Press, Cambridge, 1990.

[17] Y. Komori, Notes on the elliptic Ruijsenaars operators, *Lett. Math. Phys.*, **46** (1998), 147–155.

[18] Y. Komori, Functional equations arising from the root algebras, in preparation.

[19] Y. Komori and K. Hikami, Elliptic K-matrix associated with Belavin's symmetric R-matrix, *Nuclear Phys. B*, **494** (1997), 687–701.

[20] Y. Komori and K. Hikami, Quantum integrability of the generalized elliptic Ruijsenaars models, *J. Phys. A*, **30** (1997), 4341–4364.

[21] Y. Komori and K. Hikami, Affine R-matrix and the generalized elliptic Ruijsenaars models, *Lett. Math. Phys.*, **43** (1998), 335–346.

[22] Y. Komori and K. Hikami, Conserved operators of the generalized elliptic Ruijsenaars models, *J. Math. Phys.*, **39** (1998), 6175–6190.

[23] T. H. Koornwinder, Askey-Wilson polynomials for root systems of type BC, *Contemp. Math.*, **138** (1992), 189–204.

[24] G. Luszig, Affine-Hecke algebras and their graded version, *J. Amer. Math. Soc.*, **2** (1989), 599–635.

[25] I. G. Macdonald, Orthogonal polynomials associated with root systems, preprint.

[26] I. G. Macdonald, *Symmetric Functions and Hall Polynomials*, 2nd ed., Oxford University Press, Oxford, 1995.

[27] S. N. M. Ruijsenaars, Complete integrability of relativistic Calogero-Moser systems and elliptic function identities, *Comm. Math. Phys.*, **110** (1987), 191–213.

[28] S. N. M. Ruijsenaars, Generalized Lamé functions I: The elliptic case, *J. Math. Phys.*, **40** (1999), 1595–1626.

[29] S. N. M. Ruijsenaars, Generalized Lamé functions II: Hyperbolic and trigonometric specializations, *J. Math. Phys.*, **40** (1999), 1627–1663.

[30] Y. Shibukawa and K. Ueno, Completely $\mathbb{Z}$ symmetric R matrix, *Lett. Math. Phys.*, **25** (1992), 239–248.

[31] E. K. Sklyanin, Some algebraic structures connected with the Yang-Baxter equation. representation of quantum algebras, *Funktsional Anal. i Prilozhen*, **17** (1983), 34–48.

[32] J. F. van Diejen, Integrability of difference Calogero-Moser systems, *J. Math. Phys.*, **35** (1994), 2983–2998.

[33] J. F. van Diejen, Difference Calogero-Moser systems and finite toda chains, *J. Math. Phys.*, **36** (1995), 1299–1323.

Department of Physics
Graduate School of Science
University of Tokyo
Hongo 7-3-1
Bunkyo, Tokyo 113
Japan
komori@monet.phys.s.u-tokyo.ac.jp

A Generalization of the q-Saalschütz Sum and the Burge Transform

Anne Schilling and S. Ole Warnaar

Abstract. A generalization of the q-(Pfaff–)Saalschütz summation formula is proved. This implies a generalization of the Burge transform, resulting in an additional dimension of the "Burge tree". Limiting cases of our summation formula imply the (higher-level) Bailey lemma, provide a new decomposition of the q-multinomial coefficients, and can be used to prove the Lepowsky and Primc formula for the $A_1^{(1)}$ string functions.

Key Words. q-Saalschütz sum, Burge transform, Bailey lemma, q-multinomial coefficients, $A_1^{(1)}$ string functions

AMS Subject Classfications. Primary 33D15, 05A30, 05A10

1 Introduction

One of the most important summation formulas for basic hypergeometric functions is Jackson's q-analogue of a ${}_3F_2$ summation formula of Pfaff and Saalschütz. Employing standard notation (see, e.g., Gasper and Rahman [13]), this q-(Pfaff–)Saalschütz sum is written as

$$ {}_3\phi_2\left[\begin{matrix} a, b, q^{-n} \\ c, abq^{1-n}/c \end{matrix}; q, q\right] := \sum_{k=0}^{n} \frac{(a)_k(b)_k(q^{-n})_k\, q^k}{(q)_k(c)_k(abq^{1-n}/c)_k} = \frac{(c/a)_n(c/b)_n}{(c)_n(c/ab)_n} \qquad (1.1) $$

for $n \in \mathbb{Z}_+$. Here $(a)_n$ is the q-shifted factorial, defined for all integers n by

$$ (a;q)_\infty = (a)_\infty = \prod_{k=0}^{\infty}(1-aq^k) \qquad \text{and} \qquad (a;q)_n = (a)_n = \frac{(a)_\infty}{(aq^n)_\infty}. $$

Defining the q-binomial coefficient as

$$\begin{bmatrix} m+n \\ m \end{bmatrix} = \begin{cases} \dfrac{(q)_{m+n}}{(q)_m (q)_n} & \text{for } m,n \geq 0, \\ 0 & \text{otherwise,} \end{cases} \tag{1.2}$$

the q-Saalschütz sum is often written as the following summation formula [17, 9, 1]:

$$\sum_{i=0}^{M} q^{i(i+\ell)} \begin{bmatrix} L_1+L_2+M-i \\ M-i \end{bmatrix} \begin{bmatrix} L_1 \\ i+\ell \end{bmatrix} \begin{bmatrix} L_2 \\ i \end{bmatrix} = \begin{bmatrix} L_1+M \\ M+\ell \end{bmatrix} \begin{bmatrix} L_2+M+\ell \\ M \end{bmatrix}, \tag{1.3}$$

valid for all L_1, L_2, M, and $\ell \in \mathbb{Z}$ except when $-L_1 \leq -\ell \leq L_2 < 0 \leq M$ or $-L_2 \leq \ell \leq L_1 < 0 \leq M+\ell$. (In these cases, the left-hand side is zero, whereas the right-hand side is not.)

In this paper, we generalize the representation (1.3) of the q-Saalschütz sum to a summation formula which transforms an N-fold sum over a product of $N+2$ q-binomials to an $(N-1)$-fold sum over a product of $N+1$ q-binomials as stated in Theorem 2.1 of the next section. This generalized q-Saalschütz sum contains many important special cases and can be applied in connection with the Burge transform, the Bailey lemma, q-multinomial coefficients, and level-N $\mathrm{A}_1^{(1)}$ string functions as summarized below.

1. In [7] Burge used equation (1.3) to establish a transformation on generating functions of (restricted) partition pairs. This "Burge transform," which generalizes a special case of the Bailey lemma, can be used to derive a tree of identities for doubly bounded Virasoro characters [7, 12]. Our generalization of (1.3) adds a further dimension to the Burge tree, as discussed in Section 3.

2. Letting b tend to infinity in (1.1) yields the q-Chu–Vandermonde summation [13, equation (II.7)]. The q-binomial version of this is obtained by letting M tend to infinity in (1.3), resulting in

$$\sum_{i=0}^{L_2} q^{i(i+\ell)} \begin{bmatrix} L_1 \\ i+\ell \end{bmatrix} \begin{bmatrix} L_2 \\ i \end{bmatrix} = \begin{bmatrix} L_1+L_2 \\ L_1-\ell \end{bmatrix} \tag{1.4}$$

for $L_1, L_2, M, \ell \in \mathbb{Z}$, except when $-L_1 \leq -\ell \leq L_2 < 0$ or $-L_2 \leq \ell \leq L_1 < 0$. This identity can be viewed as a decomposition of the q-binomial and is easily understood combinatorially using the notion of the Durfee rectangle of a partition.

The q-binomials have been generalized to q-trinomials in [3], and more generally to q-multinomials in [2, 8, 18, 22, 27]. Our generalized q-Saalschütz sum implies a generalized q-Chu–Vandermonde sum, which provides a new decomposition formula for q-multinomials in terms of q-binomials (see Section 4.1).

3. When L_1 and L_2 tend to infinity in (1.3), we are left with

$$\sum_{i=0}^{M} \frac{q^{i(i+\ell)}}{(q)_{M-i}} \frac{1}{(q)_i (q)_{i+\ell}} = \frac{1}{(q)_M (q)_{M+\ell}}. \tag{1.5}$$

Let $\{\gamma\}_{L\geq 0}$ and $\{\delta\}_{L\geq 0}$ be sequences that satisfy

$$\gamma_L = \sum_{r=L}^{\infty} \frac{\delta_r}{(q)_{r-L}(aq)_{r+L}}. \tag{1.6}$$

Then the pair (γ, δ) is called a conjugate Bailey pair relative to a [5, 24]. Replacing $M \to M - L$ and $\ell \to \ell + 2L$ in equation (1.5) implies the conjugate Bailey pair

$$\gamma_L = \frac{a^L q^{L^2}}{(q)_{M-L}(aq)_{M+L}}, \qquad \delta_L = \frac{a^L q^{L^2}}{(q)_{M-L}},$$

with $a = q^\ell$.

A limit of our generalized q-Saalschütz sum yields (a special case of) the higher-level generalization of this conjugate Bailey pair from [23, 24]. For details, see Section 4.2. This paper thus provides a new proof of the higher-level Bailey lemma of [23, 24] for a special choice of one of the parameters.

4. Finally, letting L_1, L_2, and M all tend to infinity in (1.3) yields the well-known Durfee rectangle identity

$$\sum_{i=0}^{\infty} \frac{q^{i(i+\ell)}}{(q)_i (q)_{i+\ell}} = \frac{1}{(q)_\infty}.$$

This formula has many interpretations. Here we only mention that the right-hand side can be identified with the level-1 $A_1^{(1)}$ string function. Combined with the spinon formula of the string function of [4, 6, 20, 21, 25], the analogous limit of our generalized q-Saalschütz sum yields the fermionic expression for the string function due to Lepowsky and Primc [19] (see Section 4.3).

2 A generalized q-Saalschütz identity

The next theorem states the main result of this paper and provides a generalization of the q-Saalschütz summation formula (1.3). Let C be the Cartan matrix of A_{N-1} (i.e., $C_{ij} = 2\delta_{i,j} - \delta_{|i-j|,1}$ for $i, j = 1, \ldots, N-1$, where $\delta_{i,j}$ is the Kronecker delta symbol) and let $\mathcal{I} = 2I - C$ be the corresponding incidence matrix where I is the identity matrix. Furthermore, let $\boldsymbol{e}_i$, $i = 1, \ldots, N-1$, be the standard unit vectors in $\mathbb{Z}^{N-1}$, $(\boldsymbol{e}_i)_j = \delta_{i,j}$, and denote $\boldsymbol{n}C^{-1}\boldsymbol{n} = \sum_{i,j=1}^{N-1} n_i C^{-1}_{ij} n_j$ and $\boldsymbol{e}_i C^{-1}\boldsymbol{n} = (C^{-1}\boldsymbol{n})_i$ for $\boldsymbol{n} \in \mathbb{Z}^{N-1}$.

Theorem 2.1. *Let $\sigma = 0, 1$ and let N, ℓ, M, $L_1 + \frac{\ell+\sigma}{2}$, and $L_2 + \frac{\ell+\sigma}{2}$ be integers such that $\ell + \sigma N$ is even, $N \geq 1$, and $L_1, L_2 \geq 0$. Then*

$$\sum_{i=0}^{M} q^{i(i+\ell)/N} \begin{bmatrix} L_1+L_2+M-i \\ M-i \end{bmatrix} \sum_{\substack{n \in \mathbb{Z}^{N-1} \frac{2i+\ell+\sigma N}{2N}+(C^{-1}n)_1 \in \mathbb{Z}}} q^{nC^{-1}n} \begin{bmatrix} m+n \\ n \end{bmatrix} \begin{bmatrix} L_1+\frac{1}{2}m_1 \\ i+\ell \end{bmatrix} \begin{bmatrix} L_2+\frac{1}{2}m_1 \\ i \end{bmatrix}$$

$$= \sum_{\substack{\eta \in \mathbb{Z}^{N-1} \\ \frac{\ell+\sigma N}{2N}+(C^{-1}\eta)_1 \in \mathbb{Z}}} q^{\eta C^{-1}\eta} \begin{bmatrix} \mu+\eta \\ \eta \end{bmatrix} \begin{bmatrix} L_1+\frac{1}{2}(M+\mu_1) \\ M+\ell \end{bmatrix} \begin{bmatrix} L_2+\frac{1}{2}(M+\ell+\mu_{N-1}) \\ M \end{bmatrix}, \tag{2.1}$$

with

$$m+n = \frac{1}{2}(\mathcal{I}m + (2i+\ell)e_1) \tag{2.2}$$

and

$$\mu+\eta = \frac{1}{2}(\mathcal{I}\mu + (M+\ell)e_1 + Me_{N-1}). \tag{2.3}$$

The vector $m \in \mathbb{Z}^{N-1}$ on the left-hand side is determined by the (summation) variable n through the (m, n)-system (2.2). Similarly $\mu \in \mathbb{Z}^{N-1}$ is determined by (2.3). Also, $\begin{bmatrix} m+n \\ n \end{bmatrix} = \prod_{j=1}^{N-1} \begin{bmatrix} m_j+n_j \\ n_j \end{bmatrix}$ and similarly for $\begin{bmatrix} \mu+\eta \\ \eta \end{bmatrix}$. We further note that the nature of the solutions of (2.2) depends on the parity of N. When N is odd one must have

$$\begin{aligned} m_1 \equiv m_3 \equiv \cdots \equiv m_{N-2} \equiv 0 \quad &\pmod 2, \\ m_2 \equiv m_4 \equiv \cdots \equiv m_{N-1} \equiv \ell \quad &\pmod 2, \end{aligned}$$

whereas for N even one finds

$$\begin{aligned} m_1 \equiv m_3 \equiv \cdots \equiv m_{N-1} \quad &\pmod 2, \\ m_2 \equiv m_4 \equiv \cdots \equiv m_{N-2} \equiv \ell \equiv 0 \quad &\pmod 2. \end{aligned} \tag{2.4}$$

This implies that m_1 is even for N odd so that L_1 and L_2 must be integers. This indeed follows from (since N is odd) $0 \equiv \ell + \sigma N \equiv \ell + \sigma \pmod 2$. When N is even, the partity of m_1 is not fixed and there is the freedom to choose m_1 even corresponding to $\sigma = 0$ or m_1 odd corresponding to $\sigma = 1$. (Since for N even $0 \equiv \ell + \sigma N \equiv \ell \pmod 2$, ℓ is even in accordance with (2.4) and hence, since L_1 and L_2 must be integers when m_1 is even and half an odd integer when m_1 is odd, it thus follows from $L_i + (\ell+\sigma)/2 \in \mathbb{Z}$ that σ has the same parity as m_1.) A similar analysis of the solutions of the (μ, η)-system (2.3) can be carried out. The restrictions on the sums over n and η ensure that the components of m and μ are integer and have the parity as discussed above.

Equation (2.1) yields a summation formula for every $N \geq 1$. When $N = 1$, the sums over n and η drop out; on the left-hand side $m_1 = 0$ and on the right-hand side one needs to interpret $\mu_1 = M$ and $\mu_0 = M + \ell$. Then (2.1) indeed reduces to (1.3) for $N = 1$.

PROOF OF THEOREM 2.1. Note that both sides of (2.1) are zero unless $M+\ell\geq 0$ and $M\geq 0$. Furthermore, denoting the identity (2.1) by $I(L_1,L_2,M,\ell)$, it enjoys the symmetry $I(L_1,L_2,M,\ell)=I(L_2,L_1,M+\ell,-\ell)$. Hence we may assume that $\ell\geq 0$ and $M\geq 0$ in the proof below.

Throughout the proof, we use modified q-binomials defined as

$$\begin{bmatrix} m+n \\ m \end{bmatrix} = \frac{(q^{n+1})_m}{(q)_m} \quad \text{for } m\in\mathbb{Z}_+,\, n\in\mathbb{Z}, \tag{2.5}$$

and zero otherwise. Note that $\begin{bmatrix} m+n \\ m \end{bmatrix}$ is zero if $n<0$ unless $m+n<0$. Let us now show that on both sides of (2.1) the q-binomials (1.2) can be replaced by the modified q-binomials. Since $M,\ell,L_1,L_2\geq 0$, we find from (2.2) and (2.3) that $m_i+n_i\geq 0$ and $\mu_i+\eta_i\geq 0$ if $m_i,\mu_i\geq 0$ so that $\begin{bmatrix} \boldsymbol{m}+\boldsymbol{n} \\ \boldsymbol{n} \end{bmatrix}$ and $\begin{bmatrix} \boldsymbol{\mu}+\boldsymbol{\eta} \\ \boldsymbol{\eta} \end{bmatrix}$ in (2.1) can be replaced by the modified q-binomials $\begin{bmatrix} \boldsymbol{m}+\boldsymbol{n} \\ \boldsymbol{m} \end{bmatrix}$ and $\begin{bmatrix} \boldsymbol{\mu}+\boldsymbol{\eta} \\ \boldsymbol{\mu} \end{bmatrix}$, respectively. The other q-binomials can be turned into modified q-binomials since the top entries are nonnegative by the conditions on the parameters.

The proof of (2.1) makes frequent use of the following identity which is a corollary of Sears's transformation formula for a balanced ${}_4\phi_3$ series [13, equation (III.15)]

$$\begin{aligned}\sum_{i\in\mathbb{Z}} & q^{i(i-a+e+g)}\begin{bmatrix} i+a \\ a \end{bmatrix}\begin{bmatrix} b-i \\ c-i \end{bmatrix}\begin{bmatrix} d \\ i+e \end{bmatrix}\begin{bmatrix} f \\ i+g \end{bmatrix} \\ &= \sum_{i\in\mathbb{Z}} q^{i(i-a+e+g)}\begin{bmatrix} a-g \\ a-g-i \end{bmatrix}\begin{bmatrix} b-d+e \\ c-i \end{bmatrix}\begin{bmatrix} c+d-i \\ c+e \end{bmatrix}\begin{bmatrix} i+f \\ i+g \end{bmatrix},\end{aligned} \tag{2.6}$$

where $a,b,c,d,e,f,g\in\mathbb{Z}$ and the condition $a+b=c+d+f$ applies. Since we need the Sears transform (2.6) with negative entries in the q-binomials, it is essential that definition (2.5) is used here. (The above formula is not correct for all $a,\ldots,g\in\mathbb{Z}$ with the use of (1.2)).

We start by shifting $\boldsymbol{n}\to\boldsymbol{n}+i\boldsymbol{e}_1$, followed by $i\to i-n_1$. This transforms the left-hand side of (2.1) into

$$\begin{aligned}\sum_{i,\boldsymbol{n}} & q^{(i-n_1)(i-n_1-m_1+\ell)+\boldsymbol{n}C^{-1}\boldsymbol{n}} \\ &\times \begin{bmatrix} L_1+L_2+M+n_1-i \\ M+n_1-i \end{bmatrix}\begin{bmatrix} L_1+\frac{1}{2}m_1 \\ i+\ell-n_1 \end{bmatrix}\begin{bmatrix} L_2+\frac{1}{2}m_1 \\ i-n_1 \end{bmatrix}\begin{bmatrix} m_1+i \\ m_1 \end{bmatrix}\prod_{\alpha=2}^{N-1}\begin{bmatrix} m_\alpha+n_\alpha \\ m_\alpha \end{bmatrix},\end{aligned}$$

where the sum over $\boldsymbol{n}$ is restricted by

$$\frac{\ell+\sigma N}{2N}+(C^{-1}\boldsymbol{n})_1\in\mathbb{Z} \tag{2.7}$$

and the $(\boldsymbol{m},\boldsymbol{n})$-system is given by

$$\boldsymbol{m}+\boldsymbol{n}=\frac{1}{2}(\mathcal{I}\boldsymbol{m}+\ell\boldsymbol{e}_1). \tag{2.8}$$

Since the $(\boldsymbol{m},\boldsymbol{n})$-system has become i-independent, only the first four q-binomials depend on the summation variable i. Hence we may apply (2.6) with $a = m_1$, $b = L_1 + L_2 + M + n_1$, $c = M + n_1$, $d = L_1 + \frac{1}{2}m_1$, $e = \ell - n_1$, $f = L_2 + \frac{1}{2}m_1$, and $g = -n_1$ to obtain

$$\sum_{i,\boldsymbol{n}} q^{(i-n_1)(i-n_1-m_1+\ell)+\boldsymbol{n}C^{-1}\boldsymbol{n}} \\ \times \begin{bmatrix} L_1+M+n_1+\frac{1}{2}m_1-i \\ M+\ell \end{bmatrix} \begin{bmatrix} L_2+M+\ell-\frac{1}{2}m_1 \\ M+n_1-i \end{bmatrix} \begin{bmatrix} L_2+\frac{1}{2}m_1+i \\ i-n_1 \end{bmatrix} \begin{bmatrix} m_1+n_1 \\ m_1+n_1-i \end{bmatrix} \prod_{\alpha=2}^{N-1} \begin{bmatrix} m_\alpha+n_\alpha \\ m_\alpha \end{bmatrix}.$$

Shifting $\boldsymbol{n} \to \boldsymbol{n} + i(2\boldsymbol{e}_1 - \boldsymbol{e}_2)$ and $\boldsymbol{m} \to \boldsymbol{m} - 2i\boldsymbol{e}_1$, which leaves the $(\boldsymbol{m},\boldsymbol{n})$-system (2.8) and the restriction (2.7) on the summation over $\boldsymbol{n}$ invariant, yields

$$\sum_{i,\boldsymbol{n}} q^{(i+\frac{m_2-m_1}{2})^2-(\frac{m_1-\ell}{2})^2+\boldsymbol{n}C^{-1}\boldsymbol{n}} \begin{bmatrix} L_1+M+\frac{1}{2}m_1+n_1 \\ M+\ell \end{bmatrix} \begin{bmatrix} L_2+M+\ell-\frac{1}{2}m_1+i \\ M+n_1+i \end{bmatrix} \\ \times \begin{bmatrix} L_2+\frac{1}{2}m_1 \\ -i-n_1 \end{bmatrix} \begin{bmatrix} m_1+n_1 \\ m_1+n_1-i \end{bmatrix} \begin{bmatrix} m_2+n_2-i \\ m_2 \end{bmatrix} \prod_{\alpha=3}^{N-1} \begin{bmatrix} m_\alpha+n_\alpha \\ m_\alpha \end{bmatrix},$$

where we have used the $(\boldsymbol{m},\boldsymbol{n})$-system to simplify the exponent of q. Shifting $i \to n_2-i$, one can apply (2.6) with $a = m_2, b = L_2+M+\ell-\frac{1}{2}m_1+n_2, c = M+n_1+n_2$, $d = L_2 + \frac{1}{2}m_1, e = -n_1 - n_2$, $f = m_1 + n_1$, and $g = m_1 + n_1 - n_2$, observing that

$$c + d + f - a - b = 2m_1 + 2n_1 - m_2 - \ell = 0$$

thanks to (2.8). This yields

$$\sum_{i,\boldsymbol{n}} q^{(i-\frac{m_3-m_2}{2})^2-(\frac{m_1-\ell}{2})^2+\boldsymbol{n}C^{-1}\boldsymbol{n}} \begin{bmatrix} L_1+M+\frac{1}{2}m_1+n_1 \\ M+\ell \end{bmatrix} \begin{bmatrix} L_2+M+\frac{1}{2}m_1+n_1+n_2-i \\ M \end{bmatrix} \\ \times \begin{bmatrix} m_2+n_2-m_1-n_1 \\ m_2+n_2-m_1-n_1-i \end{bmatrix} \begin{bmatrix} M+\ell-m_1-n_1 \\ M+n_1+n_2-i \end{bmatrix} \begin{bmatrix} m_1+n_1+i \\ m_1+n_1-n_2+i \end{bmatrix} \prod_{\alpha=3}^{N-1} \begin{bmatrix} m_\alpha+n_\alpha \\ m_\alpha \end{bmatrix}.$$

Shifting $\boldsymbol{n} \to \boldsymbol{n}+i(\boldsymbol{e}_1+\boldsymbol{e}_2-\boldsymbol{e}_3)$ and $\boldsymbol{m} \to \boldsymbol{m}-2i(\boldsymbol{e}_1+\boldsymbol{e}_2)$, which again leaves the $(\boldsymbol{m},\boldsymbol{n})$-system (2.8) and the restriction (2.7) on the sum over $\boldsymbol{n}$ unchanged, leads to

$$\sum_{i,\boldsymbol{n}} q^{(i+\frac{m_3-m_2}{2})^2-(\frac{m_1-\ell}{2})^2+\boldsymbol{n}C^{-1}\boldsymbol{n}} \begin{bmatrix} L_1+M+n_1+\frac{1}{2}m_1 \\ M+\ell \end{bmatrix} \begin{bmatrix} L_2+M+\frac{1}{2}m_1+n_1+n_2 \\ M \end{bmatrix} \\ \times \begin{bmatrix} m_2+n_2-m_1-n_1 \\ m_2+n_2-m_1-n_1-i \end{bmatrix} \begin{bmatrix} M+\ell-m_1-n_1+i \\ M+n_1+n_2+i \end{bmatrix} \begin{bmatrix} m_1+n_1 \\ m_1+n_1-n_2-i \end{bmatrix} \begin{bmatrix} m_3+n_3-i \\ m_3 \end{bmatrix} \prod_{\alpha=4}^{N-1} \begin{bmatrix} m_\alpha+n_\alpha \\ m_\alpha \end{bmatrix}. \tag{2.9}$$

We now need the following lemma.

Lemma 2.2. *For $p = 3, \ldots, N$, let*

$$f_p = \sum_{i,\boldsymbol{n}} q^{(i+\frac{m_p-m_{p-1}}{2})^2-(\frac{m_1-\ell}{2})^2+\boldsymbol{n}C^{-1}\boldsymbol{n}} \begin{bmatrix} L_1+M+\frac{1}{2}m_1+n_1 \\ M+\ell \end{bmatrix} \begin{bmatrix} L_2+M+\frac{1}{2}m_1+n_1+n_2 \\ M \end{bmatrix}$$
$$\times \left(\prod_{\alpha=1}^{p-3} \begin{bmatrix} M+\sum_{\beta=1}^{\alpha+2} n_\beta+\sum_{\beta=1}^{\alpha}(-1)^{\alpha-\beta}(m_\beta+n_\beta) \\ M+\sum_{\beta=1}^{\alpha} n_\beta+\sum_{\beta=1}^{\alpha}(-1)^{\alpha-\beta}(m_\beta+n_\beta) \end{bmatrix} \right) \left(\prod_{\alpha=p+1}^{N-1} \begin{bmatrix} m_\alpha+n_\alpha \\ m_\alpha \end{bmatrix} \right)$$
$$\times \begin{bmatrix} \sum_{\alpha=1}^{p-1}(-1)^{p-\alpha-1}(m_\alpha+n_\alpha) \\ \sum_{\alpha=1}^{p-1}(-1)^{p-\alpha-1}(m_\alpha+n_\alpha)-i \end{bmatrix} \begin{bmatrix} M+\ell-m_1-\sum_{\alpha=1}^{p-2} n_\alpha+i \\ M+\sum_{\alpha=1}^{p-1} n_\alpha+i \end{bmatrix}$$
$$\times \begin{bmatrix} \sum_{\alpha=1}^{p-2}(-1)^{p-\alpha}(m_\alpha+n_\alpha) \\ \sum_{\alpha=1}^{p-2}(-1)^{p-\alpha}(m_\alpha+n_\alpha)-n_{p-1}-i \end{bmatrix} \begin{bmatrix} m_p+n_p-i \\ m_p \end{bmatrix},$$

with $(\boldsymbol{m},\boldsymbol{n})$*-system* (2.8) *and* $m_N = n_N = 0$. *Then* $f_p = f_{p+1}$ *for* $3 \le p < N$.

Proof. Change $i \to n_p - i$ and apply (2.6) with $a = m_p$, $b = M + \ell - m_1 - \sum_{\alpha=1}^{p-2} n_\alpha + n_p$, $c = M + \sum_{\alpha=1}^{p} n_\alpha$, $d = \sum_{\alpha=1}^{p-2}(-1)^{p-\alpha}(m_\alpha+n_\alpha)$, $e = d - n_{p-1} - n_p$, $f = \sum_{\alpha=1}^{p-1}(-1)^{p-\alpha-1}(m_\alpha + n_\alpha)$, and $g = f - n_p$, observing that

$$c+d+f-a-b = 2\sum_{\alpha=1}^{p-1} n_\alpha + m_1 + m_{p-1} - m_p - \ell = 0$$

by summing up the first $p-1$ components of the $(\boldsymbol{m},\boldsymbol{n})$-system (2.8). This leads to

$$f_p = \sum_{i,\boldsymbol{n}} q^{(n_p-i+\frac{m_p-m_{p-1}}{2})^2-(\frac{m_1-\ell}{2})^2+\boldsymbol{n}C^{-1}\boldsymbol{n}} \begin{bmatrix} L_1+M+\frac{1}{2}m_1+n_1 \\ M+\ell \end{bmatrix} \begin{bmatrix} L_2+M+\frac{1}{2}m_1+n_1+n_2 \\ M \end{bmatrix}$$
$$\times \left(\prod_{\alpha=1}^{p-3} \begin{bmatrix} M+\sum_{\beta=1}^{\alpha+2} n_\beta+\sum_{\beta=1}^{\alpha}(-1)^{\alpha-\beta}(m_\beta+n_\beta) \\ M+\sum_{\beta=1}^{\alpha} n_\beta+\sum_{\beta=1}^{\alpha}(-1)^{\alpha-\beta}(m_\beta+n_\beta) \end{bmatrix} \right) \left(\prod_{\alpha=p+1}^{N-1} \begin{bmatrix} m_\alpha+n_\alpha \\ m_\alpha \end{bmatrix} \right) \tag{2.10}$$
$$\times \begin{bmatrix} \sum_{\alpha=1}^{p}(-1)^{p-\alpha}(m_\alpha+n_\alpha) \\ \sum_{\alpha=1}^{p}(-1)^{p-\alpha}(m_\alpha+n_\alpha)-i \end{bmatrix} \begin{bmatrix} M+\ell-m_1-\sum_{\alpha=1}^{p-1} n_\alpha \\ M+\sum_{\alpha=1}^{p} n_\alpha-i \end{bmatrix}$$
$$\times \begin{bmatrix} M+\sum_{\alpha=1}^{p} n_\alpha+\sum_{\alpha=1}^{p-2}(-1)^{p-\alpha}(m_\alpha+n_\alpha)-i \\ M+\sum_{\alpha=1}^{p-2} n_\alpha+\sum_{\alpha=1}^{p-2}(-1)^{p-\alpha}(m_\alpha+n_\alpha) \end{bmatrix} \begin{bmatrix} \sum_{\alpha=1}^{p-1}(-1)^{p-\alpha-1}(m_\alpha+n_\alpha)+i \\ \sum_{\alpha=1}^{p-1}(-1)^{p-\alpha-1}(m_\alpha+n_\alpha)-n_p+i \end{bmatrix}.$$

We now carry out the transformations $\boldsymbol{n} \to \boldsymbol{n} + i(\boldsymbol{e}_1 + \boldsymbol{e}_p - \boldsymbol{e}_{p+1})$ and $\boldsymbol{m} \to \boldsymbol{m} - 2i(\boldsymbol{e}_1 + \boldsymbol{e}_2 + \cdots + \boldsymbol{e}_p)$, which leave the $(\boldsymbol{m},\boldsymbol{n})$-system unchanged. (Here $\boldsymbol{e}_N := 0$.) Using $\boldsymbol{n}C^{-1}(\boldsymbol{e}_1+\boldsymbol{e}_p-\boldsymbol{e}_{p+1}) = \sum_{\alpha=1}^{p} n_\alpha$ and $(\boldsymbol{e}_1+\boldsymbol{e}_p-\boldsymbol{e}_{p+1})C^{-1}(\boldsymbol{e}_1+\boldsymbol{e}_p-\boldsymbol{e}_{p+1}) = 2$, as well as the $(\boldsymbol{m},\boldsymbol{n})$-system, yields

$$\left(n_p - i + \frac{m_p - m_{p-1}}{2}\right)^2 - \left(\frac{m_1-\ell}{2}\right)^2 + \boldsymbol{n}C^{-1}\boldsymbol{n}$$
$$\to \left(n_p + \frac{m_p - m_{p-1}}{2}\right)^2 - \left(i - \frac{m_1-\ell}{2}\right)^2 + 2i\left(i + \sum_{\alpha=1}^{p} n_\alpha\right) + \boldsymbol{n}C^{-1}\boldsymbol{n}$$
$$= \left(i + \frac{m_{p+1} - m_p}{2}\right)^2 - \left(\frac{m_1-\ell}{2}\right)^2 + \boldsymbol{n}C^{-1}\boldsymbol{n}$$

transforming (2.10) into f_{p+1} as desired. □

Equation (2.9) corresponds to f_3 and we can thus use the above lemma to replace it with f_N. Since $m_N = 0$, the last q-binomial in f_N is 1 and we can perform the sum over i using the q-Saalschütz sum, which is the special case $a = 0$ of the Sears transformation (2.6). (When $a = 0$, the only nonvanishing term on the right-hand side of (2.6) corresponds to $i = -g$.) Specifically, we take f_N, replace i by $-i$ and apply (2.6) with the same choice of parameters as in the proof of Lemma 2.2 but with $p = N$, $n_N = 0$, and $a = m_N = 0$. Then we get

$$\begin{aligned}\sum_{\boldsymbol{n}} & q^{(\frac{m_{N-1}}{2})^2-(\frac{m_1-\ell}{2})^2+(n_{N-1}-\sum_{\alpha=1}^{N-2}(-1)^{N-\alpha}(m_\alpha+n_\alpha))(\sum_{\alpha=1}^{N-1}(-1)^{N-\alpha-1}(m_\alpha+n_\alpha))} \\ & \times q^{\boldsymbol{n}C^{-1}\boldsymbol{n}}\left(\prod_{\alpha=1}^{N-3}\begin{bmatrix}M+\sum_{\beta=1}^{\alpha+2}n_\beta+\sum_{\beta=1}^{\alpha}(-1)^{\alpha-\beta}(m_\beta+n_\beta)\\ M+\sum_{\beta=1}^{\alpha}n_\beta+\sum_{\beta=1}^{\alpha}(-1)^{\alpha-\beta}(m_\beta+n_\beta)\end{bmatrix}\right) \qquad (2.11) \\ & \times \begin{bmatrix}L_1+M+\frac{1}{2}m_1+n_1\\ M+\ell\end{bmatrix}\begin{bmatrix}M+\ell-m_1-\sum_{\alpha=1}^{N-2}n_\alpha\\ M+\ell-m_1-\sum_{\alpha=1}^{N-1}n_\alpha-\sum_{\alpha=1}^{N-1}(-1)^{N-\alpha-1}(m_\alpha+n_\alpha)\end{bmatrix} \\ & \times \begin{bmatrix}L_2+M+\frac{1}{2}m_1+n_1+n_2\\ M\end{bmatrix}\begin{bmatrix}M+\ell-m_1-\sum_{\alpha=1}^{N-1}n_\alpha\\ M+\ell-m_1-\sum_{\alpha=1}^{N-2}n_\alpha-\sum_{\alpha=1}^{N-2}(-1)^{N-\alpha}(m_\alpha+n_\alpha)\end{bmatrix}.\end{aligned}$$

All that remains to be done is to clean up the above expression. Introduce a new variable $\boldsymbol{\eta} \in \mathbb{Z}^{N-1}$ through its components as follows:

$$\begin{aligned}\eta_i &= n_{2i}+n_{2i+1} && \text{for } i=1,\dots,\lfloor N/2\rfloor-1,\\ \eta_{N-i} &= n_{2i+1}+n_{2i+2} && \text{for } i=1,\dots,\lfloor (N-1)/2\rfloor-1,\\ \eta_{\lfloor (N+1)/2\rfloor} &= \sum_{\alpha=1}^{N-2}(-1)^{N-\alpha}(m_\alpha+n_\alpha)-n_{N-1}, \\ \eta_{\lfloor (N+1)/2\rfloor\pm 1} &= \sum_{\alpha=1}^{N-1}(-1)^{N-\alpha-1}(m_\alpha+n_\alpha)+n_{N-1}\end{aligned}$$

for N even/odd. Also define $\boldsymbol{\mu}$ through the $(\boldsymbol{\mu},\boldsymbol{\eta})$-system (2.3). Eliminating $\boldsymbol{m}$ and $\boldsymbol{n}$ from (2.11) in favor of $\boldsymbol{\mu}$ and $\boldsymbol{\eta}$, we finally get the right-hand side of (2.1). We also note that $(C^{-1}\boldsymbol{\eta})_1 = \sum_{i=1}^{N-1}(N-i)\eta_i/N$ yields $(-n_1+\sum_{i=2}^{N-1}(N-i)n_i)/N = (C^{-1}\boldsymbol{n})_1 - n_1$ so that the restriction (2.7) on the sum over $\boldsymbol{n}$ translates into the restriction

$$\frac{\ell+\sigma N}{2N}+(C^{-1}\boldsymbol{\eta})_1\in\mathbb{Z}$$

for the sum over $\boldsymbol{\eta}$ as it should. □

3 The Burge transform

Perhaps the most interesting application of our generalized q-Saalschütz sum (2.1) arises when it is combined with the Burge transform [7, 12]. The Burge transform is a generalization of (a special case) of the Bailey lemma and can be utilized to derive an infinite tree (a Burge tree) of polynomial identities from a single initial identity. In this section, we show that each element of a Burge tree can be transformed using (2.1) to yield an additional infinite series of polynomial identities.

In his study of restricted partition pairs, Burge considered the polynomial

$$\begin{aligned} X^{(p,p')}_{r,s}&(M_1,L_1,M_2,L_2) \qquad\qquad (3.1)\\ &=\sum_{j=-\infty}^{\infty}\Big\{q^{j(pp'j+p'(M_{12}+r)-ps)}\begin{bmatrix}M_1+L_1-(p'-p)j\\ M_1+pj\end{bmatrix}\begin{bmatrix}M_2+L_2+(p'-p)j\\ M_2-pj\end{bmatrix}\\ &\quad -q^{(pj+M_{12}+r)(p'j+s)}\begin{bmatrix}M_1+L_1-(p'-p)j+r-s\\ M_1+pj+r\end{bmatrix}\begin{bmatrix}M_2+L_2+(p'-p)j-r+s\\ M_2-pj-r\end{bmatrix}\Big\},\end{aligned}$$

with $M_{12}=M_1-M_2$ and proved that it is the generating function of pairs of partitions (λ,μ) such that

$$0\le\lambda_1\le\cdots\le\lambda_{M_1}\le L_1,\qquad 0\le\mu_1\le\cdots\le\mu_{M_2}\le L_2,$$

and

$$\lambda_i-\mu_{i-r+1}\ge 1-s,\qquad \mu_i-\lambda_{i-p+r+1}\ge 1-p'+s.$$

Here the integers p, p', r, and s are restricted to $p,p'\ge 1$, $0\le r+M_{12}\le p$, and $0\le s-L_{12}\le p'$ with $L_{12}=L_1-L_2$. There are four exceptional cases—$r=0$, $r=p$, $r=-M_{12}$, and $r=p-M_{12}$—that demand the additional conditions $\mu_1\le s-1$, $\lambda_1\le p'-s-1$, $\lambda_{M_2}\ge L_1-s+1$, and $\mu_{M_1}\ge L_2-p'+s+1$, respectively [14, 12].

The important observation made in [7] is that

$$\begin{aligned} X^{(p,p+p')}_{r,r+s}&(M_1,L_1,M_2,L_2) \qquad\qquad (3.2)\\ &=\sum_{i\in\mathbb{Z}}q^{i(i+M_{12})}\begin{bmatrix}L_1+L_2+M_2-i\\ M_2-i\end{bmatrix}X^{(p,p')}_{r,s}(i+M_{12},L_1-i,i,L_2-M_{12}-i)\end{aligned}$$

and

$$\begin{aligned} X^{(p',p+p')}_{s-M_{12},r+s+L_{12}}&(M_1,L_1,M_2,L_2) \qquad\qquad (3.3)\\ &=\sum_{i\in\mathbb{Z}}q^{i(i+M_{12})}\begin{bmatrix}L_1+L_2+M_2-i\\ M_2-i\end{bmatrix}X^{(p,p')}_{r,s}(L_1-i,i+M_{12},L_2-M_{12}-i,i),\end{aligned}$$

where the second equation follows from the first by exploiting the symmetry

$$X^{(p,p')}_{r,s}(M_1,L_1,M_2,L_2)=X^{(p',p)}_{s-L_{12},r+M_{12}}(L_1,M_1,L_2,M_2). \qquad (3.4)$$

The proof of the Burge transform follows from the q-Saalschütz formula (1.3). In [7, 12], the defining equation (3.1) is substituted into (3.2), then the sums over i and j are interchanged, followed by the variable change $i \to i+pj$ and $i \to i+pj+r$ in the terms corresponding to the second and third line of (3.1), respectively (referred to as the positive and negative terms below). Then the q-Saalschütz sum is used with $L_1 \to L_1 + M_{12} - (p'-p)j$, $L_2 \to L_2 - M_{12} + (p'-p)j$, $M \to M_2 - pj$, and $\ell \to M_{12} + 2pj$ for the positive terms and $L_1 \to L_1 + M_{12} - (p'-p)j + r - s$, $L_2 \to L_2 - M_{12} + (p'-p)j - r + s$, $M \to M_2 - pj - r$, and $\ell \to M_{12} + 2pj + 2r$ for the negative terms. This gives the left-hand side of (3.2). However, we note that it needs to be verified that the summation (1.3) has not been employed when the variables therein lie in the ranges given just below (1.3). This means that

$$\begin{aligned} -L_1 - M_{12} + (p'-p)j - r + s &\le -M_{12} - 2pj - 2r \\ &\le L_2 - M_{12} + (p'-p)j - r + s < 0 \\ &\le M_2 - pj - r \end{aligned} \tag{3.5}$$

and

$$\begin{aligned} -L_2 + M_{12} - (p'-p)j + r - s &\le M_{12} + 2pj + 2r \\ &\le L_1 + M_{12} - (p'-p)j + r - s < 0 \\ &\le M_1 + pj + r, \end{aligned} \tag{3.6}$$

and the corresponding inequalities obtained by setting $r = s = 0$ should not hold for any $j \in \mathbb{Z}$. Eliminating j gives several conditions on the parameters in (3.2). In particular, (3.5) can hold only if

$$2pj > -M_{12} - 2r \quad \text{and} \quad 2p'j < M_{12} + L_{12} - 2s.$$

Similarly, (3.6) can hold only if

$$2pj < -M_{12} - 2r \quad \text{and} \quad 2p'j > M_{12} + L_{12} - 2s.$$

If, for example, $M_{12} = L_{12} = 0$ these conditions cannot be satisfied for any j, recalling that $0 \le r \le p$ and $0 \le s \le p'$. Hence, setting

$$X_{r,s}^{(p,p')}(M, L, M, L) = X_{r,s}^{(p,p')}(M, L),$$

the symmetric version of the Burge transform (3.2),

$$X_{r,r+s}^{(p,p+p')}(M, L) = \sum_{i=0}^{M} q^{i^2} \begin{bmatrix} 2L + M - i \\ 2L \end{bmatrix} X_{r,s}^{(p,p')}(i, L-i), \tag{3.7}$$

always holds. By the same arguments, one can show that the symmetric form of (3.3),

$$X_{s,r+s}^{(p',p+p')}(M,L) = \sum_{i=0}^{M} q^{i^2} \begin{bmatrix} 2L+M-i \\ 2L \end{bmatrix} X_{r,s}^{(p,p')}(L-i,i),$$

is true for arbitrary M and L.

By iterating the two Burge transformations, starting with an appropriate initial identity for $X_{r,s}^{(p,p')}$, one can derive an infinite tree of polynomial identities. This was mentioned in [7] and explicitly carried out in [12]. To illustrate this, we follow [12] and use the trivial result

$$X_{0,1}^{(1,2)}(M,L) = \delta_{L,0} \tag{3.8}$$

to derive the Burge tree (where a node with label $X_{r,s}^{(p,p')}$ corresponds to a polynomial identity for $X_{r,s}^{(p,p')}(M,L)$)

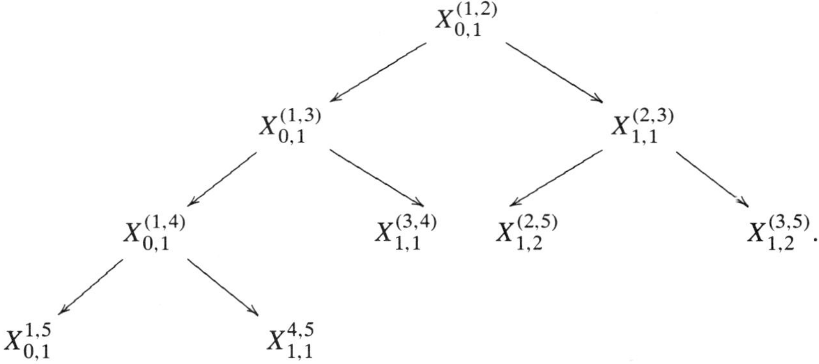

(Actually, in [12], an extension of the Burge tree was constructed by exploiting various symmetries of $X_{r,s}^{(p,p')}$.) Explicitly some of the identities in the above tree are [7, 12]

$$X_{0,1}^{(1,3)}(M,L) = q^{L^2} \begin{bmatrix} L+M \\ 2L \end{bmatrix}, \tag{3.9}$$

$$X_{1,1}^{(2,3)}(M,L) = \begin{bmatrix} 2L+M \\ 2L \end{bmatrix}, \tag{3.10}$$

$$X_{1,1}^{(3,4)}(M,L) = \sum_{\substack{m=0 \\ m \text{ even}}}^{L} q^{\frac{1}{2}m^2} \begin{bmatrix} 2L+M-\frac{1}{2}m \\ 2L \end{bmatrix} \begin{bmatrix} L \\ m \end{bmatrix}, \tag{3.11}$$

$$X_{1,2}^{(2,5)}(M,L) = \sum_{n=0}^{L} q^{n^2} \begin{bmatrix} 2L+M-n \\ 2L \end{bmatrix} \begin{bmatrix} 2L-n \\ n \end{bmatrix}. \tag{3.12}$$

Equation (3.10) is a doubly bounded version of the Euler identity, equation (3.11) is a doubly bounded analogue of the vacuum-character identity of the Ising model

$$\sum_{\substack{m=0\\ m\text{ even}}}^{\infty} \frac{q^{\frac{1}{2}m^2}}{(q)_m} = \frac{1}{2}\left\{(-q^{1/2})_\infty + (q^{1/2})_\infty\right\}$$
$$= \prod_{j=1}^{\infty} \frac{(1+q^{8j-3})(1+q^{8j-5})(1-q^{8j})}{1-q^{2j}},$$

and (3.12) is a doubly bounded version of the (first) Rogers–Ramanujan identity

$$\sum_{n=0}^{\infty} \frac{q^{n^2}}{(q)_n} = \prod_{j=1}^{\infty} \frac{1}{(1-q^{5j-1})(1-q^{5j-4})}.$$

To see how (2.1) transforms an identity in the Burge tree, let us first introduce a generalization of the polynomial $X_{r,s}^{(p,p')}(M_1, L_1, M_2, L_2)$ as follows. Let N be a positive integer, let $\sigma = 0, 1$, and let M_1, M_2, $L_1 + \frac{M_{12}+\sigma}{2}$, and $L_2 + \frac{M_{12}+\sigma}{2}$ be integers such that $M_{12} + \sigma N$ is even. Also assume that $(p' - p)/N \in \mathbb{Z}_+$ and $(r - s)/N \in \mathbb{Z}$ for r and s integers. Then

$$X_{r,s,\sigma}^{(p,p'),N}(M_1, L_1, M_2, L_2) \tag{3.13}$$
$$= \sum_{j=-\infty}^{\infty} q^{\frac{j}{N}(pp'j+p'(M_{12}+r)-ps)} \sum_{\substack{\eta\in\mathbb{Z}^{N-1}\\ \frac{M_{12}+2pj+\sigma N}{2N}+(C^{-1}\eta)_1\in\mathbb{Z}}} q^{\eta C^{-1}\eta}\begin{bmatrix}\eta+\mu\\ \eta\end{bmatrix}$$
$$\times \begin{bmatrix}M_1+L_1-(p'-p)j/N-\frac{1}{2}(M_2-pj-\mu_1)\\ M_1+pj\end{bmatrix}\begin{bmatrix}M_2+L_2+(p'-p)j/N-\frac{1}{2}(M_1+pj-\mu_{N-1})\\ M_2-pj\end{bmatrix}$$
$$- \sum_{j=-\infty}^{\infty} q^{\frac{1}{N}(pj+M_{12}+r)(p'j+s)} \sum_{\substack{\eta\in\mathbb{Z}^{N-1}\\ \frac{M_{12}+2pj+2r+\sigma N}{2N}+(C^{-1}\eta)_1\in\mathbb{Z}}} q^{\eta C^{-1}\eta}\begin{bmatrix}\eta+\mu\\ \eta\end{bmatrix}$$
$$\times \begin{bmatrix}M_1+L_1-((p'-p)j-r+s)/N-\frac{1}{2}(M_2-pj-r-\mu_1)\\ M_1+pj+r\end{bmatrix}$$
$$\times \begin{bmatrix}M_2+L_2+((p'-p)j-r+s)/N-\frac{1}{2}(M_1+pj+r-\mu_{N-1})\\ M_2-pj-r\end{bmatrix},$$

with $(\boldsymbol{\mu}, \boldsymbol{\eta})$-systems

$$\boldsymbol{\mu} + \boldsymbol{\eta} = \frac{1}{2}(\mathcal{I}\boldsymbol{\mu} + (M_1 + pj)\boldsymbol{e}_1 + (M_2 - pj)\boldsymbol{e}_{N-1})$$

for the first term of the right-hand side and

$$\boldsymbol{\mu} + \boldsymbol{\eta} = \frac{1}{2}(\mathcal{I}\boldsymbol{\mu} + (M_1 + pj + r)\boldsymbol{e}_1 + (M_2 - pj - r)\boldsymbol{e}_{N-1})$$

for the second term of the right-hand side. In Section 4.1, we will show that in the limit when M_1 and M_2 tend to infinity for fixed M_{12}, the above polynomials become proportional to the one-dimensional configuration sums of solvable lattice models of Date et al. [10, 11], which are bounded analogues of level-N $A_1^{(1)}$ branching functions.

Using (2.1), it follows that

$$X^{(p,p+Np'),N}_{r,r+Ns,\sigma}(M_1, L_1, M_2, L_2) \tag{3.14}$$
$$= \sum_{i\in\mathbb{Z}} q^{i(i+M_{12})/N} \begin{bmatrix} L_1+L_2+M_2-i \\ M_2-i \end{bmatrix} \sum_{\substack{n\in\mathbb{Z}^{N-1} \\ \frac{2i+M_{12}+\sigma N}{2N}+(C^{-1}n)_1\in\mathbb{Z}}} q^{nC^{-1}n} \begin{bmatrix} m+n \\ n \end{bmatrix}$$
$$\times X^{(p,p')}_{r,s}\Big(i+M_{12}, L_1-i+\frac{1}{2}m_1, i, L_2-M_{12}-i+\frac{1}{2}m_1\Big),$$

where on the right-hand side we assume the $(\boldsymbol{m},\boldsymbol{n})$-system

$$\boldsymbol{m}+\boldsymbol{n} = \frac{1}{2}(\mathcal{I}\boldsymbol{m} + (2i+M_{12})\boldsymbol{e}_1). \tag{3.15}$$

Because of the conditions $L_1, L_2 \geq 0$ in (2.1), a sufficiency condition for the above transformation to hold is

$$\left\lfloor \frac{L_1+M_{12}(N-1)/(2N)-s-r/N}{p'+p/N} \right\rfloor \leq \left\lfloor \frac{L_1+M_{12}+r-s}{p'-p} \right\rfloor,$$
$$\left\lfloor \frac{L_2-M_{12}(N-1)/(2N)+s+r/N}{p'+p/N} \right\rfloor \leq \left\lfloor \frac{L_2-M_{12}-r+s}{p'-p} \right\rfloor \tag{3.16}$$

together with the inequalities obtained by setting $r=s=0$, where we assumed that $p'>p$. (The kernel of $X^{(p,p+Np'),N}_{r,r+Ns,\sigma}$ and of $X^{(p,p')}_{r,s}$ on either side of (3.14) is zero unless the summation variable j lies in certain ranges. The above conditions make sure that in these ranges of j the conditions $L_1, L_2 \geq 0$ of Theorem 2.1 apply.)

Using the symmetry (3.4), one also finds

$$X^{(p',Np+p'),N}_{s-M_{12},N(r+L_{12}+M_{12})+s-M_{12},\sigma}(M_1, L_1, M_2, L_2) \tag{3.17}$$
$$= \sum_{i\in\mathbb{Z}} q^{i(i+M_{12})/N} \begin{bmatrix} L_1+L_2+M_2-i \\ M_2-i \end{bmatrix} \sum_{\substack{n\in\mathbb{Z}^{N-1} \\ \frac{2i+M_{12}+\sigma N}{2N}+(C^{-1}n)_1\in\mathbb{Z}}} q^{nC^{-1}n} \begin{bmatrix} m+n \\ n \end{bmatrix}$$
$$\times X^{(p,p')}_{r,s}\Big(L_1-i+\frac{1}{2}m_1, i+M_{12}, L_2-M_{12}-i+\frac{1}{2}m_1, i\Big),$$

where again (3.15) holds. This time a sufficient condition is that

$$\left\lfloor \frac{L_2-M_{12}(N-1)/(2N)-s-r/N}{p'+p/N} \right\rfloor \leq \left\lfloor \frac{L_1+M_{12}+r-s}{p'-p} \right\rfloor,$$
$$\left\lfloor \frac{L_1+M_{12}(N-1)/(2N)+s+r/N}{p'+p/N} \right\rfloor \leq \left\lfloor \frac{L_2-M_{12}-r+s}{p'-p} \right\rfloor \tag{3.18}$$

holds, as well as the inequalities obtained by setting $r = s = 0$, where again $p' > p$.

Again we consider the simpler case when $M_{12} = L_{12} = 0$. Setting

$$X^{(p,p'),N}_{r,s,\sigma}(M, L, M, L) = X^{(p,p'),N}_{r,s,\sigma}(M, L),$$

the generalized Burge transformations (3.14) and (3.17) simplify to

$$X^{(p,p+Np'),N}_{r,r+Ns,\sigma}(M, L) = \sum_{i=0}^{M} q^{i^2/N} \begin{bmatrix} 2L + M - i \\ 2L \end{bmatrix} \times \sum_{\substack{n\in\mathbb{Z}^{N-1} \\ \frac{2i+\sigma N}{2N}+(C^{-1}n)_1\in\mathbb{Z}}} q^{nC^{-1}n} \begin{bmatrix} m + n \\ n \end{bmatrix} X^{(p,p')}_{r,s}\Big(i, L - i + \frac{1}{2}m_1\Big) \tag{3.19}$$

and

$$X^{(p',Np+p'),N}_{s,Nr+s,\sigma}(M, L) = \sum_{i=0}^{M} q^{i^2/N} \begin{bmatrix} 2L + M - i \\ 2L \end{bmatrix} \times \sum_{\substack{n\in\mathbb{Z}^{N-1} \\ \frac{2i+\sigma N}{2N}+(C^{-1}n)_1\in\mathbb{Z}}} q^{nC^{-1}n} \begin{bmatrix} m + n \\ n \end{bmatrix} X^{(p,p')}_{r,s}\Big(L - i + \frac{1}{2}m_1, i\Big), \tag{3.20}$$

both with (m, n)-system

$$m + n = \frac{1}{2}(\mathcal{I}m + 2ie_1). \tag{3.21}$$

The sufficiency conditions (3.16) and (3.18) (and their $r = s = 0$ counterparts) reduce to the single condition

$$\left\lfloor \frac{L + s + r/N}{p' + p/N} \right\rfloor \le \left\lfloor \frac{L - r + s}{p' - p} \right\rfloor. \tag{3.22}$$

To end this section, let us give some simple examples of our extensions to the Burge transform by finding the generalizations of equations (3.9)–(3.12) to arbitrary N. First, applying (3.19) to (3.8) yields

$$X^{(1,2N+1),N}_{0,N,\sigma}(M, L) = q^{L^2} \sum_{m\in\mathbb{Z}^{N-1}} q^{\frac{1}{4}mTm} \begin{bmatrix} L + M - \frac{1}{2}m_1 \\ 2L \end{bmatrix} \begin{bmatrix} m + n \\ m \end{bmatrix},$$

with $m + n = \frac{1}{2}(\mathcal{I}_T m + 2Le_1)$ and $(\mathcal{I}_T)_{i,j} = \delta_{|i-j|,1} + \delta_{i,j}\delta_{i,1}$ the incidence matrix of the tadpole graph with $N - 1$ nodes, and $T = 2I - \mathcal{I}_T$ the corresponding Cartan-like matrix. When N is odd, $\sigma = 0$, $L \in \mathbb{Z}$ and $m \in 2\mathbb{Z}^{N-1}$. When N is even,

$m_{2i+1} \equiv 2L \equiv \sigma \pmod 2$ and $m_{2i} \equiv 0 \pmod 2$. The sufficiency condition (3.22) is satisfied. Next, applying (3.20) to (3.8) yields

$$X^{(2,N+2),N}_{1,1,\sigma}(M,L) = \begin{bmatrix} 2L+M \\ 2L \end{bmatrix} \delta_{\sigma,0}$$

which, for $\sigma = 0$, is a doubly bounded version of the Euler identity for the level-N string functions of type $A^{(1)}_1$. Our third example follows after inserting (3.9) into (3.20):

$$X^{(3,N+3),N}_{1,1,\sigma}(M,L) = \sum_{\boldsymbol{m}\in\mathbb{Z}^N} q^{\frac{1}{4}\boldsymbol{m}C\boldsymbol{m}} \begin{bmatrix} 2L+M-\frac{1}{2}m_1 \\ 2L \end{bmatrix} \begin{bmatrix} \boldsymbol{m}+\boldsymbol{n} \\ \boldsymbol{m} \end{bmatrix},$$

with $(\boldsymbol{m},\boldsymbol{n})$-system $\boldsymbol{m}+\boldsymbol{n} = \frac{1}{2}(\mathcal{I}\boldsymbol{m} + 2L\boldsymbol{e}_1) \in \mathbb{Z}^N$, where $\mathcal{I}$ is now the incidence matrix of the A_N Dynkin diagram. When N is odd, $\sigma = 0$, $L \in \mathbb{Z}$, and $\boldsymbol{m} \in 2\mathbb{Z}^{N-1}$. When N is even, $m_{2i} \equiv 2L \equiv \sigma \pmod 2$ and $m_{2i+1} \equiv 0 \pmod 2$. These identities are bounded analogues of identities for level-N $A^{(1)}_1$ branching functions isomorphic to unitary minimal Virasoro characters. Finally, we use (3.19) and (3.10) to find

$$X^{(2,3N+2),N}_{1,N+1,\sigma}(M,L) = \sum_{i=0}^{M} q^{i^2/N} \begin{bmatrix} 2L+M-i \\ 2L \end{bmatrix} \sum_{\substack{\boldsymbol{n}\in\mathbb{Z}^{N-1} \\ \frac{2i+\sigma N}{2N}+(C^{-1}\boldsymbol{n})_1\in\mathbb{Z}}} q^{\boldsymbol{n}C^{-1}\boldsymbol{n}} \begin{bmatrix} \boldsymbol{m}+\boldsymbol{n} \\ \boldsymbol{n} \end{bmatrix} \begin{bmatrix} 2L-i+m_1 \\ i \end{bmatrix},$$

where (3.21) holds. As remarked before, for $N = 1$ ($\sigma = 0$) this is a doubly bounded version of the (first) Rogers–Ramanujan identity. For $N = 2$, it becomes

$$X^{(2,8),2}_{1,3,\sigma}(M,L) = \sum_{i=0}^{M} \sum_{\substack{n=0 \\ n+i+\sigma \text{ even}}}^{i} q^{(i^2+n^2)/2} \begin{bmatrix} 2L+M-i \\ 2L \end{bmatrix} \begin{bmatrix} i \\ n \end{bmatrix} \begin{bmatrix} 2L-n \\ i \end{bmatrix},$$

which can be recognized as a doubly bounded version of

$$\sum_{n=0}^{\infty} \frac{q^{n^2}(-q;q^2)_n}{(q^2;q^2)_n} = \prod_{j=1}^{\infty} \frac{1}{(1-q^{8j-1})(1-q^{8j-4})(1-q^{8j-7})}$$

due to Slater [26] and related to the (first) Göllnitz–Gordon partition identity [15, 16].

4 Special limits of Theorem 2.1

4.1 q-multinomial coefficients. In [2, 8, 18, 22, 27], q-multinomial coefficients were introduced as q-analogues of the coefficients in the expansion

$$(1+x+x^2+\cdots+x^N)^L = \sum_{a=-\frac{NL}{2}}^{\frac{NL}{2}} \binom{L}{a}_N x^{a+\frac{NL}{2}}$$

for $L \in \mathbb{Z}_+$. The q-multinomial coefficients are the generating function of a wide class of combinatorial objects: (i) unrestricted lattice paths related to the RSOS lattice models of Date et al. with H-function statistic [10, 11]; (ii) Durfee dissection partitions [27]; and (iii) tabloids of shape (N^L) and content $(1^a 2^{NL-a})$ with the statistic "value" [8], etc.

Here we need the following explicit representation for the q-multinomials [22]

$$T_n^{(N)}(L,a) = \sum_{\substack{\boldsymbol{\eta}\in\mathbb{Z}^{N-1} \\ \frac{L}{2}+\frac{a}{N}+(C^{-1}\boldsymbol{\eta})_1\in\mathbb{Z}}} \frac{q^{\boldsymbol{\eta}C^{-1}(\boldsymbol{\eta}-\boldsymbol{e}_n)}(q)_L}{(q)_{\frac{L}{2}-\frac{a}{N}-(C^{-1}\boldsymbol{\eta})_1}(q)_{\frac{L}{2}+\frac{a}{N}-(C^{-1}\boldsymbol{\eta})_{N-1}}(q)_{\boldsymbol{\eta}}}, \tag{4.1}$$

where $L \in \mathbb{Z}_+$, $2a \in \{-NL, -NL+2, \dots, NL\}$ and $n \in \{0, 1, \dots, N-1\}$. Repeated use of Newton's binomial expansion shows that

$$\lim_{q\to 1} T_n^{(N)}(L,a) = \binom{L}{a}_N$$

so that $T_n^{(N)}(L,a)$ is indeed a q-analogue of the multinomial coefficient.

Theorem 2.1 provides a new representation of the q-multinomials when $n = 0$. To see this, we let M tend to infinity in (2.1), resulting in

$$\begin{aligned}\sum_{i=0}^{\infty} q^{i(i+\ell)/N} &\sum_{\substack{\boldsymbol{n}\in\mathbb{Z}^{N-1} \\ \frac{2i+\ell+\sigma N}{2N}+(C^{-1}\boldsymbol{n})_1\in\mathbb{Z}}} q^{\boldsymbol{n}C^{-1}\boldsymbol{n}} \begin{bmatrix}\boldsymbol{m}+\boldsymbol{n}\\ \boldsymbol{n}\end{bmatrix}\begin{bmatrix}L_1+\frac{1}{2}m_1\\ i+\ell\end{bmatrix}\begin{bmatrix}L_2+\frac{1}{2}m_1\\ i\end{bmatrix} \\ &= \sum_{\substack{\boldsymbol{\eta}\in\mathbb{Z}^{N-1} \\ \frac{\ell+\sigma N}{2N}+(C^{-1}\boldsymbol{\eta})_1\in\mathbb{Z}}} \frac{q^{\boldsymbol{\eta}C^{-1}\boldsymbol{\eta}}(q)_{L_1+L_2}}{(q)_{L_1-\frac{\ell}{2N}-\frac{\ell}{2}-(C^{-1}\boldsymbol{\eta})_1}(q)_{L_2+\frac{\ell}{2N}+\frac{\ell}{2}-(C^{-1}\boldsymbol{\eta})_{N-1}}(q)_{\boldsymbol{\eta}}}.\end{aligned}$$

If we now set $L_1 = \frac{1}{2}(L+\ell)$ and $L_2 = \frac{1}{2}(L-\ell)$ (so that $\sigma \equiv L \pmod 2$) and compare with the right-hand side of (4.1), we find that

$$\begin{aligned}&T_0^{(N)}(L,\ell/2) \\ &= \sum_{i=0}^{\infty} q^{i(i+\ell)/N} \sum_{\substack{\boldsymbol{n}\in\mathbb{Z}^{N-1} \\ \frac{L}{2}+\frac{2i+\ell}{2N}+(C^{-1}\boldsymbol{n})_1\in\mathbb{Z}}} q^{\boldsymbol{n}C^{-1}\boldsymbol{n}} \begin{bmatrix}\boldsymbol{m}+\boldsymbol{n}\\ \boldsymbol{n}\end{bmatrix}\begin{bmatrix}\frac{1}{2}(L+\ell+m_1)\\ i+\ell\end{bmatrix}\begin{bmatrix}\frac{1}{2}(L-\ell+m_1)\\ i\end{bmatrix},\end{aligned} \tag{4.2}$$

with $\boldsymbol{m}$ given by (2.2).

When $N = 1$, the above decomposition of the q-multinomial coefficients reduces to the q-Chu–Vandermonde sum (1.4) and a combinatorial interpretation is easily given as follows. The q-binomial $\begin{bmatrix}m+n\\ m\end{bmatrix}$ is the generating function of partitions that fit in a box of dimension m times n. Hence the summand on the left-hand side of (1.4) is the generating function of partitions that fit in a box of dimension $L_1 - \ell$ times

$L_2 + \ell$ which have a Durfee rectangle of size i by $i + \ell$ (maximal rectangle of the Ferrers graph that has a horizontal excess of ℓ nodes). Summing over i removes the Durfee rectangle restriction resulting in the right-hand side. It seems an interesting problem to also explain the q-multinomial decomposition (4.2) combinatorially.

There is a corresponding formula for $1 \leq n < N - 1$ which, however, is less appealing (and which we will not prove here):

$$
\begin{aligned}
T_n^{(N)}&(L, (n-\ell)/2) - q^{(\ell+1)/N} T_n^{(N)}(L, (n+\ell+2)/2) \\
&= \sum_{i=0}^{\infty} q^{i(i+\ell)/N} \sum_{\substack{n \in \mathbb{Z}^{N-1} \\ \frac{L}{2} + \frac{2i+\ell-n}{2N} + (C^{-1}\boldsymbol{n})_1 \in \mathbb{Z}}} q^{\boldsymbol{n}C^{-1}(\boldsymbol{n}-\boldsymbol{e}_{N-n})} \begin{bmatrix} \boldsymbol{m}+\boldsymbol{n} \\ \boldsymbol{n} \end{bmatrix} \\
&\quad \times \left(\begin{bmatrix} \frac{1}{2}(L+\ell+m_1) \\ i+\ell \end{bmatrix} \begin{bmatrix} \frac{1}{2}(L-\ell+m_1) \\ i \end{bmatrix} - \begin{bmatrix} \frac{1}{2}(L+\ell+2+m_1) \\ i+\ell+1 \end{bmatrix} \begin{bmatrix} \frac{1}{2}(L-\ell-2+m_1) \\ i-1 \end{bmatrix} \right),
\end{aligned}
$$

with

$$\boldsymbol{m} + \boldsymbol{n} = \frac{1}{2}(\mathcal{I}\boldsymbol{m} + (2i+\ell)\boldsymbol{e}_1 + \boldsymbol{e}_{N-n}).$$

Although this identity has the structure $f(L,\ell) - q^{(\ell+1)/N} f(L, -\ell-2) = g(L,\ell) - q^{(\ell+1)/N} g(L, -\ell-2)$, it is not true that $f(L,\ell) = g(L,\ell)$.

To conclude our discussion of the q-multinomial coefficients, let us point out that the polynomials defined in equation (3.13) are related to one-dimensional configuration sums of lattice models of Date et al. [10, 11]. Let $L \in \mathbb{Z}$ and choose

$$L_1 = \frac{1}{2}\left(L - M_{12} - \frac{r-s}{N}\right), \qquad L_2 = \frac{1}{2}\left(L + M_{12} + \frac{r-s}{N}\right)$$

so that $\sigma = 0, 1$ is fixed by the condition that $L - (r-s)/N + \sigma$ is even. Then

$$
\begin{aligned}
\lim_{\substack{M_1, M_2 \to \infty \\ M_{12}\ \text{fixed}}} &(q)_{2L} X_{r,s,\sigma}^{(p,p'),N}(M_1, L_1, M_2, L_2) \\
&= \sum_{j=-\infty}^{\infty} \left\{ q^{\frac{j}{N}(pp'j + p'(M_{12}+r) - ps)} T_0^{(N)}\left(L, \frac{1}{2}(r + M_{12} - s) + p'j\right) \right. \\
&\qquad \left. - q^{\frac{1}{N}(pj+M_{12}+r)(p'j+s)} T_0^{(N)}\left(L, \frac{1}{2}(r + M_{12} + s) + p'j\right) \right\},
\end{aligned}
$$

which, for $p' = p + N$, is proportional to the configuration sums of the models of Date et al. in the representation obtained in [22, equation (3.15)].

4.2 Bailey's lemma. In this section, we show that the limit $L_1, L_2 \to \infty$ of Theorem 2.1 gives rise to the higher-level Bailey lemma (or, more precisely, the higher-level conjugate Bailey pairs) of [23, 24].

Bailey's lemma [5] is an elegant tool to prove q-series identities such as the famous Rogers–Ramanujan identities. Let $\alpha=\{\alpha_L\}_{L\geq 0}$, $\beta=\{\beta_L\}_{L\geq 0}$ be a pair of sequences that satisfies

$$\beta_L=\sum_{i=0}^{L}\frac{\alpha_i}{(q)_{L-i}(aq)_{L+i}}.$$

Such a pair is called a Bailey pair relative to a. Recalling the definition (1.6) of a conjugate Bailey pair, it follows by a simple interchange of sums that

$$\sum_{L=0}^{\infty}\alpha_L\gamma_L=\sum_{L=0}^{\infty}\beta_L\delta_L. \tag{4.3}$$

Many known q-series identities follow from (4.3) after substitution of suitable Bailey and conjugate Bailey pairs.

Now let L_1 and L_2 tend to infinity in (2.1) and replace $i\to i-L$, $\ell\to\ell+2L$, and $M\to M-L$. This yields

$$\begin{aligned}\sum_{i=L}^{M}\frac{q^{i(i+\ell)/N}}{(q)_{i-L}(q)_{i+L+\ell}(q)_{M-i}}&\sum_{\substack{\boldsymbol{n}\in\mathbb{Z}^{N-1}\\ \frac{2i+\ell+\sigma N}{2N}+(C^{-1}\boldsymbol{n})_1\in\mathbb{Z}}}q^{\boldsymbol{n}C^{-1}\boldsymbol{n}}\begin{bmatrix}\boldsymbol{m}+\boldsymbol{n}\\ \boldsymbol{n}\end{bmatrix}\\ =\frac{q^{L(L+\ell)/N}}{(q)_{M-L}(q)_{M+L+\ell}}&\sum_{\substack{\boldsymbol{\eta}\in\mathbb{Z}^{N-1}\\ \frac{2L+\ell+\sigma N}{2N}+(C^{-1}\boldsymbol{\eta})_1\in\mathbb{Z}}}q^{\boldsymbol{\eta}C^{-1}\boldsymbol{\eta}}\begin{bmatrix}\boldsymbol{\mu}+\boldsymbol{\eta}\\ \boldsymbol{\eta}\end{bmatrix},\end{aligned}$$

with $(\boldsymbol{m},\boldsymbol{n})$-system (2.2) and $(\boldsymbol{\mu},\boldsymbol{\eta})$-system

$$\boldsymbol{\mu}+\boldsymbol{\eta}=\frac{1}{2}(\mathcal{I}\boldsymbol{\mu}+(M+L+\ell)\boldsymbol{e}_1+(M-L)\boldsymbol{e}_{N-1}). \tag{4.4}$$

Comparing with (1.6) one reads off the following conjugate Bailey pair (which is the special case $\lambda=0$ of [24, Corollary 2.1]):

$$\gamma_L=\frac{a^{L/N}q^{L^2/N}}{(q)_{M-L}(aq)_{M+L}}\sum_{\substack{\boldsymbol{\eta}\in\mathbb{Z}^{N-1}\\ \frac{2L+\ell+\sigma N}{2N}+(C^{-1}\boldsymbol{\eta})_1\in\mathbb{Z}}}q^{\boldsymbol{\eta}C^{-1}\boldsymbol{\eta}}\begin{bmatrix}\boldsymbol{\mu}+\boldsymbol{\eta}\\ \boldsymbol{\eta}\end{bmatrix},$$

$$\delta_L=\frac{a^{L/N}q^{L^2/N}}{(q)_{M-L}}\sum_{\substack{\boldsymbol{n}\in\mathbb{Z}^{N-1}\\ \frac{2L+\ell+\sigma N}{2N}+(C^{-1}\boldsymbol{n})_1\in\mathbb{Z}}}q^{\boldsymbol{n}C^{-1}\boldsymbol{n}}\begin{bmatrix}\boldsymbol{m}+\boldsymbol{n}\\ \boldsymbol{n}\end{bmatrix},$$

with $a=q^\ell$ and where (4.4) and $\boldsymbol{m}+\boldsymbol{n}=\frac{1}{2}(\mathcal{I}\boldsymbol{m}+(2L+\ell)\boldsymbol{e}_1)$ hold.

4.3 String functions. Taking the limit $L_1, L_2, M \to \infty$ in Theorem 2.1, we obtain

$$\sum_{i=0}^{\infty} \frac{q^{i(i+\ell)/N}}{(q)_i (q)_{i+\ell}} \sum_{\substack{\boldsymbol{n}\in\mathbb{Z}^{N-1} \\ \frac{2i+\ell+\sigma N}{2N}+(C^{-1}\boldsymbol{n})_1\in\mathbb{Z}}} q^{\boldsymbol{n}C^{-1}\boldsymbol{n}} \begin{bmatrix} \boldsymbol{m}+\boldsymbol{n} \\ \boldsymbol{n} \end{bmatrix} = \frac{1}{(q)_\infty} \sum_{\substack{\boldsymbol{\eta}\in\mathbb{Z}^{N-1} \\ \frac{\ell+\sigma N}{2N}+(C^{-1}\boldsymbol{\eta})_1\in\mathbb{Z}}} \frac{q^{\boldsymbol{\eta}C^{-1}\boldsymbol{\eta}}}{(q)_{\boldsymbol{\eta}}}. \tag{4.5}$$

It was shown in [4, 6, 20, 21, 25] that the left-hand side is proportional to a level-N, $\mathrm{A}_1^{(1)}$ string function $C_{m,\ell}^N$ defined as follows. Let

$$\Theta_{n,m}(z,q) = \sum_{j\in\mathbb{Z}+n/2m} q^{mj^2} z^{-mj}$$

be the classical theta function of degree m and characteristic n. The $\mathrm{A}_1^{(1)}$ character of the highest-weight module of highest-weight $(N-\ell)\Lambda_0 + \ell\Lambda_1$ (where Λ_0 and Λ_1 are the fundamental weights of $\mathrm{A}_1^{(1)}$ and $0 \le \ell \le N$) is given by

$$\chi_\ell(z,q) = \frac{\sum_{\sigma=\pm1} \sigma\Theta_{\sigma(\ell+1),N+2}(z,q)}{\sum_{\sigma=\pm1} \sigma\Theta_{\sigma,2}(z,q)}.$$

The level-N $\mathrm{A}_1^{(1)}$ string functions are defined by the expansion

$$\chi_\ell(z,q) = \sum_{m\in2\mathbb{Z}+\ell} C_{m,\ell}^N(q) q^{\frac{m^2}{4N}} z^{-\frac{1}{2}m}.$$

According to the above-cited references,

$$\begin{aligned} C_{m,\ell}^N(q) &= q^{\frac{(\ell+1)^2}{4(N+2)}-\frac{m^2}{4N}-\frac{1}{8}} \sum_{i=0}^{\infty} \frac{X_{\ell+1}^{N+2}(2i+m)}{(q)_i (q)_{i+m}} \\ &= q^{\frac{(\ell+1)^2}{4(N+2)}-\frac{\ell^2}{4N}-\frac{1}{8}} \sum_{i=0}^{\infty} \frac{q^{i(i+m)/N}}{(q)_i (q)_{i+m}} \sum_{\substack{\boldsymbol{n}\in\mathbb{Z}^{N-1} \\ \frac{2i+m+\ell}{2N}+(C^{-1}\boldsymbol{n})_1\in\mathbb{Z}}} q^{\boldsymbol{n}C^{-1}(\boldsymbol{n}-\boldsymbol{e}_\ell)} \begin{bmatrix} \boldsymbol{m}+\boldsymbol{n} \\ \boldsymbol{n} \end{bmatrix}, \end{aligned}$$

with $\boldsymbol{m}+\boldsymbol{n} = \frac{1}{2}(\mathcal{I}\boldsymbol{m} + (2i+m)\boldsymbol{e}_1 + \boldsymbol{e}_\ell)$ and $X_s^p(L)$ a one-dimensional configuration sum of the $(p-1)$-state Andrews–Baxter–Forrester model in regime I,

$$X_s^p(L) = \sum_{j=-\infty}^{\infty} q^{j(pj+s)} \left\{ \begin{bmatrix} L \\ \frac{1}{2}(L-s+1)-pj \end{bmatrix} - \begin{bmatrix} L \\ \frac{1}{2}(L-s-1)-pj \end{bmatrix} \right\}.$$

Comparing with (4.5), we obtain the following expression of the string function

$$C^N_{m,\sigma N}(q) = \frac{q^{\frac{1}{4(N+2)}-\frac{1}{8}}}{(q)_\infty} \sum_{\substack{\eta\in\mathbb{Z}^{N-1}\\ \frac{m+\sigma N}{2N}+(C^{-1}\eta)_1\in\mathbb{Z}}} \frac{q^{\eta C^{-1}\eta}}{(q)_\eta},$$

which was first derived by Lepowsky and Primc [19].

Acknowledgments. We thank Omar Foda and Trevor Welsh for discussions on the Burge transform.

The first author was supported by the Stichting Fundamenteel Onderzoek der Materie. The second author was supported by a fellowship of the Royal Netherlands Academy of Arts and Sciences.

References

[1] G. E. Andrews, *The Theory of Partitions*, Encyclopedia of Mathematics and Its Applications 2, Addison–Wesley, Reading, MA, 1976.

[2] G. E. Andrews, Schur's theorem, Capparelli's conjecture and q-trinomial coefficients, *Contemp. Math.*, **166** (1994), 141–154.

[3] G. E. Andrews and R. J. Baxter, Lattice gas generalization of the hard hexagon model III: q-Trinomial coefficients, *J. Statist. Phys.*, **47** (1987), 297–330.

[4] T. Arakawa, T. Nakanishi, K. Oshima, and A. Tsuchiya, Spectral decomposition of path space in solvable lattice model, *Comm. Math. Phys.*, **181** (1996), 157–182.

[5] W. N. Bailey, Identities of the Rogers–Ramanujan type, *Proc. London Math. Soc.* (2), **50** (1949), 1–10.

[6] P. Bouwknegt, A. W. W. Ludwig, and K. Schoutens, Spinon basis for higher level $SU(2)$ WZW models, *Phys. Lett. B*, **359** (1995), 304–312.

[7] W. H. Burge, Restricted partition pairs, *J. Combin. Theory Ser. A*, **63** (1993), 210–222.

[8] L. M. Butler, Subgroup lattices and symmetric functions, *Mem. Amer. Math. Soc.*, **112** (1994), no. 539.

[9] L. Carlitz, Remark on a combinatorial identity, *J. Combin. Theory Ser. A*, **17** (1974), 256–257.

[10] E. Date, M. Jimbo, A. Kuniba, T. Miwa, and M. Okado, Exactly solvable SOS models: Local height probabilities and theta function identities, *Nuclear Phys. B*, **290** (1987), 231–273.

[11] E. Date, M. Jimbo, A. Kuniba, T. Miwa, and M. Okado, Exactly solvable SOS models II: Proof of the star-triangle relation and combinatorial identities, *Adv. Stud. Pure Math.*, **16** (1988), 17–122.

[12] O. Foda, K. S. M. Lee, and T. A. Welsh, A Burge tree of Virasoro-type polynomial identities, *Internat. J. Modern Phys. A*, **13** (1998), 4967–5012.

[13] G. Gasper and M. Rahman, *Basic Hypergeometric Series*, Encyclopedia of Mathematics and Its Applications 35, Cambridge University Press, Cambridge, 1990.

[14] I. M. Gessel and C. Krattenthaler, Cylindric partitions, *Trans. Amer. Math. Soc.*, **349** (1997), 429–479.

[15] H. Göllnitz, Partitionen mit Differenzenbedingungen, *J. Reine Angew. Math.*, **225** (1967), 154–190.

[16] B. Gordon, Some continued fractions of the Rogers–Ramanujan type, *Duke Math. J.*, **31** (1965), 741–748.

[17] H. W. Gould, A new symmetrical combinatorial identity, *J. Combin. Theory Ser. A*, **13** (1972), 278–286.

[18] A. N. Kirillov, Dilogarithm identities, *Prog. Theoret. Phys. Suppl.*, **118** (1995), 61–142.

[19] J. Lepowsky and M. Primc, *Structure of the standard modules for the affine Lie algebra* $A_1^{(1)}$, Contemporary Mathematics 46, AMS, Providence, 1985.

[20] A. Nakayashiki and Y. Yamada, Crystallizing the spinon basis, *Comm. Math. Phys.*, **178** (1996), 179–200.

[21] A. Nakayashiki and Y. Yamada, Crystalline spinon basis for RSOS models, *Internat. J. Modern Phys. A*, **11** (1996), 395–408.

[22] A. Schilling, Multinomials and polynomial bosonic forms for the branching functions of the $\widehat{su}(2)_M \times \widehat{su}(2)_N / \widehat{su}(2)_{M+N}$ conformal coset models, *Nuclear Phys. B*, **467** (1996), 247–271.

[23] A. Schilling and S. O. Warnaar, A higher-level Bailey lemma, *Internat. J. Modern Phys. B*, **11** (1997), 189–195.

[24] A. Schilling and S. O. Warnaar, A higher level Bailey lemma: Proof and application, *Ramanujan J.*, **2** (1998), 327–349.

[25] A. Schilling and S. O. Warnaar, Conjugate Bailey pairs: From configuration sums and fractional-level string functions to Bailey's lemma, preprint math.QA/9906092.

[26] L. J. Slater, Further identities of the Rogers–Ramanujan type, *Proc. London Math. Soc.* (2), **54** (1952), 147–167.

[27] S. O. Warnaar, The Andrews-Gordon identities and q-multinomial coefficients, *Comm. Math. Phys.*, **184** (1997), 203–232.

Anne Schilling
Instituut voor Theoretische Fysica
Universiteit van Amsterdam
Valckenierstraat 65
1018 XE Amsterdam
The Netherlands
current address
Department of Mathematics
Massachusetts Institute of Technology
Cambridge MA 02139
`anne@math.mit.edu`

S. Ole Warnaar
Instituut voor Theoretische Fysica
Universiteit van Amsterdam
Valckenierstraat 65
1018 XE Amsterdam
The Netherlands
`warnaar@wins.uva.nl`

The Bethe Equation at $q = 0$, the Möbius Inversion Formula, and Weight Multiplicities I: The $\mathfrak{sl}(2)$ Case

Atsuo Kuniba and Tomoki Nakanishi

Abstract. The $U_q(\widehat{\mathfrak{sl}}(2))$ Bethe equation is studied at $q = 0$. A linear congruence equation is proposed related to the string solutions. The number of its off-diagonal solutions is expressed in terms of an explicit combinatorial formula and coincides with the weight multiplicities of the quantum space.

1 Introduction

1.1 Background. Consider the periodic spin-$\frac{1}{2}$ XXX Heisenberg Hamiltonian

$$H_{\mathrm{XXX}} = J \sum_{n=1}^{L} (\sigma_n^x \sigma_{n+1}^x + \sigma_n^y \sigma_{n+1}^y + \sigma_n^z \sigma_{n+1}^z)$$

acting on the tensor product of the L-copies of the vector representations of $\mathfrak{sl}(2)$:

$$W = \mathbb{C}^2 \otimes \cdots \otimes \mathbb{C}^2.$$

Since H_{XXX} is $\mathfrak{sl}(2)$-linear, its spectrum is degenerated within the irreducible components in the decomposition:

$$W = \bigoplus_{\lambda \in (\mathbb{Z}_{\geq 0})\Lambda_1} [W : V_\lambda] \, V_\lambda,$$

where Λ_1 is the fundamental weight and V_λ denotes the irreducible module with highest weight λ. Diagonalization of H_{XXX} was achieved by Bethe [Be] in 1931. Associated with each solution of the simultaneous equations (Bethe equation)

$$\left(\frac{u_j + \sqrt{-1}}{u_j - \sqrt{-1}} \right)^L = - \prod_{j=1}^{M} \frac{u_j - u_k + 2\sqrt{-1}}{u_j - u_k - 2\sqrt{-1}}, \quad j = 1, \ldots, M,$$

he proposed (Bethe ansatz) a vector $\psi \in W$ (Bethe vector) such that

$$H_{\mathrm{XXX}}\psi = E\psi, \quad E \in \mathbb{C},$$
$$\left(\sum_{n=1}^{L} \sigma_n^+\right)\psi = 0, \quad \sigma_n^+ = \sigma_n^x + \sqrt{-1}\sigma_n^y,$$
$$\left(\sum_{n=1}^{L} \sigma_n^z\right)\psi = (L-2M)\psi, \quad 0 \le M \le \left[\frac{L}{2}\right].$$

The second and third properties (cf. [FT]) establish that the Bethe vector is the $\mathfrak{sl}(2)$-highest of weight $(L-2M)\Lambda_1$. Therefore, in order to have completeness of the Bethe ansatz, there should exist as many solutions to the Bethe equation as the multiplicity of $V_{(L-2M)\Lambda_1}$ in W, $[W : V_{(L-2M)\Lambda_1}]$ $(= \binom{L}{M} - \binom{L}{M-1})$. Actually, ψ can vanish depending on the solutions $\{u_1, \dots, u_M\}$. In particular, it is so if $u_i = u_j$ for some $i \neq j$.

It was Bethe himself who studied completeness with the introduction of strings. (He called it *WellenKomplex*.) It is a solution in which $\{u_1, \dots, u_M\}$ are arranged as

$$\bigcup_{m\in\mathbb{N}} \bigcup_{1\le\alpha\le N_m} \bigcup_{u_{m\alpha}\in\mathbb{R}} \{u_{m\alpha} + \sqrt{-1}(m+1-2i) + \epsilon_{m\alpha i} \mid 1 \le i \le m\} \tag{1.1}$$

for each partition $M = \sum_{m\in\mathbb{N}} mN_m$ $(N_m \in \mathbb{Z}_{\ge 0})$. Here $\mathbb{N} = \mathbb{Z}_{\ge 1}$ denotes the set of positive integers and $\epsilon_{m\alpha i}$ stands for a small deviation. The m-tuple configuration (with negligible $\epsilon_{m\alpha i}$) is called the m-string with string center $u_{m\alpha}$. N_m is the number of m-strings. In general, to expect such a behavior for the solutions is called the string hypothesis. Actually, in a strict sense, it is known to invalid, as already exemplified for $M = 2$ and $L > 21$ (cf. [EKS, JD]). Nevertheless, Bethe's count of the number of string solutions led to the discovery of the identity ($M \le [\frac{L}{2}]$):

$$\sum_{N} \prod_{m\in\mathbb{N}} \binom{L - 2\sum_{k\ge1}\min(m,k)N_k + N_m}{N_m} = [W : V_{(L-2M)\Lambda_1}], \tag{1.2}$$

where $\sum_N$ runs over $N_1, N_2, \dots \in \mathbb{Z}_{\ge 0}$ such that $M = \sum_{m\ge1} mN_m$. In his count, each summand on the left side represents the number of string solutions corresponding to the prescribed values of $N_1, N_2, \dots$. The binomial coefficients originate in the fermionic restriction on the solutions $u_i \neq u_j$ $(i \neq j)$. The expression on the left side is called the fermionic formula and the above identity is called the combinatorial completeness of the string hypothesis.

Despite the gap in completeness in the rigorous sense, the above result opened a fruitful link between quantum integrable systems and representation theory. For a class of Bethe-ansatz-solvable models with Yangian symmetry $Y(X_n)$, one can set up fermionic formulas following Bethe's counting method. If combinatorial completeness holds, these formulas yield the multiplicities of irreducible X_n-modules in tensor products of a variety of finite-dimensional $Y(X_n)$-modules. The XXX chain corresponds to the $Y(\mathfrak{sl}(2))$ case. The fermionic formula associated with $Y(X_n)$ in such a sense was first written down in [KR] for general X_n, where combinatorial completeness was also announced for the classical types $X_n = A_n, B_n, C_n$, and D_n. The proof of combinatorial completeness boils down to showing recursion relations (Q-system) among classical characters of certain $Y(X_n)$-modules (cf. [HKOTY]).

1.2 Present work. The XXX chain admits an integrable q-deformation called the XXZ chain:

$$H_{\mathrm{XXZ}} = J\sum_{n=1}^{L}\left(\sigma_n^x\sigma_{n+1}^x + \sigma_n^y\sigma_{n+1}^y + \frac{q+q^{-1}}{2}\sigma_n^z\sigma_{n+1}^z\right).$$

In place of the Yangian $Y(\mathfrak{sl}(2))$, the underlying symmetry of the model is the quantum affine algebra $U_q(\widehat{\mathfrak{sl}}(2))$ as is well known. Accordingly, we regard the space W (called the quantum space) as a $U_q(\widehat{\mathfrak{sl}}(2))$-module. Under the periodic boundary condition, the spectrum is determined from the solutions of the Bethe equation ($i = 1, \ldots, M$)

$$\left(\frac{\sin\pi\left(u_i+\sqrt{-1}\hbar\right)}{\sin\pi\left(u_i-\sqrt{-1}\hbar\right)}\right)^L = -\prod_{j=1}^{M}\frac{\sin\pi\left(u_i-u_j+2\sqrt{-1}\hbar\right)}{\sin\pi\left(u_i-u_j-2\sqrt{-1}\hbar\right)},$$

where $\hbar$ is related to q by $q = e^{-2\pi\hbar}$. When the deformation parameter q tends to 1, the above equation reduces to the one in Section 1.1 by replacing u_j by $\hbar u_j$ and setting $\hbar \to 0$. A significant difference from the $q=1$ case is that the Hamiltonian is no longer invariant under $\mathfrak{sl}(2)$ or under $U_q(\mathfrak{sl}(2))$ as far as a finite chain ($L < \infty$) is considered under the periodic boundary condition. For completeness, the number of solutions to the Bethe equation should therefore coincide with the weight multiplicity of $(L-2M)\Lambda_1$, $\dim W_{(L-2M)\Lambda_1} (= \binom{L}{M})$, rather than the multiplicity of $V_{(L-2M)\Lambda_1}$. Similar facts are valid also for the generalized model in which W is replaced with

$$W(\nu) = \bigotimes_{s\geq 1}(W_s)^{\otimes \nu_s},$$

where $\nu_s \in \mathbb{Z}_{\geq 0}$ and W_s stands for the $(s+1)$-dimensional irreducible module. See (2.2) for the corresponding Bethe equation.

The purpose of this paper is to study the Bethe equation and to formulate another version of combinatorial completeness at $q=0$. This is inspired by crystal theory, where simplification at $q=0$ is known to lead to many fascinating features. In terms of the exponential variables $x_j = e^{2\pi\sqrt{-1}u_j}$, we shall consider a class of meromorphic solutions $x_j = x_j(q)$ around $q=0$ that correspond to the strings. In a sense, we are approaching the point $q=0$ within the off-critical regime $\hbar \in \mathbb{R}_{>0}$, avoiding the parity and the arithmetic complexity of strings [TS] in the critical regime $\hbar \in \sqrt{-1}\mathbb{R}$.

It is a routine calculation to reduce the Bethe equation to the one for string centers for general $\nu = (\nu_s)$ and $N = (N_m)$. At $q = 0$, the resulting string-center equation (SCE) is a linear congruence equation (2.23). As a remnant of the fermionic restriction, we seek their off-diagonal solutions (Definition 2.14). They are counted systematically by means of the Möbius inversion formula. When $P_m := \sum_{k\geq 1} \min(m,k)(\nu_k - 2N_k) \geq 0$ for any m such that $N_m > 0$, we find that the result is expressed as (cf. Theorem 3.5)

$$R(\nu, N) = \sum_{J\subset\mathbb{N}} D_J \prod_{m\in\mathbb{N}\setminus J} \binom{P_m + N_m}{N_m} \prod_{m\in J} \binom{P_m + N_m - 1}{N_m - 1},$$

$$D_J = \begin{cases} 1 & \text{if } J = \emptyset, \\ \det_{m,k\in J}(2\min(m,k) - \delta_{m,k}) & \text{otherwise.} \end{cases}$$

In the XXZ case $\nu_s = L\delta_{s,1}$, the $J = \emptyset$ term here is equal to the summand in the left side of (1.2). With this $R(\nu, N)$, combinatorial completeness at $q = 0$ is stated as (cf. Theorem 4.9)

$$\sum_N{}^{(\lambda)} R(\nu, N) = \dim W(\nu)_\lambda, \quad \lambda \in \mathbb{Z}\Lambda_1,$$

where the sum $\sum_N{}^{(\lambda)}$ runs over $N_1, N_2, \ldots \in \mathbb{Z}_{\geq 0}$ such that $\sum_{j\geq 1} j(\nu_j - 2N_j)\Lambda_1 = \lambda$. This is a nontrivial identity even when $\dim W(\nu)_\lambda = 0$ for $\lambda \in (\mathbb{Z}_{<0})\Lambda_1$. Curiously, the left side in general involves contributions from those N that break the $P_m \geq 0$ condition above. In the course of the proof, we will clarify the relationship between the known fermionic formula as in (1.2) and our $R(\nu, N)$. It is most transparently presented in terms of generating functions. See (4.25) and (4.29).

The layout of the paper is as follows. In Section 2, we study the Bethe equation at $q = 0$. We explain the relationship between solutions of the SCE and string solutions of the Bethe equation. In Section 3, we derive the formula $R(\nu, N)$ by counting the off-diagonal solutions of the SCE. In Section 4, we prove combinatorial completeness. In Section 5, we give a summary and discussion. The appendix is a summary of the Möbius inversion trick used in Section 3.

A few remarks are in order. In [KL], combinatorial completeness was investigated when q is a root of unity. Their fermionic formula for weight multiplicities is different from ours. We expect that their result describes the rich singular behavior of the meromorphic solutions (around $q = 0$) of the Bethe equation on the convergence circle $|q| = 1$. In [LS], the SCE was obtained for the XXZ case. There is a statement similar to combinatorial completeness at $q = 0$ without an explicit formula as $R(\nu, N)$. In [TV] the authors studied a deformation of the XXZ-type Bethe equation and showed completeness of the Bethe vectors for admissible off-diagonal solutions at a generic value of the deformation parameter.

Many results in this paper admit generalizations to the $U_q(X_n^{(1)})$ case for arbitrary X_n. This will be our subject in a forthcoming paper.

2 The Bethe equation at $q = 0$

In this section, we start from the Bethe equation and seek string solutions in the $q \to 0$ limit. We introduce the SCE, which is a linear congruence equation. Later sections will be devoted to studies of the SCE. Our aim here is to explain the precise relation between solutions of the SCE and string solutions of the Bethe equation. Our theorems concern mostly what we call generic string solutions.

We let α_1 and Λ_1 denote the simple root and the fundamental weight of $\mathfrak{sl}(2)$, $\alpha_1 = 2\Lambda_1$. In this paper, $U_q(\widehat{\mathfrak{sl}}(2))$ means the quantum affine algebra without the derivation operator. It is denoted by $U'_q(\widehat{\mathfrak{sl}}(2))$ in some literature.

For a meromorphic function $f(q)$ around $q = 0$, we will use the notation

$$f(q) = q^{\mathrm{ord}\,(f)}(f^0 + f^1 q + \cdots), \quad \mathrm{ord}\,(f) \in \mathbb{Z},\ f^0 \neq 0,$$
$$\tilde{f}(q) = q^{-\mathrm{ord}\,(f)} f(q) = f^0 + f^1 q + \cdots .$$

We call $\mathrm{ord}\,(f)$ the order and f^0 the leading coefficient of f. $\tilde{f}$ is a holomorphic function around $q = 0$ with nonzero constant term. Note that $f^0 = \tilde{f}^0$.

2.1 The Bethe equation. Consider a solvable vertex model associated with $U_q(\widehat{\mathfrak{sl}}(2))$. Let

$$W(\nu) = \bigotimes_{s \geq 1} (W_s)^{\otimes \nu_s} \tag{2.1}$$

be the $U_q(\widehat{\mathfrak{sl}}(2))$-module (the quantum space) on which the commuting family of row-to-row transfer matrices acts. We assume throughout this paper that only finitely many ν_ss are nonzero. Here W_s stands for an $(s+1)$-dimensional irreducible module with highest weight $s\Lambda_1$. Each W_s depends on a spectral parameter, which may be interpreted as inhomogeneity of the interaction. The Bethe equation relevant to the spectrum of the transfer matrices also depends on those spectral parameters.

In this paper, we concentrate on the regime and the situation in which the Bethe equation takes the form ($i = 1, \ldots, M$):

$$\prod_{s \geq 1} \left(\frac{\sin \pi \left(u_i + \sqrt{-1} s \hbar\right)}{\sin \pi \left(u_i - \sqrt{-1} s \hbar\right)} \right)^{\nu_s} = - \prod_{j=1}^{M} \frac{\sin \pi \left(u_i - u_j + 2\sqrt{-1}\hbar\right)}{\sin \pi \left(u_i - u_j - 2\sqrt{-1}\hbar\right)}. \tag{2.2}$$

Here $\hbar \in \mathbb{R}_{>0}$ and $M \in \mathbb{Z}_{\geq 0}$. This is a regime in which the so-called parity [TS] is irrelevant. Integer shifts of u_j do not lead to a new Bethe vector; hence one should consider $u_j \in \mathbb{C}/\mathbb{Z}$. Setting

$$q = e^{-2\pi\hbar}, \quad x_j = e^{2\pi\sqrt{-1}u_j},$$

(2.2) can be written as polynomial equations on x_js:

$$F_{i+} G_{i-} = -F_{i-} G_{i+}, \quad i = 1, \ldots, M, \tag{2.3}$$

where

$$F_{i+} = \prod_{s\geq 1}(x_i q^s - 1)^{\nu_s}, \qquad G_{i+} = \prod_{j=1}^{M}(x_i q^2 - x_j),$$

$$F_{i-} = \prod_{s\geq 1}(x_i - q^s)^{\nu_s}, \qquad G_{i-} = \prod_{j=1}^{M}(x_i - x_j q^2).$$

The equation is invariant under the permutation of the variables $x_i \leftrightarrow x_j$.

We are interested in meromorphic solutions (x_i), $x_i = x_i(q)$, of (2.3) around $q = 0$. We set $x_i(q) = q^{d_i} z_i(q)$, where $d_i = \mathrm{ord}\,(x_i)$, and $z_i(q) = \tilde{x}_i(q)$. Then the Bethe equation for $z_i(q)$ is given by (2.3) with $F_{i\pm}$, $G_{i\pm}$ now specified as

$$F_{i+} = \prod_{s\geq 1}(z_i q^{d_i+s} - 1)^{\nu_s}, \qquad G_{i+} = \prod_{j=1}^{M}(z_i q^{d_i+2} - z_j q^{d_j}), \tag{2.4}$$

$$F_{i-} = \prod_{s\geq 1}(z_i q^{d_i} - q^s)^{\nu_s}, \qquad G_{i-} = \prod_{j=1}^{M}(z_i q^{d_i} - z_j q^{d_j+2}). \tag{2.5}$$

This equation is invariant under the permutation of the variables $z_i \leftrightarrow z_j$ only when $d_i = d_j$.

2.2 String solution.

Definition 2.1. A meromorphic solution (x_i) of (2.3) is called *inadmissible* (*admissible*) if $F_{i+}G_{i-} = F_{i-}G_{i+} = 0$ for some i as a function of q around $q = 0$ (otherwise).

Let $N = (N_m)$ be an infinite sequence of nonnegative integers such that

$$M = \sum_{m\geq 1} m N_m. \tag{2.6}$$

Definition 2.2. A meromorphic solution (x_i) of (2.3) is called a *string solution of pattern* $N = (N_m)$ if

(i) (x_i) is admissible;

(ii) (x_i) can be arranged as $(x_{m\alpha i})$ with

$$m = 1, 2, \ldots, \qquad \alpha = 1, \ldots, N_m, \qquad i = 1, \ldots, m$$

such that

(a) $d_{m\alpha i} = m + 1 - 2i$ for $d_{m\alpha i} := \mathrm{ord}\,(x_{m\alpha i})$;

(b) $z^0_{m\alpha 1} = z^0_{m\alpha 2} = \cdots = z^0_{m\alpha m}$, where $z_{m\alpha i} = \tilde{x}_{m\alpha i}$.

For each $1 \le \alpha \le N_m$, $(z_{m\alpha i})_{i=1}^m$ is called an m-string. N_m is the number of m-strings. When considering string solutions, we denote the quantity in (b) by $z^0_{m\alpha}$ and call it the *string center*. We set

$$q^{\zeta_{m\alpha i}} y_{m\alpha i}(q) = z_{m\alpha i}(q) - z_{m\alpha i-1}(q), \quad 2 \le i \le m,$$

where $\zeta_{m\alpha i} = \operatorname{ord}(z_{m\alpha i} - z_{m\alpha i-1}) \in \mathbb{Z}_{\ge 1}$. For a string solution of pattern N, the Bethe equation (2.3) reads

$$F_{m\alpha i+}G_{m\alpha i-} = -F_{m\alpha i-}G_{m\alpha i+}, \tag{2.7}$$

where

$$F_{m\alpha i+} = \prod_{s\ge 1}(z_{m\alpha i}q^{d_{m\alpha i}+s} - 1)^{\nu_s}, \qquad G_{m\alpha i+} = \prod_{k\beta j}(z_{m\alpha i}q^{d_{m\alpha i}+2} - z_{k\beta j}q^{d_{k\beta j}}),$$

$$F_{m\alpha i-} = \prod_{s\ge 1}(z_{m\alpha i}q^{d_{m\alpha i}} - q^s)^{\nu_s}, \qquad G_{m\alpha i-} = \prod_{k\beta j}(z_{m\alpha i}q^{d_{m\alpha i}} - z_{k\beta j}q^{d_{k\beta j}+2}).$$

Here $\prod_{k\beta j}$ means $\prod_{k\ge 1}\prod_{\beta=1}^{N_k}\prod_{j=1}^{k}$. Let us extract the factors $y_{m\alpha i}$ from $G_{m\alpha i\pm}$ by introducing $G'_{m\alpha i\pm}$ as follows:

$$G_{m\alpha i+} = \begin{cases} G'_{m\alpha 1+}, & i = 1, \\ G'_{m\alpha i+}q^{d_{m\alpha i}+2+\zeta_{m\alpha i}}y_{m\alpha i}, & 2 \le i \le m, \end{cases}$$

$$G_{m\alpha i-} = \begin{cases} G'_{m\alpha i-}(-q^{d_{m\alpha i}+\zeta_{m\alpha i+1}}y_{m\alpha i+1}), & 1 \le i \le m-1, \\ -G'_{m\alpha m-}, & i = m. \end{cases}$$

Now the Bethe equation (2.7) takes the form

$$\tilde{F}_{m\alpha 1+}\tilde{G}'_{m\alpha 1-}y_{m\alpha 2} = \tilde{F}_{m\alpha 1-}\tilde{G}'_{m\alpha 1+}, \qquad i = 1, \tag{2.8}$$

$$\tilde{F}_{m\alpha i+}\tilde{G}'_{m\alpha i-}y_{m\alpha i+1} = \tilde{F}_{m\alpha i-}\tilde{G}'_{m\alpha i+}y_{m\alpha i}, \qquad 2 \le i \le m-1, \tag{2.9}$$

$$\tilde{F}_{m\alpha m+}\tilde{G}'_{m\alpha m-} = \tilde{F}_{m\alpha m-}\tilde{G}'_{m\alpha m+}y_{m\alpha m}, \qquad i = m. \tag{2.10}$$

In particular,

$$1 = (-1)^m \prod_{i=1}^m \frac{F_{m\alpha i+}G_{m\alpha i-}}{F_{m\alpha i-}G_{m\alpha i+}} = \prod_{i=1}^m \frac{\tilde{F}_{m\alpha i+}\tilde{G}'_{m\alpha i-}}{\tilde{F}_{m\alpha i-}\tilde{G}'_{m\alpha i+}}, \tag{2.11}$$

where the latter equality is the identity as holomorphic functions around $q = 0$.

2.3 $q \to 0$ limit of the Bethe equation. Suppose that $(x_{m\alpha i})$ is a string solution to the Bethe equation (2.7). Taking the leading coefficients of (2.8)–(2.10), we get the $q \to 0$ limit:

$$F^0_{m\alpha 1+}G'^0_{m\alpha 1-}y^0_{m\alpha 2} = F^0_{m\alpha 1-}G'^0_{m\alpha 1+}, \qquad i = 1, \tag{2.12}$$

$$F^0_{m\alpha i+}G'^0_{m\alpha i-}y^0_{m\alpha i+1} = F^0_{m\alpha i-}G'^0_{m\alpha i+}y^0_{m\alpha i}, \qquad 2 \le i \le m-1, \tag{2.13}$$

$$F^0_{m\alpha m+}G'^0_{m\alpha m-} = F^0_{m\alpha m-}G'^0_{m\alpha m+}y^0_{m\alpha m}, \qquad i = m. \tag{2.14}$$

In particular,

$$1 = (-1)^m \prod_{i=1}^{m} \frac{F^0_{m\alpha i+} G^0_{m\alpha i-}}{F^0_{m\alpha i-} G^0_{m\alpha i+}} = \prod_{i=1}^{m} \frac{F^0_{m\alpha i+} G'^0_{m\alpha i-}}{F^0_{m\alpha i-} G'^0_{m\alpha i+}} \tag{2.15}$$

holds for the leading coefficients.

2.4 Generic string solution. In order to estimate the order of the Bethe equation (2.7), we introduce

$$\begin{aligned}
\xi_{m\alpha i+} &= \sum_{s\geq 1} \nu_s \min(m+1-2i+s, 0),\\
\xi_{m\alpha i-} &= \sum_{s\geq 1} \nu_s \min(m+1-2i, s),\\
\eta_{m\alpha i+} &= \sum_{k\beta j} \min(m+3-2i, k+1-2j),\\
\eta_{m\alpha i-} &= \sum_{k\beta j} \min(m+1-2i, k+3-2j),
\end{aligned}$$

where $\sum_{k\beta j}$ is the abbreviation of $\sum_{k\geq 1}\sum_{\beta=1}^{N_k}\sum_{j=1}^{k}$. In general, one has $\xi_{m\alpha i\pm} \leq \mathrm{ord}\,(F_{m\alpha i\pm})$, $\eta_{m\alpha i+} + (1-\delta_{i,1})\zeta_{m\alpha i} \leq \mathrm{ord}\,(G_{m\alpha i+})$, and $\eta_{m\alpha i-} + (1-\delta_{i,m})\zeta_{m\alpha i+1} \leq \mathrm{ord}\,(G_{m\alpha i-})$. Let us consider the simplest situation when these inequalities are saturated.

Definition 2.3. A string solution $(x_{m\alpha i})$ to (2.3) is called *generic* if the following condition is valid:

$$\begin{aligned}
\mathrm{ord}\,(F_{m\alpha i\pm}) &= \xi_{m\alpha i\pm},\\
\mathrm{ord}\,(G_{m\alpha i+}) &= \begin{cases} \eta_{m\alpha 1+}, & i = 1,\\ \eta_{m\alpha i+} + \zeta_{m\alpha i}, & 2 \leq i \leq m, \end{cases}\\
\mathrm{ord}\,(G_{m\alpha i-}) &= \begin{cases} \eta_{m\alpha i-} + \zeta_{m\alpha i+1}, & 1 \leq i \leq m-1,\\ \eta_{m\alpha m-}, & i = m. \end{cases}
\end{aligned}$$

Our results in the rest of Section 2 concern mostly generic string solutions. Given the quantum space data $\nu = (\nu_s)$ and the string pattern $N = (N_m)$, we put

$$P_m = P_m(\nu, N) = \gamma_m - 2\sum_{k\geq 1} \min(m,k) N_k, \tag{2.16}$$

$$\gamma_m = \gamma_m(\nu) = \sum_{k\geq 1} \min(m,k)\nu_k. \tag{2.17}$$

Lemma 2.4. *For $1 \le i \le m$ we have*

$$(\xi_{m\alpha i+} + \eta_{m\alpha i-}) - (\xi_{m\alpha i-} + \eta_{m\alpha i+}) = \begin{cases} -(P_{m+1-2i} + N_{m+1-2i}), & i < \frac{m+1}{2}, \\ 0, & i = \frac{m+1}{2}, \\ P_{2i-m-1} + N_{2i-m-1}, & i > \frac{m+1}{2}. \end{cases}$$

Proposition 2.5. *A necessary condition for the existence of a generic string solution of pattern $N = (N_m)$ is as follows: If $N_m \ge 1$, then*

$$\begin{aligned} P_{m-1} + N_{m-1} \ge 1, \\ (P_{m-1} + N_{m-1}) + (P_{m-3} + N_{m-3}) \ge 1, \\ \vdots \\ (P_{m-1} + N_{m-1}) + (P_{m-3} + N_{m-3}) + \cdots + \begin{cases} (P_1 + N_1) \ge 1, & m : \textit{even}, \\ (P_2 + N_2) \ge 1, & m : \textit{odd}. \end{cases} \end{aligned} \tag{2.18}$$

PROOF. From the condition $\mathrm{ord}\,(F_{m\alpha i+}G_{m\alpha i-}) = \mathrm{ord}\,(F_{m\alpha i-}G_{m\alpha i+})$, we have

$$(\xi_{m\alpha i+} + \eta_{m\alpha i-}) - (\xi_{m\alpha i-} + \eta_{m\alpha i+}) = \begin{cases} -\zeta_{m\alpha 2}, & i = 1, \\ \zeta_{m\alpha i} - \zeta_{m\alpha i+1}, & 2 \le i \le m-1, \\ \zeta_{m\alpha m}, & i = m. \end{cases}$$

Solving this using Lemma 2.4, we get

$$\zeta_{m\alpha i} = \zeta_{m\alpha\, m+2-i} = \sum_{k=1}^{\min(i-1, m+1-i)} (P_{m+1-2k} + N_{m+1-2k}), \quad 2 \le i \le m.$$

In order to have $\zeta_{m\alpha 2}, \zeta_{m\alpha 3}, \ldots, \zeta_{m\alpha m} \ge 1$, the condition in the proposition must hold. □

2.5 The SCE.

Theorem 2.6. *Let $(x_{m\alpha i})$ be a generic string solution of pattern N. Then its string center $(z^0_{m\alpha})$ satisfies the equation*

$$\prod_{k \ge 1} \prod_{\beta=1}^{N_k} (z^0_{k\beta})^{A_{m\alpha,k\beta}} = (-1)^{P_m + N_m + 1}, \quad m \ge 1,\ 1 \le \alpha \le N_m, \tag{2.19}$$

$$A_{m\alpha,k\beta} := \delta_{mk}\delta_{\alpha\beta}(P_m + N_m) + 2\min(m,k) - \delta_{mk}. \tag{2.20}$$

We call (2.19) the string center equation (SCE). It is a linear congruence equation in the sense of (2.23).

Proof. Let us compute the ratio (2.15) explicitly.

$$\prod_{i=1}^{m} F^0_{m\alpha i\epsilon} = \begin{cases} (-1)^{\gamma_m} \prod_{s\geq 1}(f^{(s)}_{m\alpha})^{\nu_s}, & \epsilon = +, \\ (z^0_{m\alpha})^{\gamma_m} \prod_{s\geq 1}(f^{(s)}_{m\alpha})^{\nu_s}, & \epsilon = -, \end{cases}$$

$$f^{(s)}_{m\alpha} = \begin{cases} 1, & m \leq s, \\ (-z^0_{m\alpha})^{\frac{m-s}{2}}, & m > s,\ s \equiv m \bmod 2, \\ (-z^0_{m\alpha})^{\frac{m-s-1}{2}}(z^0_{m\alpha} - 1), & m > s,\ s \not\equiv m \bmod 2. \end{cases}$$

In order to calculate $\prod_{i=1}^{m}(G^0_{m\alpha i-}/G^0_{m\alpha i+})$, it is convenient first to evaluate

$$\prod_{i=1}^{m}\prod_{j=1}^{k}(z_{m\alpha i}q^{d_{m\alpha i}+1+\epsilon} - z_{k\beta j}q^{d_{k\beta j}+1-\epsilon})^0$$

$$= \begin{cases} (-z^0_{k\beta})^{2\min(m,k)-\delta_{mk}} g^{k\beta}_{m\alpha}, & \epsilon = 1, \\ (z^0_{m\alpha})^{2\min(m,k)-\delta_{mk}}(-1)^{(m-1)\delta_{mk}\delta_{\alpha\beta}} g^{k\beta}_{m\alpha}, & \epsilon = -1, \end{cases}$$

$$g^{k\beta}_{m\alpha} = \begin{cases} (-z^0_{m\alpha}z^0_{k\beta})^{\frac{mk}{2}-\min(m,k)}, & m \not\equiv k \bmod 2, \\ (-z^0_{m\alpha}z^0_{k\beta})^{\frac{mk}{2}-\frac{3}{2}\min(m,k)+\delta_{mk}} \\ \quad \times (z^0_{m\alpha} - z^0_{k\beta})^{\min(m,k)-\delta_{mk}}, & m\alpha \neq k\beta,\ m \equiv k \bmod 2, \\ (-z^0_{m\alpha}z^0_{k\beta})^{\frac{mk}{2}-\frac{3}{2}\min(m,k)+\delta_{mk}} \\ \quad \times y^0_{m\alpha 2}\cdots y^0_{m\alpha m}, & m\alpha = k\beta. \end{cases}$$

The factors $\prod_{s\geq 1}(f^{(s)}_{m\alpha})^{\nu_s}$ and $g^{k\beta}_{m\alpha}$ are all nonzero for a generic string solution $(x_{m\alpha i})$. They are canceled in the ratio (2.15) and we find

$$1 = (-1)^m \prod_{i=1}^{m} \frac{F^0_{m\alpha i+}G^0_{m\alpha i-}}{F^0_{m\alpha i-}G^0_{m\alpha i+}} = (-1)^{P_m+N_m+1}\prod_{k\beta}(z^0_{k\beta})^{-A_{m\alpha,k\beta}}. \tag{2.21}$$

The theorem is proved. □

Remark 2.7. From the condition $\prod_{s\geq 1}(f^{(s)}_{m\alpha})^{\nu_s}$, $g^{k\beta}_{m\alpha} \neq 0$ in the above proof, we see that a string solution $(x_{m\alpha i})$ is generic if and only if

$$\begin{aligned} \prod\nolimits_{1\leq s<m,\, s\not\equiv m(2)}(z^0_{m\alpha} - 1)^{\nu_s} &\neq 0, \\ \prod\nolimits_{k\beta(\neq m\alpha),\, k\equiv m(2)}(z^0_{m\alpha} - z^0_{k\beta})^{\min(m,k)-\delta_{mk}} &\neq 0 \end{aligned} \tag{2.22}$$

for any $m \geq 1$, $1 \leq \alpha \leq N_m$. In the latter, the power $\min(m,k) - \delta_{mk}$ implies that the collision $z^0_{m\alpha} = z^0_{k\beta}$ is allowed only when $m = k = 1$.

Definition 2.8. A solution to the SCE (2.19) is called *generic* if it satisfies condition (2.22).

By this definition, the solutions to the SCE (2.19) arising from the string centers of generic string solutions as in Theorem 2.6 are generic.

The SCE (2.19) becomes the linear congruence equation in terms of the variables $u_{k\beta} \in \mathbb{R}/\mathbb{Z}$ defined by $z^0_{k\beta} = \exp(2\pi\sqrt{-1}u_{k\beta})$:

$$\sum_{k\geq 1}\sum_{\beta=1}^{N_k} A_{m\alpha,k\beta}u_{k\beta} \equiv \frac{P_m + N_m + 1}{2} \quad \mathrm{mod}\ \mathbb{Z}. \tag{2.23}$$

This will also be called the SCE. In the limit $\hbar \to \infty$, the asymptotic behavior of the original variable $u_j = \frac{1}{2\pi\sqrt{-1}}\log(x_{k\beta i})$ in (2.2) is $u_{k\beta} + \sqrt{-1}\hbar(k + 1 - 2i)$.

2.6 Lifting generic solutions of the SCE to generic string solutions. In Section 2.5, we have seen that string centers of a generic string solution to the Bethe equation (2.7) yield a generic solution to the SCE. Here we show the inverse. Let $A = (A_{m\alpha,k\beta})$ be the matrix of size $N_1 + N_2 + \cdots$ defined by (2.20).

Theorem 2.9. *Suppose that* $N = (N_m)$ *satisfies the conditions in* (2.18) *and that* $\det A \neq 0$. *Let* $(z^0_{m\alpha})$ *be a generic solution to the SCE* (2.19). *Then there exists a unique generic string solution* $(x'_{m\alpha i}(q))$ *to the Bethe equation* (2.3) *such that* $z'^0_{m\alpha} = z^0_{m\alpha}$.

Define the variables (holomorphic functions of q) $w_{m\alpha i}$ by

$$w_{m\alpha i} = \begin{cases} z_{m\alpha i}, & i = 1, \\ y_{m\alpha i}, & 2 \leq i \leq m. \end{cases}$$

Then

$$z_{m\alpha i} = w_{m\alpha 1} + q^{\zeta_{m\alpha 2}}w_{m\alpha 2} + \cdots + q^{\zeta_{m\alpha i}}w_{m\alpha i}, \quad 1 \leq i \leq m. \tag{2.24}$$

Denote the ith equation of (2.8)–(2.10) by $L_{m\alpha i} = R_{m\alpha i}$. Let $J = (J_{m\alpha i,k\beta j})$ be the matrix of size $N_1 + 2N_2 + \cdots (= M)$ defined by $J_{m\alpha i,k\beta j} = \frac{\partial}{\partial w_{k\beta j}}(\frac{L_{m\alpha i}}{R_{m\alpha i}} - 1)$.

Lemma 2.10. *Suppose that* $N = (N_m)$ *satisfies the conditions in* (2.18) *and that* $\det A \neq 0$. *Then* $\det J$ *is nonzero at* $q = 0$ *(i.e.,* $\det(J^0_{m\alpha i,k\beta j}) \neq 0$).

Proof. Owing to the assumption (2.18), we have $\forall\, \zeta_{m\alpha i} \geq 1$. Since $L^0_{m\alpha i} = R^0_{m\alpha i} \neq 0$, it suffices to show that $\det \mathcal{J} \neq 0$ for $\mathcal{J}_{m\alpha i,k\beta j} = \frac{\partial}{\partial w_{k\beta j}}\log\frac{L_{m\alpha i}}{R_{m\alpha i}}$. From (2.24), both $\frac{\partial \tilde{F}_{m\alpha i\pm}}{\partial w_{k\beta j}}$ and $\frac{\partial \tilde{G}'_{m\alpha i\pm}}{\partial w_{k\beta j}}$ for $j \neq 1$ are zero at $q = 0$. Thus among $\mathcal{J}^0_{m\alpha i,k\beta j}$s, the nonvanishing ones are only $\mathcal{J}^0_{m\alpha i,k\beta 1}$ $(1 \leq i \leq m)$, $\mathcal{J}^0_{m\alpha i,m\alpha i} = -1/y^0_{m\alpha i}$ $(2 \leq i \leq m)$, and $\mathcal{J}^0_{m\alpha i,m\alpha i+1} = 1/y^0_{m\alpha i+1}$ $(1 \leq i \leq m-1)$. Let $\vec{\mathcal{J}}^0_{m\alpha i} = (\mathcal{J}^0_{m\alpha i,k\beta j})_{k\beta j}$ be a row vector of the matrix $\mathcal{J}$. In view of the above result, the linear dependence $\sum_{m\alpha i} c_{m\alpha i}\vec{\mathcal{J}}^0_{m\alpha i} = 0$ can possibly hold only when $c_{m\alpha i}$ is independent of i. Consequently, we consider the equation $\sum_{m\alpha} c_{m\alpha} \sum_{i=1}^{m} \vec{\mathcal{J}}^0_{m\alpha i} = 0$. The $(k\beta 1)$st component of the vector $\sum_{i=1}^{m} \vec{\mathcal{J}}^0_{m\alpha i}$ is given by

$$\lim_{q\to 0} \frac{\partial}{\partial w_{k\beta 1}} \log \prod_{i=1}^{m} \frac{\tilde{F}_{m\alpha i+} \tilde{G}'_{m\alpha i-}}{\tilde{F}_{m\alpha i-} \tilde{G}'_{m\alpha i+}} = \frac{\partial}{\partial z^0_{k\beta}} \log \prod_{i=1}^{m} \frac{-F^0_{m\alpha i+} G^0_{m\alpha i-}}{F^0_{m\alpha i-} G^0_{m\alpha i+}},$$

where we have taken (2.15), (2.24), and $\forall\, \zeta_{m\alpha i} \geq 1$ into account. Due to (2.21), the last expression is equal to $-A_{m\alpha,k\beta}/z^0_{k\beta}$. Thus the equation $\sum_{m\alpha} c_{m\alpha} \sum_{i=1}^{m} \vec{\mathcal{J}}^0_{m\alpha i} = 0$ is equivalent to $\sum_{m\alpha} c_{m\alpha} A_{m\alpha,k\beta} = 0$ for any $k\beta$. This admits only the trivial solution for $c_{m\alpha}$ if $\det A \neq 0$. □

Proof of Theorem 2.9. The Bethe equations (2.8)–(2.10) are simultaneous equations on the variables $(w_{m\alpha i}, q)$. At $q = 0$, (2.8)–(2.10) reduce to (2.12)–(2.14). The latter fix $(y^0_{m\alpha i})$ unambiguously once a generic solution $(z^0_{m\alpha})$ to the SCE is given. Denote the resulting value of $w_{m\alpha i}$ by $w^0_{m\alpha i}$. Thus (2.8)–(2.10) are valid at $(w_{m\alpha i}, q) = (w^0_{m\alpha i}, 0)$. From the implicit function theorem, there uniquely exist functions $\tilde{w}_{m\alpha i}(q)$ satisfying (2.8)–(2.10) and $\tilde{w}_{m\alpha i}(0) = w^0_{m\alpha i}$ if the Jacobian $\det J$ at $(w_{m\alpha i}, q) = (w^0_{m\alpha i}, 0)$ is nonzero. By Lemma 2.10, this has been guaranteed. □

From Theorems 2.6 and 2.9, we have the following.

Corollary 2.11. *Suppose that $N = (N_m)$ satisfies the conditions in* (2.18) *and that* $\det A \neq 0$. *Then there is a one-to-one correspondence between generic string solutions to the Bethe equation* (2.3) *and generic solutions to the SCE* (2.19).

Remark 2.12. In the next section, we assume a further condition (3.1) for $N = (N_m)$. The condition $\det A \neq 0$ follows from (3.1), as we will show in Corollary 3.4. In general, the conditions in (2.18) and (3.1) are simultaneously satisfied if $\sum_m m N_m$ is sufficiently smaller than $\sum_s \nu_s$. In other words, for a module $W(\nu)$ with $\sum_s \nu_s$ large enough, these conditions hold at least "near the highest weight."

2.7 Off-diagonal solution. To restrict oneself to the admissible solutions to the Bethe equation in Corollary 2.11 is natural because otherwise the associated Bethe vectors vanish [TV]. On the other hand, the limitation to the generic case was made for a technical reason. In fact the most essential constraint on the solution of the Bethe equation (2.3) is that $x_1, \ldots, x_M$ are all distinct. Otherwise, the Bethe vectors again vanish. To make this precise in the present context, we introduce the following.

Definition 2.13. A meromorphic solution (x_i) of (2.3) is called *diagonal* (*off-diagonal*) if $x_i = x_j$ for some $i \neq j$ as a function of q around $q = 0$ (otherwise).

Definition 2.14. A solution $(z^0_{m\alpha})$ of the SCE is called *diagonal* (*off-diagonal*) if $z^0_{m\alpha} = z^0_{m\beta}$ for some $\alpha \neq \beta$ (otherwise).

According to the above definition, the string solution $(z_{m\alpha i})$ of the Bethe equation is diagonal (off-diagonal) if $z_{m\alpha i}(q) = z_{k\beta j}(q)$ and $d_{m\alpha i} = d_{k\beta j}$ for some $m\alpha i \neq k\beta j$ (otherwise). We expect that the number of off-diagonal string solutions of the

Bethe equation is equal to the number of off-diagonal solutions of the SCE under a certain condition for $N = (N_m)$ like (2.18) and $\det A \neq 0$. Thus far, we have been unable to solve the discrepancy between "generic" and "off-diagonal."

Remark 2.15. In [LS], the SCE was given for the XXZ case ($\nu_s = L\delta_{1s}$). It is claimed that the SCE is satisfied regardless of whether a string solution is generic. Moreover, each off-diagonal solution of the SCE gives rise to an off-diagonal string solution.

3 Counting of off-diagonal solutions

This section is devoted to an expository proof of Theorem 3.5. It provides an explicit combinatorial formula that counts the number of off-diagonal solutions to the SCE (2.19) in the sense of Section 2.7. We assume that the string pattern $N = (N_m)$ satisfies the condition

$$P_m(\nu, N) \geq 0 \text{ whenever } N_m \geq 1 \tag{3.1}$$

throughout Section 3.

3.1 Rule of counting. We shall work with the logarithmic form of the SCE (2.23), presented as

$$A\vec{u} \equiv \vec{c} \mod \mathbb{Z}^d, \tag{3.2}$$

where $d = N_1 + N_2 + \cdots$. Here $\vec{u} = (u_{k\beta})$ is the unknown and $\vec{c}$ is some constant vector. The d-dimensional matrix $A = (A_{m\alpha,k\beta})$ is specified by (2.20). It has a block structure according to the string pattern. For example, if only N_1, N_2 and N_3 are nonzero, it looks as follows ($\mathcal{P}_i = P_i + N_i$):

$$\begin{pmatrix}
\mathcal{P}_1+1 & \cdots & 1 & 2 & \cdots & 2 & 2 & \cdots & 2 \\
\vdots & \ddots & \vdots & \vdots & & \vdots & \vdots & & \vdots \\
1 & \cdots & \mathcal{P}_1+1 & 2 & \cdots & 2 & 2 & \cdots & 2 \\
2 & \cdots & 2 & \mathcal{P}_2+3 & \cdots & 3 & 4 & \cdots & 4 \\
\vdots & & \vdots & \vdots & \ddots & \vdots & \vdots & & \vdots \\
2 & \cdots & 2 & 3 & \cdots & \mathcal{P}_2+3 & 4 & \cdots & 4 \\
2 & \cdots & 2 & 4 & \cdots & 4 & \mathcal{P}_3+5 & \cdots & 5 \\
\vdots & & \vdots & \vdots & \ddots & \vdots & \vdots & & \vdots \\
2 & \cdots & 2 & 4 & \cdots & 4 & 5 & \cdots & \mathcal{P}_3+5
\end{pmatrix},$$

which consists of nine submatrices of size $N_i \times N_j$ ($1 \leq i, j \leq 3$).

We will seek the number of off-diagonal solutions to the SCE (3.2) in the sense of Section 2.7. Let us remember the three essential rules for counting it. First, there is no change in (2.2) under any integer shift of u_j. Accordingly, we should not distinguish the solutions $\vec{u}$ and $\vec{u}'$ to (3.2) if $\vec{u} - \vec{u}' \in \mathbb{Z}^d$. In other words, we consider $u_{k\beta} \in \mathbb{R}/\mathbb{Z}$

rather than $\mathbb{R}$. Second, the original Bethe equation (2.2) is symmetric with respect to $u_1, \ldots, u_M$, but their permutation does not lead to a new Bethe vector, as is well known. Consequently, for a solution $\vec{u} = (u_{k\beta})$ of pattern $N = (N_m)$, we should regard

$$(u_{k1}, u_{k2}, \ldots, u_{kN_k}) \in (\mathbb{R}/\mathbb{Z})^{N_k} / \mathfrak{S}_{N_k}$$

for each $k \in \mathbb{N}$. Here $\mathfrak{S}_{N_k}$ stands for the symmetric group permuting the N_k components. Last, but most important, Definition 2.14 postulates that $u_{k\beta} \neq u_{k\beta'} \in \mathbb{R}/\mathbb{Z}$ if $1 \leq \beta \neq \beta' \leq N_k$ for each $k \in \mathbb{N}$. To summarize these rules, we start with a fixed string pattern (N_m) and specify A and $\vec{c}$ by (2.23). Then we are to count the number of solutions $\vec{u} = (u_{k\beta})$ to the SCE (3.2) such that

$$(u_{k1}, u_{k2}, \ldots, u_{kN_k}) \in \left((\mathbb{R}/\mathbb{Z})^{N_k} - \Delta_{N_k}\right)/\mathfrak{S}_{N_k},$$
$$\Delta_n = \{(v_1, \ldots, v_n) \in (\mathbb{R}/\mathbb{Z})^n \mid v_\alpha = v_\beta \text{ for some } 1 \leq \alpha \neq \beta \leq n\}$$

for each $k \in \mathbb{N}$. In practice, one has simply to find the number of solutions such that $(u_{k1}, u_{k2}, \ldots, u_{kN_k}) \in (\mathbb{R}/\mathbb{Z})^{N_k} - \Delta_{N_k}$ and divide afterwards by $N_k!$ for each k.

3.2 Example. Before treating the general case, let us illustrate how to enumerate off-diagonal solutions with the simplest example, $\nu_s = L\delta_{s,1}$. This corresponds to the spin-$\frac{1}{2}$ XXZ model with L-sites. Thus the character of the quantum space is expanded as ($x = e^{\Lambda_1}$)

$$(x + x^{-1})^L = x^L + \binom{L}{1}x^{L-2} + \binom{L}{2}x^{L-4} + \binom{L}{3}x^{L-6} + \cdots$$

according to the "magnon number" $M = 0, 1, 2, 3, \ldots$. Here and in what follows, the symbol $\binom{\cdot}{\cdot}$ will always denote the generalized binomial coefficient

$$\binom{\xi}{n} = \begin{cases} \xi(\xi-1)\cdots(\xi-n+1)/n! & \text{if } n \in \mathbb{Z}_{\geq 1}, \\ 1 & \text{if } n = 0, \\ 0 & \text{otherwise}, \end{cases} \qquad \xi \in \mathbb{C}.$$

Counting the off-diagonal solutions sketched below up to $M = 3$ is useful for gaining an idea for treating the general case. The result coincides with the weight multiplicity $\binom{L}{M}$ that appears in the above expansion. The proof of the coincidence in the general case will be given in Section 4. According to (3.1), we assume that $P_m = L - 2\sum_{k\geq 1} \min(m, k)N_k \geq 0$ whenever $N_m > 0$. We use the following fact.

Proposition 3.1. *Let B be an n by n integer matrix with* $\det B \neq 0$. *Then for any $\vec{b} \in \mathbb{R}^n$, the equation $B\vec{x} \equiv \vec{b}$* mod $\mathbb{Z}^n$ *has exactly* $|\det B|$ *solutions $\vec{x}$ in* $(\mathbb{R}/\mathbb{Z})^n$.

Proof. Because of linearity, it is enough to show the statement for $\vec{b} = \vec{0}$. Consider a pair of lattices $L \supset M$, where L consists of the solutions $\vec{x} \in \mathbb{R}^n$ for the homogeneous equation $B\vec{x} \equiv \vec{0} \bmod \mathbb{Z}^n$, and $M = \mathbb{Z}^n$. The column vectors $\vec{f}_1, \ldots, \vec{f}_n$ of B^{-1} form a basis of L. It is well known [C, 1.2.2, Lemma 1] that $|L/M|$ is equal to $|\det(\vec{f}_1, \ldots, \vec{f}_n)|^{-1}$, which is $|\det B|$. □

As it turns out, condition (3.1) assures $\det B > 0$ not only for $B = A$ but also for all the relevant matrices B coming into the game (cf. Lemma 3.3)

The $M = 0$ case. Equation (3.2) is void. We simply define the number of off-diagonal solutions to be 1, in agreement with $\binom{L}{0}$.

The $M = 1$ case. The only string pattern (2.6) is $N_1 = 1$. Then (3.2) is a scalar equation $Lu_{1,1} \equiv c \bmod \mathbb{Z}$ for some $c \in \mathbb{R}$. Thus the number of off-diagonal solutions is L, in agreement with $\binom{L}{1}$.

The $M = 2$ case. There are two string patterns—(i) $N_2 = 1$ and (ii) $N_1 = 2$—that satisfy (2.6). (i) Equation (3.2) is $Lu_{2,1} \equiv c$ for some $c \in \mathbb{R}$. Thus there are L off-diagonal solutions. (ii) Equation (3.2) reads

$$A\vec{u} = \begin{pmatrix} L-1 & 1 \\ 1 & L-1 \end{pmatrix} \begin{pmatrix} u_{1,1} \\ u_{1,2} \end{pmatrix} \equiv \vec{c} \mod \mathbb{Z}^2$$

for some $\vec{c}$. From the assumption $P_1 = L - 4 \geq 0$, there are $\det A = L(L-2)$ solutions to this by Proposition 3.1. However, they include the diagonal solutions $u_{1,1} = u_{1,2}$. Under this constraint, the above matrix equation reduces to $Lu_{1,1} \equiv c'$ with some c', establishing that the number of diagonal solutions is L. Therefore, the nondiagonal solutions from (ii) is enumerated as $(L(L-2) - L)/2$ by recalling the $\mathfrak{S}_2$ redundancy. Collecting the contributions from the string patterns (i) and (ii), one finds the number of off-diagonal solutions:

$$L + \frac{1}{2}(L(L-2) - L) = \binom{L}{2}.$$

The $M = 3$ case. There are three string patterns—(i) $N_3 = 1$, (ii) $N_1 = N_2 = 1$, and (iii) $N_1 = 3$—that satisfy (2.6). (i) Equation (3.2) is $Lu_{3,1} \equiv c$ for some $c \in \mathbb{R}$. Thus there are L off-diagonal solutions. (ii) Equation (3.2) reads

$$A\vec{u} = \begin{pmatrix} L-2 & 2 \\ 2 & L-2 \end{pmatrix} \begin{pmatrix} u_{1,1} \\ u_{2,1} \end{pmatrix} \equiv \vec{c} \mod \mathbb{Z}^2$$

for some $\vec{c}$. From the assumption $P_1 = L - 4$, $P_2 = L - 6 \geq 0$, this has $\det A = L(L-4)$ solutions by Proposition 3.1. In this case, there is no permutation redundancy to remove. (iii) Equation (3.2) reads

$$A^{1/2/3}\vec{u} = \begin{pmatrix} L-2 & 1 & 1 \\ 1 & L-2 & 1 \\ 1 & 1 & L-2 \end{pmatrix} \begin{pmatrix} u_{1,1} \\ u_{1,2} \\ u_{1,3} \end{pmatrix} \equiv \vec{c} \mod \mathbb{Z}^3$$

for some $\vec{c}$. Here we have written A as $A^{1/2/3}$ to match the notation in Section 3.3. Due to the assumption $P_1 = L - 6 \geq 0$, this has $\det A^{1/2/3} = L(L-3)^2$ solutions in $(\mathbb{R}/\mathbb{Z})^3$ by Proposition 3.1. However, they include various diagonal solutions. For example, under the condition $u_{1,1} = u_{1,2}$, the above equation reduces to

$$A^{12/3}\vec{u} = \begin{pmatrix} L-1 & 1 \\ 2 & L-2 \end{pmatrix} \begin{pmatrix} u_{1,1} \\ u_{1,3} \end{pmatrix} \equiv \vec{c}' \mod \mathbb{Z}^2$$

for some $\vec{c}'$, which has $\det A^{12/3} = L(L-3)$ solutions. Similarly, there are diagonal solutions counted by $\det A^{13/2}$ and $\det A^{23/1}$, which are both equal to $L(L-3)$. Finally, there is the completely diagonal one $u_{1,1} = u_{1,2} = u_{1,3}$ satisfying $A^{123}u_{1,1} = Lu_{1,1} \equiv c'' \bmod \mathbb{Z}$ for some c''. By the inclusion–exclusion principle, we can now compute the number of off-diagonal solutions in (iii) as

$$\begin{aligned}&\frac{1}{3!}\left(\det A^{1/2/3} - \det A^{12/3} - \det A^{13/2} - \det A^{23/1} + 2\det A^{123}\right)\\ &\qquad = \frac{1}{6}L(L-4)(L-5),\end{aligned} \tag{3.3}$$

where we have removed the 3!-fold $\mathfrak{S}_3$ redundancy. Assembling the contributions from (i), (ii), and (iii) we get

$$L + L(L-4) + \frac{1}{6}L(L-4)(L-5) = \binom{L}{3}$$

as desired.

3.3 General case. Let us proceed to the general case where both the quantum space data $\nu = (\nu_s)$ and the magnon number $M \in \mathbb{Z}_{\geq 0}$ are arbitrary. The examples in Section 3.2 already elucidate the essential feature in our counting. We dare to do some overcounting by classifying diagonal solutions in terms of their patterns like 12/3 and finally subtract them via a kind of inclusion–exclusion principle. The most natural framework for systematizing such a process is the partition of sets and the Möbius inversion trick, summarized in the appendix. Compare the coefficients 1, −1, −1, −1, and 2 appearing in (3.3) with (A.1). Below we will use the terminology and notation therein.

We start with the SCE (3.2). Given $M \in \mathbb{N}$, fix a string pattern (N_m) satisfying (2.6). Set

$$\mathcal{J} = \{j \in \mathbb{N} \mid N_j > 0\}, \qquad j_0 = \max \mathcal{J}. \tag{3.4}$$

Thus $d = \sum_{j\in\mathcal{J}} N_j$. Take an element $\pi = (\pi^{(1)}, \pi^{(2)}, \ldots, \pi^{(j_0)}) \in L_{N_1} \times \cdots \times L_{N_{j_0}}$ in the sense of the appendix. Thus for each k, $\pi^{(k)} = (\pi^{(k)}_1, \ldots, \pi^{(k)}_l)$ is a partition of the set $\{1, \ldots, N_k\}$ into the blocks

$$\{1, \ldots, N_k\} = \pi^{(k)}_1 \sqcup \cdots \sqcup \pi^{(k)}_l$$

for some l. Here and in what follows, if $N_i = 0$ for $1 \leq i < j_0$, the corresponding ith component should simply be dropped. For example, $L_{N_1} \times \cdots \times L_{N_{j_0}}$ actually means the product of the set L_{N_j} over $j \in \mathcal{J}$, and $(\pi^{(1)}, \ldots, \pi^{(j_0)})$ is in fact $(\pi^{(j)})_{j\in\mathcal{J}}$. To classify the diagonal solutions, we introduce

$$\begin{aligned}\mathrm{Sol}'_\pi &= \{\vec{u} = (u_{k\beta}) \mid u_{k\alpha} = u_{k\beta} \text{ if } \alpha \text{ and } \beta \text{ belong}\\ &\qquad\qquad \text{to the same block of } \pi^{(k)} \text{ for each } k\},\\[1ex] \mathrm{Sol}_\pi &= \{\vec{u} = (u_{k\beta}) \mid u_{k\alpha} = u_{k\beta} \text{ if and only if } \alpha \text{ and } \beta \text{ belong}\\ &\qquad\qquad \text{to the same block of } \pi^{(k)} \text{ for each } k\}.\end{aligned}$$

By this definition, it follows that ($|\cdot|$ here denotes cardinality)

$$|\mathrm{Sol}'_\pi| = \sum_{\pi' \le \pi} |\mathrm{Sol}_{\pi'}|$$

in terms of the partial order $\le$ of the poset $L_{N_1} \times \cdots \times L_{N_{j_0}}$, introduced in the appendix in Section A.5. By means of the Möbius inversion formula, this is equivalent to

$$|\mathrm{Sol}_\pi| = \sum_{\pi' \le \pi} \mu(\pi', \pi)|\mathrm{Sol}'_{\pi'}| \quad \text{for any } \pi \in L_{N_1} \times \cdots \times L_{N_{j_0}}.$$

The off-diagonal solutions in question are counted by setting $\pi = \pi_{\max}$, which is the maximal element of $L_{N_1} \times \cdots \times L_{N_{j_0}}$, explained in the appendix in Section A.5. Removing the $\mathfrak{S}_{N_m}$ redundancy for $m \in \mathcal{J}$, they are enumerated as

$$\frac{|\mathrm{Sol}_{\pi_{\max}}|}{\prod_{m \in \mathcal{J}} N_m!} = \frac{\sum_{\pi \in L_{N_1} \times \cdots \times L_{N_{j_0}}} \mu(\pi, \pi_{\max})|\mathrm{Sol}'_\pi|}{\prod_{m \in \mathcal{J}} N_m!}. \tag{3.5}$$

Let us evaluate $|\mathrm{Sol}'_\pi|$. In the original SCE (3.2), impose the constraint $u_{k\alpha} = u_{k\beta}$ on $\vec{u} = (u_{k\beta})$ if α and β belong to the same block of $\pi^{(k)}$ for each $k \in \mathcal{J}$. As we obtained $A^{12/3}$ from $A = A^{1/2/3}$ in Section 3.2, the result has the form

$$A^\pi \vec{u}_\pi \equiv \vec{c}_\pi \mod \mathbb{Z}^{l(\pi)} \quad \text{for some } \vec{c}_\pi \in \mathbb{R}^{l(\pi)}. \tag{3.6}$$

Here $l(\pi)$ denotes the length of $\pi = (\pi^{(1)}, \ldots, \pi^{(j_0)})$, as specified in the appendix in Section A.5. In the new unknown $\vec{u}_\pi = (u_{k\beta})$, β is now labeled by the blocks of $\pi^{(k)}$ for each $k \in \mathcal{J}$. A^π is an $l(\pi)$ by $l(\pi)$ integer matrix obtained by a reduction of the matrix A, which was d by d originally ($d = \sum_{j \in \mathcal{J}} N_j$). It is formed by summing the $(k\beta)$ columns of A over those β belonging to the same block of $\pi^{(k)}$ and discarding all but one row for each block. For example, suppose $\mathcal{J} = \{1, 2\}$. Thus only N_1 and N_2 are nonzero and $\pi = (\pi^{(1)}, \pi^{(2)}) \in L_{N_1} \times L_{N_2}$. Let $\pi^{(1)} = (\pi_1^{(1)}, \pi_2^{(1)}, \pi_3^{(1)})$ and $\pi^{(2)} = (\pi_1^{(2)}, \pi_2^{(2)})$ so that $l(\pi^{(1)}) = 3$, $l(\pi^{(2)}) = 2$, and $l(\pi) = 5$. Denote the number of elements in the block $\pi_i^{(1)}$ by λ_i (resp., $|\pi_i^{(2)}|$ by μ_i). By definition, $\lambda_1 + \lambda_2 + \lambda_3 = N_1$, $\mu_1 + \mu_2 = N_2$, and the matrix A^π reads ($\mathcal{P}_i = P_i + N_i$)

$$\begin{pmatrix} \mathcal{P}_1 + \lambda_1 & \lambda_2 & \lambda_3 & 2\mu_1 & 2\mu_2 \\ \lambda_1 & \mathcal{P}_1 + \lambda_2 & \lambda_3 & 2\mu_1 & 2\mu_2 \\ \lambda_1 & \lambda_2 & \mathcal{P}_1 + \lambda_3 & 2\mu_1 & 2\mu_2 \\ 2\lambda_1 & 2\lambda_2 & 2\lambda_3 & \mathcal{P}_2 + 3\mu_1 & 3\mu_2 \\ 2\lambda_1 & 2\lambda_2 & 2\lambda_3 & 3\mu_1 & \mathcal{P}_2 + 3\mu_2 \end{pmatrix}.$$

This is easily seen from the example of A in Section 3.1. In general, A^π is not symmetric. Its matrix element is given by $A^\pi_{(m,i),(k,j)} = \delta_{m,k}\delta_{i,j}(P_m + N_m) + (2\min(m,k) - \delta_{m,k})|\pi_j^{(k)}|$ for $m, k \in \mathcal{J}$, $1 \le i \le l(\pi^{(m)})$, and $1 \le j \le l(\pi^{(k)})$. Note that $A = A^{\pi_{\max}}$.

By definition, $|\mathrm{Sol}'_\pi|$ counts the number of all the solutions to (3.6). Therefore, from Proposition 3.1, we have

$$|\mathrm{Sol}'_\pi| = |\det A^\pi|. \tag{3.7}$$

It is straightforward to show the following.

Lemma 3.2. *For any* $\pi \in L_{N_1} \times \cdots \times L_{N_{j_0}}$, *we have*

$$\det A^\pi = \det_{m,k\in\mathcal{J}}(F_{m,k}) \prod_{m\in\mathcal{J}} (P_m + N_m)^{l(\pi^{(m)})-1}, \tag{3.8}$$

$$F_{m,k} = \delta_{m,k} P_m + 2\min(m,k) N_k. \tag{3.9}$$

The lemma holds without assuming (3.1). The dependence on π enters only through $l(\pi^{(m)})$. Like A^π, the matrix F is nonsymmetric in general:

$$F = \begin{pmatrix} P_1 + 2N_1 & 2N_2 & 2N_3 & \cdots & 2N_{j_0} \\ 2N_1 & P_2 + 4N_2 & 4N_3 & \cdots & 4N_{j_0} \\ 2N_1 & 4N_2 & P_3 + 6N_3 & \cdots & \vdots \\ \vdots & \vdots & \vdots & \ddots & 2(j_0-1)N_{j_0} \\ 2N_1 & 4N_2 & 6N_3 & \cdots & P_{j_0} + 2j_0 N_{j_0} \end{pmatrix}.$$

We remark that

$$\sum_{k\geq 1} F_{m,k} = \gamma_m \tag{3.10}$$

for any m, where γ_m is defined in (2.17).

Lemma 3.3. *If* $P_m \geq 0$ *for any* $m \in \mathcal{J}$, *then* $\det A^\pi > 0$.

Proof. Denote $\det_{m,k\in\mathcal{J}}(F_{m,k})$ by $\det_{\mathcal{J}} F$. By Lemma 3.2, it suffices to verify that $\det_{\mathcal{J}} F > 0$. We do this by a double induction on $|\mathcal{J}|$ and $\sum_m P_m$, regarding P_m as nonnegative variables independent of $\{N_m\}$. First, let $\mathcal{J} = \{j_1 < \cdots < j_l\}$ be arbitrary and $\sum_m P_m = 0$. Thus $P_m = 0$ for any $m \in \mathcal{J}$; hence $\det_{\mathcal{J}} F = (\prod_{j\in\mathcal{J}} 2N_j) \det_{m,k\in\mathcal{J}}(\min(m,k)) = (\prod_{j\in\mathcal{J}} 2N_j) j_1(j_2 - j_1)\cdots(j_l - j_{l-1}) > 0$. Next, let $\mathcal{J} = \{j\}$. Then $\det_{\mathcal{J}} F = P_j + 2jN_j > 0$ because of the assumption $P_j \geq 0$. Finally, let $\mathcal{J}$ and $\sum_m P_m > 0$ be arbitrary. Then there exists $i \in \mathcal{J}$ such that $P_i > 0$. Setting $P'_j = P_j - \delta_{j,i}$ and $\mathcal{J}' = \mathcal{J} \setminus \{i\}$, one can expand the determinant as $\det_{\mathcal{J}} F(\{P_m\}) = \det_{\mathcal{J}'} F(\{P_m\}) + \det_{\mathcal{J}} F(\{P'_m\})$. By induction, the two terms on the right-hand side are both positive. □

Although it is more direct to expand $\det_{m,k\in\mathcal{J}}(F_{m,k})$ from the beginning, we have presented a proof in the above form because it generalizes to an arbitrary simple Lie algebra case that will be treated in our subsequent paper.

The specialization $\pi = \pi_{\max}$ in the above leads to the following.

Corollary 3.4. *If $P_m \geq 0$ for any $m \in \mathcal{J}$, then* $\det A > 0$.

Given the quantum space data $\nu = (\nu_s)$ and the string pattern $N = (N_m)$, we define

$$R(\nu, N) = \det_{m,k \in \mathcal{J}} (F_{m,k}) \prod_{m \in \mathcal{J}} \frac{1}{N_m} \binom{P_m + N_m - 1}{N_m - 1}, \tag{3.11}$$

when $\mathcal{J} \neq \emptyset$. Here $P_m = P_m(\nu, N)$, $F_{m,k}$, and $\mathcal{J}$ are given by (2.16), (3.9), and (3.4), respectively. When $\mathcal{J} = \emptyset$, namely, $\forall\, N_m = 0$, we set $R(\nu, N) = 1$ regardless of ν. In the definition itself, we do not need to assume that $P_m \geq 0$ for those $m \in \mathcal{J}$, and (ν_s) can be arbitrary complex parameters.

Theorem 3.5. *Assume that $P_m \geq 0$ for any $m \in \mathcal{J}$. Then the number of off-diagonal solutions to the SCE* (3.2) *is equal to* $R(\nu, N)$.

Proof. By the assumption of the theorem, the number of off-diagonal solutions has already been obtained in (3.5). By virtue of (3.7) and Lemma 3.3, this number is equal to

$$\begin{aligned} &\frac{1}{\prod_{m \in \mathcal{J}} N_m!} \sum_{\pi \in L_{N_1} \times \cdots \times L_{N_{j_0}}} \mu(\pi, \pi_{\max}) \det A^{\pi} \\ &= \det_{m,k \in \mathcal{J}} (F_{m,k}) \sum_{\pi \in L_{N_1} \times \cdots \times L_{N_{j_0}}} \mu(\pi, \pi_{\max}) \prod_{m \in \mathcal{J}} \frac{(P_m + N_m)^{l(\pi^{(m)}) - 1}}{N_m!}, \end{aligned}$$

where we have substituted (3.8). By means of (A.2), the π-sum can be taken, leading to

$$\det_{m,k \in \mathcal{J}} (F_{m,k}) \prod_{m \in \mathcal{J}} \frac{(P_m + N_m)_{N_m}}{N_m!(P_m + N_m)} = R(\nu, N). \qquad \square$$

Under the specialization $\nu_s = L\delta_{1,s}$, the above $R(\nu, N)$ reproduces the number of off-diagonal solutions for each string pattern exemplified in Section 3.2.

Expanding the determinant in (3.11), one can rewrite $R(\nu, N)$ as follows:

$$R(\nu, N) = \sum_{J \subset \mathbb{N}} D_J \prod_{m \in \mathbb{N} \setminus J} \binom{P_m + N_m}{N_m} \prod_{m \in J} \binom{P_m + N_m - 1}{N_m - 1}, \tag{3.12}$$

$$D_J = \begin{cases} 1 & \text{if } J = \emptyset, \\ \det_{m,k \in J} (2 \min(m, k) - \delta_{m,k}) & \text{otherwise.} \end{cases} \tag{3.13}$$

In deriving this, we have used $\binom{*}{0} = 1$ and $\binom{*}{-1} = 0$. From this expression, $R(\nu, N) \in \mathbb{Z}$ is manifest if $\forall\, \nu_s \in \mathbb{Z}$. In Section 4, we will work mainly with (3.12) rather than (3.11).

4 $R(\nu, N)$ as weight multiplicity

Set

$$K(\nu, N) = \prod_{m \geq 1} \binom{P_m + N_m}{N_m}.$$

This is a generalization of Bethe's fermionic formula in (1.2) corresponding to the quantum space (2.1) [K]. The nature of our $R(\nu, N)$ becomes most transparent by a parallel analysis on $K(\nu, N)$. It contains $K(\nu, N)$ as the summand in (3.12) corresponding to $J = \emptyset$.

In this section, we fix $l \in \mathbb{N}$. It plays a role of "cutoff" similar to that of j_0 in (3.4) and has nothing to do with the length function of partitions. We will introduce various functions indexed with l, which tend to the quantities in our problem in the limit $l \to \infty$. In particular, P_m and γ_m in Sections 4.1–4.3 stand for the truncations of (2.16)–(2.17) by l:

$$P_m = P_m(\nu, N) = \gamma_m - 2\sum_{k=1}^{l} \min(m, k) N_k, \tag{4.1}$$

$$\gamma_m = \gamma_m(\nu) = \sum_{k=1}^{l} \min(m, k)\nu_k. \tag{4.2}$$

We do not introduce new symbols for them as they will be used only in the sections indicated. We set $\mathbb{N}_l = \{1, 2, \ldots, l\}$. The binomial coefficient is that specified in the beginning of Section 3.2.

4.1 $R_l(\nu, N)$ and $K_l(\nu, N)$. Let $\nu = (\nu_s)$, $\nu_1, \ldots, \nu_l \in \mathbb{C}$ and $N = (N_m)$, $N_1, \ldots, N_l \in \mathbb{Z}_{\geq 0}$, be arbitrary. Define

$$R_l(\nu, N) = \sum_{J \subset \mathbb{N}_l} D_J \prod_{m \in \mathbb{N}_l \setminus J} \binom{P_m + N_m}{N_m} \prod_{m \in J} \binom{P_m + N_m - 1}{N_m - 1}, \tag{4.3}$$

$$K_l(\nu, N) = \prod_{m \in \mathbb{N}_l} \binom{P_m + N_m}{N_m}, \tag{4.4}$$

where D_J is specified by (3.13). When $N = 0$ (i.e., $\forall\, N_m = 0$), we have $R_l(\nu, 0) = K_l(\nu, 0) = 1$ regardless of ν. Obviously, we have $R(\nu, N) = \lim_{l\to\infty} R_l(\nu, N)$ and $K(\nu, N) = \lim_{l\to\infty} K_l(\nu, N)$, where the limits render no subtlety. We will utilize two other expressions of $R_l(\nu, N)$. The first is the analogue of (3.11):

$$R_l(\nu, N) = \left(\det_{m,k \in \{i \in \mathbb{N}_l | N_i \neq 0\}} F_{m,k}\right) \prod_{m \in \mathbb{N}_l, N_m \neq 0} \frac{1}{N_m} \binom{P_m + N_m - 1}{N_m - 1}. \tag{4.5}$$

To match (4.3), the right side of this equation should be understood as 1 when $N = 0$. To deduce the second expression, for $J \subset \mathbb{N}_l$, we introduce

$$\nu[J] = (\nu[J]_s), \qquad \nu[J]_s = \nu_s - 2\theta(s \in J), \tag{4.6}$$

$$N[J] = (N[J]_m), \qquad N[J]_m = N_m - \theta(m \in J), \tag{4.7}$$

where $\theta(\text{true}) = 1$ and $\theta(\text{false}) = 0$. With the aid of

$$P_m(\nu[J], N[J]) = P_m(\nu, N) =: P[J]_m,$$

we can rewrite (4.3) as

$$R_l(\nu, N) = \sum_{J \subset \mathbb{N}_l} D_J \prod_{m \in \mathbb{N}_l} \binom{P[J]_m + N[J]_m}{N[J]_m}. \tag{4.8}$$

4.2 Generating functions. Let us introduce the generating functions

$$R_l(\nu|w) = \sum_N R_l(\nu, N) w_1^{N_1} \cdots w_l^{N_l}, \tag{4.9}$$

$$K_l(\nu|w) = \sum_N K_l(\nu, N) w_1^{N_1} \cdots w_l^{N_l}, \tag{4.10}$$

where $w = (w_1, \ldots, w_l)$ and $\sum_N$ extends over $N_1, \ldots, N_l \in \mathbb{Z}_{\geq 0}$.

Proposition 4.1. *When $\nu = 0$ (i.e., $\forall\, \nu_s = 0$), we have*

$$R_l(0|w) = 1.$$

Proof. We show that $R_l(0, N) = 0$ for any $N \neq 0$. Note that $\nu = 0$ implies $\forall\, \gamma_m = 0$. Therefore, the assertion follows from expressions (4.5) and (3.10). □

This simple observation will eventually lead to the nontrivial consequence (4.24), whose derivation is analogous to the "denominator formula." In contrast, $K_l(0|w)$ is not a simple function. See (4.19).

4.3 Analytic formula for generating functions. Consider the variables $\{z_{j,i-1} \mid 1 \leq i \leq j \leq l\}$ related via

$$z_{j,i} = z_{j,i-1} \left(1 - z_{i,i-1}\right)^{-2(j-i)}, \quad 1 \leq i < j \leq l. \tag{4.11}$$

For $1 \leq i \leq l$, we define the function ψ_i by

$$\psi_i = \prod_{j=i}^{l} \left(1 - z_{j,j-1}\right)^{-\beta_j - 1}, \quad 1 \leq i \leq l, \tag{4.12}$$

where $\beta_1, \ldots, \beta_l \in \mathbb{C}$ are parameters.

Lemma 4.2. *ψ_i has a formal power series expansion*

$$\psi_i = \sum_{\{N_j\}} \prod_{j=i}^{l} \binom{\beta_j + 2\sum_{j<k\le l}(k-j)N_k + N_j}{N_j} \left(z_{j,i-1}\right)^{N_j}, \quad 1 \le i \le l,$$

where the sum $\sum_{\{N_j\}}$ extends over $N_i, \dots, N_l \in \mathbb{Z}_{\ge 0}$.

Proof. We proceed by induction on i. The case $i = l$ is due to the formula

$$(1-z)^{-\beta-1} = \sum_{N=0}^{\infty} \binom{\beta+N}{N} z^N. \tag{4.13}$$

Assume that ψ_{i+1} has the above expansion. Then from (4.12), ψ_i is

$$(1 - z_{i,i-1})^{-\beta_i-1} \sum_{\{N_j\}}{}' \prod_{j=i+1}^{l} \binom{\beta_j + 2\sum_{j<k\le l}(k-j)N_k + N_j}{N_j} \left(z_{j,i}\right)^{N_j},$$

where the sum $\sum'_{\{N_j\}}$ is over $N_{i+1}, \dots, N_l \in \mathbb{Z}_{\ge 0}$. Upon substituting (4.11), the right-hand side becomes

$$\sum_{\{N_j\}}{}' \left\{ (1 - z_{i,i-1})^{-\beta_i - 2\sum_{i<k\le l}(k-i)N_k - 1} \times \prod_{j=i+1}^{l} \binom{\beta_j + 2\sum_{j<k\le l}(k-j)N_k + N_j}{N_j} \left(z_{j,i-1}\right)^{N_j} \right\}.$$

Applying (4.13) again, we obtain the desired expansion. □

This lemma is originally due to [K]. Here we have quoted the version reproduced in [HKOTY].

In what follows, we will work only with the variables $z_j = z_{j,0}$, $v_j = z_{j,j-1}$, and w_j $(1 \le j \le l)$:

$$v_j = z_j \prod_{k=1}^{j-1} (1 - v_k)^{-2(j-k)}, \qquad w_j = z_j \prod_{k=1}^{l} (1 - v_k)^{-2j}, \tag{4.14}$$

where the former relation is due to (4.11) while the latter is the definition of w_j. (z_i here should not be confused with the $z_i(q)$ in Section 2.) Note that $v_i = w_i \prod_{k=1}^{l} (1 - v_k)^{2\min(i,k)}$. Let $z = (z_1, \dots, z_l)$ and $w = (w_1, \dots, w_l)$. We will denote $z_1 = \cdots = z_l = 0$ simply by $z = 0$, $dz_1 \wedge \cdots \wedge dz_l / z_1 \cdots z_l$ by dz/z, etc. The variables z are holomorphic functions of w around $w = 0$. This is due to $w = 0$ and $\partial w_i/\partial z_j = \delta_{ij}$ at $z = 0$. Setting $i = 1$ in Lemma 4.2, we have

$$\prod_{j=1}^{l} \binom{\beta_j + 2\sum_{j<k\le l}(k-j)N_k + N_j}{N_j} = \operatorname*{Res}_{z=0} \left(\prod_{j=1}^{l} (1 - v_j)^{-\beta_j - 1} z_j^{-N_j} \right) \frac{dz}{z}. \tag{4.15}$$

Under a further specialization to $\beta_j = \gamma_j - 2\sum_{k=1}^{l} kN_k$, this becomes

$$\begin{aligned} \prod_{m\in\mathbb{N}_l} \binom{P_m + N_m}{N_m} &= \operatorname*{Res}_{z=0} \left(\prod_{j=1}^{l} (1 - v_j)^{-\gamma_j - 1} w_j^{-N_j} \right) \frac{dz}{z} \\ &= \operatorname*{Res}_{w=0} \left(\prod_{j=1}^{l} (1 - v_j)^{-\gamma_j - l(l+1) - 1} w_j^{-N_j} \right) \frac{\partial z}{\partial w} \frac{dw}{w}, \end{aligned} \tag{4.16}$$

where $\partial z/\partial w$ represents the Jacobian $\det_{i,j\in\mathbb{N}_l}(\partial z_i/\partial w_j)$. In (4.16), replace N_m by $N[J]_m$, P_m by $P_m(\nu[J], N[J])$ (defined in (4.6)–(4.7)), and γ_j by $\gamma_j - 2\sum_{i\in J}\min(j, i)$. The result reads

$$\begin{aligned} &\prod_{m\in\mathbb{N}_l} \binom{P[J]_m + N[J]_m}{N[J]_m} \\ &= \operatorname*{Res}_{w=0} \left(\prod_{j=1}^{l} (1 - v_j)^{-\gamma_j - l(l+1) - 1} w_j^{-N_j} \right) \left(\prod_{i\in J} v_i \right) \frac{\partial z}{\partial w} \frac{dw}{w}. \end{aligned} \tag{4.17}$$

Notice that the left-hand side constitutes the summand in (4.8). In terms of the generating functions, the results (4.16)–(4.17) are stated as follows.

Proposition 4.3.

$$K_l(\nu|w) = K_l(0|w) \prod_{j=1}^{l} (1 - v_j)^{-\gamma_j}, \tag{4.18}$$

$$K_l(0|w) = \frac{\partial z}{\partial w} \prod_{j=1}^{l} (1 - v_j)^{-l(l+1)-1}, \tag{4.19}$$

$$R_l(\nu|w) = R_l(0|w) \prod_{j=1}^{l} (1 - v_j)^{-\gamma_j}, \tag{4.20}$$

$$R_l(0|w) = K_l(0|w) \left(\sum_{J\subset\mathbb{N}_l} D_J \prod_{i\in J} v_i \right). \tag{4.21}$$

Combining this with Proposition 4.1, we find the following.

Theorem 4.4.

$$R_l(\nu|w) = \frac{K_l(\nu|w)}{K_l(0|w)} = \prod_{j=1}^{l}(1-v_j)^{-\gamma_j}, \tag{4.22}$$

$$R_l(\nu|w)R_l(\nu'|w) = R_l(\nu+\nu'|w), \tag{4.23}$$

$$\frac{\partial w}{\partial z}\prod_{j=1}^{l}(1-v_j)^{l(l+1)+1} = \sum_{J\subset \mathbb{N}_l} D_J \prod_{i\in J} v_i, \tag{4.24}$$

where $\nu+\nu' = (\nu_s+\nu'_s)_{s=1}^{l}$ *and* γ_j *is the truncated one* (4.2).

The factorization property (4.23) is enjoyed only by R_l and not by K_l. It is due to (4.22) and $\gamma(\nu+\nu')_j = \gamma(\nu)_j + \gamma(\nu')_j$.

4.4 $l \to \infty$ limit. Let $R(\nu|w) = \lim_{l\to\infty} R_l(\nu|w)$ and $K(\nu|w) = \lim_{l\to\infty} K_l(\nu|w)$ be formal power series in infinitely many variables $w = (w_j)_{j\geq 1}$. They can also be viewed as the series in $(v_j)_{j\geq 1}$ upon the substitution $w_i = v_i \prod_{k\geq 1}(1-v_k)^{-2\min(i,k)}$. See the remark after (4.14). In the $l \to \infty$ limit, Theorem 4.4 yields the following.

Theorem 4.5.

$$R(\nu|w) = \frac{K(\nu|w)}{K(0|w)} = \prod_{j\geq 1}(1-v_j)^{-\gamma_j}, \tag{4.25}$$

$$R(\nu|w)R(\nu'|w) = R(\nu+\nu'|w), \tag{4.26}$$

where $\nu+\nu' = (\nu_s+\nu'_s)_{s\geq 1}$ *and* γ_j *is defined by* (2.17).

We specialize $R(\nu|w)$ and $K(\nu|w)$ as follows:

$$R(\nu) := e^{\gamma_\infty(\nu)\Lambda_1} R(\nu|w)|_{w_j=e^{-j\alpha_1}} = \sum_N R(\nu,N)x^{\gamma_\infty(\nu)-2\sum_{j\geq 1} jN_j}, \tag{4.27}$$

$$K(\nu) := e^{\gamma_\infty(\nu)\Lambda_1} K(\nu|w)|_{w_j=e^{-j\alpha_1}} = \sum_N K(\nu,N)x^{\gamma_\infty(\nu)-2\sum_{j\geq 1} jN_j}, \tag{4.28}$$

where the sum $\sum_N$ runs over $N_1, N_2, \ldots \in \mathbb{Z}_{\geq 0}$ and α_1 and Λ_1 are the simple root and the fundamental weight, respectively. $x = e^{\Lambda_1}$ is a formal variable and $\gamma_\infty(\nu) = \sum_{s\geq 1} s\nu_s$ in accordance with (2.17). From Theorem 4.5 it follows that

$$R(\nu) = \frac{K(\nu)}{K(0)}, \tag{4.29}$$

$$R(\nu)R(\nu') = R(\nu+\nu'). \tag{4.30}$$

We remark that Proposition 4.3 and Theorems 4.4 and 4.5 are all valid for $\nu_s \in \mathbb{C}$. The specialization $w_j = e^{-j\alpha_1}$ also induces an effect for v_j (4.14) and $\prod_{j\geq 1}(1-v_j)^{-\gamma_j}$ in (4.25). This will be worked out in Section 4.5.

4.5 Combinatorial completeness. Henceforth, we assume that $\forall\, \nu_s \in \mathbb{Z}_{\geq 0}$. For $m \in \mathbb{Z}_{\geq 0}$, we define

$$\begin{aligned} \delta_m &= (\nu_s), \qquad \nu_s = \delta_{s,m}, \\ Q_m &= R(\delta_m) \in x^m \mathbb{C}[[x^{-2}]]. \end{aligned} \tag{4.31}$$

From the decomposition $\nu = (\nu_s) = \sum_{s\geq 1} \nu_s \delta_s$ and (4.30), we have

$$R(\nu) = \prod_{s\geq 1} Q_s^{\nu_s} \tag{4.32}$$

for general $\nu = (\nu_s)$.

Proposition 4.6.

$$R(\lambda) = R(\mu) + R(\nu),$$

where $\lambda = (\lambda_s)$, $\mu = (\mu_s)$ *and* $\nu = (\nu_s)$ *are related as* $(s \in \mathbb{N})$

$$\lambda_s = \nu_s + 2\delta_{s,k}, \qquad \mu_s = \nu_s + \delta_{s,k+1} + \delta_{s,k-1}$$

for some $k \in \mathbb{N}$.

Proof. Set $N' = (N'_m)$, $N'_m = N_m - \delta_{m,k}$. Then it is easy to check that

$$\gamma_\infty(\lambda) - 2\sum_{j\geq 1} jN_j = \gamma_\infty(\mu) - 2\sum_{j\geq 1} jN_j = \gamma_\infty(\nu) - 2\sum_{j\geq 1} jN'_j, \tag{4.33}$$

$$P_m(\lambda, N) = P_m(\nu, N') = P_m(\mu, N) + \delta_{m,k}. \tag{4.34}$$

From (4.27) and (4.33), we are to show that $R(\lambda, N) = R(\mu, N) + R(\nu, N')$. By expanding a binomial coefficient in (3.12), the $R(\lambda, N)$ is expressed as ($P_m = P_m(\lambda, N)$)

$$\begin{aligned} &\sum_{J\subset\mathbb{N},\, k\notin J} D_J\,(A+B) \prod_{m\in\mathbb{N}\setminus J,\, m\neq k} \binom{P_m + N_m}{N_m} \prod_{m\in J} \binom{P_m + N_m - 1}{N_m - 1} \\ &\quad + \sum_{J\subset\mathbb{N},\, k\in J} D_J \prod_{m\in\mathbb{N}\setminus J} \binom{P_m + N_m}{N_m} (C + D) \prod_{m\in J,\, m\neq k} \binom{P_m + N_m - 1}{N_m - 1}, \end{aligned}$$

where $A = \binom{P_k+N_k-1}{N_k}$, $B = \binom{P_k+N_k-1}{N_k-1}$, $C = \binom{P_k+N_k-2}{N_k-1}$, and $D = \binom{P_k+N_k-2}{N_k-2}$. By (4.34), these can also be written as $A = \binom{P_k(\mu,N)+N_k}{N_k}$, $B = \binom{P_k(\nu,N')+N'_k}{N'_k}$, $C = \binom{P_k(\mu,N)+N_k-1}{N_k-1}$, and $D = \binom{P_k(\nu,N')+N'_k-1}{N'_k-1}$. Thus the contributions containing A and C (resp., B and D) amount to $R(\mu, N)$ (resp., $R(\nu, N')$). □

Let $\mathbb{Q}((x))$ denote the field of the formal Laurent series in x over $\mathbb{Q}$ with finitely many negative powers. Clearly, $Q_m \in \mathbb{Q}((x^{-1}))$.

Proposition 4.7.

(i) *Q_m satisfies*

(a) (*recursion relation*)

$$Q_0 = 1, \quad Q_k^2 = Q_{k+1}Q_{k-1} + 1, \quad k \in \mathbb{N},$$

(b) (*asymptotic property*)

$$\lim_{k\to\infty} \frac{Q_{k+1}}{Q_k} = x.$$

(ii) *Conversely, properties* (a) *and* (b) *above characterize the series* $Q_m \in \mathbb{Q}((x^{-1}))$.

Proof. (i). (a) Set $\nu = 0$ in Proposition 4.6 and apply (4.32). (b) It is enough to show that the limit $\lim_{k\to\infty} x^{-k} Q_k$ exists in $\mathbb{Q}[[x^{-1}]]$. Note that $P_m(\delta_k, N) = P_m(\delta_{k+1}, N) - \theta(m \geq k+1)$ from (2.16). In the series $x^{-k}Q_k = x^{-k}R(\delta_k)$ in (4.27), those $N = (N_m)$ containing $N_j > 0$ with $j \geq k+1$ make contributions in order higher than $2k+1$. It follows that $x^{-k}Q_k \equiv x^{-k-1}Q_{k+1} \mod x^{-2k-2}\mathbb{Q}[[x^{-1}]]$. Then we have

$$x^{-k}Q_k \equiv x^{-k-1}Q_{k+1} \equiv x^{-k-2}Q_{k+2} \equiv \cdots \mod x^{-2k-2}\mathbb{Q}[[x^{-1}]],$$

which means that $\lim_{k\to\infty} x^{-k}Q_k$ exists.

(ii). Suppose that $\tilde{Q}_m$ satisfies (a) and (b). Setting $v_j = 1 - \frac{\tilde{Q}_{j-1}\tilde{Q}_{j+1}}{\tilde{Q}_j^2}$, we find that

$$\prod_{j=1}^{l}(1-v_j)^{-\gamma_j} = \left(\frac{\tilde{Q}_l}{\tilde{Q}_{l+1}}\right)^{\gamma_l} \prod_{j=1}^{l} \tilde{Q}_j^{\nu_j}, \qquad w_j = \left(\frac{\tilde{Q}_l}{\tilde{Q}_{l+1}}\right)^{2j}$$

by (4.14). (γ_j here is the truncated one (4.2).) Therefore, (4.22) specializes to

$$\sum_N R_l(\nu, N) \prod_{j=1}^{l} \left(\frac{\tilde{Q}_l}{\tilde{Q}_{l+1}}\right)^{2jN_j} = \left(\frac{\tilde{Q}_l}{\tilde{Q}_{l+1}}\right)^{\gamma_l} \prod_{j=1}^{l} \tilde{Q}_j^{\nu_j},$$

where $\sum_N$ is over $N_1, \ldots, N_l \in \mathbb{Z}_{\geq 0}$. By taking the limit $l \to \infty$ using (b) for $\tilde{Q}_m$, this leads to $R(\nu) = \prod_{j\geq 1} \tilde{Q}_j^{\nu_j}$. Since ν_js are arbitrary, we obtain $\tilde{Q}_m = R(\delta_m)$. Comparing this with (4.31), we conclude that $\tilde{Q}_m = Q_m$. □

It is immediate to check that the character (with respect to the classical Cartan subalgebra) of the $(m+1)$-dimensional irreducible $U_q(\widehat{\mathfrak{sl}}(2))$-module W_m

$$\mathrm{ch}\, W_m = \frac{x^{m+1} - x^{-m-1}}{x - x^{-1}}, \quad x = e^{\Lambda_1},$$

fulfills properties (a) and (b) in Proposition 4.7. Thus from (ii), we have the following.

Proposition 4.8.

$$Q_m = \operatorname{ch} W_m, \quad m \in \mathbb{Z}_{\geq 0}.$$

Our main result in Section 4 is the following.

Theorem 4.9 (Combinatorial completeness). *Let $W(\nu)$ be the quantum space in* (2.1), $W(\nu) = \bigotimes_{s\geq 1}(W_s)^{\otimes \nu_s}$.

(i)

$$R(\nu) = \operatorname{ch} W(\nu),$$
$$\sum_N {}^{(\lambda)} R(\nu, N) = \dim W(\nu)_\lambda, \quad \lambda \in \mathbb{Z}\Lambda_1.$$

Here the sum $\sum_N^{(\lambda)}$ extends over $N_1, N_2, \ldots \in \mathbb{Z}_{\geq 0}$ such that $\sum_{j\geq 1} j(\nu_j - 2N_j)\Lambda_1 = \lambda$, and $\dim W(\nu)_\lambda$ *denotes the multiplicity of the weight λ. In particular, $R(\nu)$ is invariant under the Weyl group.*

(ii) (*Kirillov* [K])

$$K(\nu) = (1 - e^{-\alpha_1}) \operatorname{ch} W(\nu),$$
$$\sum_N {}^{(\lambda)} K(\nu, N) = [W(\nu) : V_\lambda], \quad \lambda \in (\mathbb{Z}_{\geq 0})\Lambda_1.$$

Here $[W(\nu) : V_\lambda]$ denotes the multiplicity of the irreducible $U_q(\mathfrak{sl}(2))$-module V_λ with highest weight λ. The sum $\sum_N^{(\lambda)}$ is the same as (i). *In particular $e^{\Lambda_1} K(\nu)$ is skew-invariant under the Weyl group.*

PROOF. (i) In view of (4.27), the two equalities are equivalent. The first is due to (4.32) and $\operatorname{ch} W(\nu) = \prod_{s\geq 1} Q_s^{\nu_s}$ by Proposition 4.8.

(ii) In view of (4.28), the two equalities are again equivalent. To be self-contained, let us include a quick proof of the first, although this was done in [K]. Let γ_j be as in (4.2) and $\mu \in \mathbb{Z}_{\geq 0}$. In the expansion of $\prod_{j=1}^{l}(1 - v_j)^{-\beta_j - 1}$ by means of (4.15), specialize the variables as $v_j = 1 - \frac{Q_{j-1}Q_{j+1}}{Q_j^2}$ (hence $z_j = Q_1^{-2j}$) and $\beta_j = \gamma_j - \mu$. The result reads

$$Q_1^{-\mu+1}\left(\frac{Q_l}{Q_{l+1}}\right)^{\gamma_l - \mu + 1} \prod_{s=1}^{l} Q_s^{\nu_s}$$
$$= \sum_N Q_1^{-2\sum_{i=1}^{l} iN_i} \prod_{j=1}^{l} \binom{\gamma_j - \mu + 2\sum_{j<k\leq l}(k-j)N_k + N_j}{N_j},$$

where $\sum_N$ is taken over $N_1, N_2, \ldots, \in \mathbb{Z}_{\geq 0}$. Picking up the coefficient of $Q_1^{-\mu}$, we get

$$\sum_{N:2\sum_{i=1}^{l} iN_i=\mu} K_l(\nu, N) = (-1)^l \operatorname*{Res}_{Q_1=\infty} \left(Q_1 \left(\frac{Q_l}{Q_{l+1}} \right)^{\gamma_l-\mu+1} \prod_{s=1}^{l} Q_s^{\nu_s} \right) \frac{dQ_1}{Q_1}$$
$$= (-1)^l \operatorname*{Res}_{x=\infty} \left(x(1-x^{-2}) \left(\frac{Q_l}{Q_{l+1}} \right)^{\gamma_l-\mu+1} \prod_{s=1}^{l} Q_s^{\nu_s} \right) \frac{dx}{x},$$

where $Q_1 = x + x^{-1}$ is used. In the limit $l \to \infty$, this is equivalent to $K(\nu) = (1 - e^{-\alpha_1}) \operatorname{ch} W(\nu)$ due to (4.28) and property (b) in Proposition 4.7. □

It is curious that in general the sum $\sum_N^{(\lambda)}$ involves the contributions from those N that do not satisfy the assumption in Theorem 3.5.

5 Discussion

In this paper, we have proposed the SCE relevant to the string solutions of the Bethe equation at $q = 0$. The number of off-diagonal solutions to the SCE is identified with weight multiplicities of the quantum space by constructing an explicit combinatorial formula $R(\nu, N)$.

It is quite common to reduce the Bethe equation to that for string centers. Indeed, such analyses have been done extensively at $q = 1$ and have led to the well-known fermionic formula $K(\nu, N)$ [K]. However, at $q = 0$, systematic counting of the number of solutions have been left untouched. The result in this paper reveals another aspect of combinatorial completeness of the string hypothesis. The fermionic form $K(\nu, N)$ is relevant to $q = 1$ and the multiplicity of irreducible components, while our $R(\nu, N)$ is relevant to $q = 0$ and weight multiplicities. Their generating functions are simply related as (4.25) and (4.29).

In this paper, we have treated the $U_q(\widehat{\mathfrak{sl}}(2))$ case exclusively. Many results here admit straightforward generalizations to $U_q(X_n^{(1)})$, which will be the subject of our subsequent paper. In place of (3.12)–(3.13), our main formula is ($\nu = (\nu_s^{(a)})$, $N = (N_m^{(a)})$)

$$R(\nu, N) = \sum_{J \subset \mathbb{N}^n} D_J \prod_{(a,m)\in\mathbb{N}^n\setminus J} \binom{P_m^{(a)} + N_m^{(a)}}{N_m^{(a)}} \prod_{(a,m)\in J} \binom{P_m^{(a)} + N_m^{(a)} - 1}{N_m^{(a)} - 1},$$
$$D_J = \begin{cases} 1 & \text{if } J = \emptyset, \\ \det_{(a,m),(b,k)\in J} \left((\alpha_a|\alpha_b) \min(t_b m, t_a k) - \delta_{a,b}\delta_{m,k} \right) & \text{otherwise,} \end{cases}$$

where $\mathbb{N}^n = \{(a, m) \mid 1 \le a \le n, m \in \mathbb{N}\}$ and the other notation is the same as in [HKOTY] under the identification of $(P_j^{(a)}, N_j^{(a)})$ here with $(p_j^{(a)}, m_j^{(a)})$ there. As the $\mathfrak{sl}(2)$ case, the above $R(\nu, N)$ contains the fermionic form in [KR] as the summand corresponding to $J = \emptyset$. With this $R(\nu, N)$, Theorems 3.5, 4.5, and 4.9 and Propositions 4.7 and 4.8 can be generalized.

Another direction of the generalization is to seek a q-analogue of $R(\nu, N)$ that expresses the unrestricted one-dimensional configuration sums (1dsums) over the quantum space W in the sense of [HKOTY]. Thus far, we have jointly with G. Hatayama, M. Okado, and T. Takagi obtained only a conjecture for the XXZ case.

Appendix Möbius function $\mu(\pi, \pi')$

Let us explain a minimum about the partition of sets and the Möbius function on it. For a more extensive treatment, see [A, B, S].

A.1 Partition of a set. Let $N \in \mathbb{N}$. By definition, $\pi = (\pi_1, \ldots, \pi_l)$ is called a *partition* of a set $\{1, \ldots, N\}$ if

$$\{1, \ldots, N\} = \pi_1 \sqcup \cdots \sqcup \pi_l$$

is a disjoint union decomposition. Here the ordering of $\pi_1, \pi_2, \ldots, \pi_l$ does not matter, e.g., $(\pi_1, \pi_2, \ldots, \pi_l)$ and $(\pi_2, \pi_1, \ldots, \pi_l)$ are the same partition. Each π_i is called a *block* of π and l is called a *length* of π. Let L_N denote the set of partitions of $\{1, \ldots, N\}$. Here are the first three:

$$\begin{aligned} &L_1 = \{1\}, \\ &L_2 = \{12,\ 1/2\}, \\ &L_3 = \{123,\ 12/3,\ 13/2,\ 23/1,\ 1/2/3\}, \end{aligned}$$

where, for example, $23/1$ stands for the partition $\pi = (\pi_1, \pi_2)$ of length $l(\pi) = 2$ consisting of the blocks $\pi_1 = \{2, 3\}$ and $\pi_2 = \{1\}$.

A.2 Poset structure. One can endow a natural partial order "$\leq$" with the set L_N. Given two partitions $\pi, \pi' \in L_N$, we say that $\pi \leq \pi'$ if each block of π' is contained in a block of π. For L_2 in the above, we have $12 \leq 1/2$, and for L_3,

$$\begin{matrix} & 12/3 & \\ 123 \leq & 13/2 & \leq 1/2/3, \\ & 23/1 & \end{matrix}$$

where there is no order among the middle three. Sometimes π' is called a *refinement* of π when $\pi \leq \pi'$. The partition $\pi_{\max} = 1/2/\cdots/N$ (resp., $\pi_{\min} = 12\ldots N$) is the unique maximal (resp., minimal) element in L_N of length $l(\pi_{\max}) = N$ (resp., $l(\pi_{\min}) = 1$). Clearly, the following three axioms hold:

1. *Reflexivity*: For any $\pi \in L_N, \pi \leq \pi$.

2. *Antisymmetry*: If $\pi \leq \pi'$ and $\pi' \leq \pi$, then $\pi = \pi'$.

3. *Transitivity*: If $\pi \leq \pi'$ and $\pi' \leq \pi''$, then $\pi \leq \pi''$.

Thus L_N equipped with $\leq$ is a partially ordered set (poset) in the sense of [S].

A.3 Möbius function. Consider an $|L_N|$ by $|L_N|$ matrix ζ defined by

$$\zeta = \big(\zeta(\pi,\pi')\big)_{\pi,\pi'\in L_N}, \quad \zeta(\pi,\pi') = \begin{cases} 1 & \text{if } \pi \le \pi', \\ 0 & \text{otherwise.} \end{cases}$$

This matrix is upper triangular with all the diagonal elements being 1. Thus it has the inverse

$$\zeta\mu = 1_{L_N}, \quad \mu = \big(\mu(\pi,\pi')\big)_{\pi,\pi'\in L_N}.$$

The matrix elements $\mu(\pi,\pi') \in \mathbb{Z}$ are called the Möbius function of the poset L_N. Note from the definition that $\mu(\pi,\pi) = 1$ for any π and $\mu(\pi,\pi') = 0$ unless $\pi \le \pi'$. For example, in the $N = 2$ and 3 cases in the above they are explicitly given by

$$\mu = \begin{pmatrix} 1 & -1 \\ 0 & 1 \end{pmatrix}, \qquad \mu = \begin{pmatrix} 1 & -1 & -1 & -1 & 2 \\ 0 & 1 & 0 & 0 & -1 \\ 0 & 0 & 1 & 0 & -1 \\ 0 & 0 & 0 & 1 & -1 \\ 0 & 0 & 0 & 0 & 1 \end{pmatrix}. \tag{A.1}$$

For general N, an explicit formula of $\mu(\pi,\pi')$ is available [A, B], but we do not need it in this paper.

A.4 Möbius inversion formula. Given any function $f : L_N \to \mathbb{C}$, define another function $g : L_N \to \mathbb{C}$ by

$$g(\pi) = \sum_{\pi'\le\pi} f(\pi').$$

This is a composition with ζ, introduced previously. In vector-matrix notation, it is expressed as $g = f\zeta$; hence it is equivalent to $f = g\mu$:

$$f(\pi) = \sum_{\pi'\le\pi} g(\pi')\mu(\pi',\pi),$$

which is the Möbius inversion formula. Here is a simple example of its application.

Proposition A.1. *Let X be an indeterminate. For any $\pi \in L_N$, we have*

$$X^{l(\pi)} = \sum_{\pi'\le\pi} (X)_{l(\pi')},$$

$$(X)_{l(\pi)} = \sum_{\pi'\le\pi} \mu(\pi',\pi) X^{l(\pi')},$$

where $(X)_l = X(X-1)\cdots(X-l+1)$.

PROOF. It suffices to show the former, assuming that X is any positive integer. Notice that $X^{l(\pi)}$ is the number of maps $\phi : \{1, \ldots, N\} \to \{1, \ldots, X\}$ such that $\phi(i) = \phi(j)$ if i and j belong to the same block of π. Similarly, $(X)_{l(\pi)}$ is the number of maps $\phi : \{1, \ldots, N\} \to \{1, \ldots, X\}$ such that $\phi(i) = \phi(j)$ if and only if i and j belong to the same block of π. Since the "if" case consists of the disjoint union of "if and only if" cases labeled by π' $(\leq \pi)$, the former relation holds. □

A.5 Product poset. Consider the product set $L_{N_1} \times \cdots \times L_{N_m}$ for any positive integers $N_1, \ldots, N_m$. Denote its elements by $\pi = (\pi^{(1)}, \ldots, \pi^{(m)})$, where $\pi^{(i)} \in L_{N_i}$. We can equip $L_{N_1} \times \cdots \times L_{N_m}$ with a poset structure by introducing the partial order as

$$(\pi^{(1)}, \ldots, \pi^{(m)}) \leq (\pi^{(1)'}, \ldots, \pi^{(m)'}) \overset{\text{def}}{\Longleftrightarrow} \pi^{(i)} \leq \pi^{(i)'} \quad \text{for any } 1 \leq i \leq m.$$

The length function is also introduced as $l\big((\pi^{(1)}, \ldots, \pi^{(m)})\big) = l(\pi^{(1)}) + \cdots + l(\pi^{(m)})$. The unique maximal element in $L_{N_1} \times \cdots \times L_{N_m}$ is $\pi_{\max} = (\pi^{(1)}_{\max}, \ldots, \pi^{(m)}_{\max})$, where $\pi^{(i)}_{\max} = 1/2/\cdots/N_i$ is the maximal one in L_{N_i}. Obviously, the Möbius function of this poset is the direct product of that for each component:

$$\mu\big((\pi^{(1)}, \ldots, \pi^{(m)}), (\pi^{(1)'}, \ldots, \pi^{(m)'})\big) = \prod_{i=1}^{m} \mu_{N_i}(\pi^{(i)}, \pi^{(i)'}),$$

where we have written the Möbius function of L_{N_i} as μ_{N_i}. Combining this with Proposition A.1, we get

$$\sum_{\pi \in L_{N_1} \times \cdots \times L_{N_m}} \mu(\pi, \pi_{\max}) X_1^{l(\pi^{(1)})} \cdots X_m^{l(\pi^{(m)})} = \prod_{i=1}^{m} (X_i)_{l(\pi^{(i)}_{\max})} = \prod_{i=1}^{m} (X_i)_{N_i}, \tag{A.2}$$

where $X_1, \ldots, X_m$ are indeterminates. In the main text, we use the Möbius inversion formula for the poset $L_{N_1} \times \cdots \times L_{N_m}$ and (A.2).

Acknowledgments. The authors thank G. Hatayama, M. Okado, and T. Takagi for stimulating discussion and collaboration on a generalization of the present work. They also thank M. T. Batchelor and V. O. Tarasov for useful correspondence. In addition, thanks are due to Z. Tsuboi for a critical reading of this manuscript.

REFERENCES

[A] G. E. Andrews, *The Theory of Partitions*, Encyclopedia of Mathematics and Its Applications 2, Addison–Wesley, Reading, MA, 1976.

[B] C. Berge, *Principes de Combinatoire*, Dunod, Paris, 1968.

[Be] H. A. Bethe, Zur Theorie der Metalle I: Eigenwerte und Eigenfunktionen der linearen Atomkette, *Z. Phys.*, **71** (1931), 205–231.

[C] J. W. S. Cassels, *An Introduction to the Geometry of Numbers*, Springer-Verlag, Berlin, New York, Heidelberg, 1971.

[EKS] H. L. Eßler, V. E. Korepin, and K. Schoutens, Fine structure of the Bethe ansatz equations for the isotropic spin-$\frac{1}{2}$ Heisenberg XXX model, *J. Phys. A*, **25** (1992), 4115–4126.

[FT] L. D. Faddeev and L. A. Takhtadzhyan, Spectrum and scattering of excitations in the one-dimensional isotropic Heisenberg model, *J. Soviet Math.*, **24** (1984), 241–246.

[HKOTY] G. Hatayama, A. Kuniba, M. Okado, T. Takagi, and Y. Yamada, Remarks on fermionic formula, in N. Jing and K. C. Misra, eds., *Recent Developments in Quantum Affine Algebras and Related Topics*, AMS, Providence, RI, 1999.

[JD] G. Jüttner and B. D. Dörfel, New solutions of the Bethe ansatz equations for the isotropic and anisotropic spin-$\frac{1}{2}$ Heisenberg chain, *J. Phys. A*, **26** (1993), 3105–3120.

[K] A. N. Kirillov, Combinatorial identities and completeness of states for the Heisenberg magnet, *J. Soviet Math.*, **30** (1985), 2298–3310.

[KL] A. N. Kirillov and N. A. Liskova, Completeness of Bethe's states for the generalized XXZ model, *J. Phys. A*, **30** (1997), 1209–1226.

[KR] A. N. Kirillov and N. Yu. Reshetikhin, Representations of Yangians and multiplicity of occurrence of the irreducible components of the tensor product of representations of simple Lie algebras, *J. Soviet Math.*, **52** (1990), 3156–3164.

[LS] R. P. Langlands and Y. Saint-Aubin, Algebro-geometric aspects of the Bethe equations, in A. Gülen, C. Saclioglu, and M. Daroglu, eds., *Strings and Symmetries*, Lecture Notes in Physics 447, Springer-Verlag, Berlin, New York, Heidelberg, 1995, 40–53.

[S] R. P. Stanley, *Enumerative Combinatorics*, Vol. 1, Cambridge University Press, Cambridge, 1997.

[TS] M. Takahashi and M. Suzuki, One-dimensional anisotropic Heisenberg model at finite temperature, *Prog. Theoret. Phys.*, **48** (1972), 2187–2209.

[TV] V. Tarasov and A. Varchenko, Completeness of Bethe vectors and difference equations with regular singular points, *Internat. Math. Res. Notices*, 1995, 637–669.

Atsuo Kuniba
Institute of Physics
University of Tokyo
Tokyo 153-8902
Japan
atsuo@gokutan.c.u-tokyo.ac.jp

Tomoki Nakanishi
Graduate School of Mathematics
Nagoya University
Nagoya 464-8602
Japan
nakanisi@math.nagoya-u.ac.jp

Hidden E-Type Structures in Dilute A Models

J. Suzuki

Abstract. The hidden $E_7(E_6)$ structure has been conjectured for the minimal model $\mathcal{M}_{4,5}(\mathcal{M}_{6,7})$ perturbed by $\Phi_{1,2}$ in the context of conformal field theory (CFT). Motivated by this, we examine the dilute $A_{4,6}$ models, which are expected to be corresponding lattice models. Thermodynamics of the equivalent one-dimensional quantum systems is analyzed via the quantum transfer matrix approach. Appropriate auxiliary functions, related to kinks in the theory, play a role in constructing functional relations among transfer matrices. We successfully recover the universal Y-systems and thereby the thermodynamic Bethe ansatz equations for $E_{6,7}$ from the dilute $A_{6,4}$ model, respectively.

1 Introduction

The impact of perturbed conformal field theory (CFT) has many aspects [57, 58]. In this communication we explore one of its predictions, the "trinity" of minimal unitary CFT: $\mathcal{M}_{p,p+1}$, $p = 3, 4, 6$, perturbed by $\Phi_{1,2}$, lattice models of criticality (the Ising model in a field, the tricritical Ising model of the critical temperature, and the tricritical three-state Potts model), and the dilute $A_{3,4,6}$ models. Many results have already been accumulated on equivalence in universality [1]–[3], [5, 10, 11, 17], [34]–[36], [40], [52]–[56]. The scaling exponents of the dilute A_3 model in periodic or open boundary conditions have been evaluated analytically [2, 3, 5], [53]–[55]. These results agree with numerical results for the Ising model in a magnetic field. Masses for eight elementary excitations of the dilute A_3 model are found to be proportional to components of the (largest) eigenvector of the Cartan matrix for E_8 [3, 36]. Vertex operators of $A_2^{(2)}$, which is the symmetry of the dilute A_3 model at criticality, satisfy a set of relations that indicate the hidden E_8 structure [16].

In particular, we want to call attention to thermodynamics of a one-dimensional (1D) system related to the dilute A_3 model in [9]. A set of solutions to the eigenvalue problem of the 1D Hamiltonian has been identified in exquisite "string" forms [9, 13, 14]. Nine of these are expected to contribute nontrivially in the thermodynamic limit. This observation leads to a set of integral equations, the thermodynamic Bethe ansatz (TBA), which determines the free energy. Remarkably, the TBA exhibits the underlying E_8 structure [8].

In [44], we attacked the same problem in a different setting. By following general frameworks, one represents the free energy by the largest eigenvalue of the "quantum transfer matrix" (QTM) acting on a virtual space [12, 28], [46]–[50]. We have managed to solve the (single) eigenvalue problem of commuting the QTM by introducing auxiliary functions related to fusion of the QTM [22, 24, 32]. We will simply call them fusion QTMs. Eight fusion QTMs are found to satisfy a closed set of functional relations related to E_8. A quantum analogue of the Jacobi–Trudi formula [7, 31] as well as combinatorial aspects in terms of the "Yangian analogue" of Young tableaux [7, 31, 33, 43, 51] play a fundamental part in the proof of the relations. Nice analytic properties of fusion QTMs allow for the transformation of functional relations into coupled integral equations. The resultant TBA equation yields a direct evaluation of free energy. Again, it coincides with a hypothetical TBA for E_8 theory.

As promised in [44], we carry out this program for the dilute A_4 and A_6 models. A novel feature lies in the fact that "a box" of the Young tableaux is no longer the fundamental constituent. This may be natural in view of representation theory; the vector representation is no longer minimal. In the language of S-matrix theory, boxes present breathers rather than kinks. We conjecture explicit forms of QTMs related to these kinks. The investigation on these kink QTMs reveals connections between $U_q(A_2^{(2)})$ modules of symmetric tensors of the vector representation and $U_q(\widehat{E_6})$, $U_q(\widehat{E_7})$ modules when q equals a proper root of unity. With help of this observation, closed functional relations among fusion QTMs are also found for the dilute $A_{4,6}$ models. Quite parallel to the dilute A_3 model, one recovers the TBAs expected for $E_{6,7}$ [8].

The paper is organized as follows. Because this subject may not be very familiar to readers, in Section 2, we present a brief survey on the QTM approach, together with the sketch of the idea of analytic Bethe ansatz. The dilute A_L model is briefly described in Section 3. The sl_3 fusion structure of the model is presented in view of analytic Bethe ansatz, the Yangian analogue of Young tableaux, and the quantum Jacobi–Trudi formula in Section 4. We concentrate on the dilute A_4 model in Sections 5 and 6. The explicit form of the QTM related to the "kink" is proposed in Section 5. Yangian homomorphisms among $Y(E_7)$ modules serve as a useful guide in search of the form. The remaining QTMs are defined and their functional relations are examined in Section 6. This result coincides with a prediction in [30]. Similar results for the dilute A_6 model are given in Section 7. With some information on analyticity of these QTMs, supported by numerics, the desired TBAs are recovered in Section 8. We conclude the paper with a short summary in Section 9.

2 Survey of the QTM approach

Exact evaluation of physical quantities at finite temperatures poses serious difficulties even for integrable models. One has to go far beyond mere diagonalization of a Hamiltonian; summation over eigenspectra must be performed.

The string hypothesis brought the first breakthrough and success. It postulates dominant solutions to the Bethe ansatz equation (BAE) in the thermodynamic limit. In a sense, the method tackles the combinatorial aspect of the problem directly.

2.1 The QTM. The QTM method takes a different route. It utilizes the famous mapping between the Hamiltonian $\mathcal{H}_M$ of a 1D quantum system and the row-to-row transfer matrix $T_{RTR}(u)$ of the corresponding two-dimensional (2D) classical model [12, 28], [46]–[50]. In the present context, the latter is given by

$$(T_{\mathrm{RTR}}(u))_{\{a\}}^{\{b\}} = \prod_{j=1}^{M} {}^{b_j}_{a_j} \boxed{u} {}^{b_{j+1}}_{a_{j+1}}.$$

Here the box represents the RSOS weight and M is the number of sites. See the appendix for the explicit weights for the dilute A_L models. The parameter u represents the anisotropy of interactions between horizontal and vertical directions and is called the spectral parameter. The explicit relation between $\mathcal{H}_M$ and $T_{RTR}(u)$ is

$$T_{RTR}(u) \sim T_{RTR}(0)\left(1 + \frac{u}{\epsilon}\mathcal{H}_M + O(u^2)\right),$$

where ϵ is the normalization parameter of the Hamiltonian. The essential idea in the QTM approach is encoded in the identity

$$\begin{aligned} \exp(-\beta\mathcal{H}_M) &= \lim_{N\to\infty}\left(1 + \frac{u}{\epsilon}\mathcal{H}_M\right)^N\Bigg|_{u\to-\beta\epsilon/N} = \lim_{N\to\infty}\left(T'_{RTR}(u = -\beta\epsilon/N)\right)^N, \\ T'_{RTR}(u) &:= \left(T_{RTR}(u)(T_{RTR}(0))^{-1}\right)^N. \end{aligned}$$

Namely, the partition function of the original problem is transformed into that of the 2D classical models on $M \times N$ sites. The fictitious dimension N is sometimes referred to as the Trotter number. We can interpret $T'_{RTR}(u)$ as the row-to-row transfer matrix in the "vertical" direction. Similarly, one can construct a transfer matrix propagating in the "horizontal" direction $T'_{QTM}(u)$, which acts on N sites. Therefore, we have

$$\mathrm{Tr}\,\exp(-\beta\mathcal{H}_M) = \lim_{N\to\infty} \mathrm{Tr}\, T'_{QTM}(u = -\beta\epsilon/N)^M.$$

It may be better to rewrite this in the form

$$\lim_{M\to\infty}\frac{1}{M}\log\left(\mathrm{Tr}\,\exp(-\beta\mathcal{H}_M)\right) = \lim_{N\to\infty}\lim_{M\to\infty}\frac{1}{M}\left(\mathrm{Tr}\,T'_{QTM}(u = -\beta\epsilon/N)^M\right).$$

The exchangeability of two limits is proven in [48].

A gap opens up between the largest and the second largest eigenvalues of $T'_{QTM}(u)$. In the thermodynamic limit $M \to \infty$, we have to deal with only the largest eigenvalue of the QTM. This contrasts strongly with the spectra of $T'_{RTR}(u)$. One observes almost degenerate low-lying excitations in the latter case as $M \to \infty$. The evaluation of free energy per site of the 1D quantum system is thus reduced to the largest eigenvalue problem of $T'_{QTM}(u)$. We are free from the summation problem.

This is, unfortunately, not the happy ending of the story. The Trotter number should approach infinity at the end. The diagonalization of the QTM is accomplished by the application of the Bethe ansatz method. The BAE depends nontrivially on N, which originates from the local interaction parameter u. Thus we cannot resort to the simple-minded application of the usual scheme of converting the transcendental equation into the integral equation. This makes the extrapolation $N \to \infty$ quite nontrivial.

2.2 Commuting the QTM. Instead of dealing with the BAE roots directly, we employ a different idea. The integrable structure of the underlying model allows for the introduction of a one-parameter family of commuting QTMs, which is labeled by a novel complex parameter x. For an explicit demonstration of this, we adopt a more sophisticated approach [24, 25] than the one presented above. We introduce a "staggered-manner" QTM to avoid a $(T_{RTR}(0))^{-1}$ factor in the definition of $T'_{RTR}(u)$.

$$(T_{\mathrm{QTM}}(u,x))^{\{b\}}_{\{a\}} = \prod_{j=1}^{N/2} \;{}^{b_{2j-1}}_{a_{2j-1}}\boxed{u+ix}\,{}^{b_{2j}}_{a_{2j}} \quad {}^{a_{2j}}_{a_{2j+1}}\boxed{u-ix}\,{}^{b_{2j}}_{b_{2j+1}}.$$

The following relation is still valid:

$$\begin{aligned}\beta f &= -\lim_{M\to\infty}\frac{1}{M}\ln \mathrm{Tr}\,\exp(-\beta\mathcal{H}_\epsilon)\\ &= -\lim_{N\to\infty}\ln\left(\text{the largest eigenvalue of } T_{\mathrm{QTM}}\left(u=-\epsilon\frac{\beta}{N}, x=0\right)\right).\end{aligned}$$

As emphasized above, we find an intriguing fact, commutativity of QTMs:

$$[T_{\mathrm{QTM}}(u,x), T_{\mathrm{QTM}}(u,x')] = 0.$$

Generally, one can construct a "higher-spin" QTM $T_{\mathrm{fusion}}(u,x)$ via the fusion procedure. Yang–Baxter integrability also assures commutativity among these generalized QTMs:

$$[T_{\mathrm{fusion}}(u,x), T_{\mathrm{fusion'}}(u,x')] = 0.$$

Note that the factor u is common. Henceforth, we will sometimes drop this common factor in commuting QTMs. We also sometimes use same notation for transfer matrices and their eigenvalues since we are considering them on the identical eigenspace.

We utilize the existence of the complex x-plane in which QTMs are simultaneously diagonalizable. There exist functional relations among these fusion QTMs in the x-plane. Our idea is to utilize these functional relations in place of the BAE. See [6, 27] for a discussion of the usual case of row-to-row transfer matrices. Our motivation is simple. The number of roots is of order N, the Trotter number. All of these locations change with N, while functional relations depend weakly on N. This dependence can be summarized in the known scalar factors in functional relations. One may expect the tractable limit $N \to \infty$ for functional relations. The problem of combinatorics (summation over eigenspectra) is then reduced to the study of functional relations[1] and analytic structures of fusion QTMs, as will be discussed below.

2.3 The analytic Bethe ansatz and functional relations. QTMs should not possess singularities in the x-plane, as Boltzmann weights are regular functions of x. The BAE can be interpreted as the pole-free condition of QTMs in the complex x-plane. Conversely, the analyticity requirement imposes restrictions on explicit eigenvalues of QTMs.

The analytic Bethe ansatz was proposed in [38] as a tool in deriving expressions of eigenvalues of transfer matrices. It starts with a simple observation: the eigenvalue at the "vacuum sector" is determined by diagonal elements of the R matrix, which are referred to as vacuum expectation values. In general sectors, eigenvalues should be modified such that each vacuum expectation value is "dressed" by appropriate combination of Baxter's Q operators. See (4) for a typical example. The combination is determined by requiring analyticity of the transfer matrix. We call the resultant expression the dressed vacuum form (DVF).

A "universal" BAE was proposed in [38, 39]. One can write down the BAE for the model based on $U_q(\hat{\mathfrak{g}})$ using only algebraic data of $U_q(\mathfrak{g})$. Starting from a properly chosen "highest-weight term," we can construct a pole-free set of functions under the BAE. In [33], a similarity was pointed out between the above procedure and the construction of the highest-weight module of the Lie algebra. This leads to the assumption that there exists a set of functions, pole-free under the BAE, that correspond to an irreducible module of the quantum affine Lie algebra. The set is naturally identified with the eigenvalue of the transfer matrix, the trace of which is taken over the irreducible module. This was promoted as an axiom in [33] and the subsequent papers [31, 51], producing fruitful results. We take sl_2 as the simplest example. By $V_m(x)$ and $T_m(x)$, we mean the $m+1$-dimensional sl_2 module and the associated transfer matrix, respectively. The DVF of $T_1(x)$ consists of two terms. We do not specify their forms; instead we represent them by boxes with letters 1 and 2,

$$T_1(x) = \boxed{1}_x + \boxed{2}_x.$$

Each box carries spurious poles, which are actually canceled due to the BAE. We represent this situation graphically as

$$\boxed{1}_x \to \boxed{2}_x.$$

[1] One can adopt different auxiliary functions from the fusion hierarchy. For results on other choices of auxiliary functions for several models, see [19]–[21], [23, 25, 26, 41, 45].

The eigenvalue of a fusion QTM can apparently be represented by the sum over products of boxes with various letters and spectral parameters. For example, one can construct a transfer matrix in which auxiliary space acts on the symmetric subspace of $V_1 \times V_1$. We associate with this the set of glued boxes $\boxed{i_1}_{x-i}\ \boxed{i_2}_{x+i}$, $(i_1 \leq i_2)$. The difference in spectral parameters is fixed to match the singularity of the R matrix. The cancellation of spurious singularities is again depicted as

$$\boxed{1\,|\,1} \to \boxed{1\,|\,2} \to \boxed{2\,|\,2},$$

where we omit spectral parameters. The eigenvalue of $T_2(x)$ is given by the sum of the three diagrams above. The extension to general T_m is now obvious. Starting from the highest-weight term, $\boxed{1\,|\,1\,|\,\cdots\,|\,1}$, we have the cancellation diagram

$$\boxed{1\,|\,1\,|\,\cdots\,|\,1} \to \boxed{1\,|\,1\,|\,\cdots\,|\,2} \to \cdots \to \boxed{2\,|\,2\,|\,\cdots\,|\,2}.$$

The sum over these, regarded as expressions, yields the eigenvalue of $T_m(x)$. One can easily identify the diagram with the crystal graph of the m-fold tensor of $U_q(sl_2)$ that represents the irreducible module.

Suppose that we have a short exact sequence among tensor products of irreducible modules of the quantum affine Lie algebra:

$$0 \to W_0 \otimes W_1 \to W_2 \otimes W_3 \to W_4 \otimes W_5 \to 0.$$

Then a functional relation follows:

$$0 = T_{W_0}T_{W_1} - T_{W_2}T_{W_3} + T_{W_4}T_{W_5}.$$

We remark that spectral parameter dependencies are implicit in W_is.[2] The desired functional relations are derived as a consequence of relations among affine modules.

Even if exact sequences are not available, one can still check the validity of hypothetical functional relations using explicit forms of transfer matrices, which can be derived by applying the analytic Bethe ansatz. Indeed, functional relations for sl_2 are easily derived without knowledge of exact sequences. By using the box representation above, one can graphically derive

$$T_m(x-i)T_m(x+i) = g_m(x) + T_{m-1}(x)T_{m+1}(x), \quad m \geq 1, \tag{1}$$

where $g_m(x)$ is a known scalar function which depends on N.

Such functional relations are sometimes referred to as the T-system.

[2]To be precise, we first consider a vertex model in which quantum space (auxiliary space) is given by W_i and denote the transfer matrix by T_{W_i}. In a later section, we use the same notation for the transfer matrix of the corresponding RSOS model.

2.4 Functional relations and the TBA. Unfortunately, functional relations alone do not provide enough information on the explicit eigenvalues. This can easily be seen from the fact that excited states' eigenvalues satisfy the same algebraic relations. We need additional information on analyticities of fusion QTMs. Let us again demonstrate this for the sl_2 case. We conveniently rewrite the T-system (1) in terms of $Y_m(x) = T_{m-1}(x)T_{m+1}(x)/g_m(x)$:

$$Y_m(x-i)Y_m(x+i) = (1+Y_{m-1}(x))(1+Y_{m+1}(x)), \tag{2}$$

where a known property of the g-function, $g_m(x+i)g_m(x-i) = g_{m-1}(x)g_{m+1}(x)$, is used.

Consider functions in the largest eigenvalue sector of $T_1(x)$. We have convincing numerical evidence for the conjecture that zeros of $T_m(x)$ lie approximately on the curve $\Im x \sim \pm(m+1)$. Then both sides of (2) are analytic, nonzero, and asymptotically constant (ANZC) within the strip $\Im x \in [-1, 1]$. Strictly speaking, we must modify the left-hand side for $m = 1$. We will not go into such detail in this introductory section.

This information is now sufficient to transform the algebraic relations into integral equations that enable the explicit evaluation of Y_m and then T_m.

Take the logarithmic derivatives of both sides of (2) and perform Fourier transformations. Let $\tilde{dl}Y_m[k]$ be the Fourier transformation of the logarithmic derivative of $Y_m(x)$. Thanks to the ANZC property, the Cauchy theorem applies. The resultant equation is given simply by

$$2\cosh k\tilde{dl}Y_m[k] = \tilde{dl}(1+Y_{m-1})[k] + \tilde{dl}(1+Y_{m+1})[k], \quad m \geq 2.$$

Remarkably, both sides contain only functions with the same Fourier mode. Dividing both sides by $2\cosh k$, performing an inverse Fourier transformation, and integrating once over x, we reach the integral equation, which is identical to the TBA:

$$\log Y_m(x) = \int_{-\infty}^{\infty} K(x-y)\log(1+Y_{m-1})(1+Y_{m+1})(y)dy, \quad m \geq 2. \tag{3}$$

$K(x)$ denotes the Fourier transformation of $1/2\cosh k$. Though we have omitted it above, for $m = 1$, the right-hand side of (3) has a nontrivial scalar factor originating from the g-function. However, it brings an N-dependency by the combination uN in the $N \to \infty$ limit. Remembering that u is inversely proportional to N, we can send $N \to \infty$ analytically! The resultant drive term depends only on β.

In this way, we take a completely different route from the string hypothesis but reach the same conclusion. In the absence of appropriate conjectures on dominant patterns of roots, our method has an explicit advantage in attacking the problem. This is the case with the dilute $A_{4,6}$ models.[3]

[3]Although as yet unpublished, there is also progress in view of the string hypothesis for these cases. The author thanks V. V. Bazhanov and O. Warnaar for this information.

In the rest of this paper, we shall extensively apply the above ideas to these cases. Before this, we repeat some lessons discussed above. To find functional relations is not enough. One must find functional relations that have the ANZC property in appropriate domains of the complex x-plane. This is the most crucial step in the present approach.

3 The dilute A_L model

The dilute A_L model is proposed in [53, 54] as an elliptic extension of the Izergin–Korepin model [18]. (See [29] for an elliptic extension of a different type.) The model is of the restricted SOS type with local variables $\in \{1, 2, \dots, L\}$. The variables $\{a, b\}$ on neighboring sites should satisfy the adjacency condition $|a - b| \leq 1$. The solvable weights contain parameters u, q, and λ. We supplement their explicit forms in the appendix. The model exhibits four different physical regimes depending on parameters:

- Regime 1. $0 < u < 3$, $\lambda = \frac{\pi L}{4(L+1)}$, $L \geq 2$.
- Regime 2. $0 < u < 3$, $\lambda = \frac{\pi(L+2)}{4(L+1)}$, $L \geq 3$.
- Regime 3. $3 - \frac{\pi}{\lambda} < u < 0$, $\lambda = \frac{\pi(L+2)}{4(L+1)}$, $L \geq 3$.
- Regime 4. $3 - \frac{\pi}{\lambda} < u < 0$, $\lambda = \frac{\pi L}{4(L+1)}$, $L \geq 2$.

We are interested in Regimes 2 and 3. As in Section 2, we defines the Hamiltonian of the associated 1D quantum chain by

$$\mathcal{H}_\epsilon = \epsilon \frac{\partial}{\partial u} \ln T_{\mathrm{RTR}}(u)|_{u=0}$$

as in [9]. $\epsilon = -1$ (respectively, 1) corresponds to Regime 2 (respectively, 3).

The one-particle excitations for the dilute A_L case have been examined in [4, 42] for $L = 3$, 4, and 6. (See also [36] for another derivation for $L = 3$.) Eight, seven, and six particles are identified, respectively, and their masses are summarized by a single formula in the trigonometric limit

$$m_j \sim \sum_a \sin\left(\frac{a\pi}{g^\vee}\right),$$

where $g^\vee$ is 30, 18, 12 for $L = 3$, 4, 6 and is simply the dual Coxeter number for E_8, E_7, and E_6, respectively. We present sets of allowed as in Tables 1 and 2 for $L = 4$, 6, in which we are currently interested.

A number k in the bracket means that it corresponds to the kth light particle. Note that vectors of the form $(m_1, m_2, \dots)$ coincide with the eigenvectors of Cartan matrices for E_7 and E_6, respectively. The exponents $\{a\}$ will reappear in a novel context later. The leftmost numbers, which have simply been indices thus far, will be connected to indices for nodes in the Dynkin diagrams of E_7 and E_6. (See Figure 1.)

j	set of allowed as for A_4
1 (2)	$\{1, 7\}$
2 (5)	$\{2, 6, 8\}$
3 (7)	$\{3, 5, 7, 9\}$
4 (6)	$\{4, 6, 8\}$
5 (4)	$\{5, 7\}$
6 (1)	$\{6\}$
7 (3)	$\{4, 8\}$

Table 1.

j	set of allowed as for A_6
1, 5 (1)	$\{4\}$
2, 4 (4)	$\{3, 5\}$
6 (3)	$\{1, 5\}$
3 (6)	$\{2, 4, 6\}$

Table 2.

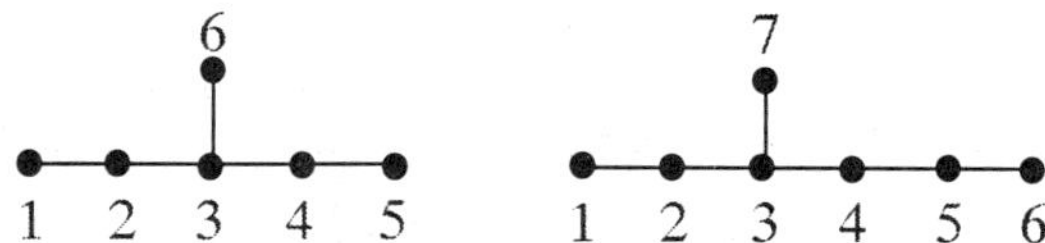

Figure 1. Dynkin diagrams for E_7 and E_6.

4 The sl_3 fusion structure and the quantum Jacobi–Trudi formula

The sl_3-type fusion structure in the dilute A_L model was discussed in [15]. This comes from the singularity of RSOS weights at $u = \pm 3$; the face operator becomes a projector related to sl_3 at these points. One obtains a desired subspace from the tensor products of spaces using these projectors. The adjacency conditions of local states are described by the combinatorics of tableaux.

We are interested in eigenvalues of fusion QTMs. Then the most relevant is the fact that these eigenvalues are again expressible in terms of Young tableaux depending on spectral parameters, as exemplified in Section 2.3 for the sl_2 case.

Explicitly, the eigenvalue $T_1(u, x)$ of $T_{\rm QTM}(u, x)$ is given by

$$\begin{aligned} T_1(u,x) &= w\phi\left(x+\frac{3}{2}i\right)\phi\left(x+\frac{1}{2}i\right)\frac{Q(x-5/2i)}{Q(x-1/2i)} \\ &\quad +\phi\left(x+\frac{3}{2}i\right)\phi\left(x-\frac{3}{2}i\right)\frac{Q(x-3/2i)\,Q(x+3/2i)}{Q(x-1/2i)\,Q(x+1/2i)} \end{aligned}$$

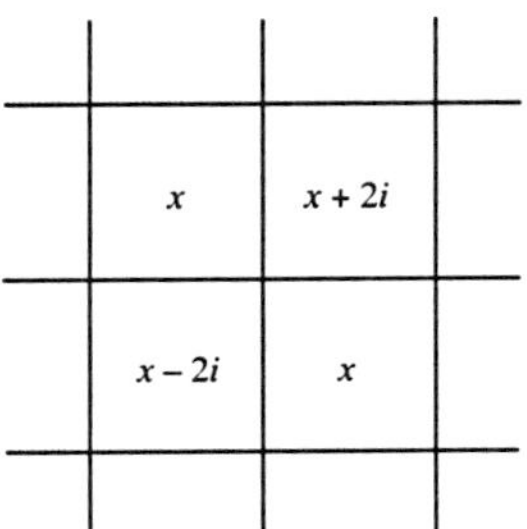

Figure 2. Assignment of the spectral parameter.

$$+w^{-1}\phi\left(x-\frac{3}{2}i\right)\phi\left(x-\frac{1}{2}i\right)\frac{Q(x+5/2i)}{Q(x+1/2i)},$$

$$Q(x) := \prod_{j=1}^{N} h[x-x_j],$$

$$\phi(x) := \left(\frac{h[x+(3/2-u)i]h[x-(3/2-u)i]}{h[2i]h[3i]}\right)^{N/2}, \quad h[x] := \theta_1(ix), \tag{4}$$

and $w = \exp(i\pi\frac{\ell}{L+1})$, where $\ell = 1$ for the largest eigenvalue sector. $\theta_1(ix)$ is defined in the appendix. The parameters, $\{x_j\}$ are solutions to the BAE:

$$w\frac{\phi(x_j+i)}{\phi(x_j-i)} = \frac{Q(x_j-i)Q(x_j+2i)}{Q(x_j+i)Q(x_j-2i)}, \quad j=1,\ldots,N. \tag{5}$$

As in Section 2, we represent these three terms by three boxes with letters 1, 2, and 3:

$$T_1(u,x) = \boxed{1}_x + \boxed{2}_x + \boxed{3}_x. \tag{6}$$

One infers from the sl_2 example that a combinatorial aspect may also appear. This turns out to be true. The eigenvalues of fusion QTMs are given by the sum of combinations of boxes, which can be identified with semistandard Young tableaux for sl_3. On each diagram, the spectral parameter changes $+2i$ from left to right and $-2i$ from top to bottom. (See Figure 2.)

For now, we restrict ourselves to rectangular-shaped diagrams. First, we note that the QTM associated with the $2\times m$ $(3\times m)$ Young diagram can be reduced to that associated with $1\times m$ (or just scalars). This is due to the identities

$$\begin{array}{|c|}\hline 1 \\ \hline 2 \\ \hline\end{array}\begin{matrix} {}_{x+i} \\ {}_{x-i}\end{matrix} = \phi\left(x+\frac{5}{2}i\right)\phi\left(x-\frac{5}{2}i\right)\boxed{1}_x,$$

$$\begin{array}{|c|}\hline 1 \\ \hline 3 \\ \hline\end{array}\begin{matrix} {}_{x+i} \\ {}_{x-i}\end{matrix} = \phi\left(x+\frac{5}{2}i\right)\phi\left(x-\frac{5}{2}i\right)\boxed{2}_x,$$

$$\begin{array}{|c|}\hline 2 \\ \hline 3 \\ \hline\end{array}\begin{matrix} {}_{x+i} \\ {}_{x-i}\end{matrix} = \phi\left(x+\frac{5}{2}i\right)\phi\left(x-\frac{5}{2}i\right)\boxed{3}_x,$$

$$\begin{array}{|c|l}\hline 1 & _{x+2i} \\ \hline 2 & _{x} \\ \hline 3 & _{x-2i} \\ \hline \end{array} = \prod_{j=1}^{3} \phi\left(x+\left(\frac{9}{2}-j\right)i\right)\phi\left(x-\left(\frac{9}{2}-j\right)i\right).$$

Second, eigenvalues of $1 \times m$ fusion QTMs have "duality" in the following sense. Let us denote a renormalized $1 \times m$ fusion QTM by $T_m(x)$:

$$T_m(x) = \frac{1}{f_m(x)} \sum_{i_1 \le i_2 \le \cdots \le i_m} \begin{array}{|c|c|c|c|}\hline i_1 & i_2 & \cdots & i_m \\ \hline \end{array}. \tag{7}$$

The spectral parameters are assigned $x - i(m-1), \ldots, x + i(m-1)$ from left to right. The renormalization factor, common to all expressions of length m tableaux, is given by

$$f_m(x) := \prod_{j=1}^{m-1} \phi\left(x \pm i\left(\frac{2m-1}{2} - j\right)\right).$$

Henceforth, we sometimes denote $f(x \pm iy) := f(x+iy)f(x-iy)$.

The resultant T_ms are all of degree $2N$ with respect to $h[x + \text{shift}]$. Obviously, we have a periodicity due to Boltzmann weights:

$$T_m\left(x + \frac{10}{3}i\right) = T_m(x) \qquad \left(T_m\left(x + \frac{14}{4}i\right) = T_m(x)\right) \tag{8}$$

for the dilute A_4 (A_6) model. From the sl_3 structure together with the above property, one can prove the following functional relations:

$$\begin{aligned} T_m(x-i)T_m(x+i) &= g_m(x)T_m(x) + T_{m+1}(x)T_{m-1}(x), \quad m \ge 1, \\ g_m(x) &= \phi\left(x \pm i\left(m + \frac{3}{2}\right)\right), \\ T_{-1}(x) &:= 0, \\ T_0(x) &:= f_2(x). \end{aligned} \tag{9}$$

The periodicity for the dilute A_4 model, $\phi(x+10/3i) = \phi(x)$, leads to $g_{m+10}(x) = g_m(x)$, $m \ge 0$, and $g_{7-m}(x) = g_m(x)$, $0 \le m \le 7$. From the adjacency matrix, we conclude that $T_{8,9}(x) = 0$.

Thus functional relations are invariant under the transformation $T_m(x) \to T_{7-m}(x)$, $m = 0, \ldots, 7$, or $T_m(x) \to T_{m+10}(x)$, $m \ge -1$. In general, the symmetry of functional relations is not necessarily inherited by its solution. However, we verify duality,

$$T_m(x) = T_{7-m}(x), \qquad m = 0, \ldots, 7, \tag{10}$$

and $T_{m+10}(x) = T_m(x)$, $m \ge -1$ numerically for the largest eigenvalue sector, the only one in which we are interested. These duality relations are also observed and numerically verified for A_6 models with a change in period; see Section 7. However, functional relations among these do not possess the desired analytical property, as

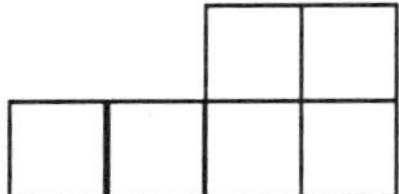

Figure 3. An example of a (4, 4)–(2) skew Young table.

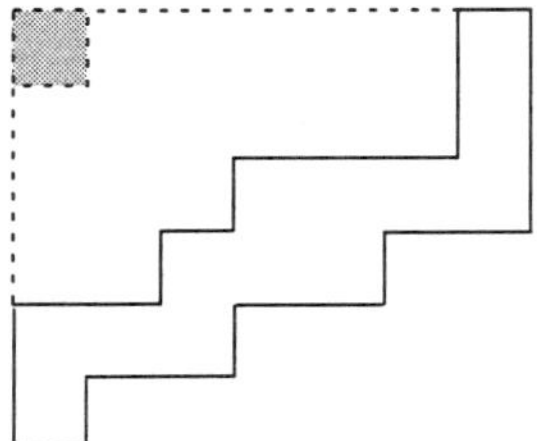

Figure 4. The spectral parameter $x + i(\mu_1' - \mu_1)$ is assigned to the hatched place.

discussed in [44]. We thus introduce another class of QTMs related to skew Young diagrams.

Let μ and λ be a pair of Young diagrams that satisfy $\mu_i \geq \lambda_i \; \forall i$. We subtract a diagram λ from μ. We call the result a skew Young diagram $\mu - \lambda$; it consists of $(\mu_1 - \lambda_1, \mu_2 - \lambda_2, \dots)$ boxes. (See Figure 3.) In the theory of symmetric polynomials, the Jacobi–Trudi formula tells us that a complex Schur function associated with a skew Young diagram can be expressed by a determinant of a matrix in which elements are given by elementary Schur functions associated with one-row or one-column diagrams. A quite parallel formula holds for the present situation, which we call the quantum Jacobi–Trudi formula [7, 31, 33].

Consider a set of semistandard skew Young tableaux of $\mu - \lambda$ shape. We assign an expression to each table. The spectral parameter of the top-left box is fixed to $x + i(\mu_1' - \mu_1)$,[4] where μ_1' denotes the depth of the tableaux. (See Figure 4.)

We identify each box in a table with an expression under rule (6) with the shift of the spectral parameter. Then the product over all constitutent boxes yields the desired expression for the table.

Theorem 1. *Let $\mathcal{T}_{\mu/\lambda}(x)$ be the sum over the resultant expressions divided by a common factor,* $\prod_{j=1}^{\mu_1'} f_{\mu_j - \lambda_j}(x + i(\mu_1' - \mu_1 + \mu_j + \lambda_j - 2j + 1))$. *Then the following equality holds:*

$$\mathcal{T}_{\mu/\lambda}(x) = \det_{1 \leq j,k \leq \mu_1'} (T_{\mu_j - \lambda_k - j + k}(x + i(\mu_1' - \mu_1 + \mu_j + \lambda_k - j - k + 1))), \quad (11)$$

where $T_{m<0} := 0$.

[4]We take this opportunity to note misprints in [44] in the fourth row of the second paragraph of Section 5 and the caption of Figure 2.

The proof of Theorem 1 is quite similar to that of Young tableaux [31, 33]. One must only keep in mind that the allowed position of a box is restricted by its spectral parameter.

The crucial fact is that the $\mathcal{T}_{\mu/\lambda}(x)$ defined thusly is an analytic function of x due to the BAE and contains $T_1(x)$ as a special case. The former is not so obvious from the original definition by the tableaux, but it follows trivially from the quantum Jacobi–Trudi formula.

In the same spirit, we introduce $\Lambda_{\mu/\lambda}(x)$, which is analytic under the BAE, by setting $T_{m\geq 8}(x) = 0$ in $\mathcal{T}_{\mu/\lambda}(x)$:

$$\Lambda_{\mu/\lambda}(x) := \mathcal{T}_{\mu/\lambda}(x)/.\{T_{m\geq 8}(x) = 0\} \tag{12}$$

for the dilute A_4 model. The pole-free property of $\Lambda_{\mu/\lambda}(x)$ is obvious from (11). For the dilute A_6 model, we define $\Lambda_{\mu/\lambda}(x)$ by setting $T_{m\geq 11}(x) = 0$.

5 The kink transfer matrix

In following few sections, we restrict our discussion to the dilute A_4 model. We will summarize our results for the dilute A_6 model in Section 7.

For the E_8 case, the vector representation is minimal and other representations are constructed by fusion with it. Eigenvalues of associated transfer matrices of the dilute A_3 model are thus derived from products of $T_1(x + \text{some shift})$.

This is no longer true for the dilute A_4 model. Let $W_a(x)$ be the Yangian highest-weight module associated with the node a in Figure 1. From several pieces of evidence, we identify $T_1(x)$ of the model with the transfer matrix connected to $W_1(x)$, which is not minimal. The most fundamental object in E_7 is $W_6(x)$ rather than $W_1(x)$. Any other object may be constructed from $T^{(6)}(x)$, the transfer matrix of $W_6(x)$, of which the explicit form is not known from the dilute A_4 model. Determination of the explicit form of $T^{(6)}(x)$ is thus vital to the present approach. For this purpose, homomorphisms of $U_q(\hat{\mathfrak{g}})$ modules deserve our attention. These lead to nontrivial algebraic relations among $T^{(6)}(x)$ and other QTMs. Although such information is not available for $U_q(\hat{E}_7)$, there exists a list of homomorphisms of $Y(E_7)$ modules in [37]:

$$W_{6-a}(x) \hookrightarrow W_{7-a}\left(x - \frac{1}{6}i\right) \otimes W_6\left(x + \frac{a}{6}i\right), \quad a = 1, 2, 3, \tag{13}$$

$$W_1(x) \hookrightarrow W_6\left(x + \frac{5}{6}i\right) \otimes W_6\left(x - \frac{5}{6}i\right), \tag{14}$$

$$C \hookrightarrow W_6\left(x + \frac{3}{2}i\right) \otimes W_6\left(x - \frac{3}{2}i\right), \tag{15}$$

$$W_2(x) \hookrightarrow W_1\left(x + \frac{1}{6}i\right) \otimes W_1\left(x - \frac{1}{6}i\right), \tag{16}$$

$$W_7(x) \hookrightarrow W_1\left(x + \frac{1}{2}i\right) \otimes W_6\left(x - \frac{2}{3}i\right). \tag{17}$$

Note that the normalization of the spectral parameter is different from [37].

For example, the second relation implies

$$T^{(6)}\left(x+i\frac{5}{6}\right)T^{(6)}\left(x-i\frac{5}{6}\right)=T^{(1)}(x)+\cdots.$$

After trial and error, we find the following ansatz compatible with the above homomorphisms:

$$\begin{aligned}T^{(6)}(x) = \frac{1}{\sqrt{2}}\Bigg(& w^2\phi\left(x+\frac{2}{3}i\right)\frac{Q(x+\frac{2}{3}i)}{Q(x-\frac{1}{3}i)} + w\phi\left(x-\frac{4}{3}i\right)\frac{Q(x-\frac{4}{3}i)Q(x+\frac{5}{3}i)}{Q(x-\frac{1}{3}i)Q(x+i)} \\ &+\phi(x)\frac{Q(x+\frac{5}{3}i)Q(x)}{Q(x+i)Q(x-i)} + \frac{1}{w}\phi\left(x+\frac{4}{3}i\right)\frac{Q(x+\frac{4}{3}i)Q(x-\frac{5}{3}i)}{Q(x+\frac{1}{3}i)Q(x-i)} \\ &+\frac{1}{w^2}\phi\left(x-\frac{2}{3}i\right)\frac{Q(x-\frac{2}{3}i)}{Q(x+\frac{1}{3}i)}\Bigg). \end{aligned} \tag{18}$$

Since explicit RSOS weights are not yet derived, it may be inappropriate to call $T^{(6)}(x)$ the eigenvalue of the transfer matrix. In the following discussion, however, we do not use the assumption that it coincides with the actual eigenvalue of the transfer matrix of W_6. Rather, we simply use the following facts: (i) it is pole free under the BAE; (ii) it satisfies the desired relations expected from (14) and (15). The reader should understand this terminology as merely a "nickname."

The following functional relations between $T^{(6)}(x)$ and $T_m(x)$ will facilitate discussions in later sections.

Lemma 1.

$$T^{(6)}\left(x+\frac{1}{6}i\right)T^{(6)}\left(x-\frac{1}{6}i\right) = T_2\left(x+\frac{10}{6}i\right)+T_0(x), \tag{19}$$

$$T^{(6)}\left(x+\frac{3}{6}i\right)T^{(6)}\left(x-\frac{3}{6}i\right) = T_3(x), \tag{20}$$

$$T^{(6)}\left(x+\frac{5}{6}i\right)T^{(6)}\left(x-\frac{5}{6}i\right) = T_1(x)+T_1\left(x+\frac{10}{6}i\right), \tag{21}$$

$$T^{(6)}\left(x+\frac{7}{6}i\right)T^{(6)}\left(x-\frac{7}{6}i\right) = T_3\left(x+\frac{10}{6}i\right), \tag{22}$$

$$T^{(6)}\left(x+\frac{9}{6}i\right)T^{(6)}\left(x-\frac{9}{6}i\right) = T_2(x)+T_0\left(x+\frac{10}{6}i\right). \tag{23}$$

Proof. The duality relation (10) plays a fundamental role in the proof. The direct substitution of (18) in the left-hand sides of (20), (21), and (23) yields $1/2(T_3(x)+T_4(x))$, $1/2(T_1(x+i\frac{10}{6})+T_6(x+\frac{10}{6}i)+2T_1(x))$, and $1/2(T_2(x)+T_5(x))+T_0(x+\frac{10}{6}i)$, respectively. We remark that nontrivial cancellation of terms occurs due to $w^5=-1$.

These results coincide with the right-hand side due to the dualities $T_3 = T_4$, $T_1 = T_6$, and $T_2 = T_5$. The rest of the proof follows from these results by $x \to x + i\frac{10}{6}$. □

These relations suggest the underlying homomorphisms between $U_q(\hat{E}_7)$ and $U_{q'}(A_2^{(2)})$ modules at $q = \exp(i\pi/20)$, $q' = \exp(i3\pi/10)$. This may be an interesting but an independent subject from the present problem thus we will not go into detail here.

6 Fusion quantum transfer matrices and the T-system for the dilute A_4 model

Having defined the "kink" QTM $T^{(6)}$, we are in position to introduce other QTMs and explore functional relations among them.

First, we present QTMs defined by skew Young tableaux.

Definition 1.

$$T^{(1)}(x) = \Lambda_{(1)}(x) \quad (= T_1(x)), \tag{24}$$

$$T^{(2)}(x) = \frac{1}{\phi(x - \frac{5}{3}i)} \Lambda_{(6,1)}\left(x - \frac{5}{6}i\right), \tag{25}$$

$$T^{(3)}(x) = \frac{1}{\phi(x \pm \frac{3}{2}i)} \Lambda_{(11,6,6)/(5,5)}(x), \tag{26}$$

$$T^{(5)}(x) = \Lambda_{(2)}\left(x + \frac{5}{3}i\right) \quad \left(= T_2\left(x + \frac{5}{3}i\right)\right). \tag{27}$$

We have two comments.

First, one can rewrite (25) and (26) equivalently in terms of $\Lambda_{(6,6)/(5)}$ or $\Lambda_{(6,6,1)/(5)}(x)$ using the relations

$$\Lambda_{(6,6)/(5)}(x) = \Lambda_{(6,1)}(x + 5i), \tag{28}$$

$$\Lambda_{(6,6,1)/(5)}(x) = \Lambda_{(11,6,6)/(5,5)}(x). \tag{29}$$

These are outcomes of the quantum Jacobi–Trudi formula and duality (10).

The second comment concerns the complex conjugate property. In the largest eigenvalue sector of the QTM, we confirm $Q(x) = \overline{Q(\bar{x})}$ numerically. The explicit forms of the DVFs in (4) and (18) then conclude $T^{(1)}(x) = \overline{T^{(1)}(\bar{x})}$, $T^{(6)}(x) = \overline{T^{(6)}(\bar{x})}$. By remembering $5/3i$ equals to the half-period of our elliptic function $\theta_1(x)$, one also verifies that $T^{(5)}(x) = \overline{T^{(5)}(\bar{x})}$. The conjugate property of $T^{(2)}(x)$ and $T^{(3)}(x)$ is less obvious. Nevertheless, one can show this using the quantum Jacobi–Trudi formula.

$T^{(6)}$ comes into expressions of the remaining QTMs. The Yangian homomorphisms turn out to be useful in deriving their explicit forms. For instance, we take

$T^{(4)}(x)$, related to $W_4(x)$. As argued in Section 2.3, we identify an analytic set with an affine irreducible module. Thus (13) for $a = 2$ implies that

$$T^{(5)}(x - i/6)T^{(6)}(x + i/3) \sim T^{(4)}(x) + T'(x).$$

That is, the product of DVFs $T^{(5)}(x - I/6)T^{(6)}(x + i/3)$ decomposes into two (or more) subsets and each subset is analytic within itself. We look at the explicit DVF of the left-hand side and find that it contains an analytic subset given by $\phi(x + \frac{4}{3}i)\phi(x - i)T^{(6)}(x - \frac{1}{3}i)$. The sum of the remaining terms must be analytic under the BAE. We identify them as $T^{(4)}(x)$. In a similar way, we deduce $T^{(7)}(x)$.

Definition 2.

$$T^{(4)}(x) \tag{30}$$
$$= \left(T^{(5)}\left(x - \frac{1}{6}i\right) T^{(6)}\left(x + \frac{1}{3}i\right) - \phi\left(x + \frac{2}{3}i\right)\phi\left(x - \frac{1}{3}i\right) T^{(6)}\left(x - \frac{1}{3}i\right)\right),$$

$$T^{(7)}(x) \tag{31}$$
$$= \frac{1}{\phi(x - \frac{4}{3}i)}\left(T^{(1)}\left(x + \frac{1}{2}i\right) T^{(6)}\left(x - \frac{2}{3}i\right) - \phi(x)\phi(x - i)T^{(6)}\left(x + \frac{4}{3}i\right)\right).$$

The common factor $1/\phi(x - \frac{4}{3}i)$ is divided out for $T^{(7)}(x)$. We note that the conjugate property, discussed for other $T^{(a)}$s, is also verified for $T^{(7)}(x)$ when it is written in terms of its explicit DVF. On the other hand, this property is not so apparent for $T^{(4)}(x)$, although one can prove it by a different route. See the discussion after the proof of (36).

We are now ready to describe the statement as to functional relations among QTMs.

Proposition 1. *The above QTMs enjoy the following T-system:*

$$T^{(1)}\left(x - \frac{1}{6}i\right) T^{(1)}\left(x + \frac{1}{6}i\right) = \phi\left(x - \frac{5}{3}i\right) T^{(2)}(x) + T_0\left(x \pm \frac{5}{6}i\right), \tag{32}$$

$$T^{(2)}\left(x - \frac{1}{6}i\right) T^{(2)}\left(x + \frac{1}{6}i\right) = T_0(x) T_0\left(x \pm \frac{4}{6}i\right) + T^{(1)}(x)T^{(3)}(x), \tag{33}$$

$$T^{(3)}\left(x - \frac{1}{6}i\right) T^{(3)}\left(x + \frac{1}{6}i\right) = T_0\left(x \pm \frac{1}{2}i\right) T_0\left(x \pm \frac{1}{6}i\right) + T^{(2)}(x)T^{(4)}(x)T^{(7)}(x), \tag{34}$$

$$T^{(4)}\left(x - \frac{1}{6}i\right) T^{(4)}\left(x + \frac{1}{6}i\right) = T_0(x) T_0\left(x \pm \frac{1}{3}i\right) + T^{(3)}(x)T^{(5)}(x), \tag{35}$$

$$T^{(5)}\left(x - \frac{1}{6}i\right) T^{(5)}\left(x + \frac{1}{6}i\right) = T_0\left(x \pm \frac{1}{6}i\right) + T^{(4)}(x)T^{(6)}(x), \tag{36}$$

$$T^{(6)}\left(x - \frac{1}{6}i\right) T^{(6)}\left(x + \frac{1}{6}i\right) = T_0(x) + T^{(5)}(x), \tag{37}$$

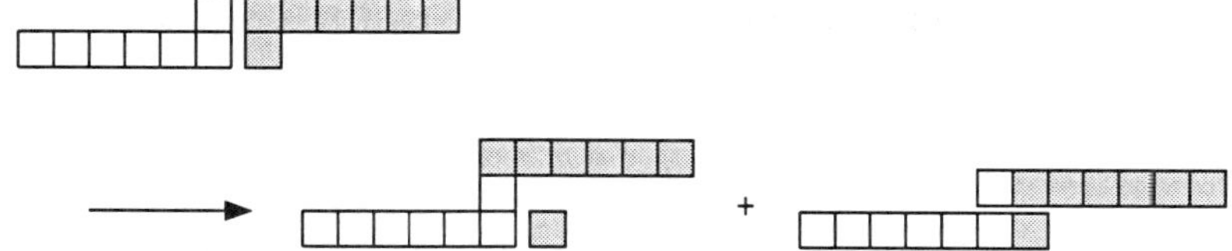

Figure 5. The decomposition of $\Lambda_{(6,6)/(5)}(x)\Lambda_{(6,1)}(x+12i)$.

$$T^{(7)}\left(x-\frac{1}{6}i\right)T^{(7)}\left(x+\frac{1}{6}i\right)=T_0\left(x\pm\frac{1}{3}i\right)+T^{(3)}(x). \tag{38}$$

These coincide with the T-system for E_7 proposed in [30] in a different context. Note that (37) has already been proven in (19) using definition (27).

PROOF OF (32) AND (33). We take the simpler case (32) first. We consider the decomposition of the product $T_1(x-6i)T_6(x+i)$. Quite similar to the combinatorics of a semistandard tableaux, we have

$$T_1(x-6i)T_6(x+i)=\Lambda_{(6,1)}(x)+T_7(x)T_0(x-5i),$$

which can also be verified using the quantum Jacobi–Trudi formula. By utilizing the dualities $T_6(x)=T_1(x)$ and $T_7(x)=T_0(x)$ and periodicity, one recovers (32) after the shift $x\to x-\frac{5}{6}i$ in both sides.

Equation (33) follows by considering the decomposition of $\Lambda_{(6,6)/(5)}(x)\Lambda_{(6,1)}(x+12i)$. (See Figure 5.) The left-hand side of the arrow contains two Young diagrams corresponding to $\Lambda_{(6,6)/(5)}(x)$ and $\Lambda_{(6,1)}(x+12i)$ (hatched) in the relative position, compatible with the shift in the spectral parameter. We then employ the "recombination" of boxes, as in usual Young diagrams. Two rules need particular attention. First, we must place the box so as to match the spectral parameter. Second, due to the condition $T_{m\geq 8}(x)=0$ in the definition of $\Lambda_{\mu/\lambda}(x)$, the width of a column in the resultant Young diagram must be less than or equal to 7. Then two terms result from the recombination as depicted on the right-hand side of the arrow, $\Lambda_{(11,6,6)/(5,5)}(x-8i)T_1(x+6i)$ and $T_7(x)T_7(x+12i)$. The boxes that once belonged to $\Lambda_{(6,1)}(x+12i)$ retain their hatching. Note that the figure represents equality of the DVFs in terms of boxes. Thus it needs proper normalization factors, viewed as the relation between $\Lambda_{\mu/\lambda}(x)$ or $T_m(x)$, due to scalar factors in the definitions (7) of $T_m(x)$ and $\mathcal{T}_{\mu/\lambda}(x)$ (and then $\Lambda_{\mu/\lambda}(x)$) in Theorem 1. Then, using property (28), one finds

$$\begin{aligned}\Lambda_{(6,1)}(x-5i)\Lambda_{(6,1)}(x+12i)&=\Lambda_{(11,6,6)/(5,5)}(x-8i)T_1(x+6i)\\&+\phi\left(x-\frac{9}{2}i\right)\phi\left(x-\frac{11}{2}i\right)\phi\left(x+\frac{13}{2}i\right)\phi\left(x+\frac{15}{2}i\right)T_7(x)T_7(x+12i),\end{aligned}$$

where the factor in front of $T_7(x)T_7(x+12i)$ comes from $f_7(x)f_7(x+12i)/f_6(x+i)f_6(x+13i)$. Finally, let us shift the spectral parameter by $+i4/6$ and use the periodicity (8). Then (33) follows from the definitions (25), (26), and $T_7=T_0$. □

PROOF OF (36). The proof utilizes (19) as follows. Consider the product $T^{(4)}(x)T^{(6)}(x)$. Substituting (30), we have

$$\begin{aligned} &T^{(4)}(x)T^{(6)}(x) \\ &= T^{(5)}\left(x-\frac{1}{6}i\right)T^{(6)}(x)T^{(6)}\left(x+\frac{1}{3}i\right) - T_0\left(x+\frac{1}{6}i\right)T^{(6)}(x)T^{(6)}\left(x-\frac{1}{3}i\right). \end{aligned}$$

We apply rule (19) to the product of two $T^{(6)}$s by shifting $x \pm i/6$. The result leads to

$$T^{(4)}(x)T^{(6)}(x) = T^{(5)}\left(x-\frac{1}{6}i\right)T^{(5)}\left(x+\frac{1}{6}i\right) - T_0\left(x \pm \frac{1}{6}i\right),$$

which coincides with (36). □

As mentioned previously, the complex conjugate property of $T^{(4)}(x)$ is not obvious from its DVF. Instead, we can now show it from the established relation (36) with the help of the conjugate property of $T^{(5)}(x)$ and $T^{(6)}(x)$, discussed below (27).

To prove the remaining relations, we need to present further lemmas.

Lemma 2. *The following decompositions are valid:*

$$T^{(5)}\left(x+\frac{1}{3}i\right)T^{(5)}\left(x-\frac{1}{3}i\right) = T_0(x)T_4(x) + T^{(3)}(x), \tag{39}$$

$$\begin{aligned} T^{(5)}\left(x+\frac{1}{3}i\right)T^{(5)}\left(x-\frac{1}{3}i\right) &= T_4(x)(T_0(x)+T^{(5)}(x)) - T_0\left(x\pm\frac{1}{3}i\right) \\ &- \left(T_0\left(x+\frac{1}{3}i\right)T^{(5)}\left(x-\frac{1}{3}i\right) + T_0\left(x-\frac{1}{3}i\right)T^{(5)}\left(x+\frac{1}{3}i\right)\right). \end{aligned} \tag{40}$$

PROOF. Relation (39) is checked by comparing the DVFs of both sides directly. To prove (40), we rewrite $T^{(5)}$ in the left-hand side in terms of $T^{(6)}$ by using (19):

$$\begin{aligned} &T^{(5)}\left(x+\frac{1}{3}i\right)T^{(5)}\left(x-\frac{1}{3}i\right) \\ &= T_0\left(x\pm\frac{1}{3}i\right) + T^{(6)}\left(x-\frac{1}{6}i\right)T^{(6)}\left(x+\frac{1}{6}i\right)T^{(6)}\left(x-\frac{1}{2}i\right)T^{(6)}\left(x+\frac{1}{2}i\right) \\ &\quad - \left(T_0\left(x+\frac{1}{3}i\right)T^{(6)}\left(x-\frac{1}{6}i\right)T^{(6)}\left(x-\frac{1}{2}i\right)\right. \\ &\quad \left. + T_0\left(x-\frac{1}{3}i\right)T^{(6)}\left(x+\frac{1}{6}i\right)T^{(6)}\left(x+\frac{1}{2}i\right)\right). \end{aligned}$$

By applying (19) and (20), we reach the right-hand side of (40). □

For later use, we rearrange the sum of (39) and (40):

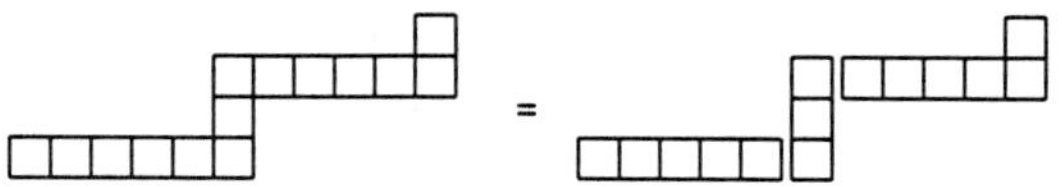

Figure 6. Decomposition of the skew diagram (11,11,6,6)/(10,5,5). The height-3 piece in the middle of the right-hand side reduces to a known scalar factor.

$$\begin{aligned}&\Big(2T_0(x)T_4(x) + T^{(3)}(x) + T_4(x)T^{(5)}(x) - T_0\left(x+\frac{1}{3}i\right)T^{(5)}\left(x-\frac{1}{3}i\right)\\&\quad - T_0\left(x-\frac{1}{3}i\right)T^{(5)}\left(x+\frac{1}{3}i\right) - 2T^{(5)}\left(x-\frac{1}{3}i\right)T^{(5)}\left(x+\frac{1}{3}i\right)\Big)\\&\quad = T_0\left(x\pm\frac{1}{3}i\right).\end{aligned} \tag{41}$$

Lemma 3.

$$\begin{aligned}&T_0(x\pm i)T_3(x) + T_1(x-2i)T_1(x)T_1(x+2i)\\&\quad T_0(x-i)T_1(x+2i)T_2(x-i) - T_0(x+i)T_1(x-2i)T_2(x+i)\\&\quad = \phi\left(x\pm\frac{7}{2}i\right)\phi\left(x\pm\frac{5}{2}i\right)\phi\left(x\pm\frac{3}{2}i\right).\end{aligned}$$

This is the analogue of the relation

□□□ + □ ⊗ □ ⊗ □ − □ ⊗ □□ − □□ ⊗ □ = (column of three boxes)

and can be shown in a similar manner.

To prove (34), we need the decomposition of a huge QTM.

Lemma 4.

$$\begin{aligned}\Lambda_{(11,11,6,6)/(10,5,5)}(x) &= \phi\left(x+\frac{1}{2}i\right)\phi\left(x-\frac{5}{2}i\right)\\&\quad\times\big(T_1(x+3i)T_2(x+i)T_2(x-3i) - T_0(x+2i)T_1(x-2i)T_2(x+i)\\&\qquad - T_0(x-i)T_1(x+3i)T_3(x-i) + T_0(x-i)T_0(x+2i)T_4(x)\big).\end{aligned}$$

PROOF. For convenience, we go back to the original definition of $\mathcal{T}_{(11,11,6,6)/(10,5,5)}(x)$. Thanks to the semistandard condition, it decomposed into pieces. We consequently have (see Figure 6)

$$\begin{aligned}f_6(x+6i)f_6(x-8i)\mathcal{T}_{(11,11,6,6)/(10,5,5)}(x) &= \frac{f_5(x-i)f_5(x-9i)f_5(x+7i)}{f_2(x-i)}\\&\quad\times T_5(x-9i)\big(T_1(x+13i)T_5(x+5i) - T_0(x+12i)T_6(x+8i)\big).\end{aligned} \tag{42}$$

On the other hand, the quantum Jacobi–Trudi formula relates the same quantity to

$$\begin{aligned}&f_6(x+6i)f_6(x-8i)\Lambda_{(11,11,6,6)/(10,5,5)}(x)\\&\quad + f_{13}(x-i)T_1(x+13i)T_{13}(x-i) - f_{14}(x)T_{14}(x).\end{aligned} \tag{43}$$

Then the equality (42) = (43) with properties $T_{m+10}(x) = T_m(x)$ and $T_{7-m}(x) = T_m(x)$ leads to Lemma 4 after renormalization. □

Our final lemma concerns the decomposition of $T^{(4)}(x)T^{(7)}(x)$.

Lemma 5.

$$\begin{aligned}
&\phi\left(x-\frac{4}{3}i\right)T^{(4)}(x)T^{(7)}(x)\\
&\quad = T_2\left(x+\frac{11}{6}i\right)\left\{T_1\left(x+\frac{1}{2}i\right)T_2\left(x+\frac{7}{6}i\right)-T_0\left(x+\frac{17}{6}i\right)T_1\left(x-\frac{7}{6}i\right)\right\}\\
&\qquad - T_0\left(x-\frac{1}{6}i\right)T_1\left(x+\frac{1}{2}i\right)T_3\left(x-\frac{1}{6}i\right)\\
&\qquad + T_0\left(x-\frac{1}{6}i\right)T_0\left(x-\frac{1}{2}i\right)T_3\left(x+\frac{5}{6}i\right).
\end{aligned}$$

PROOF. We use a trick; the complex conjugate property, established shortly after the proof of (36), allows us to replace $T^{(4)}(x)$ by $\overline{T^{(4)}(x)}$ in the left-hand side. After this, we substitute definitions (30) and (31) into the left-hand side:

$$\begin{aligned}
&\phi\left(x-\frac{4}{3}i\right)\overline{T^{(4)}(x)}T^{(7)}(x)\\
&= T_2\left(x+\frac{11}{6}i\right)T_1\left(x+\frac{1}{2}i\right)T^{(6)}\left(x-\frac{1}{2}i+\frac{1}{6}i\right)T^{(6)}\left(x-\frac{1}{2}i-\frac{1}{6}i\right)\\
&\quad - \phi(x)\phi(x-i)T_2\left(x+\frac{11}{6}i\right)T^{(6)}\left(x+\frac{1}{2}i+\frac{5}{6}i\right)T^{(6)}\left(x+\frac{1}{2}i-\frac{5}{6}i\right)\\
&\quad - \phi\left(x+\frac{1}{3}i\right)\phi\left(x-\frac{2}{3}i\right)T_1\left(x+\frac{1}{2}i\right)T^{(6)}\left(x-\frac{1}{6}i+\frac{1}{2}i\right)T^{(6)}\left(x-\frac{1}{6}i-\frac{1}{2}i\right)\\
&\quad + \phi(x-i)\phi\left(x-\frac{2}{3}i\right)\phi(x)\phi\left(x+\frac{1}{3}i\right)T^{(6)}\left(x+\frac{5i}{6}+\frac{1}{2}i\right)T^{(6)}\left(x+\frac{5i}{6}-\frac{1}{2}i\right),
\end{aligned}$$

where we have used definitions (24) and (27) to rewrite $T^{(1)}$ and $T^{(5)}$ by T_1 and T_2. Thanks to this trick, the differences in arguments of products of $T^{(6)}$ are such that one can apply (19), (20), and (21). The result of the application then agrees with the right-hand side of the equality in Lemma 5. □

By comparing the right-hand sides of Lemmas 4 and 5, using the duality $T_3 = T_4$, we notice the following equality.

Lemma 6.

$$\Lambda_{(11,11,6,6)/(10,5,5)}\left(x+\frac{5}{6}i\right) = \phi\left(x\pm\frac{8}{6}i\right)\phi\left(x-\frac{10}{6}i\right)T^{(4)}(x)T^{(7)}(x).$$

With these preparations, we prove the remaining relations.

PROOF OF (35). Let us rewrite the left-hand side by substituting definition (30). In doing so, we employ a trick similar to the one above: we replace $T^{(4)}(x+\frac{1}{6}i)$ in the left-hand side by $\overline{T^{(4)}(x - \frac{1}{6}i)}$. Then it follows that

$$
\begin{aligned}
&T^{(4)}\left(x+\frac{1}{6}i\right)T^{(4)}\left(x-\frac{1}{6}i\right)\\
&= T^{(5)}\left(x-\frac{1}{3}i\right)T^{(5)}\left(x+\frac{1}{3}i\right)T^{(6)}\left(x-\frac{1}{6}i\right)T^{(6)}\left(x+\frac{1}{6}i\right)\\
&\quad - T_0(x)T^{(5)}\left(x-\frac{1}{3}i\right)T^{(6)}\left(x+\frac{1}{3}i-\frac{1}{6}i\right)T^{(6)}\left(x+\frac{1}{3}i+\frac{1}{6}i\right)\\
&\quad - T_0(x)T^{(5)}\left(x+\frac{1}{3}i\right)T^{(6)}\left(x-\frac{1}{3}i-\frac{1}{6}i\right)T^{(6)}\left(x-\frac{1}{3}i+\frac{1}{6}i\right)\\
&\quad + (T_0(x))^2T^{(6)}\left(x-\frac{1}{2}i\right)T^{(6)}\left(x+\frac{1}{2}i\right).
\end{aligned}
$$

In this form, (19) applies to the first three terms of the right-hand side and (20) can be used in the fourth. Also, we use (39) the first term of the right-hand side. The result reads

$$
\begin{aligned}
&T^{(4)}\left(x+\frac{1}{6}i\right)T^{(4)}\left(x-\frac{1}{6}i\right)\\
&= T^{(3)}(x)T^{(5)}(x) + T_0(x)\Bigg(2T_0(x)T_4(x) + T^{(3)}(x) + T_4(x)T^{(5)}(x)\\
&\quad - T_0\left(x+\frac{1}{3}i\right)T^{(5)}\left(x-\frac{1}{3}i\right) - T_0\left(x-\frac{1}{3}i\right)T^{(5)}\left(x+\frac{1}{3}i\right)\\
&\quad - 2T^{(5)}\left(x-\frac{1}{3}i\right)T^{(5)}\left(x+\frac{1}{3}i\right)\Bigg),
\end{aligned}
$$

where we represent $T_2(x)$ by $T^{(5)}(x + \frac{5}{3}i)$ and use the duality $T_3 = T_4$. One notices that the content of the main parenthesis on the right-hand side reduces to $T_0(x \pm \frac{1}{3}i)$ due to (41), which completes the proof. □

Proof of (38). In the same manner, we start from the equivalent expression $\overline{T^{(7)}(x - \frac{1}{6}i)}T^{(7)}(x - \frac{1}{6}i)$ and substitute (31). After rearrangement using (21), (22), and (23), we find that this expression is equal to

$$
\begin{aligned}
&T^{(7)}\left(x+\frac{1}{6}i\right)T^{(7)}\left(x-\frac{1}{6}i\right) = \frac{1}{\phi(x\pm\frac{3}{2}i)}\Bigg(T_0\left(x\pm\frac{2}{3}i\right)T_3\left(x+\frac{10}{6}i\right)\\
&+ T_1\left(x-\frac{1}{3}i\right)T_1\left(x+\frac{1}{3}i\right)T_1\left(x+\frac{10}{6}i\right) + T_1\left(x-\frac{1}{3}i\right)T_1\left(x+\frac{1}{3}i\right)T_1(x)\\
&- T_0\left(x+\frac{2}{3}i\right)T_0(x+i)T_1\left(x+\frac{1}{3}i\right) - T_0\left(x+\frac{2}{3}i\right)T_1\left(x+\frac{1}{3}i\right)T_2\left(x+\frac{2}{3}i\right)\\
&- T_0\left(x-\frac{2}{3}i\right)T_0(x-i)T_1\left(x-\frac{1}{3}i\right) - T_0\left(x-\frac{2}{3}i\right)T_1\left(x-\frac{1}{3}i\right)T_2\left(x-\frac{2}{3}i\right)\Bigg).
\end{aligned}
$$

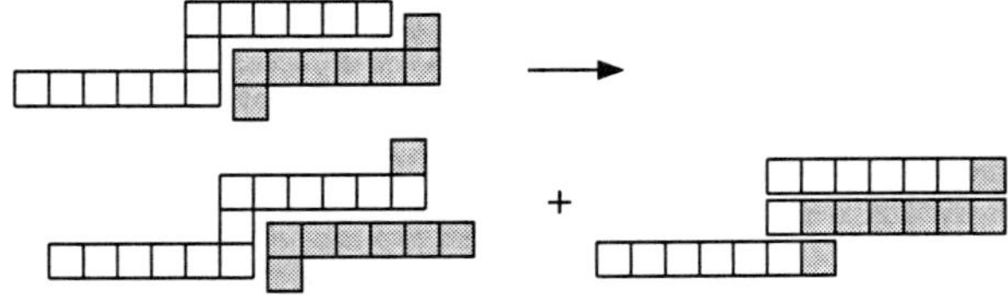

Figure 7. Graphical rule for the decomposition of two diagrams $(11, 6, 6)/(5, 5)$ and $(6, 6, 1)/(5)$.

Let us subtract $T^{(3)}(x)$ from the above. Note that $T^{(3)}(x)$ should be understood as the result of the application of (12) to (26) and of the duality $T_7 \to T_0$. Then we have

$$T^{(7)}\Big(x+\frac{1}{6}i\Big)T^{(7)}\Big(x-\frac{1}{6}i\Big) - T^{(3)}(x) = \frac{1}{\phi(x \pm \frac{3}{2}i)}\Big(T_0\Big(x \pm \frac{2}{3}i\Big)T_3\Big(x+\frac{10}{6}i\Big)$$
$$+ T_1\Big(x-\frac{1}{3}i\Big)T_1\Big(x+\frac{1}{3}i\Big)T_1\Big(x+\frac{10}{6}i\Big) - T_0\Big(x+\frac{2}{3}i\Big)T_1\Big(x+\frac{1}{3}i\Big)T_2\Big(x+\frac{2}{3}i\Big)$$
$$- T_0\Big(x-\frac{2}{3}i\Big)\,T_1\Big(x-\frac{1}{3}i\Big)\,T_2\Big(x-\frac{2}{3}i\Big)\Big).$$

The content of the main parentheses is identical to the left-hand side of Lemma 3 with the shift $x \to x + \frac{10}{6}i$ by noticing the periodicity (8). One immediately verifies that the right-hand side reduces to $T_0(x \pm \frac{1}{3}i)$, and (38) is proven. □

Proof of (34). The decomposition of $\Lambda_{(11,6,6)/(5,5)}(x)\Lambda_{(6,6,1)/(5)}(x + 7i)$ can be done using (12). Equivalently, we can argue it in a graphical manner, as shown in Figure 7 (just as in Figure 5.)

$$\begin{aligned}
&\Lambda_{(11,6,6)/(5,5)}(x)\Lambda_{(6,6,1)/(5)}(x+7i)\\
&\quad = \Lambda_{(6,1)}(x+6i)\Lambda_{(11,11,6,6)/(10,5,5)}(x+i)\\
&\qquad + T_0(x \pm i)T_0(x \pm 6i)T_0(x+8i)T_0(x+13i),
\end{aligned}$$

where T_7 is replaced by T_0.

By using (26) and (29) and taking normalization factors into account, one finds

$$T^{(3)}\Big(x-\frac{1}{6}i\Big)\,T^{(3)}\Big(x+\frac{1}{6}i\Big) = T_0\Big(x \pm \frac{1}{2}i\Big)\,T_0\Big(x \pm \frac{1}{6}i\Big)$$
$$+ \frac{\phi(x-\frac{10}{6}i)}{\phi(x \pm \frac{8}{6}i)\phi(x \pm \frac{10}{6}i)}T^{(2)}(x)\Lambda_{(11,11,6,6)/(10,5,5)}\Big(x+\frac{5}{6}i\Big).$$

Then the equivalent statement to (38) is

$$\Lambda_{(11,11,6,6)/(10,5,5)}\Big(x+\frac{5}{6}i\Big) = \phi\Big(x \pm \frac{8}{6}i\Big)\,\phi\Big(x-\frac{10}{6}i\Big)\,T^{(4)}(x)T^{(7)}(x).$$

This is simply Lemma 6. We thus complete the proof of Proposition 1. □

In the next section, we summarize similar results for the dilute A_6 model without proof.

7 The T-system for the dilute A_6 model

We firstly comment on the duality property in the dilute A_6 model:

$$T_m(x) = T_{11-m}(x), \quad m = 0, \ldots, 11,$$

and $T_{12}(x) = T_{13}(x) = 0$. This is again compatible with the symmetry of functional relations but is still a conjecture supported by numerics.

Yangian representation theory asserts that irreducible modules of $Y(E_6)$ are made by tensoring the minimal objects $W_1(x)$ and $W_5(x')$. On the other hand, we assume

$$T_1(x) = T^{(6)}(x)$$

to be the QTM for $W_6(x)$. Thus the situation is similar to the E_7 case; one must figure out eigenvalues of the QTMs associated with $W_1(x)$ and $W_5(x')$ independently from knowledge of the dilute A_6 model. We conjecture that the eigenvalues of the QTMs for these are the same and that its explicit form reads

$$\begin{aligned}
T^{(1)}(x) &= T^{(5)}(x) \\
&= \frac{1}{2}\left(\omega^3\phi\left(x+\frac{3}{4}i\right)\frac{Q(x+\frac{3}{4}i)}{Q(x-\frac{1}{4}i)} + \omega^2\phi\left(x-\frac{5}{4}i\right)\frac{Q(x-\frac{5}{4}i)Q(x-\frac{7}{4}i)}{Q(x-\frac{1}{4}i)Q(x+\frac{5}{4}i)}\right. \\
&+\omega\phi\left(x+\frac{1}{4}i\right)\frac{Q(x+\frac{1}{4}i)Q(x-\frac{7}{4}i)}{Q(x-\frac{3}{4}i)Q(x+\frac{5}{4}i)} + \phi\left(x-\frac{7}{4}i\right)\frac{Q(x-\frac{7}{4}i)^2}{Q(x-\frac{3}{4}i)Q(x+\frac{3}{4}i)} \\
&+\frac{1}{\omega}\phi\left(x-\frac{1}{4}i\right)\frac{Q(x-\frac{1}{4}i)Q(x+\frac{7}{4}i)}{Q(x+\frac{3}{4}i)Q(x-\frac{5}{4}i)} + \frac{1}{\omega^2}\phi\left(x+\frac{5}{4}i\right)\frac{Q(x+\frac{5}{4}i)Q(x+\frac{7}{4}i)}{Q(x+\frac{1}{4}i)Q(x-\frac{5}{4}i)} \\
&\left.+\frac{1}{\omega^3}\phi\left(x-\frac{3}{4}i\right)\frac{Q(x-\frac{3}{4}i)}{Q(x+\frac{1}{4}i)}\right).
\end{aligned} \tag{44}$$

The following relations hold in parallel to (19)–(23), which are key ingredients in the proof of the T-system.

$$(T^{(1)}(x))^2 = T_2\left(x+i\frac{7}{4}\right), \tag{45}$$

$$T^{(1)}\left(x+i\frac{1}{4}\right)T^{(1)}\left(x-i\frac{1}{4}\right) = \frac{1}{2}\big(T_4(x)+T_0(x)\big), \tag{46}$$

$$T^{(1)}\left(x+i\frac{1}{2}\right)T^{(1)}\left(x-i\frac{1}{2}\right) = \frac{1}{2}T_5\left(x+i\frac{7}{4}\right), \tag{47}$$

$$T^{(1)}\left(x+i\frac{3}{4}\right)T^{(1)}\left(x-i\frac{3}{4}\right) = T_1(x)+\phi(x)T^{(1)}\left(x+i\frac{7}{4}\right). \tag{48}$$

(We omit relations obtained by $x \to x + \frac{7}{4}i$.)

Let the other QTMs be

$$T^{(2)}(x) = T^{(4)}(x) = T^{(1)}\left(x - \frac{1}{4}i\right) T^{(1)}\left(x + \frac{1}{4}i) - T_0(x\right),$$

$$T^{(3)}(x) = \frac{1}{\phi\left(x + i\frac{7}{4}\right)}\left(T_1\left(x - \frac{1}{4}i\right) T_1\left(x + \frac{1}{4}i\right) - T_0\left(x \pm \frac{3i}{4}\right)\right).$$

Then the following T-system is valid.

Proposition 2.

$$T^{(1)}\left(x - \frac{1}{4}i\right) T^{(1)}\left(x + \frac{1}{4}i\right) = T_0(x) + T^{(2)}(x), \tag{49}$$

$$T^{(5)}\left(x - \frac{1}{4}i\right) T^{(5)}\left(x + \frac{1}{4}i\right) = T_0(x) + T^{(4)}(x), \tag{50}$$

$$T^{(6)}\left(x - \frac{1}{4}i\right) T^{(6)}\left(x + \frac{1}{4}i\right) = T_0\left(x \pm \frac{3i}{4}\right) + \phi\left(x + i\frac{7}{4}\right) T^{(3)}(x), \tag{51}$$

$$T^{(2)}\left(x - \frac{1}{4}i\right) T^{(2)}\left(x + \frac{1}{4}i\right) = T_0\left(x \pm \frac{1}{4}i\right) + T^{(1)}(x) T^{(3)}(x), \tag{52}$$

$$T^{(3)}\left(x - \frac{1}{4}i\right) T^{(3)}\left(x + \frac{1}{4}i\right) = T_0(x) T_0\left(x \pm \frac{1}{2}i\right) + (T^{(2)}(x))^2 T^{(6)}(x), \tag{53}$$

$$T^{(4)}\left(x - \frac{1}{4}i\right) T^{(4)}\left(x + \frac{1}{4}i\right) = T_0\left(x \pm \frac{1}{4}i\right) + T^{(3)}(x) T^{(5)}(x). \tag{54}$$

The first three relations are trivial rewritings of definitions. The last three equations have nontrivial proofs, which we omit for brevity.

8 The Y-systems and the E-type thermodynamic Bethe ansatz

We define Y-functions by combinations of $T^{(a)}$s and transform the T-systems into equivalent but desired forms. In the dilute A_4 case, they explicitly read as follows.

Definition 3.

$$Y^{(1)}(x) := \frac{\phi(\frac{x}{6} + \frac{10}{6}i) T^{(2)}(\frac{x}{6})}{T_0(\frac{x}{6} \pm \frac{5}{6}i)}, \qquad Y^{(2)}(x) := \frac{T^{(1)}(\frac{x}{6}) T^{(3)}(\frac{x}{6})}{T_0(\frac{x}{6}) T_0(\frac{x}{6} \pm \frac{4}{6}i)},$$

$$Y^{(3)}(x) := \frac{T^{(2)}(\frac{x}{6}) T^{(4)}(\frac{x}{6}) T^{(7)}(\frac{x}{6})}{T_0(\frac{x}{6} \pm \frac{1}{6}i) T_0(\frac{x}{6} \pm \frac{1}{2}i)}, \qquad Y^{(4)}(x) := \frac{T^{(3)}(\frac{x}{6}) T^{(5)}(\frac{x}{6})}{T_0(\frac{x}{6}) T_0(\frac{x}{6} \pm \frac{1}{3}i)},$$

$$Y^{(5)}(x) := \frac{T^{(4)}(\frac{x}{6}) T^{(6)}(\frac{x}{6})}{T_0(\frac{x}{6} \pm \frac{1}{6}i)}, \qquad Y^{(6)}(x) := \frac{T^{(5)}(\frac{x}{6})}{T_0(\frac{x}{6})},$$

$$Y^{(7)}(x) := \frac{T^{(3)}(\frac{x}{6})}{T_0(\frac{x}{6} \pm \frac{1}{3}i)}.$$

Similarly for the dilute A_6 model, we have the following.

Definition 4.

$$
\begin{aligned}
Y^{(1)}(x) &= Y^{(5)}(x) := \frac{T^{(2)}(\frac{x}{4})}{T_0(\frac{x}{4})}, \\
Y^{(2)}(x) &= Y^{(4)}(x) := \frac{T^{(1)}(\frac{x}{4})T^{(3)}(\frac{x}{4})}{T_0(\frac{x}{4} \pm \frac{1}{4}i)}, \\
Y^{(3)}(x) &:= \frac{(T^{(2)}(\frac{x}{4}))^2 T^{(6)}(\frac{x}{4})}{T_0(\frac{x}{4})T_0(\frac{x}{4} \pm \frac{1}{2}i)}, \\
Y^{(6)}(x) &:= \frac{\phi(\frac{x}{4} + i\frac{7}{4})T^{(3)}(\frac{x}{4})}{T_0(\frac{x}{4} \pm \frac{3i}{4})}.
\end{aligned}
$$

Then new sets of functional relations (the Y-systems) follow from the T-systems.

Theorem 2. *Functional relations among Y-functions exhibit the $E_{6,7}$ structure in the following form:*

$$
Y^{(a)}(x-i)Y^{(a)}(x+i) = \prod_{b \sim a}(1 + Y^{(b)}(x)), \quad a = 1, \ldots, a_{\max}.
$$

Here $a_{\max} = 6$ and 7 for the dilute A_6 and A_4 models, respectively. We denote $a \sim b$ if a and b are adjacent nodes in the E_6 or E_7 Dynkin diagram.

This coincides with the $E_{6,7}$ case of the universal Y-system in [59].

The derivation of the TBA from the Y-system requires some information on the analytic structures of $Y^{(a)}(x)$, $1+Y^{(a)}(x)$. As we stressed in the survey section, only the Y-system with nice analytic (ANZC) properties yields an explicit algorithm in the evaluation of free energy.

We employ numerical calculations for some fixed values of N and β for this purpose. This is relatively facile since one has to deal with only the largest eigenvalue sector. Though we have performed the numerical calculation for small values of N, it already reveals intriguing patterns for zeros of $T^{(a)}(x)$, which are also observed for the dilute A_3 model. Namely, imaginary parts of coordinates of zeros show remarkable coincidence with exponents related to mass spectra in Table 1. We state it as a conjecture for *arbitrary* N.

Conjecture 1. *Zeros of $T^{(a)}$ distribute approximately along lines: $\Im x \sim \pm\frac{1}{6}(a_j+1)$ for the dilute A_4 model and $\pm\frac{1}{4}(a_j+1)$ for the dilute A_6 model. The set $\{a_j\}$ agrees with $\{a\}$ for the particle j in Table 1 (Table 2).*

Therefore, we have a lemma parallel to the dilute A_3 case.

Lemma 7. *Assume that Conjecture 1 is valid. Then $\widetilde{Y^{(a)}}(x)$ and $1 + Y^{(a)}(x)$ are analytic and nonzero and have constant asymptotic behavior (ANZC) in strips $\Im x \in [-1, 1]$, $[-0^+, 0^+]$, respectively.*

$\widetilde{Y^{(a)}}(x)$ is defined by

$$\widetilde{Y^{(a)}}(x) = \begin{cases} Y^{(a)}(x)/\{\kappa(x+i(1+\tilde{u}))\kappa(x-i(1+\tilde{u}))\} & \text{for } a = 1(6), u < 0, \\ Y^{(a)}(x)\kappa(x+i(1-\tilde{u}))\kappa(x-i(1-\tilde{u})) & \text{for } a = 1(6), u > 0, \\ Y^{(a)}(x) & \text{otherwise} \end{cases}$$

and $\tilde{u} = 6u$ $(4u)$ for the dilute A_4 (A_6) model. The renormalization factor is given by

$$\kappa(x) = \left(\frac{\vartheta_1(i\pi v/4, \tau')}{\vartheta_2(i\pi v/4, \tau')}\right)^{N/2},$$

where $\tau' = 5\tau(\frac{7}{2}\tau)$ for the dilute A_4 (A_6) model.

The significance of the above property is clear when one considers these relations (to be precise, logarithmic derivatives of them) in Fourier space, or "k" space. Cauchy's theorem assures that all quantities satisfy algebraic equations at the same k, i.e., without mixing of modes. Thus they can be solved in an elementary way. We omit the explicit procedure since it has been given for other models [22, 32, 44]. The resultant coupled integral equations read

$$\begin{aligned} \ln Y^{(a)}(x) &= -\epsilon\delta_{a,t}\widetilde{\beta}s(x) + \mathcal{C}_{a,b} * \ln(1+Y^{(b)})(x), \\ s(x) &= \frac{\delta}{2\pi}\sum_n e^{ik_n x}\frac{1}{2\cosh k_n}, \\ \mathcal{C}_{a,b}(x) &= s(x)(2I - C^{\mathfrak{g}})_{a,b}, \end{aligned} \tag{55}$$

where $t = 1(6)$, $\widetilde{\beta} = 12\pi\beta(8\pi\beta)$, and $C^{\mathfrak{g}}$ denotes the Cartan matrix for E_7 (E_6). δ also depends on whether we are dealing with the dilute A_4 model or the dilute A_6 model through $\delta = \pi^2/(2\tau')$ and $k_n = n\delta$. We also adopt the abbreviation $A * B(x) := \int_{-2\tau'/\pi}^{2\tau'/\pi} A(x-x')B(x')dx'$.

This is simply the conjectured TBA for the $E_{6,7}$ RSOS model at level 2 [8].

The free energy is expressed via Y-functions with the aid of (32). We shall give the result only for $\epsilon = 1$.

$$\begin{aligned} -\beta f &= -\beta e_0 - \widetilde{\beta}b_1 * s(0) + s * \ln(1+Y^{(1)})(0), \\ e_0 &:= \begin{cases} \lambda[\ln(\vartheta_1(\pi/10)\vartheta_1(4\pi/10))]' & \text{for } A_4, \\ \lambda[\ln(\vartheta_1(\pi/7)/\vartheta_1(3\pi/7))]' & \text{for } A_3, \end{cases} \\ \widehat{b_1}(x) &: = \begin{cases} \frac{\sinh 3x + \sinh 9x}{\sinh 10x} & \text{for } A_4, \\ \frac{\sinh 6x}{\sinh 7x} & \text{for } A_3. \end{cases} \end{aligned} \tag{56}$$

9 Conclusion

We have seen that the $E_{6,7,8}$ structure appears in the dilute $A_{6,4,3}$ model: exponents of mass scale, zeros of the QTMs, and the TBAs. These results strongly support the underlying E-type symmetry in the dilute A_L model.

A Yangian analogue of Young tableaux arises in the proof of the T-system. The combinatorial aspects provide interesting problems on their own. We thus believe that the subject is worth further research.

Appendix

The RSOS weights for the dilute A_L model are given by

$$
\begin{aligned}
\begin{smallmatrix} a & & a \\ & \boxed{u} & \\ a & & a \end{smallmatrix} &= \frac{\theta_1(6-u)\theta_1(3+u)}{\theta_1(6)\theta_1(3)} - \frac{\theta_1(u)\theta_1(3-u)}{\theta_1(6)\theta_1(3)} \\
&\quad \times \left(\frac{S_{a+1}}{S_a}\frac{\theta_4(2a-5)}{\theta_4(2a+1)} + \frac{S_{a-1}}{S_a}\frac{\theta_4(2a+5)}{\theta_4(2a-1)} \right), \\
\begin{smallmatrix} a\pm 1 & & a \\ & \boxed{u} & \\ a & & a \end{smallmatrix} &= \begin{smallmatrix} a & & a \\ & \boxed{u} & \\ a & & a\pm 1 \end{smallmatrix} = \frac{\theta_1(3-u)\theta_4(\pm 2a+1-u)}{\theta_1(3)\theta_4(\pm 2a+1)}, \\
\begin{smallmatrix} a & & a \\ & \boxed{u} & \\ a\pm 1 & & a \end{smallmatrix} &= \begin{smallmatrix} a & & a\pm 1 \\ & \boxed{u} & \\ a & & a \end{smallmatrix} = \left(\frac{S_{a\pm 1}}{S_a}\right)^{1/2} \frac{\theta_1(u)\theta_4(\pm 2a-2+u)}{\theta_1(3)\theta_4(\pm 2a+1)}, \\
\begin{smallmatrix} a & & a\pm 1 \\ & \boxed{u} & \\ a & & a\pm 1 \end{smallmatrix} &= \begin{smallmatrix} a\pm 1 & & a\pm 1 \\ & \boxed{u} & \\ a & & a \end{smallmatrix} \\
&= \left(\frac{\theta_4(\pm 2a+3)\theta_4(\pm 2a-1)}{\theta_4^2(\pm 2a+1)} \right)^{1/2} \frac{\theta_1(u)\theta_1(3-u)}{\theta_1(2)\theta_1(3)}, \\
\begin{smallmatrix} a\pm 1 & & a \\ & \boxed{u} & \\ a & & a\mp 1 \end{smallmatrix} &= \frac{\theta_1(2-u)\theta_1(3-u)}{\theta_1(2)\theta_1(3)}, \\
\begin{smallmatrix} a & & a\mp 1 \\ & \boxed{u} & \\ a\pm 1 & & a \end{smallmatrix} &= -\left(\frac{S_{a-1}S_{a+1}}{S_a^2}\right)^{1/2} \frac{\theta_1(u)\theta_1(1-u)}{\theta_1(2)\theta_1(3)}, \\
\begin{smallmatrix} a & & a\pm 1 \\ & \boxed{u} & \\ a\pm 1 & & a \end{smallmatrix} &= \frac{\theta_1(3-u)\theta_1(\pm 4a+2+u)}{\theta_1(3)\theta_1(\pm 4a+2)} \\
&\quad + \frac{S_{a\pm 1}}{S_a}\frac{\theta_1(u)\theta_1(\pm 4a-1+u)}{\theta_1(3)\theta_1(\pm 4a+2)} \qquad \text{for } \pm 4a+2 \neq 0, \\
&= \frac{\theta_1(3+u)\theta_1(\pm 4a-4+u)}{\theta_1(3)\theta_1(\pm 4a-4)} + \left(\frac{S_{a\mp 1}\theta_1(4)}{S_a\theta_1(2)} - \frac{\theta_4(\pm 2a-5)}{\theta_4(\pm 2a+1)} \right) \\
&\quad \times \frac{\theta_1(u)\theta_1(\pm 4a-1+u)}{\theta_1(3)\theta_1(\pm 4a-4)} \qquad \text{otherwise.}
\end{aligned}
\tag{57}
$$

Here $\theta_{1,4}(x) = \vartheta_{1,4}(\lambda x, \tau)$,

$$
\vartheta_1(x,\tau) = 2q^{1/4}\sin x \prod_{n=1}^{\infty}(1-2q^{2n}\cos 2x + q^{4n})(1-q^{2n}),
$$

$$
\vartheta_4(x,\tau) = \prod_{n=1}^{\infty}(1-2q^{2n-1}\cos 2x + q^{4n-2})(1-q^{2n}),
$$

and $q = \exp(-\tau)$. λ is a parameter of the model specified in Section 3 and S_a denotes

$$
S_a = (-1)^a \frac{\theta_1(4a)}{\theta_4(2a)}.
$$

Acknowledgments. The author thanks the organizers of and participants in the "Physical Combinatorics" conference, where he enjoyed many discussions. He also thanks V. V. Bazhanov, M. Jimbo, J. Shiraishi, and O. Warnaar for discussion and comments.

References

[1] C. Acerbi and G. Mussardo, Form-factors and correlation functions of the stress-energy tensors in massive deformation of the minimal models $(E_n)_1 \otimes (E_n)_1/(E_n)_2$, *Internat. J. Modern Phys. A*, **11** (1996), 5327–5364.

[2] M. T. Batchelor, V. Fridkin, and Y. K. Zhou, An Ising model in a magnetic field with a boundary, *J. Phys. A*, **29** (1996), L61–L67.

[3] M. T. Batchelor and K. A. Seaton, Magnetic correlation length and universal amplitude of the lattice E_8 Ising model, *J. Phys. A*, **30** (1997), L479–L484.

[4] M. T. Batchelor and K. A. Seaton, Excitations in the dilute A_L lattice model: E_6, E_7 and E_8 mass spectra, *European Phys. J. B*, **5** (1998), 719–725.

[5] M. T. Batchelor and K. A. Seaton, Correlation lengths and E_8 mass spectrum of the dilute A_3 lattice model, *Nuclear Phys. B*, **520** (1998), 697–744.

[6] R. J. Baxter and P. A. Pearce, Hard hexagons: Interfacial tension and correlation length, *J. Phys. A*, **15** (1982), 897–910.

[7] V. V. Bazhanov and N. Yu. Reshetikhin, Restricted solid on solid models connected with simply laced Lie algebra, *J. Phys. A*, **23** (1990), 1477–1492.

[8] V. V. Bazhanov and N. Yu. Reshetikhin, Scattering amplitudes in off-critical models and RSOS integrable models, *Prog. Theoret. Phys. Suppl.*, **102** (1990), 301–318.

[9] V. V. Bazhanov, O. Warnaar, and B. Nienhuis, Lattice Ising model in a field: E_8 scattering theory, *Phys. Lett. B*, **322** (1994), 198–206.

[10] P. Christe and G. Mussardo, Integrable systems away from criticality: The Toda fields theory and S matrix of the tricritical Ising model, *Nuclear Phys. B*, **330** (1990), 465.

[11] G. Delfino and G. Mussardo, Spin-spin correlation function in the two dimensional Ising model in a magnetic field at $T = T_c$, *Nuclear Phys. B*, **455** (1995), 724–758.

[12] C. Destri and H. J. de Vega, New thermodynamic Bethe ansatz equations without strings, *Phys. Rev. Lett.*, **69** (1992), 2313–2317.

[13] U. Grimm and B. Nienhuis, Scaling properties of the Ising model in a field, in M. L. Ge and F. Y. Wu, eds., *Symmetry, Statistical Mechanical Models and Applications: Proceedings of the Seventh Nankai Workshop* (*Tianjin* 1995), World Scientific, Singapore, 1996, 384–393.

[14] U. Grimm and B. Nienhuis, Scaling limit of the Ising model in a field, *Phys. Rev. E*, **55** (1997), 5011–5025.

[15] U. Grimm, P. A. Pearce, and Y. K. Zhou, Fusion of dilute A_L lattice models, *Physica A*, **222** (1995), 261–306.

[16] Y. Hara, M. Jimbo, H. Konno, S. Odake, and J. Shiraishi, Free field approach to the dilute A_L models, preprint math/9902150.

[17] M. Henkel and H. Saleur, The two dimensional Ising model in a magnetic field: Numerical check of Zamolodchikov's conjecture, *J. Phys. A*, **22** (1989), L513–L518.

[18] A. G. Izergin and V. E. Korepin, The inverse scattering method approach to the quantum Shabat-Mikhailov model, *Comm. Math. Phys.*, **79** (1981), 303–316.

[19] G. Jüttner and A. Klümper, Exact calculation of thermodynamical properties of the integrable $t - J$ model, *European Phys. Lett.*, **37** (1997), 335–340.

[20] G. Jüttner, A. Klümper, and J. Suzuki, Thermodynamics of correlated electrons with bond charge and Hubbard interaction in one dimension, *J. Phys. A*, **30** (1997), 1181–1886.

[21] G. Jüttner, A. Klümper, and J. Suzuki, Exact thermodynamics and Luttinger liquid properties of the integrable $t - J$ model, *Nuclear Phys. B*, **486** (1997), 650–674.

[22] G. Jüttner, A. Klümper, and J. Suzuki, From fusion hierarchy to excited TBA, *Nuclear Phys. B*, **512** (1998), 581–600.

[23] G. Jüttner, A. Klümper, and J. Suzuki, The Hubbard chain at finite temperatures: Ab-initio calculations of Tomonaga-Luttinger liquid properties, *Nuclear Phys. B*, **522** (1998), 471–502.

[24] A. Klümper, Free energy and correlation lengths of quantum chains related to restricted solid-on-solid lattice models, *Ann. Phys.*, **1** (1992), 540.

[25] A. Klümper, Thermodynamics of the anisotropic spin 1/2 Heisenberg chain and related quantum spins, *Z. Phys. B*, **91** (1993), 507–519.

[26] A. Klümper and R. Z. Bariev, Exact thermodynamics of the Hubbard chain: Frere energy and correlation lengths, *Nuclear Phys. B*, **458** (1996), 623–639.

[27] A. Klümper and P. A. Pearce, Conformal weights of RSOS lattice models and their fusion hierarchies, *Physica A*, **183** (1992), 304–350.

[28] T. Koma, Thermal Bethe-Ansatz method for the one dimensional Heisenberg model, *Prog. Theoret. Phys.*, **78** (1987), 1213–1218.

[29] A. Kuniba, Exact solution of solid-on-solid models for twisted affine Lie algebras $A^{(2)}_{2n}$ and $A^{(2)}_{2n-1}$, *Nuclear Phys. B*, **355** (1991), 801–821.

[30] A. Kuniba, T. Nakanishi, and J. Suzuki, Functional relations in solvable lattice models I and II, *Internat. J. Modern Phys. A*, **9** (1994), 5215–5266 and 5267-5312.

[31] A. Kuniba, Y. Ohta, and J. Suzuki, Quantum Jacobi-Trudi and Giambelli formulae for $U_q(B^{(1)}_r)$ from the analytic Bethe ansatz, *J. Phys. A*, **28** (1995), 6211–6226.

[32] A. Kuniba, K. Sakai, and J. Suzuki, Continued fraction TBA and functional relations in XXZ model at root of unity, *Nuclear Phys. B*, **525** (1998), 597–626.

[33] A. Kuniba and J. Suzuki, Analytic Bethe ansatz for fundamental representations of Yangians, *Comm. Math. Phys.*, **173** (1995), 225–264.

[34] M. Lassig, G. Mussardo, and J. L. Cardy, The scaling regions of the tricritical Ising model in two dimensions, *Nuclear Phys. B*, **348** (1990), 591–618.

[35] P. G. Lauwers and V. Rittenberg, The critical 2D Ising model in a magnetic field: A Monte Carlo study using Swendsen-Wang algorithm, *Phys. Lett. B*, **233** (1989), 197–200.

[36] B. McCoy and P. Orrick, Single particle excitations in the lattice E_8 Ising model, *Phys. Lett. A*, **230** (1997), 24–32.

[37] T. Nakanishi, Fusion, mass and representation theory of the Yangian algebra, *Nuclear Phys. B*, **439** (1995), 441–459.

[38] N. Yu. Reshetikhin, Integrable models of quantum one dimensional magnets with $O(n)$ and $Sp(2k)$ symmetry, *Theoret. Math. Phys.*, **63** (1985), 555–569.

[39] N. Yu. Reshetikhin, The spectrum of the transfer matrices connected with Kac-Moody algebras, *Lett. Math. Phys.*, **14** (1987), 235–246.

[40] I. R. Sagdeev and A. B. Zamolodchikov, *Modern Phys. Lett. B*, **3** (1989), 1375.

[41] K. Sakai, M. Shiroishi, J. Suzuki, and Y. Umeno, Commuting quantum transfer-matrix approach to intrinsic fermion system: Correlation length of a spinless fermion model, *Phys. Rev. B*, **60** (1999), 5186–5201.

[42] K. A. Seaton and M. T. Batchelor, E_8, E_7, and E_6 symmetries in the dilute A_L lattice model, in S. P. Corney, R. Delbourgo, and P. D. Jarvis, eds., *Group22: Proceedings of the* XXII *International Colloquium on Group Theoretical Methods in Physics*, International Press, Cambridge, MA, 1998.

[43] J. Suzuki, Fusion $U_q(G_2^{(1)})$ vertex models and analytic Bethe ansätze, *Phys. Lett. A*, **195** (1994), 190–197.

[44] J. Suzuki, Quantum Jacobi-Trudi formula and E_8 structure in the Ising model in a field, *Nuclear Phys. B*, **528** (1998), 683–700.

[45] J. Suzuki, Spinons in magnetic chains of arbitrary spins at finite temperatures, *J. Phys. A*, **32** (1999), 2341–2359.

[46] J. Suzuki, Y. Akutsu, and M. Wadati, A new approach to quantum spin chains at finite temperature, *J. Phys. Soc. Japan*, **59** (1990), 2667–2680.

[47] J. Suzuki, T. Nagao, and M. Wadati, Exactly solvable models and finite size corrections, *Internat. J. Modern Phys. B*, **6** (1992), 1119–1180.

[48] M. Suzuki, Transfer matrix method and Monte Carlo simulation in quantum spin system, *Phys. Rev. B*, **31** (1985), 2957–2965.

[49] M. Suzuki and M. Inoue, The ST-transformation approach to analytic solutions of quantum system I, *Prog. Theoret. Phys.*, **78** (1987), 787-799.

[50] M. Takahashi, Correlation length and free energy of the $S = 1/2$ XYZ chain, *Phys. Rev. B*, **43** (1991), 5788–5797.

[51] Z. Tsuboi, *Physica A*, **252** (1998), 565.

[52] G. von Gehlen, Off critical behavior of the Blume-Capel quantum chain as a check of Zamolodchikov's conjecture, *Nuclear Phys. B*, **330** (1990), 741–756.

[53] S. O. Warnaar, B. Nienhuis and K. A. Seaton, New construction of solvable lattice models including an Ising model in a field, *Phys. Rev. Lett.*, **69** (1992), 710–713.

[54] S. O. Warnaar, B. Nienhuis, and K. A. Seaton, A critical Ising model in a magnetic field, *Internat. J. Modern Phys. B*, **7** (1993), 3727–3736.
[55] S. O. Warnaar, P. A. Pearce, B. Nienhuis, and K. A. Seaton, Order parameters of the dilute A models, *J. Statist. Phys.*, **74** (1994), 469–532.
[56] V. P. Yurov and Al. B. Zamolodchikov, *Internat. J. Modern Phys. A*, **6** (1991), 4556.
[57] A. B. Zamolodchikov, Integrable field theory from conformal field theory, *Adv. Stud. Pure Math.*, **19** (1989), 641–674.
[58] A. B. Zamolodchikov, Integrals of motion and S-matrix of the scaled $T = T_c$ Ising model with magnetic field, *Internat. J. Modern Phys. A*, **4** (1989), 4235–4248.
[59] Al. B. Zamolodchikov, On the thermodynamic Bethe ansatz equations for reflectionless ADE scattering theories, *Phys. Lett. B*, **253** (1991), 391–394.

Department of Physics
Faculty of Science
Shizuoka University
Ohya 836, Shizuoka
Japan
sjsuzuk@ipc.shizuoka.ac.jp

Canonical Bases of Higher-Level q-Deformed Fock Spaces and Kazhdan–Lusztig Polynomials

Denis Uglov

Abstract. The aim of this paper is to generalize some aspects of the recent work of Leclerc–Thibon and Varagnolo–Vasserot on the canonical bases of the level 1 q-deformed Fock spaces of Hayashi. Namely, we define canonical bases for the higher-level q-deformed Fock spaces of Jimbo–Misra–Miwa–Okado and establish a relation between these bases and (parabolic) Kazhdan–Lusztig polynomials for the affine Weyl group of type $A^{(1)}_{r-1}$. As an application, we derive an inversion formula for a subfamily of these polynomials.

1 Introduction

For any symmetrizable Kac–Moody Lie algebra $\mathfrak{g}$, Kashiwara introduces in [11] the notion of a lower global crystal basis of an integrable module M of the universal quantum enveloping algebra $U_q(\mathfrak{g})$. Furthermore, he proves the existence and uniqueness of such a basis when M is irreducible.

The lower global crystal bases of irreducible modules were recently recognized to be closely related with the representation theory of Hecke algebras. It was observed in [16] that the vacuum irreducible module of $\widehat{\mathfrak{sl}}_n$ at level 1 can be identified with the direct sum $\oplus_m K_m$, where K_m is the complexified Grothendieck group of finitely generated projective modules of the Hecke algebra $H_m(v)$ at v, a complex nth root of unity. Furthermore, it was conjectured that, under this identification, the specialization of the lower global crystal basis at $q = 1$ corresponds to the basis of K formed by the indecomposable summands of the Hecke algebra. A proof of this conjecture and of a more general result providing a similar interpretation for the lower global crystal basis for any irreducible integrable module of $\widehat{\mathfrak{sl}}_n$ was given in [2].

If an integrable module of a quantum enveloping algebra is not irreducible, it can have more than one lower global crystal basis. An example is the q-deformed Fock space of Hayashi [9, 20], which is a reducible integrable module of $U_q(\widehat{\mathfrak{sl}}_n)$ at level 1 spanned by the set of all partitions. One may then ask whether among the many lower global crystal bases of the q-deformed Fock space, one can single out a canonical one with particularly favorable properties.

Leclerc and Thibon gave a definition of such a canonical basis in [17]. The essential point of their definition was to introduce a bar-involution of the q-deformed Fock space by using the semiinfinite wedge construction of the latter given by Stern [S, 13]. With the involution in hand, the canonical basis is defined as the unique bar-invariant basis with a certain congruence property with respect to the $\mathbf{Q}[q]$-lattice spanned by partitions.

It was conjectured in [17] that the specialization at $q = 1$ of the transition matrix between the canonical basis of the q-deformed Fock space and the basis formed by partitions coincides with the decomposition matrix of the Weyl modules of the v-Schur algebra at v a complex nth root of unity. This conjecture was proved in [25], where it was shown that the entries of the transition matrix are given by certain parabolic Kazhdan–Lusztig polynomials for affine Weyl groups of type $A^{(1)}_{r-1}$. An excellent review of these developments can be found in [18].

The higher-level q-deformed Fock spaces, generalizing those of Hayashi, were introduced in [10] to compute the crystal graph of an arbitrary irreducible integrable module of $U_q(\widehat{\mathfrak{sl}}_n)$. For each sequence of l integers called the charge, there is a q-deformed Fock space which is an integrable module of $U_q(\widehat{\mathfrak{sl}}_n)$ at level l having as a basis the set of all l-tuples of partitions.

The main aim of this article is to give a construction of canonical bases for these q-deformed Fock spaces, generalizing the construction of Leclerc and Thibon for level 1.

As in that case, the first step of this construction is to give a semiinfinite wedge realization of each q-deformed Fock space. This was already done in [23], and we follow that paper except in some minor details. The wedge realization allows us to define a natural bar-involution on a q-deformed Fock space. Then the definition of the canonical basis proceeds exactly as in the level 1 case. In particular, as in [25], we find that the entries of the transition matrix between the canonical basis and the basis formed by the l-tuples of partitions are parabolic Kazhdan–Lusztig polynomials, the type of parabolic subgroup being determined by the charge of the q-deformed Fock space.

For a q-deformed Fock space of charge $(s_1, \ldots, s_l)$, the $U_q(\widehat{\mathfrak{sl}}_n)$-submodule M generated by the l-tuple of empty partitions is isomorphic to the irreducible module with the highest-weight $\sum_{b=1}^{l} \Lambda_{s_b \bmod n}$. The canonical basis of the Fock space is, as in the case of level 1, a lower global crystal basis and contains the lower global crystal basis of M as a subset.

The definition of the canonical bases given in this article is constructive. It provides one with a simple algorithm for computation of the transition matrices between these bases and the bases formed by l-tuples of partitions. Due to the main result of [2], the submatrix of this transition matrix that corresponds to the expansion of the lower global crystal basis of M is known to be a q-analogue of the decomposition matrix of a certain Ariki–Koike algebra at the nth root of unity. We therefore get an algorithm for computing this decomposition matrix.

Now let us give an outline of the present article. In Section 2, we give the definitions of the q-deformed Fock spaces, of their crystal bases, and of the lower global crystal bases of their irreducible submodules. This part is entirely expository and mostly follows the work of [5]. In Section 3, after giving the necessary background on affine Weyl groups and affine Hecke algebras, we introduce wedge products and their canonical bases. We also establish the connection between these canonical bases and parabolic Kazhdan–Lusztig polynomials. Most of the results of this part are straightforward generalizations of results of [13] and of a small subset of results in [25]. In the first part of Section 4, we give the realizations of the q-deformed Fock spaces as subspaces of the semiinfinite wedge products. Here we follow [23], with some deviations. In the second part of this section, we introduce the bar-involution and define canonical bases. The content of this part is very close to that of [17]. In Section 5, we describe a symmetry of the bar-involution and derive by using this symmetry an inversion formula for certain parabolic Kazhdan–Lusztig polynomials. When the level l equals 1, this formula has already been established in [18].

Since most of the results in the present article are to be found, in the special case of level 1, in [18], we tried to organize the exposition so that it parallels the relevant parts of that work.

The preliminary version of this article appeared as the preprint [24].

2 The q-deformed Fock spaces

2.1 Definitions. Let $\mathfrak{h} = (\oplus_{i=0}^{n-1}\mathbf{Q}h_i) \oplus \mathbf{Q}\partial$ be the Cartan subalgebra of $\widehat{\mathfrak{sl}}_n$ and let $\mathfrak{h}^* = (\oplus_{i=0}^{n-1}\mathbf{Q}\Lambda_i) \oplus \mathbf{Q}\delta$ be its dual. Here $\Lambda_0, \ldots, \Lambda_{n-1}$ and δ are, respectively, the fundamental weights and the null root defined in terms of the pairing between $\mathfrak{h}^*$ and $\mathfrak{h}$ by

$$\langle \Lambda_i, h_j \rangle = \delta_{ij}, \qquad \langle \Lambda_i, \partial \rangle = \langle \delta, h_i \rangle = 0, \qquad \langle \delta, \partial \rangle = 1.$$

The space $\mathfrak{h}^*$ is equipped with a bilinear symmetric form defined by

$$(\Lambda_i|\Lambda_j) = \min(i, j) - \frac{ij}{n}, \qquad (\Lambda_i|\delta) = 1, \qquad (\delta|\delta) = 0.$$

For $\Lambda \in \mathfrak{h}^*$, we shall write $|\Lambda|^2$ to mean $(\Lambda|\Lambda)$. It will be convenient to extend the index set of the fundamental weights to all integers by setting $\Lambda_i = \Lambda_{i \bmod n}$ for $i \in \mathbf{Z}$. Then the simple roots are defined for all integers i as $\alpha_i = 2\Lambda_i - \Lambda_{i+1} - \Lambda_{i-1} + \delta_{i\equiv 0 \bmod n}\delta$, where for a statement S we set $\delta_S = 1$ if S is true and $\delta_S = 0$ otherwise.

Let $U_q(\widehat{\mathfrak{sl}}_n)$ be the q-deformed universal enveloping algebra of $\widehat{\mathfrak{sl}}_n$. This is an algebra over $\mathbf{K} = \mathbf{Q}(q)$ with generators e_i, f_i, t_i, and $(t_i)^{-1}$ $(i = 0, \ldots, n-1)$ and ∂. The relations between e_i, f_i, t_i, and $(t_i)^{-1}$ are standard (see, e.g., [14]). We define the relations between the degree generator ∂ and the rest of the generators by

$$[\partial, e_i] = \delta_{i,0} e_i, \qquad [\partial, f_i] = -\delta_{i,0} f_i, \qquad [\partial, t_i] = 0.$$

For $l \in \mathbf{Z}$, a module M of $U_q(\widehat{\mathfrak{sl}}_n)$ is said to have *level* l if the canonical central element $t_0 t_1 \cdots t_{n-1}$ of $U_q(\widehat{\mathfrak{sl}}_n)$ acts on M as the multiplication by q^l. Let $U'_q(\widehat{\mathfrak{sl}}_n)$ be the subalgebra of $U_q(\widehat{\mathfrak{sl}}_n)$ generated by e_i, f_i, t_i, and $(t_i)^{-1}$.

For a nonnegative integer k, let Π_k be the set of partitions of k, i.e., the set of all nondecreasing sequences of nonnegative integers $\lambda = (\lambda_1, \lambda_2, \ldots)$ summing to k. Let $\Pi = \sqcup_{k \geq 0} \Pi_k$ be the set of all partitions. For $l \in \mathbf{N}$, an element $\boldsymbol{\lambda}_l = (\lambda^{(1)}, \ldots, \lambda^{(l)})$ of Π^l is called an *l-multipartition*. It will be convenient to identify a multipartition $\boldsymbol{\lambda}_l$ with its diagram defined as the set

$$\{(i, j, b) \in \mathbf{N}^3 \mid 1 \leq b \leq l,\ 1 \leq j \leq \lambda_i^{(b)}\}.$$

An element of the diagram of a multipartition $\boldsymbol{\lambda}_l$ is called a *node* of $\boldsymbol{\lambda}_l$, and the total number of nodes of $\boldsymbol{\lambda}_l$ is denoted by $|\boldsymbol{\lambda}_l|$.

Let $\boldsymbol{s}_l = (s_1, \ldots, s_l)$ be a sequence of l integers. With any such sequence one associates the *q-deformed Fock space* $\mathbf{F}_q[\boldsymbol{s}_l]$ defined as

$$\mathbf{F}_q[\boldsymbol{s}_l] = \bigoplus_{\boldsymbol{\lambda}_l \in \Pi^l} \mathbf{K}\, |\boldsymbol{\lambda}_l, \boldsymbol{s}_l\rangle.$$

In other words, $\mathbf{F}_q[\boldsymbol{s}_l]$ is a $\mathbf{K}$-linear space with a distinguished basis $|\boldsymbol{\lambda}_l, \boldsymbol{s}_l\rangle$ labeled by the set of all l-multipartitions. The number l is called *the level* of $\mathbf{F}_q[\boldsymbol{s}_l]$, and the sequence $\boldsymbol{s}_l$ is called the *charge* of $\mathbf{F}_q[\boldsymbol{s}_l]$.

It was shown in [10] that $\mathbf{F}_q[\boldsymbol{s}_l]$ can be endowed with a structure of an integrable $U_q(\widehat{\mathfrak{sl}}_n)$-module. We shall describe this structure following the exposition given in [5]. To do this we introduce some notation. For a node $\gamma = (i, j, b)$ of a multipartition $\boldsymbol{\lambda}_l$, one defines its *n-residue* as $\mathrm{res}_n(\gamma) = (s_b + j - i) \bmod n$. For i between 0 and $n-1$, we say that $\gamma \in \boldsymbol{\lambda}_l$ is an *i-node* of $\boldsymbol{\lambda}_l$ if $\mathrm{res}_n(\gamma) = i$. Given two nodes $\gamma = (i, j, b)$ and $\gamma' = (i', j', b')$ of a multipartition $\boldsymbol{\lambda}_l$, we write $\gamma < \gamma'$ if either $(s_b + j - i) < (s_{b'} + j' - i')$ or $(s_b + j - i) = (s_{b'} + j' - i')$ and $b < b'$. If $\boldsymbol{\mu}_l$ and $\boldsymbol{\lambda}_l$ are two multipartitions such that $\boldsymbol{\mu}_l \supset \boldsymbol{\lambda}_l$, and $\gamma = \boldsymbol{\mu}_l \setminus \boldsymbol{\lambda}_l$ is an i-node of $\boldsymbol{\mu}_l$, we say that γ is a removable i-node of $\boldsymbol{\mu}_l$ and is an addable i-node of $\boldsymbol{\lambda}_l$. In this case, we define

$$\begin{aligned} N_i^{>}(\boldsymbol{\lambda}_l, \boldsymbol{\mu}_l | \boldsymbol{s}_l, n) &= \sharp\{\text{addable } i\text{-nodes } \gamma' \text{ of } \boldsymbol{\lambda}_l \text{ such that } \gamma' > \gamma\} \\ &\quad - \sharp\{\text{removable } i\text{-nodes } \gamma' \text{ of } \boldsymbol{\lambda}_l \text{ such that } \gamma' > \gamma\}, \\ N_i^{<}(\boldsymbol{\lambda}_l, \boldsymbol{\mu}_l | \boldsymbol{s}_l, n) &= \sharp\{\text{addable } i\text{-nodes } \gamma' \text{ of } \boldsymbol{\lambda}_l \text{ such that } \gamma' < \gamma\} \\ &\quad - \sharp\{\text{removable } i\text{-nodes } \gamma' \text{ of } \boldsymbol{\lambda}_l \text{ such that } \gamma' < \gamma\}. \end{aligned}$$

Also, for a multipartition $\boldsymbol{\lambda}_l$ and i between 0 and $n-1$, we define

$$N_i(\boldsymbol{\lambda}_l|\boldsymbol{s}_l, n) = \sharp\{\text{addable } i\text{-nodes of } \boldsymbol{\lambda}_l\} - \sharp\{\text{removable } i\text{-nodes of } \boldsymbol{\lambda}_l\},$$
$$M_i(\boldsymbol{\lambda}_l|\boldsymbol{s}_l, n) = \sharp\{i\text{-nodes of } \boldsymbol{\lambda}_l\},$$

and for $\boldsymbol{s}_l = (s_1, \ldots, s_l) \in \mathbf{Z}^l$, we set

$$\Delta(\boldsymbol{s}_l|n) = \frac{1}{2}\sum_{b=1}^{l} |\Lambda_{s_b}|^2 + \frac{1}{2}\sum_{b=1}^{l}\left(\frac{s_b^2}{n} - s_b\right).$$

Now we can state the following.

Theorem 2.1 ([10, 5]). *The following formulas define on $\mathbf{F}_q[\boldsymbol{s}_l]$ a structure of an integrable $U_q(\widehat{\mathfrak{sl}}_n)$-module.*

$$f_i|\boldsymbol{\lambda}_l, \boldsymbol{s}_l\rangle = \sum_{\mathrm{res}_n(\boldsymbol{\mu}_l/\boldsymbol{\lambda}_l)=i} q^{N_i^{>}(\boldsymbol{\lambda}_l, \boldsymbol{\mu}_l|\boldsymbol{s}_l, n)} |\boldsymbol{\mu}_l, \boldsymbol{s}_l\rangle,$$
$$e_i|\boldsymbol{\mu}_l, \boldsymbol{s}_l\rangle = \sum_{\mathrm{res}_n(\boldsymbol{\mu}_l/\boldsymbol{\lambda}_l)=i} q^{-N_i^{<}(\boldsymbol{\lambda}_l, \boldsymbol{\mu}_l|\boldsymbol{s}_l, n)} |\boldsymbol{\lambda}_l, \boldsymbol{s}_l\rangle,$$
$$t_i|\boldsymbol{\lambda}_l, \boldsymbol{s}_l\rangle = q^{N_i(\boldsymbol{\lambda}_l|\boldsymbol{s}_l, n)}|\boldsymbol{\lambda}_l, \boldsymbol{s}_l\rangle,$$
$$\partial|\boldsymbol{\lambda}_l, \boldsymbol{s}_l\rangle = -(\Delta(\boldsymbol{s}_l|n) + M_0(\boldsymbol{\lambda}_l|\boldsymbol{s}_l, n))|\boldsymbol{\lambda}_l, \boldsymbol{s}_l\rangle.$$

Remark 2.2. Our labeling of the basis vectors differs from that of [5] by the transformation reversing the order of components in $\boldsymbol{\lambda}_l = (\lambda^{(1)}, \ldots, \lambda^{(l)})$ and $\boldsymbol{s}_l = (s_1, \ldots, s_l)$. Also, Theorem 2.1 as well as Theorem 2.4 below are stated in [5] only for $\boldsymbol{s}_l$ such that $n > s_1 \geq s_2 \geq \cdots \geq s_l \geq 0$. Generalizations for all $\boldsymbol{s}_l \in \mathbf{Z}^l$ are straightforward.

Note that the vector $|\varnothing_l, \boldsymbol{s}_l\rangle$, where $\varnothing_l$ denotes the l-tuple of empty partitions, is a highest-weight vector of $\mathbf{F}_q[\boldsymbol{s}_l]$. Since $\mathbf{F}_q[\boldsymbol{s}_l]$ is an integrable module, it follows that

$$\mathbf{F}_q[\boldsymbol{s}_l] = U_q(\widehat{\mathfrak{sl}}_n)\,|\varnothing_l, \boldsymbol{s}_l\rangle$$

is an irreducible submodule of $\mathbf{F}_q[\boldsymbol{s}_l]$. Computing the weight of $|\varnothing_l, \boldsymbol{s}_l\rangle$ in accordance with Theorem 2.1, we see that $\mathbf{F}_q[\boldsymbol{s}_l]$ is isomorphic to the irreducible $U_q(\widehat{\mathfrak{sl}}_n)$-module $V_q(\Lambda)$ with highest-weight $\Lambda = -\Delta(\boldsymbol{s}_l|n)\delta + \Lambda_{s_1} + \cdots + \Lambda_{s_l}$.

2.2 Crystal bases. The q-deformed Fock spaces were introduced in [10] in order to compute the crystal graphs of irreducible integrable modules of $U_q(\widehat{\mathfrak{sl}}_n)$. We have seen that any such module is embedded into a q-deformed Fock space as the component generated by the highest-weight vector labeled by the empty multipartition. From crystal base theory, it follows that the crystal graph of an irreducible module is embedded into the crystal graph of the corresponding Fock space. The last crystal graph was described in [10]. To recall how the arrows of this graph are obtained, we introduce, following [5], the notion of a *good node* of a multipartition $\boldsymbol{\lambda}_l$.

First, observe that for each i between 0 and $n-1$ the relation $\gamma < \gamma'$ defines a total order on the set of all i-addable and i-removable nodes.

Example 2.3. Let $n = 3$, $l = 4$ and $s_l = (5, 0, 2, 1)$. Then, marking the 0-addable and the 0-removable nodes of the multipartition

$$\lambda_l = ((5, 3^2, 1), (3, 2), (4, 3, 1), (2^3, 1))$$

on the diagram of λ_l by $\bullet$, we get

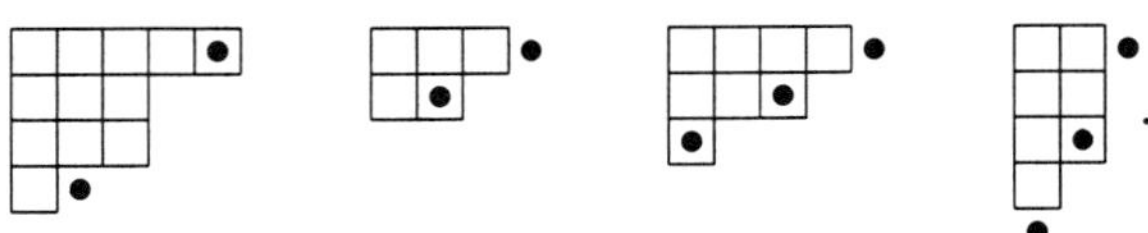

Thus these nodes are ordered as

$$A_{-3,4} < R_{0,2} < R_{0,3} < R_{0,4} < A_{3,1} < A_{3,2} < R_{3,3} < A_{3,4} < A_{6,3} < R_{9,1},$$

where $A_{d,b}$ ($R_{d,b}$) denotes an addable (removable) node (i, j, b) with $s_b + j - i = d$.

Next, for a multipartition λ_l, write the sequence of its addable and removable i-nodes ordered as explained above. Then remove from this sequence recursively all pairs RA until no such pairs remain. The resulting sequence then has the form $A \dots AR \dots R$. The rightmost R-node in this sequence is called the *good removable i-node* of λ_l, and the leftmost A-node in this sequence is called the *good addable i-node* of λ_l. Clearly, there can be at most one of each. For instance, for the multipartition considered in Example 2.3, the nodes $A_{-3,4}$ and $R_{9,1}$ are good 0-nodes.

Let $A \subset \mathbf{Q}(q)$ be the ring of rational functions without pole at $q = 0$. Let $\mathcal{L}[s_l] = \oplus_{\lambda_l \in \Pi^l} A|\lambda_l, s_l\rangle$ and let $\mathcal{B}[s_l]$ be the $\mathbf{Q}$-basis of $\mathcal{L}[s_l]/q\mathcal{L}[s_l]$ given by $\mathcal{B}[s_l] = \{|\lambda_l, s_l\rangle \bmod q\mathcal{L}[s_l] \mid \lambda_l \in \Pi^l\}$.

Theorem 2.4 ([10, 5]). *The pair $(\mathcal{L}[s_l], \mathcal{B}[s_l])$ is a lower crystal basis of $\mathbf{F}_q[s_l]$ at $q = 0$. Moreover, the crystal graph $\mathcal{B}[s_l]$ contains the arrow*

$$|\lambda_l, s_l\rangle \bmod q\mathcal{L}[s_l] \xrightarrow{i} |\mu_l, s_l\rangle \bmod q\mathcal{L}[s_l]$$

if and only if μ_l is obtained from λ_l by adding a good i-node.

Let $\Pi^l(s_l)$ be the subset of Π^l such that $\mathcal{B}[s_l]^\circ = \{|\lambda_l, s_l\rangle \bmod q\mathcal{L}[s_l] \mid \lambda_l \in \Pi^l(s_l)\}$ is the set of vertices in the connected component of $|\varnothing_l, s_l\rangle \bmod q\mathcal{L}[s_l]$ in the crystal graph $\mathcal{B}[s_l]$ of $\mathbf{F}_q[s_l]$. Then [11, Theorem 3] implies that $\mathcal{B}[s_l]^\circ$ is isomorphic to the crystal graph of the irreducible submodule $\mathbf{F}_q[s_l]$ of $\mathbf{F}_q[s_l]$.

Let us now briefly review the notion of the global crystal base of an irreducible module $\mathbf{F}_q[s_l]$. First, recall the involution $x \mapsto \overline{x}$ of $U'_q(\widehat{\mathfrak{sl}}_n)$ defined as the unique algebra automorphism satisfying

$$\overline{q} = q^{-1}, \qquad \overline{t_i} = (t_i)^{-1}, \qquad \overline{e_i} = e_i, \qquad \overline{f_i} = f_i.$$

Now each vector v of $\mathbf{F}_q[s_l]$ can be written as $v = x|\varnothing_l, s_l\rangle$ for some $x \in U'_q(\widehat{\mathfrak{sl}}_n)$. Then we set $\overline{v} = \overline{x}|\varnothing_l, s_l\rangle$. Finally, denote by $U^-_{\mathbf{Q}}$ the $\mathbf{Q}[q, q^{-1}]$-subalgebra of $U'_q(\widehat{\mathfrak{sl}}_n)$ generated by the q-divided differences $f_i^k/[k]!$ and let $\mathbf{F}_q[s_l]_{\mathbf{Q}} = U^-_{\mathbf{Q}}|\varnothing_l, s_l\rangle$. (Here, $[k]!$ denotes the q-factorial, that is, $[k] = (q^k - q^{-k})/(q - q^{-1})$ and $[k]! = [k][k-1]\cdots[1]$.)

Theorem 2.5 ([11]). *There exists a unique* $\mathbf{Q}[q, q^{-1}]$*-basis* $\{\mathcal{G}(\boldsymbol{\lambda}_l, s_l) \mid \boldsymbol{\lambda}_l \in \Pi^l(s_l)\}$ *of* $\mathbf{F}_q[s_l]_{\mathbf{Q}}$ *such that*

$$\text{(i)} \qquad \overline{\mathcal{G}(\boldsymbol{\lambda}_l, s_l)} = \mathcal{G}(\boldsymbol{\lambda}_l, s_l),$$

$$\text{(ii)} \qquad \mathcal{G}(\boldsymbol{\lambda}_l, s_l) \equiv |\boldsymbol{\lambda}_l, s_l\rangle \mod q\mathcal{L}[s_l].$$

The basis $\{\mathcal{G}(\boldsymbol{\lambda}_l, s_l) \mid \boldsymbol{\lambda}_l \in \Pi^l(s_l)\}$ is called the *lower global crystal basis* of $\mathbf{F}_q[s_l]$.

3 Canonical bases of wedge products

3.1 Affine Weyl group. Let $\mathfrak{t}^* = \oplus_{i=1}^r \mathbf{C}\varepsilon_i$ be the dual space of the Cartan subalgebra of $\mathfrak{gl}_r$. Let $\hat{\mathfrak{t}}^* = \mathfrak{t}^* \oplus \mathbf{C}\Lambda_0 \oplus \mathbf{C}\delta$ be the dual space of the Cartan subalgebra of $\widehat{\mathfrak{gl}}_r$. The space $\hat{\mathfrak{t}}^*$ is equipped with the bilinear symmetric form defined by $(\varepsilon_i, \varepsilon_j) = \delta_{ij}$, $(\varepsilon_i, \Lambda_0) = (\varepsilon_i, \delta) = (\delta, \delta) = (\Lambda_0, \Lambda_0) = 0$, $(\Lambda_0, \delta) = 1$. The systems of roots R, positive roots R^+, and simple roots Π of type A_{r-1} are the subsets of $\mathfrak{t}^*$ defined by

$$\begin{aligned} R &= \{\alpha_{ij} = \varepsilon_i - \varepsilon_j \mid i \neq j\}, \\ R^+ &= \{\alpha_{ij} \mid i < j\}, \\ \Pi &= \{\alpha_1, \ldots, \alpha_{r-1}\} \quad (\alpha_i := \alpha_{ii+1}). \end{aligned}$$

The systems of roots $\widehat{R}$, positive roots $\widehat{R}^+$, and simple roots $\widehat{\Pi}$ of type $A_{r-1}^{(1)}$ are the subsets of $\hat{\mathfrak{t}}^*$ defined by

$$\begin{aligned} \widehat{R} &= \{\alpha + k\delta \mid \alpha \in R,\ k \in \mathbf{Z}\}, \\ \widehat{R}^+ &= \{\alpha + k\delta \mid \alpha \in R^+,\ k \geq 0\} \sqcup \{-\alpha + k\delta \mid \alpha \in R^+,\ k > 0\}, \\ \widehat{\Pi} &= \{\alpha_0 := \delta - (\varepsilon_1 - \varepsilon_r)\} \sqcup \Pi. \end{aligned}$$

The Weyl group W of $\mathfrak{gl}_r$ is isomorphic to the symmetric group $\mathfrak{S}_r$ and has a realization as the group generated by the reflections $s_\alpha(\xi) = \xi - (\alpha, \xi)\alpha$ ($\alpha \in R$) of $\mathfrak{t}^*$. Let $Q = \oplus_{i=1}^{r-1} \mathbf{Z}\alpha_i$ and $P = \oplus_{i=1}^r \mathbf{Z}\varepsilon_i$ be, respectively, the root and the weight lattices of $\mathfrak{gl}_r$. They both are preserved by W.

The affine Weyl group is defined as the semidirect product

$$\widehat{W} = W \ltimes P$$

with relations $wt_\eta = t_{w(\eta)}w$, where w and t_η are elements of $\widehat{W}$ that correspond to $w \in W$, $\eta \in P$. The group $\widehat{W}$ contains the Weyl group $\widetilde{W} = W \ltimes Q$ of type $A_{r-1}^{(1)}$ as a subgroup. The group $\widehat{W}$ acts on $\hat{\mathfrak{t}}^*$ by

$$\begin{aligned} s_\alpha(\zeta) &= \zeta - (\alpha, \zeta)\alpha & &(\zeta \in \hat{\mathfrak{t}}^*,\ \alpha \in R), \\ t_\eta(\zeta) &= \zeta + (\delta, \zeta)\eta - \left((\eta, \zeta) + \frac{1}{2}(\eta, \eta)(\delta, \zeta)\right)\delta & &(\zeta \in \hat{\mathfrak{t}}^*,\ \eta \in P). \end{aligned}$$

This action preserves the root system $\widehat{R}$, and the bilinear form on $\hat{\mathfrak{t}}^*$ is invariant with respect to this action.

For an affine root $\hat{\alpha} = \alpha + k\delta$ ($\alpha \in R, k \in \mathbf{Z}$), define the corresponding affine reflection as $s_{\hat{\alpha}} = t_{-k\alpha}\, s_\alpha$, and set $s_i = s_{\alpha_i}$ $(i = 0, 1, \ldots, r-1)$, $\pi = t_{\varepsilon_1} s_1 \cdots s_{r-1}$. The group $\widehat{W}$ is generated by $\pi, \pi^{-1}, s_0, s_1, \ldots, s_{r-1}$ and is defined by the relations

$$\begin{aligned} s_i s_{i+1} s_i &= s_{i+1} s_i s_{i+1}, \\ s_i s_j &= s_j s_i \quad (i - j \neq \pm 1), \\ s_i^2 &= 1, \qquad \pi s_i = s_{i+1}\pi, \end{aligned}$$

where the subscripts are understood to be modulo r. In this presentation, $\widetilde{W}$ is the Coxeter subgroup generated by $s_0, s_1, \ldots, s_{r-1}$ and $\widehat{W} \cong \Omega \ltimes \widetilde{W}$, where $\Omega \cong \mathbf{Z}$ is the subgroup of $\widehat{W}$ generated by π, π^{-1}.

For $w \in \widehat{W}$, let $S(w) = \widehat{R}^+ \cap w^{-1}(\widehat{R}^-)$, where $\widehat{R}^- = \widehat{R} \setminus \widehat{R}^+$ is the set of negative roots. The length $l(w)$ of w is defined as the number $\sharp S(w)$ of elements in $S(w)$. The length of w is zero if and only if $w \in \Omega$. A partial order on $\widehat{W}$ is defined by $\pi^k w \preceq \pi^{k'} w'$ ($w, w' \in \widetilde{W}$) if $k = k'$ and $w \preceq w'$ in the Bruhat order of $\widetilde{W}$.

The next lemma follows immediately from the definition of $S(w)$.

Lemma 3.1.
(i) *For* $w \in \widehat{W}$, $S(w^{-1}) = -w(S(w))$.
(ii) *For* $u, v \in \widetilde{W}$, $S(u) = S(v)$ *implies* $u = v$.
(iii) *For* $w \in W$, $\lambda \in P$,

$$l(w\, t_\lambda) = \sum_{\alpha \in R^+,\ w(\alpha) \in R^+} |(\lambda, \alpha)| + \sum_{\alpha \in R^+,\ w(\alpha) \in R^-} |1 + (\lambda, \alpha)|.$$

A corollary to (iii) above is the equality $l(t_\lambda) = l(t_\mu)$ for $\lambda, \mu \in P$ such that $\lambda = w(\mu)$ for some $w \in W$.

The following lemma is contained in [1] as Definition and Proposition 2.2.2.

Lemma 3.2. *For* $\lambda \in P$, *let* w *be the shortest element of* W *such that* $w(\lambda) \in P^+$. *Then*

$$S(w) = \{\alpha \in R^+ \mid (\lambda, \alpha) < 0\}.$$

Proposition 3.3. *For every* $x \in \widehat{W}$, *there is a unique factorization of the form* $x = u\, t_\lambda v$, *where* $u, v \in W$, $\lambda \in P^+$, *and* $S(v) = \{\alpha \in R^+ \mid (\lambda, v(\alpha)) < 0\}$. *Moreover,* $l(x) = l(u) + l(t_\lambda) - l(v)$.

Proof. Every $x \in \widehat{W}$ can be factorized as $x = w\, t_\mu$, where $w \in W$, $\mu \in P$. Let $v \in W$ be the shortest element such that $v(\mu) \in P^+$. By Lemma 3.2, $S(v) = \{\alpha \in R^+ \mid (\mu, \alpha) < 0\}$. The desired factorization is afforded by $x = u\, t_\lambda v$, with $u = wv^{-1}$, $\lambda = v(\mu)$.

Assume that $x = u_1 t_{\lambda_1} v_1 = u_2 t_{\lambda_2} v_2$, where u_i, v_i, and λ_i satisfy the conditions listed in the statement of the proposition. Set $\mu_i = v_i^{-1}(\lambda_i)$ so that $x = u_i v_i t_{\mu_i}$. The presentation of x in the form $w\,t_\mu$, $(w \in W,\ \mu \in P)$ is unique; hence $\mu_1 = \mu_2$, $u_1 v_1 = u_2 v_2$. The equality $\mu_1 = \mu_2$ implies $S(v_1) = S(v_2)$, whence $v_1 = v_2$, and, therefore, $u_1 = u_2$. The factorization is unique.

It remains to show the relation $l(x) = l(u) + l(t_\lambda) - l(v)$. The length formula of Lemma 3.1 together with $S(v) = \{\alpha \in R^+ \mid (\mu, \alpha) < 0\}$ give

$$\begin{aligned} l(x) = l(w\,t_\mu) &= \sum_{\alpha \in R^+ \setminus S(v),\ w(\alpha) \in R^-} 1 - \sum_{\alpha \in S(v),\ w(\alpha) \in R^-} 1 \\ &\quad + \sum_{\alpha \in R^+ \setminus S(v)} (\mu, \alpha) - \sum_{\alpha \in S(v)} (\mu, \alpha). \end{aligned}$$

On the other hand,

$$\begin{aligned} l(v) &= \sum_{\alpha \in S(v)} 1 = \sum_{\alpha \in S(v),\ w(\alpha) \in R^+} 1 + \sum_{\alpha \in S(v),\ w(\alpha) \in R^-} 1, \\ l(u) &= l(wv^{-1}) = \sum_{\alpha \in R^+ \setminus S(v),\ w(\alpha) \in R^-} 1 + \sum_{\alpha \in S(v),\ w(\alpha) \in R^+} 1, \\ l(t_\lambda) &= l(t_\mu) = \sum_{\alpha \in R^+ \setminus S(v)} (\mu, \alpha) - \sum_{\alpha \in S(v)} (\mu, \alpha). \end{aligned}$$

The relation $l(x) = l(u) + l(t_\lambda) - l(v)$ follows. □

3.1.1 *A right action of $\widehat{W}$ on P.* Let n be a positive integer, and define a right action of $\widehat{W}$ on P by

$$\begin{aligned} \zeta \cdot s_i &= s_i(\zeta) \qquad (\zeta \in P,\ 1 \le i < r), \\ \zeta \cdot t_\mu &= \zeta + n\mu \quad (\zeta \in P,\ \mu \in P). \end{aligned} \tag{1}$$

In coordinates $(\zeta_1, \ldots, \zeta_r)$ of $\zeta = \sum_{i=1}^r \zeta_i \varepsilon_i$, this action looks as follows:

$$\begin{aligned} (\zeta_1, \ldots, \zeta_r) \cdot s_i &= (\ldots, \zeta_{i+1}, \zeta_i, \ldots) \quad (1 \le i < r), \\ (\zeta_1, \ldots, \zeta_r) \cdot t_{\varepsilon_i} &= (\ldots, \zeta_i + n, \ldots) \quad (1 \le i \le r). \end{aligned}$$

Hence

$$\begin{aligned} (\zeta_1, \ldots, \zeta_r) \cdot \pi &= (\zeta_2, \ldots, \zeta_r, \zeta_1 + n), \\ (\zeta_1, \ldots, \zeta_r) \cdot s_0 &= (\zeta_r - n, \zeta_2, \ldots, \zeta_{r-1}, \zeta_1 + n). \end{aligned}$$

Define $A^n \subset P$ by

$$A^n = \{a = (a_1, \ldots, a_r) \in P \mid 1 \le a_1 \le a_2 \le \cdots \le a_r \le n\}.$$

Then A^n is a fundamental domain of the action given by (1). For $a \in A^n$, denote by W_a the stabilizer of a. The inequality $a_r - a_1 < n$ implies that $W_a \subset W$. Let ${}^a\widehat{W}$ (resp., ${}^aW_a$) be the set of minimal length representatives in the cosets $W_a \setminus \widehat{W}$ (resp., $W_a \setminus W$).

Lemma 3.4. *Let $x = u\,t_\lambda v$ be the factorized presentation of $x \in \widehat{W}$ given by Proposition 3.3. Then $x \in {}^a\widehat{W}$ if and only if $u \in {}^aW_a$.*

Proof. For any $w \in W_a$ the factorized presentation of wx is $wx = (wu)\,t_\lambda v$. It now follows from the length relation of Proposition 3.3 that $l(wx) \geq l(x) \iff l(wu) \geq l(u)$, i.e., x is the shortest element of its coset if and only if u is the shortest element of *its* coset. □

Lemma 3.5 ([1]). *For $a \in A^n$, let $R_a^+ = \{\varepsilon_i - \varepsilon_j \in R^+ \mid a_i = a_j\}$. Then*

$$ {}^aW_a = \{u \in W \mid S(u^{-1}) \subset R^+ \setminus R_a^+\}. $$

For $w \in W$, let a map $w : \{1, \dots, r\} \to \{1, \dots, r\}$ be defined by $\varepsilon_{w^{-1}(i)} = w(\varepsilon_i)$. Note that for $u, v \in W$, $u(v(i)) = vu(i)$.

Lemma 3.6. *Let $u \in {}^aW_a$, and let $c = (c_1, \dots, c_r) = a \cdot u$. Then*

$$ S(u) = \{\varepsilon_i - \varepsilon_j \in R^+ \mid c_i > c_j\}. $$

Proof. Observe that $c_i = a_{u^{-1}(i)}$ for all $i = 1, \dots, r$. Also, $\varepsilon_i - \varepsilon_j \in S(u)$ if and only if $i < j, u^{-1}(i) > u^{-1}(j)$. Since the sequence a is nondecreasing, $i < j, a_{u^{-1}(i)} > a_{u^{-1}(j)}$ implies $i < j, u^{-1}(i) > u^{-1}(j)$, i.e., $\varepsilon_i - \varepsilon_j \in S(u)$. Conversely, $\varepsilon_i - \varepsilon_j \in S(u)$ implies, by Lemmas 3.1(i) and 3.5, that $u(\varepsilon_j - \varepsilon_i) = \varepsilon_{u^{-1}(j)} - \varepsilon_{u^{-1}(i)} \in R^+ \setminus R_a^+$. This gives $a_{u^{-1}(i)} > a_{u^{-1}(j)}$, and the lemma follows. □

Proposition 3.7. *For $a \in A^n$ and $x \in {}^a\widehat{W}$, let $a \cdot x = (c_1 + n\mu_1, \dots, c_r + n\mu_r)$, where $c_i \in \{1, \dots, n\}$ and $\mu_i \in \mathbf{Z}$. Then*

$$ \begin{aligned} l(x) &= \sharp\{i < j \mid c_i > c_j,\ \mu_i \geq \mu_j\} + \sharp\{i < j \mid c_i < c_j,\ \mu_i < \mu_j\} \\ &\quad + \sum_{i<j,\ \mu_i > \mu_j} (\mu_i - \mu_j) + \sum_{i<j,\ \mu_i < \mu_j} (\mu_j - \mu_i - 1). \end{aligned} $$

Proof. Let $x = u\,t_\lambda v$ be the factorized presentation of Proposition 3.3. The expression for $l(x)$ follows from $l(x) = l(u) + l(t_\lambda) - l(v)$, Lemmas 3.4 and 3.6, and the length formula of Lemma 3.1(iii) □

Proposition 3.8. *For $a \in A^n$, let $x \in {}^a\widehat{W}$, and let $\zeta = (\zeta_1, \dots, \zeta_r) = a \cdot x$. Set $\zeta_0 = \zeta_r - n$. Then for each $i = 0, 1, \dots, r-1$, one has the following complete set of alternatives:*

(i) $\zeta_i = \zeta_{i+1} \iff xs_i \notin {}^a\widehat{W}$,

(ii) $\zeta_i > \zeta_{i+1} \iff xs_i \in {}^a\widehat{W},\quad l(xs_i) = l(x) - 1$,

(iii) $\zeta_i < \zeta_{i+1} \iff xs_i \in {}^a\widehat{W},\quad l(xs_i) = l(x) + 1$.

Moreover, in case (i), $xs_i = s_jx$, *where* $s_j \in W_a$.

PROOF. First, we show (i). Let $\zeta_i = \zeta_{i+1}$, then $\zeta \cdot s_i = \zeta$. Assuming that $xs_i \in {}^a\widehat{W}$, the length formula of Proposition 3.7 is applicable and immediately gives $l(xs_i) = l(x)$, which is impossible. Hence $\zeta_i = \zeta_{i+1} \Longrightarrow xs_i \notin {}^a\widehat{W}$. Now let $xs_i \notin {}^c\widehat{W}$. By [4, Lemma 2.1(iii)], in this case $xs_i = s_j x$ for some $s_j \in W_a$, which implies $\zeta \cdot s_i = \zeta$; hence $\zeta_i = \zeta_{i+1}$. Thus $\zeta_i = \zeta_{i+1} \Longleftarrow xs_i \notin {}^a\widehat{W}$.

Let $\zeta_i > \zeta_{i+1}$ (resp., $\zeta_i < \zeta_{i+1}$). Then $xs_i \in {}^a\widehat{W}$ and one may use the length formula of Proposition 3.7 to show $l(xs_i) = l(x) - 1$ (resp., $l(xs_i) = l(x) + 1$). Since (i) is already established, this proves (ii) and (iii).

Finally, [4, Lemma 2.1(iii)] states that in case (i), $xs_i = s_j x$, where $s_j \in W_a$. □

3.1.2 *A left action of $\widehat{W}$ on P.* Let l be a positive integer, and define a left action of $\widehat{W}$ on P by

$$\begin{aligned} s_i \cdot \eta &= s_i(\eta) \qquad (\eta \in P,\ 1 \le i < r), \\ t_\mu \cdot \eta &= \eta + l\mu \quad (\eta \in P,\ \mu \in P). \end{aligned} \tag{2}$$

In coordinates $(\eta_1, \dots, \eta_r)$ of $\eta = \sum_{i=1}^r \eta_i \varepsilon_i$, this action looks as follows:

$$\begin{aligned} s_i \cdot (\eta_1, \dots, \eta_r) &= (\dots, \eta_{i+1}, \eta_i, \dots) \quad (1 \le i < r), \\ t_{\varepsilon_i} \cdot (\eta_1, \dots, \eta_r) &= (\dots, \eta_i + l, \dots) \qquad (1 \le i \le r). \end{aligned}$$

Hence

$$\begin{aligned} \pi \cdot (\eta_1, \dots, \eta_r) &= (\eta_r + l, \eta_1, \dots, \eta_{r-1}), \\ s_0 \cdot (\eta_1, \dots, \eta_r) &= (\eta_r + l, \eta_2, \dots, \eta_{r-1}, \eta_1 - l). \end{aligned}$$

Define $B^l \subset P$ by

$$B^l = \{b = (b_1, \dots, b_r) \in P \mid l \ge b_1 \ge b_2 \ge \cdots \ge b_r \ge 1\}.$$

Then B^l is a fundamental domain of the action given by (2). For $b \in B^l$, denote by W_b the stabilizer of b. The inequality $b_1 - b_r < l$ implies that $W_b \subset W$. Let $\widehat{W}^b$ (resp., W^b) be the set of minimal length representatives in the cosets $\widehat{W}/W_b$ (resp., W/W_b).

Lemma 3.9 ([1]). *For $b \in B^l$, let $R_b^+ = \{\varepsilon_i - \varepsilon_j \in R^+ \mid b_i = b_j\}$. Then*

$$W^b = \{v \in W \mid S(v) \subset R^+ \setminus R_b^+\}.$$

Lemma 3.10. *Let $v \in W^b$ and let $d = (d_1, \dots, d_r) = v \cdot b$. Then*

$$S(v^{-1}) = \{\varepsilon_i - \varepsilon_j \in R^+ \mid d_i < d_j\}.$$

Proposition 3.11. *For $b \in B^l$ and $x \in \widehat{W}^b$, let $x \cdot b = (d_1 + l\mu_1, \dots, d_r + l\mu_r)$, where $d_i \in \{1, \dots, l\}$ and $\mu_i \in \mathbf{Z}$. Then*

$$\begin{aligned} l(x) \;=\; & \sharp\{i < j \mid d_i < d_j,\ \mu_i \le \mu_j\} + \sharp\{i < j \mid d_i > d_j,\ \mu_i > \mu_j\} \\ & + \sum_{i<j,\ \mu_i > \mu_j} (\mu_i - \mu_j - 1) + \sum_{i<j,\ \mu_i < \mu_j} (\mu_j - \mu_i). \end{aligned}$$

Proposition 3.12. *For $b \in B^l$, let $x \in \widehat{W}^b$, and let $\eta = (\eta_1, \dots, \eta_r) = x \cdot b$. Set $\eta_0 = \eta_r + l$. Then for each $i = 0, 1, \dots, r-1$, one has the following complete set of alternatives:*

$$\text{(i)} \qquad \eta_i = \eta_{i+1} \Longleftrightarrow s_i x \notin \widehat{W}^b,$$

$$\text{(ii)} \qquad \eta_i > \eta_{i+1} \Longleftrightarrow s_i x \in \widehat{W}^b, \quad l(s_i x) = l(x) + 1,$$

$$\text{(iii)} \qquad \eta_i < \eta_{i+1} \Longleftrightarrow s_i x \in \widehat{W}^b, \quad l(s_i x) = l(x) - 1.$$

Moreover, in case (i), $s_i x = x s_j$, *where* $s_j \in W_b$.

We omit proofs of Lemma 3.10 and Propositions 3.11 and 3.12 because they are almost identical to the proofs of Lemma 3.6 and Propositions 3.7 and 3.8.

3.2 Affine Hecke algebra. The Hecke algebra $\widehat{H}$ of the Weyl group $\widehat{W}$ is the algebra over $\mathbf{K} = \mathbf{Q}(q)$ with basis T_x $(x \in \widehat{W})$ and relations

$$T_x T_y = T_{xy} \quad \text{whenever } l(xy) = l(x) + l(y),$$

$$(T_{s_i} - q^{-1})(T_{s_i} + q) = 0 \qquad \text{for all } i = 0, 1, \dots, r-1.$$

The subalgebra H of $\widehat{H}$ generated by T_w $(w \in W)$ is isomorphic to the Hecke algebra of the finite Weyl group W. A system of generators of $\widehat{H}$ is afforded by elements T_π, $T_{\pi^{-1}}$ and $T_0, T_1, \dots, T_{r-1}$, where, for simplicity, we put $T_i := T_{s_i}$.

Another system of generators is obtained as follows. For $\lambda \in P$, write $\lambda = \mu - \nu$, where $\mu, \nu \in P^+$, and define

$$X^\lambda = T_{t_\nu} T_{t_\mu}^{-1}.$$

Note that Lemma 3.1(iii) implies that $l(t_\mu t_\nu) = l(t_\mu) + l(t_\nu)$ for $\mu, \nu \in P^+$. From this it follows that X^λ does not depend on the choice of $\mu, \nu \in P^+$, and

$$X^\lambda X^\mu = X^{\lambda+\mu} \quad \text{for } \lambda, \mu \in P.$$

A proof of the following result is contained, e.g., in [15].

Lemma 3.13.

(i) *The elements $X^\lambda, \lambda \in P$, $T_1, \dots, T_{r-1}$ generate $\widehat{H}$.*

(ii) *For $\lambda \in P$, $i = 1, \dots, r-1$,*

$$X^\lambda T_i = T_i X^{s_i(\lambda)} + (q - q^{-1}) \frac{X^{s_i(\lambda)} - X^\lambda}{1 - X^{\alpha_i}},$$

$$T_i X^\lambda = X^{s_i(\lambda)} T_i + (q - q^{-1}) \frac{X^{s_i(\lambda)} - X^\lambda}{1 - X^{\alpha_i}}.$$

Following [21], let us now briefly recall the notions of the canonical bases and the Kazhdan–Lusztig polynomials of $\widehat{W}$.

First, we recall that there is a canonical involution $h \mapsto \overline{h}$ of $\widehat{H}$ defined as the unique algebra automorphism such that $\overline{T_x} = (T_{x^{-1}})^{-1}$ and $\overline{q} = q^{-1}$. A proof of the following lemma is straightforward.

Lemma 3.14. *For $u, v \in W$ and $\lambda \in P$,*

$$\overline{T_u\, X^{\lambda}\, (T_{v^{-1}})^{-1}} = T_{u\omega}\, X^{\omega(\lambda)}\, (T_{(\omega v)^{-1}})^{-1},$$

where ω is the longest element of W.

Let L^+ (resp., L^-) be the lattice spanned over $\mathbf{Z}[q]$ (resp., $\mathbf{Z}[q^{-1}]$) by T_x $(x \in \widehat{W})$. The canonical bases C'_x, C_x $(x \in \widehat{W})$ are the unique bases of $\widehat{H}$ with the properties

$$\overline{C'_x} = C'_x, \qquad \overline{C_x} = C_x,$$
$$C'_x \equiv T_x \bmod qL^+, \qquad C_x \equiv T_x \bmod q^{-1}L^-.$$

Let

$$C'_x = \sum_y \mathcal{P}^+_{y,x} T_y, \qquad C_x = \sum_y \mathcal{P}^-_{y,x} T_y.$$

The coefficients $\mathcal{P}^{\pm}_{y,x}$ are called the Kazhdan–Lusztig polynomials of $\widehat{W}$. They are nonzero only if $y \preceq x$, that is, only if $x = \pi^k \tilde{x}$, $y = \pi^k \tilde{y}$ for some $k \in \mathbf{Z}$, $\tilde{x}, \tilde{y} \in \widetilde{W}$ such that $\tilde{y} \preceq \tilde{x}$. In this case,

$$\mathcal{P}^+_{y,x} = q^{l(x)-l(y)} P_{\tilde{y},\tilde{x}}, \qquad \mathcal{P}^-_{y,x} = (-q)^{l(y)-l(x)} \overline{P_{\tilde{y},\tilde{x}}},$$

where $P_{\tilde{y},\tilde{x}} \in \mathbf{Z}_{\geq 0}[q^{-2}]$ are the Kazhdan–Lusztig polynomials of the Coxeter group $\widetilde{W}$.

3.2.1 *A right representation of $\widehat{H}$.* For $a \in A^n$, let H_a be the parabolic subalgebra of $\widehat{H}$ generated by T_w $(w \in W_a)$. Let $\mathbf{K1}^+_a$ be a one-dimensional right representation of H_a defined by

$$\mathbf{1}^+_a \cdot T_i = q^{-1}\, \mathbf{1}^+_a \quad (s_i \in W_a).$$

The induced right representation

$$\mathbf{K1}^+_a \otimes_{H_a} \widehat{H}$$

of $\widehat{H}$ has as its basis $\mathbf{1}^+_a \otimes_{H_a} T_x$ $(x \in {}^a\widehat{W})$. For $x \in {}^a\widehat{W}$, define

$$(\zeta| := \mathbf{1}^+_a \otimes_{H_a} T_x, \quad \text{where } \zeta = a \cdot x,\ \zeta \in P.$$

Then $(\zeta|$ $(\zeta \in a \cdot \widehat{W})$ is a basis of $\mathbf{K1}^+_a \otimes_{H_a} \widehat{H}$ and $(\zeta|$ $(\zeta \in P)$ is a basis of $\oplus_{a\in A^n} \mathbf{K1}^+_a \otimes_{H_a} \widehat{H}$. Proposition 3.8 allows us to describe the action of the affine Hecke algebra in the basis $(\zeta|$ explicitly. We have

$$(\zeta| \cdot T_i = \begin{cases} (\zeta \cdot s_i| & \text{if } \zeta_i < \zeta_{i+1}, \\ q^{-1}(\zeta| & \text{if } \zeta_i = \zeta_{i+1} \quad (0 \leq i < r), \\ (\zeta \cdot s_i| - (q - q^{-1})(\zeta| & \text{if } \zeta_i > \zeta_{i+1}, \end{cases} \tag{3}$$
$$(\zeta| \cdot T_\pi = (\zeta \cdot \pi|.$$

Define a canonical involution $v \mapsto \overline{v}$ of $\mathbf{K}\mathbf{1}_a^+ \otimes_{H_a} \widehat{H}$ by

$$\overline{\mathbf{1}_a^+ \otimes_{H_a} h} = \mathbf{1}_a^+ \otimes_{H_a} \overline{h} \quad (h \in \widehat{H})$$

and two lattices by

$$L_a^+ := \bigoplus_{\zeta \in a \cdot \widehat{W}} \mathbf{Z}[q]\,(\zeta|, \qquad L_a^- := \bigoplus_{\zeta \in a \cdot \widehat{W}} \mathbf{Z}[q^{-1}]\,(\zeta|.$$

Theorem 3.15 ([4]). *There are unique bases $C_\zeta^\pm$ $(\zeta \in a \cdot \widehat{W})$ of $\mathbf{K}\mathbf{1}_a^+ \otimes_{H_a} \widehat{H}$, such that*

(i) $\overline{C_\zeta^\pm} = C_\zeta^\pm,$

(ii) $C_\zeta^\pm \equiv (\zeta| \bmod q^{\pm 1} L_a^\pm.$

Moreover, if

$$C_\zeta^\pm = \sum_\eta P_{\eta,\zeta}^\pm \,(\eta|,$$

then

$$P_{\eta,\zeta}^+ = \mathcal{P}_{\omega_a y, \omega_a x}^+, \qquad P_{\eta,\zeta}^- = \sum_{u \in W_a} q^{-l(u)}\, \mathcal{P}_{uy,x}^-,$$

where x and y are unique elements of ${}^a\widehat{W}$ such that $a \cdot x = \zeta$, $a \cdot y = \eta$, and ω_a is the longest element of W_a.

In the proof of Theorem 3.26 below, we shall use the following relation [18, formula (31)]:

$$P_{\eta \cdot s_i, \zeta}^- = -q^{-1} P_{\eta,\zeta}^- \quad \text{if } \zeta_i > \zeta_{i+1},\ \eta_i > \eta_{i+1}. \tag{4}$$

Here $i = 0, 1, \ldots, r-1$.

3.2.2 *A left representation of $\widehat{H}$.* For $b \in B^l$, let H_b be the parabolic subalgebra of $\widehat{H}$ generated by T_w $(w \in W_b)$. Let $\mathbf{K}\mathbf{1}_b^-$ be a one-dimensional left representation of H_b defined by

$$T_i \cdot \mathbf{1}_b^- = -q\, \mathbf{1}_b^- \quad (s_i \in W_b).$$

The induced left representation

$$\widehat{H} \otimes_{H_b} \mathbf{K}\mathbf{1}_b^-$$

of $\widehat{H}$ has as its basis $(T_{x^{-1}})^{-1} \otimes_{H_b} \mathbf{1}_b^-$ $(x \in \widehat{W}^b)$. For $x \in \widehat{W}^b$, define

$$|\eta) := (T_{x^{-1}})^{-1} \otimes_{H_b} \mathbf{1}_b^-, \quad \text{where } \eta = x \cdot b,\ \eta \in P.$$

Then $|\eta)$ ($\eta \in \widehat{W} \cdot b$) is a basis of $\widehat{H} \otimes_{H_b} \mathbf{K1}_b^-$ and $|\eta)$ ($\eta \in P$) is a basis of $\oplus_{b \in B^l} \widehat{H} \otimes_{H_b} \mathbf{K1}_b^-$. Proposition 3.12 allows us to describe the action of the affine Hecke algebra in the basis $|\eta)$ explicitly:

$$T_i \cdot |\eta) = \begin{cases} |s_i \cdot \eta) & \text{if } \eta_i < \eta_{i+1}, \\ -q\,|\eta) & \text{if } \eta_i = \eta_{i+1} \\ |s_i \cdot \eta) - (q - q^{-1})\,|\eta) & \text{if } \eta_i > \eta_{i+1}, \end{cases} \qquad (0 \le i < r), \tag{5}$$

$$T_\pi \cdot |\eta) = |\pi \cdot \eta).$$

3.3 Wedge product. For $a \in A^n$ and $b \in B^l$, define a vector space $\Lambda^r(a, b)$ by

$$\Lambda^r(a, b) := \mathbf{K1}_a^+ \otimes_{H_a} \widehat{H} \otimes_{H_b} \mathbf{K1}_b^-.$$

Note that the maps

$$\mathbf{K1}_a^+ \otimes_{H_a} \widehat{H} \otimes_H H \otimes_{H_b} \mathbf{K1}_b^- \to \Lambda^r(a, b) \ : \ \mathbf{1}_a^+ \otimes \hat{h} \otimes h \otimes \mathbf{1}_b^- \mapsto \mathbf{1}_a^+ \otimes \hat{h}h \otimes \mathbf{1}_b^-, \tag{6}$$

$$\mathbf{K1}_a^+ \otimes_{H_a} H \otimes_H \widehat{H} \otimes_{H_b} \mathbf{K1}_b^- \to \Lambda^r(a, b) \ : \ \mathbf{1}_a^+ \otimes h \otimes \hat{h} \otimes \mathbf{1}_b^- \mapsto \mathbf{1}_a^+ \otimes h\hat{h} \otimes \mathbf{1}_b^- \tag{7}$$

are isomorphisms of vector spaces. Let

$$\Lambda^r(a) := \bigoplus_{b \in B^l} \Lambda^r(a, b) \quad (a \in A^n),$$

$$\Lambda^r(b) := \bigoplus_{a \in A^n} \Lambda^r(a, b) \quad (b \in B^l),$$

$$\Lambda^r := \bigoplus_{b \in B^l} \Lambda^r(b).$$

Let $v_1, \ldots, v_n$ (resp., $\dot{v}_1, \ldots, \dot{v}_l$) be a basis of $\mathbf{K}^n$ (resp., $\mathbf{K}^l$). With a sequence

$$v_{c_1} X^{\mu_1} \dot{v}_{d_1}, \ldots, v_{c_r} X^{\mu_r} \dot{v}_{d_r}, \quad \text{where } v_{c_i} X^{\mu_i} \dot{v}_{d_i} \in (\mathbf{K}^n \otimes \mathbf{K}^l)[X, X^{-1}],$$

we associate unique $a \in A^n$, $b \in B^l$, and unique $u \in {}^aW_a$, $v \in W^b$ such that

$$c = (c_1, \ldots, c_r) = a \cdot u, \quad d = (d_1, \ldots, d_r) = v \cdot b$$

and define the following vector of $\Lambda^r(a, b)$ (here ω_b is the longest element of W_b):

$$(v_{c_1} X^{\mu_1} \dot{v}_{d_1}) \wedge \cdots \wedge (v_{c_r} X^{\mu_r} \dot{v}_{d_r}) := (-q^{-1})^{l(\omega_b)}\, \mathbf{1}_a^+ \otimes_{H_a} T_u X^\mu (T_{v^{-1}})^{-1} \otimes_{H_b} \mathbf{1}_b^-. \tag{8}$$

Note that, using isomorphism (6) to identify the vector spaces, we have

$$(v_{c_1} X^{\mu_1} \dot{v}_{d_1}) \wedge \cdots \wedge (v_{c_r} X^{\mu_r} \dot{v}_{d_r}) = (-q^{-1})^{l(\omega_b)}\, (c| \cdot X^\mu \otimes_H |d), \tag{9}$$

and, using isomorphism (7), we have

$$(v_{c_1} X^{\mu_1} \dot{v}_{d_1}) \wedge \cdots \wedge (v_{c_r} X^{\mu_r} \dot{v}_{d_r}) = (-q^{-1})^{l(\omega_b)} (c| \otimes_H X^\mu \cdot |d). \tag{10}$$

We shall call a vector of the form (8) a *wedge* and the vector space Λ^r the *wedge product*. In what follows, it will often be convenient to use a slightly different indexation of wedges: in the notation of (8), set $u_{k_i} := v_{c_i} X^{\mu_i} \dot{v}_{d_i}$, where $k_i := c_i + n(d_i - 1) - nl\mu_i$. Since the integers c_i (resp., d_i) range from 1 to n (resp., from 1 to l), a wedge (hence c, d, μ, a, b, u, v) is completely determined by the sequence $\boldsymbol{k} = (k_1, k_2, \ldots, k_r)$. To emphasize this, we write

$$c = c(\boldsymbol{k}),\quad d = d(\boldsymbol{k}),\quad \mu = \mu(\boldsymbol{k}),\quad a = a(\boldsymbol{k}),\quad b = b(\boldsymbol{k}),\quad u = u(\boldsymbol{k}),\quad v = v(\boldsymbol{k}). \tag{11}$$

Denote the left-hand side of (8) by

$$u_{\boldsymbol{k}} = u_{k_1} \wedge \cdots \wedge u_{k_r}.$$

Then $u_{\boldsymbol{k}}$ ($\boldsymbol{k} \in \mathbf{P} := \mathbf{Z}^r$) is a spanning set of Λ^r. However, the vectors of this set are not linearly independent. Indeed, using, e.g., (9), it follows that there are relations among these vectors that come from

$$(c| \cdot X^\mu T_i \otimes_H |d) = (c| \cdot X^\mu \otimes_H T_i \cdot |d) \quad (i = 1, 2, \ldots, r-1). \tag{12}$$

The exchange formula for X^μ and T_i of Lemma 3.13(ii) and the formulas for the action of T_i on $(c|$ and $|d)$ given, respectively, by (3) and (5) allow us to compute the relations among the wedges explicitly. Note that the relations for general r follow from those for $r = 2$.

Let us call a wedge $u_{\boldsymbol{k}}$ *ordered* if $\boldsymbol{k} \in \mathbf{P}^{++} := \{\boldsymbol{k} \in \mathbf{P} \mid k_1 > k_2 > \cdots > k_r\}$.

Proposition 3.16. (i) *Let $r = 2$. For integers k_1 and k_2 such that $k_1 \le k_2$, let $c_i \in \{1, \ldots, n\}$, $d_i \in \{1, \ldots, l\}$, and $\mu_i \in \mathbf{Z}$ be the unique numbers satisfying $k_i = c_i + n(d_i - 1) - nl\mu_i$. Let γ (resp., δ) be the residue of $c_2 - c_1$ (resp., $n(d_2 - d_1)$) modulo nl. Then*

$$u_{k_1} \wedge u_{k_2} = -\, u_{k_2} \wedge u_{k_1} \qquad \text{if } \gamma = 0,\ \delta = 0, \tag{R1}$$

$$\begin{aligned} u_{k_1} \wedge u_{k_2} = &-q^{-1} u_{k_2} \wedge u_{k_1} \\ &+ (q^{-2} - 1) \sum_{m \ge 0} q^{-2m} u_{k_2 - \gamma - nlm} \wedge u_{k_1 + \gamma + nlm} \\ &- (q^{-2} - 1) \sum_{m \ge 1} q^{-2m+1} u_{k_2 - nlm} \wedge u_{k_1 + nlm} \qquad \text{if } \gamma > 0,\ \delta = 0, \end{aligned} \tag{R2}$$

$$\begin{aligned} u_{k_1} \wedge u_{k_2} = &\, q u_{k_2} \wedge u_{k_1} \\ &+ (q^2 - 1) \sum_{m \ge 0} q^{2m} u_{k_2 - \delta - nlm} \wedge u_{k_1 + \delta + nlm} \end{aligned} \tag{R3}$$

$$+ (q^2 - 1) \sum_{m \geq 1} q^{2m-1} u_{k_2 - nlm} \wedge u_{k_1 + nlm} \qquad \textit{if } \gamma = 0,\ \delta > 0,$$

$$u_{k_1} \wedge u_{k_2} = u_{k_2} \wedge u_{k_1} \tag{R4}$$

$$+ (q - q^{-1}) \sum_{m \geq 0} \frac{(q^{2m+1} + q^{-2m-1})}{(q + q^{-1})} u_{k_2 - \delta - nlm} \wedge u_{k_1 + \delta + nlm}$$

$$+ (q - q^{-1}) \sum_{m \geq 0} \frac{(q^{2m+1} + q^{-2m-1})}{(q + q^{-1})} u_{k_2 - \gamma - nlm} \wedge u_{k_1 + \gamma + nlm}$$

$$+ (q - q^{-1}) \sum_{m \geq 1} \frac{(q^{2m} - q^{-2m})}{(q + q^{-1})} u_{k_2 + nl - \gamma - \delta - nlm} \wedge u_{k_1 - nl + \gamma + \delta + nlm}$$

$$+ (q - q^{-1}) \sum_{m \geq 1} \frac{(q^{2m} - q^{-2m})}{(q + q^{-1})} u_{k_2 - nlm} \wedge u_{k_1 + nlm}$$

$$\textit{if } \gamma > 0,\ \delta > 0,$$

where summations continue as long as wedges appearing under the sums remain ordered.

(ii) *Let $r > 2$. Then the relations of* (i) *hold in every pair of adjacent factors of $u_{k_1} \wedge u_{k_2} \wedge \cdots \wedge u_{k_r}$.*

It follows from this proposition that ordered wedges span Λ^r. Relations (R1)–(R4) can then be thought of as ordering rules that allow us to straighten an arbitrary wedge as a linear combination of ordered wedges.

Remark 3.17. (i) In order to compute the ordering rules of the wedges one can use—instead of isomorphism (6) and relations (12)—isomorphism (7) and relations $(c| \cdot T_i \otimes_H X^\mu \cdot |d) = (c| \otimes_H T_i X^\mu \cdot |d)$. The result is easily seen to be the same.

(ii) The ordering rules given in Proposition 3.16 differ from those used in [23, 24]. This difference is due to a different definition of wedges adopted here. In the present notation, the wedge of [24] is $(-1)^{l(v)} u_{\mathbf{k}}$, where v is the same as in (8).

The next lemma follows easily from Proposition 3.16.

Lemma 3.18. *Let $k \geq m$. Then*

(i) $$u_m \wedge u_k \wedge u_{k-1} \wedge \cdots \wedge u_m = 0,$$

(ii) $$u_k \wedge u_{k-1} \wedge \cdots \wedge u_m \wedge u_k = 0.$$

Now our aim is to show that ordered wedges form a basis of the wedge product. To this end, for $b \in B^l$ and $\zeta \in P$, we define $[\zeta]_b \in \Lambda^r(b)$ by

$$[\zeta]_b := (-q^{-1})^{l(\omega_b)} (\zeta| \otimes_{H_b} \mathbf{1}_b^- .$$

Then (3) implies that

$$[\zeta]_b = \begin{cases} 0 & \text{if } \zeta_i = \zeta_{i+1},\ b_i = b_{i+1}, \\ -q^{-1}[\zeta \cdot s_i]_b & \text{if } \zeta_i < \zeta_{i+1},\ b_i = b_{i+1}. \end{cases} \tag{13}$$

From this it follows that $[\zeta]_b$ ($\zeta \in P_b^{++}$), where $P_b^{++} := \{\zeta \in P \mid (\alpha_i, \zeta) > 0$ if $b_i = b_{i+1}\}$, span $\Lambda^r(b)$. For b such that $P_b^{++} = P^{++}$, a proof of the following lemma is given in [18]; a proof for general b is completely similar.

Lemma 3.19. *For each $b \in B^l$, the set $\{[\zeta]_b \mid \zeta \in P_b^{++}\}$ is a basis of $\Lambda^r(b)$.*

For $\boldsymbol{k} \in \mathbf{P}$, we define $\zeta(\boldsymbol{k}) \in P$ by

$$\zeta(\boldsymbol{k}) := v(\boldsymbol{k})^{-1} \cdot (c(\boldsymbol{k}) - n\mu(\boldsymbol{k})),$$

where $v(\boldsymbol{k}), c(\boldsymbol{k})$ and $\mu(\boldsymbol{k})$ are defined in (11).

Proposition 3.20. *Suppose $\boldsymbol{k} \in \mathbf{P}^{++}$. Then $\zeta(\boldsymbol{k}) \in P_{b(\boldsymbol{k})}^{++}$ and $u_{\boldsymbol{k}} = [\zeta(\boldsymbol{k})]_{b(\boldsymbol{k})}$. Conversely, for $b \in B^l$ and $\zeta \in P_b^{++}$, there is $\boldsymbol{k} \in \mathbf{P}^{++}$ such that $b = b(\boldsymbol{k})$, $\zeta = \zeta(\boldsymbol{k})$.*

PROOF. First, let us show that $\boldsymbol{k} \in \mathbf{P}^{++}$ implies that $u_{\boldsymbol{k}} = [\zeta(\boldsymbol{k})]_{b(\boldsymbol{k})}$. We set $\zeta := \zeta(\boldsymbol{k})$ and use the notation of (8) and (11). Let $\lambda := -\mu$. Observe that $\boldsymbol{k} \in P^{++}$ implies that

$$\lambda \in P^+, \tag{14}$$
$$i < j,\ d_i < d_j \Longrightarrow \lambda_i > \lambda_j, \tag{15}$$
$$i < j,\ d_i = d_j \Longrightarrow c_i + n\lambda_i > c_j + n\lambda_j. \tag{16}$$

From the dominance of λ, it follows that $X^\mu = T_{t_\lambda}$. Using $v \in W^b$, Lemma 3.10, and (15), we get $\alpha \in S(v^{-1}) \Longrightarrow (\alpha, \lambda) > 0$. Since $S(v^{-1}) = -v(S(v))$, this gives $\alpha \in S(v) \Longrightarrow (v(\alpha), \lambda) < 0$, i.e., $S(v) \subset \{\alpha \in R^+ \mid (\alpha, \nu) < 0\}$, where we set $\nu := v^{-1}(\lambda)$. Let $w \in W$ be the shortest element such that $w(\nu) \in P^+$; clearly, $w(\nu) = \lambda$. By Lemma 3.2, $S(w) = \{\alpha \in R^+ \mid (\alpha, \nu)\}$; hence $S(v) \subset S(w)$. On the other hand, the length of w does not exceed the length of v. This is possible only if $S(v) = S(w)$; hence $v = w$. Thus $S(v) = \{\alpha \in R^+ \mid (v(\alpha), \lambda) < 0\}$. Now, Proposition 3.3 implies $l(u\, t_\lambda v) = l(u) + l(t_\lambda) - l(v)$. This, $l(u) + l(t_\lambda) = l(u\, t_\lambda)$, and $X^\mu = T_{t_\lambda}$ give

$$T_u\, X^\mu (T_{v^{-1}})^{-1} = T_{u t_\lambda v}.$$

Since $u \in {}^aW_a$, by Lemma 3.4, we have $u\, t_\lambda v \in {}^a\widehat{W}$; hence

$$u_{\boldsymbol{k}} = [a \cdot u\, t_\lambda v]_b.$$

It remains to observe that

$$a \cdot u\, t_\lambda v = (c + n\lambda) \cdot v = v^{-1} \cdot (c + n\lambda) = \zeta.$$

Next, we show that $\boldsymbol{k} \in \mathbf{P}^{++}$ implies that $\zeta \in P_b^{++}$. It follows from $v \in W^b$ and Lemma 3.9 that for $i < j$,

$$b_i = b_j \Rightarrow v^{-1}(i) < v^{-1}(j); \quad \text{equivalently,} \quad d_{v^{-1}(i)} = d_{v^{-1}(j)} \Rightarrow v^{-1}(i) < v^{-1}(j).$$

Therefore, using $c_{v^{-1}(i)} + n\lambda_{v^{-1}(i)} = \zeta_i$ and (16), we get $\zeta \in P_b^{++}$.

Finally, $u_{\boldsymbol{k}}$ with $\boldsymbol{k} \in \mathbf{P}^{++}$ span Λ^r. Hence for $b \in B^l$ and $\zeta \in P_b^{++}$, we have

$$[\zeta]_b = \sum_{l \in \mathbf{P}^{++}} e_l u_l \quad (e_l \in \mathbf{K}).$$

However, $\{[\zeta]_b | b \in B^l, \zeta \in P_b^{++}\}$ is a basis of Λ^r. Therefore, writing u_l as $[\zeta(\boldsymbol{l})]_{b(\boldsymbol{l})}$, we get $[\zeta]_b = u_{\boldsymbol{k}}$, where $\boldsymbol{k} \in \mathbf{P}^{++}$ is such that $b = b(\boldsymbol{k})$ and $\zeta = \zeta(\boldsymbol{k})$. □

This proposition and Lemma 3.19 immediately imply the following.

Proposition 3.21. *$\{u_{\boldsymbol{k}} \mid \boldsymbol{k} \in \mathbf{P}^{++}\}$ is a basis of Λ^r.*

3.4 Canonical bases of the wedge product. Define an involution $x \mapsto \overline{x}$ of Λ^r by

$$\overline{\mathbf{1}_a^+ \otimes_{H_a} h \otimes_{H_b} \mathbf{1}_b^-} = \mathbf{1}_a^+ \otimes_{H_a} \overline{h} \otimes_{H_b} \mathbf{1}_b^-, \quad \overline{q} = q^{-1} \quad (h \in \widehat{H}).$$

Lemma 3.22. *Let $u \in {}^aW_a$, $v \in W^b$, and $\mu \in P$. Then $\omega_a u \omega \in {}^aW_a$, $\omega v \omega_b \in W^b$, and*

$$\overline{T_u X^\mu (T_{v^{-1}})^{-1}} = T_{\omega_a} T_{\omega_a u \omega} X^{\omega(\mu)} (T_{(\omega v \omega_b)^{-1}})^{-1} (T_{\omega_b})^{-1}.$$

Proof. Since for $w \in W$, we have $l(w\omega) = l(\omega) - l(w)$; $u\omega$ is the longest element of the coset $W_a u\omega$. Hence $\omega_a u\omega$ is the shortest element of this coset, i.e., $\omega_a u \omega \in {}^aW_a$. Therefore, $l(u\omega) = l(\omega) + l(\omega_a u \omega)$ and

$$T_{u\omega} = T_{\omega_a} T_{\omega_a u \omega}.$$

In a completely similar fashion, we get $\omega v \omega_b \in W^b$ and

$$(T_{(\omega v)^{-1}})^{-1} = (T_{(\omega v \omega_b)^{-1}})^{-1} (T_{\omega_b})^{-1}.$$

Lemma 3.14 implies the remaining statement. □

Proposition 3.23. *For $u_{k_1} \wedge u_{k_2} \wedge \cdots \wedge u_{k_r} \in \Lambda^r(a, b)$, we have*

$$\overline{u_{k_1} \wedge u_{k_2} \wedge \cdots \wedge u_{k_r}} = (-q)^{l(\omega_b)} q^{-l(\omega_a)} u_{k_r} \wedge \cdots \wedge u_{k_2} \wedge u_{k_1}.$$

Proof. Using (9) in the right-hand side of (8), we have, by Lemma 3.22,

$$\overline{\mathbf{1}_a^+ \otimes_{H_a} T_u X^\mu (T_{v^{-1}})^{-1} \otimes_{H_b} \mathbf{1}_b^-} = q^{-l(\omega_a)} (-q)^{-l(\omega_b)} (a \cdot u\omega| \cdot X^{\omega(\mu)} \otimes_H |\omega v \cdot b).$$

The result follows. □

Remark 3.24. It is easy to see that in the notation of (8), we have

$$l(\omega_a) = \sharp\{i < j \mid c_i = c_j\}, \qquad l(\omega_b) = \sharp\{i < j \mid d_i = d_j\}.$$

For $\boldsymbol{k} \in \mathbf{P}^{++}$, set

$$\overline{u_{\boldsymbol{k}}} = \sum_{\boldsymbol{l} \in \mathbf{P}^{++}} R_{\boldsymbol{k},\boldsymbol{l}}(q)\, u_{\boldsymbol{l}}.$$

By Proposition 3.16, the entries of matrix $\| R_{\boldsymbol{k},\boldsymbol{l}}(q) \|$ are Laurent polynomials in q with integral coefficients, and by Remark 3.24, we have $R_{\boldsymbol{k},\boldsymbol{k}}(q) = 1$.

We define a partial order on $\mathbf{P}^{++}$ by

$$\boldsymbol{k} \geq \boldsymbol{l} \quad \text{iff} \quad \left| \begin{array}{l} \sum_{i=1}^{j} k_i \geq \sum_{i=1}^{j} l_i \quad \text{for all } j = 1, \dots, r, \\ \sum_{i=1}^{r} k_i = \sum_{i=1}^{r} l_i, \\ k_r \leq l_r. \end{array} \right.$$

The ordering rules of Proposition 3.16 imply that the matrix $\| R_{\boldsymbol{k},\boldsymbol{l}}(q) \|$ is lower triangular with respect to this order. That is, $R_{\boldsymbol{k},\boldsymbol{l}}(q)$ is not zero only if $\boldsymbol{k} \geq \boldsymbol{l}$.

Define two lattices of Λ^r by

$$\mathcal{L}^+ := \bigoplus_{\boldsymbol{k} \in \mathbf{P}^{++}} \mathbf{Q}[q]\, u_{\boldsymbol{k}}, \qquad \mathcal{L}^- := \bigoplus_{\boldsymbol{k} \in \mathbf{P}^{++}} \mathbf{Q}[q^{-1}]\, u_{\boldsymbol{k}}.$$

The unitriangularity of $\| R_{\boldsymbol{k},\boldsymbol{l}}(q) \|$ implies, by the standard argument going back to Kazhdan and Lusztig, the following.

Theorem 3.25. *There are unique bases $\{G_{\boldsymbol{k}}^+ \mid \boldsymbol{k} \in \mathbf{P}^{++}\}$, $\{G_{\boldsymbol{k}}^- \mid \boldsymbol{k} \in \mathbf{P}^{++}\}$ of Λ^r such that*

(i) $\qquad \overline{G_{\boldsymbol{k}}^+} = G_{\boldsymbol{k}}^+, \qquad \overline{G_{\boldsymbol{k}}^-} = G_{\boldsymbol{k}}^-,$

(ii) $\qquad G_{\boldsymbol{k}}^+ \equiv u_{\boldsymbol{k}} \bmod q\mathcal{L}^+, \qquad G_{\boldsymbol{k}}^- \equiv u_{\boldsymbol{k}} \bmod q^{-1}\mathcal{L}^-.$

Set

$$G_{\boldsymbol{k}}^+ = \sum_{\boldsymbol{l} \in \mathbf{P}^{++}} \Delta_{\boldsymbol{k},\boldsymbol{l}}^+(q)\, u_{\boldsymbol{l}}, \qquad G_{\boldsymbol{k}}^- = \sum_{\boldsymbol{l} \in \mathbf{P}^{++}} \Delta_{\boldsymbol{k},\boldsymbol{l}}^-(q)\, u_{\boldsymbol{l}}.$$

It is clear that $\Delta_{\boldsymbol{k},\boldsymbol{l}}^+(q)$ or $\Delta_{\boldsymbol{k},\boldsymbol{l}}^-(q)$ is nonzero only if $u_{\boldsymbol{k}}$ and $u_{\boldsymbol{l}}$ belong to the same subspace $\Lambda^r(a,b)$. That is, in the notation of (11), only if $a(\boldsymbol{k}) = a(\boldsymbol{l})$ and $b(\boldsymbol{k}) = b(\boldsymbol{l})$.

Theorem 3.26. *For $\boldsymbol{k}, \boldsymbol{l} \in \mathbf{P}^{++}$ such that $a(\boldsymbol{k}) = a(\boldsymbol{l})$, $b(\boldsymbol{k}) = b(\boldsymbol{l})$, set $\xi = \zeta(\boldsymbol{k})$, $\eta = \zeta(\boldsymbol{l})$ and $a = a(\boldsymbol{k})$, $b = b(\boldsymbol{k})$. Then*

(i) $\qquad \Delta_{\boldsymbol{k},\boldsymbol{l}}^-(q) = P_{\eta,\xi}^-,$

(ii) $\qquad \Delta_{\boldsymbol{k},\boldsymbol{l}}^+(q) = \sum_{v \in W_b} (-q)^{l(v)} P_{\eta\cdot\omega_b v,\, \xi\cdot\omega_b}^+,$

where $P_{\eta,\xi}^-$ and $P_{\eta,\xi}^+$ are the parabolic Kazhdan–Lusztig polynomials associated with $\mathbf{1}_a^+ \otimes_{H_a} \widehat{H}$.

PROOF. Set $D_\xi = C^-_\xi \otimes \mathbf{1}^-_b$. Then $\overline{D_\xi} = D_\xi$. Using (4) and (13), we obtain

$$D_\xi = (-q)^{l(\omega_b)} \sum_{\eta \in P} P^-_{\eta,\xi}\,[\eta]_b = z_b \sum_{\eta \in P^{++}_b} P^-_{\eta,\xi}\,[\eta]_b,$$

where $z_b = (-q)^{l(\omega_b)} \sum_{v \in W_b} q^{-2l(v)}$. Since $\overline{z_b} = z_b$, we have $D_\xi = z_b\, G^-_{\boldsymbol{k}}$ and (i) follows.

Next, let $E_\xi = C^-_{\xi \cdot \omega_b} \otimes \mathbf{1}^-_b$. Then $\overline{E_\xi} = E_\xi$, and

$$E_\xi = \sum_{\eta \in P^{++}_b} \sum_{v \in W_b} (-q)^{l(\omega_b)-l(v)} P^+_{\eta \cdot v, \xi \cdot \omega_b}\,[\eta]_b = \sum_{\eta \in P^{++}_b} \sum_{v \in W_b} (-q)^{l(v)} P^+_{\eta \cdot \omega_b v, \xi \cdot \omega_b}\,[\eta]_b.$$

Hence $E_\xi = G^+_{\boldsymbol{k}}$, which implies (ii). □

Remark 3.27. For $\xi \in a \cdot \widehat{W}$, let $x(\xi)$ be the unique element of ${}^a\widehat{W}$ such that $\xi = a \cdot x(\xi)$. It follows from Proposition 3.8 that for $\xi \in P^{++}_b$ we have $x(\xi \cdot v) = x(\xi)v$ for all $v \in W_b$. Hence, one can rewrite the formulas of Theorem 3.26 in terms of (ordinary) Kazhdan–Lusztig polynomials for $\widehat{H}$ as

(i) $$\Delta^-_{\boldsymbol{k},\boldsymbol{l}}(q) = \sum_{u \in W_a} q^{-l(u)} \mathcal{P}^-_{u\,x(\eta),\, x(\xi)},$$

(ii) $$\Delta^+_{\boldsymbol{k},\boldsymbol{l}}(q) = \sum_{v \in W_b} (-q)^{l(v)} \mathcal{P}^+_{\omega_a\, x(\eta)\, \omega_b\, v,\, \omega_a\, x(\xi)\, \omega_b}.$$

3.5 Actions of quantum affine algebras on the wedge product. It is easy to verify that the formulas

$$\begin{aligned}
e_i(v_c X^m) &= \delta_{i+1 \equiv c \bmod n}\; v_{c-1} X^{m+\delta_{i0}},\\
f_i(v_c X^m) &= \delta_{i \equiv c \bmod n}\; v_{c+1} X^{m-\delta_{i0}},\\
t_i(v_c X^m) &= q^{\delta_{i \equiv c \bmod n} - \delta_{i+1 \equiv c \bmod n}}\; v_c X^m,\\
\partial(v_c X^m) &= m v_c X^m
\end{aligned}$$

define a level zero action of $U_q(\widehat{\mathfrak{sl}}_n)$ on $\mathbf{K}^n[X, X^{-1}]$. Here it is understood that $v_0 = v_n$, $v_{n+1} = v_1$. Also, for a statement S we set $\delta_S = 1$ (resp., 0) if S is true (resp., false). Using the coproduct

$$\begin{aligned}
\Delta(e_i) &= e_i \otimes t_i^{-1} + 1 \otimes e_i, & \Delta(t_i) &= t_i \otimes t_i,\\
\Delta(f_i) &= f_i \otimes 1 + t_i \otimes f_i, & \Delta(\partial) &= \partial \otimes 1 + 1 \otimes \partial,
\end{aligned}$$

we extend this action on the tensor product $(\mathbf{K}^n[X, X^{-1}])^{\otimes r}$. Next, using the isomorphism

$$(\mathbf{K}^n[X, X^{-1}])^{\otimes r} \xrightarrow{\sim} \bigoplus_{a \in A^n} \mathbf{K}\mathbf{1}^+_a \otimes_{H_a} \widehat{H}, \quad v_{c_1} X^{\mu_1} \otimes \cdots \otimes v_{c_r} X^{\mu_r} \mapsto (c| \cdot X^\mu,$$

where $c=(c_1,\dots,c_r)$ and $\mu=(\mu_1,\dots,\mu_r)$, to identify the involved vector spaces, we obtain an action of $U_q(\widehat{\mathfrak{sl}}_n)$ on the $\widehat{H}$–module $M_R=\oplus_{a\in A^n}\mathbf{K1}_a^+\otimes_{H_a}\widehat{H}$. The following proposition is due to [8]. It is easily verified by reducing to the case $r=2$ (cf. [18, proof of Proposition 7.1]).

Proposition 3.28. *The actions of $U_q(\widehat{\mathfrak{sl}}_n)$ and $H\subset\widehat{H}$ on M_R commute.*

By isomorphism (6), for any $b\in B^l$, the commutativity of $U_q(\widehat{\mathfrak{sl}}_n)$ and H allows us to restrict the action of $U_q(\widehat{\mathfrak{sl}}_n)$ on $\Lambda^r(b)$, which then gives an action of $U_q(\widehat{\mathfrak{sl}}_n)$ on Λ^r. In terms of the wedge vectors $u_{\boldsymbol{k}}=u_{k_1}\wedge\cdots\wedge u_{k_r}$ ($\boldsymbol{k}=(k_1,\dots,k_r)\in\mathbf{P}$), this action is written as

$$e_i(u_{\boldsymbol{k}})=\sum_{j=1}^{r}u_{k_1}\wedge\cdots\wedge u_{k_{j-1}}\wedge e_i(u_{k_j})\wedge t_i^{-1}(u_{k_{j+1}})\wedge\cdots\wedge t_i^{-1}(u_{k_r}),\tag{17}$$

$$f_i(u_{\boldsymbol{k}})=\sum_{j=1}^{r}t_i(u_{k_1})\wedge\cdots\wedge t_i(u_{k_{j-1}})\wedge f_i(u_{k_j})\wedge u_{k_{j+1}}\wedge\cdots\wedge u_{k_r},\tag{18}$$

$$t_i(u_{\boldsymbol{k}})=t_i(u_{k_1})\wedge\cdots\wedge t_i(u_{k_r}),\tag{19}$$

$$\partial(u_{\boldsymbol{k}})=\sum_{j=1}^{r}u_{k_1}\wedge\cdots\wedge u_{k_{j-1}}\wedge\partial(u_{k_j})\wedge u_{k_{j+1}}\wedge\cdots\wedge u_{k_r}.\tag{20}$$

Here we set $e_i(u_{k_j})=e_i(v_{c_j}X^{\mu_j}\dot{v}_{d_j}):=e_i(v_{c_j}X^{\mu_j})\dot{v}_{d_j}$ and similarly for the rest of the generators.

In a completely similar fashion, we define on the wedge product an action of the quantum affine algebra $U_p(\widehat{\mathfrak{sl}}_l)$, where $p:=-q^{-1}$. In order to distinguish between this action and the action of $U_q(\widehat{\mathfrak{sl}}_n)$, we put dots over the generators of $U_p(\widehat{\mathfrak{sl}}_l)$. Define a level-zero action of $U_p(\widehat{\mathfrak{sl}}_l)$ on $\mathbf{K}^l[X,X^{-1}]$ by

$$\dot{e}_i(X^m\dot{v}_d)=\delta_{i+1\equiv d \bmod l}\,X^{m+\delta_{i0}}\dot{v}_{d-1},\qquad \dot{t}_i(X^m\dot{v}_d)=p^{\delta_{i\equiv d \bmod l}-\delta_{i+1\equiv d\bmod l}}\,X^m\dot{v}_d,$$
$$\dot{f}_i(X^m\dot{v}_d)=\delta_{i\equiv d\bmod l}\,X^{m-\delta_{i0}}\dot{v}_{d+1},\qquad \dot{\partial}(X^m\dot{v}_d)=mX^m\dot{v}_d.$$

Here it is understood that $\dot{v}_0=\dot{v}_l$, $\dot{v}_{l+1}=\dot{v}_1$. Using the coproduct

$$\Delta(\dot{e}_i)=\dot{e}_i\otimes\dot{t}_i^{-1}+1\otimes\dot{e}_i,\qquad \Delta(\dot{t}_i)=\dot{t}_i\otimes\dot{t}_i,$$
$$\Delta(\dot{f}_i)=\dot{f}_i\otimes 1+\dot{t}_i\otimes\dot{f}_i,\qquad \Delta(\dot{\partial})=\dot{\partial}\otimes 1+1\otimes\dot{\partial},$$

we extend this action on the tensor product $(\mathbf{K}^l[X,X^{-1}])^{\otimes r}$. Next, using the isomorphism

$$(\mathbf{K}^l[X,X^{-1}])^{\otimes r}\ \xrightarrow{\sim}\ \bigoplus_{b\in B^l}\widehat{H}\otimes_{H_b}\mathbf{K1}_b^-,$$
$$X^{\mu_1}\dot{v}_{d_1}\otimes\cdots\otimes X^{\mu_r}\dot{v}_{d_r}\ \mapsto\ X^{\mu}\cdot|d)\,(-q^{-1})^{l(\omega_b)},$$

where $d=(d_1,\dots,d_r)$, $\mu=(\mu_1,\dots,\mu_r)$, and b is the unique point of B^l in the orbit $W\cdot d$, to identify the involved vector spaces, we obtain an action of $U_p(\widehat{\mathfrak{sl}}_l)$

on the $\widehat{H}$-module $M_L = \oplus_{b \in B^l} \widehat{H} \otimes_{H_b} \mathbf{K1}_b^-$. The following result is an analogue of Proposition 3.28 and is verified similarly.

Proposition 3.29. *The actions of $U_p(\widehat{\mathfrak{sl}}_l)$ and $H \subset \widehat{H}$ on M_L commute.*

By the isomorphism (7), for any $a \in A^n$, the commutativity of $U_p(\widehat{\mathfrak{sl}}_l)$ and H allows us to restrict the action of $U_p(\widehat{\mathfrak{sl}}_l)$ on $\Lambda^r(a)$, which gives then an action of $U_p(\widehat{\mathfrak{sl}}_l)$ on Λ^r. In terms of the wedge vectors $u_{\boldsymbol{k}} = u_{k_1} \wedge \cdots \wedge u_{k_r}$ ($\boldsymbol{k} = (k_1, \ldots, k_r) \in \mathbf{P}$), this action is written as

$$\dot{e}_i(u_{\boldsymbol{k}}) = \textstyle\sum_{j=1}^r u_{k_1} \wedge \cdots \wedge u_{k_{j-1}} \wedge \dot{e}_i(u_{k_j}) \wedge \dot{t}_i^{-1}(u_{k_{j+1}}) \wedge \cdots \wedge \dot{t}_i^{-1}(u_{k_r}), \tag{21}$$

$$\dot{f}_i(u_{\boldsymbol{k}}) = \textstyle\sum_{j=1}^r \dot{t}_i(u_{k_1}) \wedge \cdots \wedge \dot{t}_i(u_{k_{j-1}}) \wedge \dot{f}_i(u_{k_j}) \wedge u_{k_{j+1}} \wedge \cdots \wedge u_{k_r}, \tag{22}$$

$$\dot{t}_i(u_{\boldsymbol{k}}) = \dot{t}_i(u_{k_1}) \wedge \cdots \wedge \dot{t}_i(u_{k_r}), \tag{23}$$

$$\dot{\partial}(u_{\boldsymbol{k}}) = \textstyle\sum_{j=1}^r u_{k_1} \wedge \cdots \wedge u_{k_{j-1}} \wedge \dot{\partial}(u_{k_j}) \wedge u_{k_{j+1}} \wedge \cdots \wedge u_{k_r}. \tag{24}$$

Here we put $\dot{e}_i(u_{k_j}) = \dot{e}_i(v_{c_j} X^{\mu_j} \dot{v}_{d_j}) := v_{c_j} \dot{e}_i(X^{\mu_j} \dot{v}_{d_j})$ and similarly for the rest of the generators.

A well-known result of Bernstein (see, e.g., [15]) says that the centre $Z(\widehat{H})$ of $\widehat{H}$ is generated by symmetric Laurent polynomials in $X_i := X^{\varepsilon_i}$. It follows by either using (6) or (7) that $Z(\widehat{H})$ acts on the wedge product Λ^r. This action may be computed in terms of the wedge vectors $u_{\boldsymbol{k}} = u_{k_1} \wedge \cdots \wedge u_{k_r}$ ($\boldsymbol{k} = (k_1, \ldots, k_r) \in \mathbf{P}$) by using either (9) or (10). In particular, for $B_m := \sum_{i=1}^r X_i^m$ ($m \in \mathbf{Z}^*$), we get

$$B_m(u_{\boldsymbol{k}}) = \sum_{j=1}^r u_{k_1} \wedge \cdots \wedge u_{k_{j-1}} \wedge u_{k_j - nlm} \wedge u_{k_{j+1}} \wedge \cdots \wedge u_{k_r}. \tag{25}$$

Proposition 3.30. *The actions of $U_q'(\widehat{\mathfrak{sl}}_n)$, $U_p'(\widehat{\mathfrak{sl}}_l)$, and $Z(\widehat{H})$ on Λ^r are pairwise mutually commutative.*

PROOF. The commutativity of $U_q'(\widehat{\mathfrak{sl}}_n)$ (resp., $U_p'(\widehat{\mathfrak{sl}}_l)$) with $Z(\widehat{H})$ is immediate by Proposition 3.28 (resp., Proposition 3.29). The commutativity of $U_q'(\widehat{\mathfrak{sl}}_n)$ with $U_p'(\widehat{\mathfrak{sl}}_l)$ follows from (17)–(19) and (21)–(23). □

The actions of $U_q(\widehat{\mathfrak{sl}}_n)$, $U_p(\widehat{\mathfrak{sl}}_l)$ and $Z(\widehat{H})$ on Λ^r are compatible with the bar involution in the following sense.

Proposition 3.31. *For $v \in \Lambda^r$*

$$\overline{e_i(v)} = e_i(\overline{v}), \quad \overline{f_i(v)} = f_i(\overline{v}), \quad \overline{t_i(v)} = t_i^{-1}(\overline{v}), \quad \overline{\partial(v)} = \partial(\overline{v}),$$
$$\overline{\dot{e}_j(v)} = \dot{e}_j(\overline{v}), \quad \overline{\dot{f}_j(v)} = \dot{f}_j(\overline{v}), \quad \overline{\dot{t}_j(v)} = \dot{t}_j^{-1}(\overline{v}), \quad \overline{\dot{\partial}(v)} = \dot{\partial}(\overline{v}),$$
$$\overline{B_m(v)} = B_m(\overline{v})$$

for $i = 0, 1, \ldots, n-1$, $j = 0, 1, \ldots, l-1$, and $m \in \mathbf{Z}^*$.

Proof. It is enough to show this for $v = u_{\boldsymbol{k}}$ ($\boldsymbol{k} = (k_1, \dots, k_r) \in \mathbf{P}$). As in Section 3.3, we set $k_i = c_i + n(d_i - 1) - nl\mu_i$ with $c_i \in \{1, \dots, n\}$, $d_i \in \{1, \dots, l\}$, and $\mu_i \in \mathbf{Z}$. By Proposition 3.23 and Remark 3.24, we have

$$\overline{u_{\boldsymbol{k}}} = q^{-\kappa(c)}(-q)^{\kappa(d)} u_{k_r} \wedge \cdots \wedge u_{k_1} = p^{-\kappa(d)}(-p)^{\kappa(c)} u_{k_r} \wedge \cdots \wedge u_{k_1},$$

where $\kappa(c)$ (resp., $\kappa(d)$) is the number of pairs (i, j) such that $c_i = c_j$ (resp., $d_i = d_j$). Using (22), we have

$$\overline{\dot{f}_0(u_{\boldsymbol{k}})} = (-p)^{\kappa(c)} \sum_{j,\, d_j = l} p^{-\kappa(d'_{(j)}) + \sum_{i<j} \delta_{d_i=1} - \delta_{d_i=l}} \; u_{k_r} \wedge \cdots \wedge (v_{c_j} X^{\mu_j - 1} \dot{v}_1) \wedge \cdots \wedge u_{k_1}$$

and

$$\dot{f}_0(\overline{u_{\boldsymbol{k}}}) = (-p)^{\kappa(c)} \sum_{j,\, d_j = l} p^{-\kappa(d) + \sum_{i>j} \delta_{d_i=l} - \delta_{d_i=1}} \; u_{k_r} \wedge \cdots \wedge (v_{c_j} X^{\mu_j - 1} \dot{v}_1) \wedge \cdots \wedge u_{k_1},$$

where

$$\kappa(d'_{(j)}) = \kappa(d + \varepsilon_j - l\varepsilon_j) = \kappa(d) + \sum_{i \neq j} \delta_{d_i=1} - \delta_{d_i=l}.$$

This gives $\overline{\dot{f}_0(u_{\boldsymbol{k}})} = \dot{f}_0(\overline{u_{\boldsymbol{k}}})$. The relations for the rest of the generators of $U_p(\widehat{\mathfrak{sl}}_l)$ and for the generators of $U_q(\widehat{\mathfrak{sl}}_n)$ are verified similarly. Finally, $\overline{B_m(v)} = B_m(\overline{v})$ follows from Lemma 3.14. □

4 Canonical bases of the q-deformed Fock spaces

4.1 Semiinfinite wedge product. For each integer s, we define a vector space $\Lambda^{s+\frac{\infty}{2}}$ as the inductive limit $\lim_{\to} \Lambda^r$, where the maps $\Lambda^r \to \Lambda^t$ ($t > r$) are given by

$$v \mapsto v \wedge u_{s-r} \wedge u_{s-r-1} \wedge \cdots \wedge u_{s-t+1}.$$

The vector space $\Lambda^{s+\frac{\infty}{2}}$ will be called the semiinfinite wedge product of charge s. For $v \in \Lambda^r$, we shall use the semiinfinite expression

$$v \wedge u_{s-r} \wedge u_{s-r-1} \wedge u_{s-r-2} \wedge \cdots$$

to denote the image of v with respect to the canonical map from Λ^r to $\lim_{\to} \Lambda^r$. Let $\mathbf{P}(s)$ be the set of semiinfinite sequences $\boldsymbol{k} = (k_1, k_2, \dots) \in \mathbf{Z}^\infty$ such that $k_i = s - i + 1$ for all $i \gg 1$. By definition, $\Lambda^{s+\frac{\infty}{2}}$ is spanned by $u_{\boldsymbol{k}} := u_{k_1} \wedge u_{k_2} \wedge \cdots$ with $\boldsymbol{k} \in \mathbf{P}(s)$. We shall call a semiinfinite wedge $u_{\boldsymbol{k}}$ *ordered* if $\boldsymbol{k} \in \mathbf{P}^{++}(s) := \{(k_1, k_2, \dots) \in \mathbf{P}(s) \mid k_1 > k_2 > \cdots\}$.

Proposition 4.1 ([23]). *Ordered wedges form a basis of* $\Lambda^{s+\frac{\infty}{2}}$.

In what follows, we shall use, besides the indexation by $\mathbf{P}^{++}(s)$, three other indexations of the basis formed by ordered wedges. The first of these is the obvious indexation by the set Π of all partitions. Namely, with $\boldsymbol{k} = (k_1, k_2, \dots) \in \mathbf{P}^{++}(s)$, we associate a partition $\lambda = (\lambda_1, \lambda_2, \dots)$ by taking $\lambda_i = k_i - s + i - 1$ for all $i \in \mathbf{N}$, and we set $|\lambda, s\rangle := u_{\boldsymbol{k}}$.

Another indexation is by the set of pairs $(\boldsymbol{\lambda}_l, \boldsymbol{s}_l)$, where $\boldsymbol{\lambda}_l = (\lambda^{(1)}, \dots, \lambda^{(l)})$ is an l-multipartition and $\boldsymbol{s}_l = (s_1, \dots, s_l)$ is a sequence of integers summing up to s. For any $d \in \{1, \dots, l\}$ the partition $\lambda^{(d)}$ and the integer s_d are defined as follows. For each $i \in \mathbf{N}^*$ write $k_i = c_i + n(d_i - 1) - nlm_i$, where $c_i \in \{1, \dots, n\}$, $d_i \in \{1, \dots, l\}$, and $m_i \in \mathbf{Z}$. Then let $k_1^{(d)}$ be equal $c_i - nm_i$, where i is the smallest such that $d_i = d$, let $k_2^{(d)}$ be equal to $c_j - nm_j$, where j is the second smallest such that $d_j = d$, and so on. In this way, we obtain a strictly decreasing semiinfinite sequence $(k_1^{(d)}, k_2^{(d)}, \dots)$ such that $k_i^{(d)} = s_d - i + 1$ $(i \gg 1)$ for some unique integer s_d. Now we define the partition $\lambda^{(d)} = (\lambda_1^{(d)}, \lambda_2^{(d)}, \dots)$ by $\lambda_i^{(d)} = k_i^{(d)} - s_d + i - 1$. It is easy to check that the s_d obtained in this way satisfies $s_1 + \dots + s_l = s$.

In a completely similar fashion, we associate with each $\boldsymbol{k} \in \mathbf{P}^{++}(s)$ a pair $(\boldsymbol{\lambda}_n, \boldsymbol{s}_n)$, where $\boldsymbol{\lambda}_n$ is an n-multipartition and $\boldsymbol{s}_n$ is a sequence of n integers summing up to s. The $(\boldsymbol{\lambda}_n, \boldsymbol{s}_n)$ is obtained by the same procedure as the $(\boldsymbol{\lambda}_l, \boldsymbol{s}_l)$, reversing everywhere the roles of n and l and the roles of c_i and d_i.

For any $s \in \mathbf{Z}$, let $\mathbf{Z}^l(s)$ be the set of l-tuples of integers summing to s. Define the map $\tau_l^s : \Pi \to \Pi^l \times \mathbf{Z}^l(s)$ (resp., the map $\tau_n^s : \Pi \to \Pi^n \times \mathbf{Z}^n(s)$) by

$$\tau_l^s : \lambda \mapsto (\boldsymbol{\lambda}_l, \boldsymbol{s}_l), \quad \text{(resp., by)} \quad \tau_n^s : \lambda \mapsto (\boldsymbol{\lambda}_n, \boldsymbol{s}_n). \tag{26}$$

It is not difficult to see that for each s, the maps τ_l^s and τ_n^s are bijections. Hence, setting $|\boldsymbol{\lambda}_l, \boldsymbol{s}_l\rangle := |\lambda, s\rangle$ (resp., $|\boldsymbol{\lambda}_n, \boldsymbol{s}_n\rangle := |\lambda, s\rangle$) if $(\boldsymbol{\lambda}_l, \boldsymbol{s}_l) = \tau_l^s(\lambda)$ (resp., if $(\boldsymbol{\lambda}_n, \boldsymbol{s}_n) = \tau_n^s(\lambda)$), we obtain that $B(s) := \{|\boldsymbol{\lambda}_l, \boldsymbol{s}_l\rangle \mid \boldsymbol{\lambda}_l \in \Pi^l,\ \boldsymbol{s}_l \in \mathbf{Z}^l(s)\} = \{|\boldsymbol{\lambda}_n, \boldsymbol{s}_n\rangle \mid \boldsymbol{\lambda}_n \in \Pi^n,\ \boldsymbol{s}_n \in \mathbf{Z}^n(s)\} = \{|\lambda, s\rangle \mid \lambda \in \Pi\}$ is a basis of $\Lambda^{s+\frac{\infty}{2}}$.

Remark 4.2. (i) The pair $(\boldsymbol{\lambda}_n = (\lambda^{(1)}, \dots, \lambda^{(n)}), \boldsymbol{s}_n = (s_1, \dots, s_n))$ can be read off the diagram of the partition λ in the following manner. For $r \in \{0, 1, \dots, n-1\}$, let R_r (resp., C_r) be the set of all rows (resp., columns) of λ that have a node of residue r as their rightmost (resp., bottom) node. Then for each $c \in \{1, \dots, n\}$, the diagram of $\lambda^{(c)}$ is embedded into the diagram of λ by

$$\lambda^{(c)} = \lambda \cap C_c \cap R_{c-1},$$

where we set $C_n = C_0$. Hence $\boldsymbol{\lambda}_n$ is the n-quotient of λ. On the other hand, the sequence $\boldsymbol{s}_n$ is obtained as follows. Let $N_r(\lambda|s, n)$ be the number of addable nodes of residue r minus the number of removable nodes of residue r. Then for each $c \in \{1, \dots, n-1\}$, we have $s_c - s_{c+1} = N_c(\lambda|s, n)$. These equalities, together with $\sum_{a=1}^{n} s_a = s$, determine $\boldsymbol{s}_n$ completely. It follows that $\boldsymbol{s}_n$ is a particular parameterization of the n-core (cf. [19]) of λ.

(ii) In a similar manner one can describe the pair $(\boldsymbol{\lambda}_l, \boldsymbol{s}_l)$ corresponding to λ and s. In this case, we first associate with λ and s another partition, which we denote by

$\sigma^s(\lambda)$. Let $\boldsymbol{k} = (k_1, k_2, \dots)$ be the element of $\mathbf{P}^{++}(s)$ defined by (λ, s). For each $i \in \mathbf{N}$, we set $k_i = e_i - nlm_i$, where $e_i \in \{1, \dots, nl\}$, $m_i \in \mathbf{Z}$. Next, for every $e \in \{1, \dots, nl\}$, we define $(k_1^{(e)}, k_2^{(e)}, \dots)$ to be the semiinfinite strictly decreasing sequence $(1 - m_i \mid e_i = e)$. This sequence stabilizes, for $i \gg 1$, to $t_e - i + 1$, where t_e is a uniquely defined integer. Then we define a partition $\mu^{(e)} = (\mu_1^{(e)}, \mu_2^{(e)}, \dots)$ by $\mu_i^{(e)} = k_i^{(e)} - t_e + i - 1$. In this way, we get an nl-multipartition $(\mu^{(1)}, \dots, \mu^{(nl)})$ and a sequence of nl integers $(t_1, \dots, t_{nl})$ summing to s. Of course, these are simply the nl-quotient and the nl-core of λ. Let σ be the permutation of $\{1, \dots, nl\}$ defined by

$$\sigma : d + l(c-1) \mapsto c + n(d-1) \quad (c \in \{1, \dots, n\},\ d \in \{1, \dots, l\}).$$

Now $\sigma^s(\lambda)$ is defined to be the unique partition with the nl-quotient $(\mu^{(\sigma(1))}, \dots, \mu^{(\sigma(nl))})$ and the nl-core $(t_{\sigma(1)}, \dots, t_{\sigma(nl)})$. Finally, $\boldsymbol{\lambda}_l$ and s_l can be read off the diagram of $\sigma^s(\lambda)$ in exactly the same way as $(\boldsymbol{\lambda}_n, s_n)$ are read off the diagram of λ, i.e., as explained in (i) above, but reversing everywhere the roles of n and l.

Example 4.3. Table 1 illustrates, for partitions λ of size 7, the three indexations defined above. Here $n = 3$, $l = 2$, and $s = 0$.

λ	$\sigma^s(\lambda)$	$\boldsymbol{\lambda}_l$	s_l	$\boldsymbol{\lambda}_n$	s_n
(7)	(7)	$((3), \varnothing)$	$(1, -1)$	$((2), \varnothing, \varnothing)$	$(1, 0, -1)$
$(6, 1)$	$(6, 1^2)$	$(\varnothing, (3, 1))$	$(0, 0)$	$(\varnothing, \varnothing, (1))$	$(0, -1, 1)$
$(5, 2)$	$(4, 2, 1)$	$(\varnothing, (3))$	$(1, -1)$	$(\varnothing, (2), \varnothing)$	$(1, 0, -1)$
$(5, 1^2)$	$(4, 1^4)$	$(\varnothing, (2, 1^2))$	$(0, 0)$	$(\varnothing, (1), \varnothing)$	$(-1, 1, 0)$
$(4, 3)$	$(3^2, 1)$	$((1), (2))$	$(1, -1)$	$((1), (1), \varnothing)$	$(1, 0, -1)$
$(4, 2, 1)$	$(2^2, 1^3)$	$(\varnothing, (2, 1))$	$(1, -1)$	$((1^2), \varnothing, \varnothing)$	$(1, 0, -1)$
$(4, 1^3)$	$(2, 1)$	$(\varnothing, \varnothing)$	$(-1, 1)$	$((1), \varnothing, (1))$	$(1, 0, -1)$
$(3^2, 1)$	$(5, 4, 1^2)$	$((1^2), (1))$	$(2, -2)$	$(\varnothing, \varnothing, (1))$	$(-1, 1, 0)$
$(3, 2^2)$	$(5, 2^3, 1)$	$((1), (2))$	$(2, -2)$	$((1), \varnothing, \varnothing)$	$(0, -1, 1)$
$(3, 2, 1^2)$	$(5, 2)$	$((2, 1), \varnothing)$	$(1, -1)$	$(\varnothing, \varnothing, (2))$	$(1, 0, -1)$
$(3, 1^4)$	$(5, 1^3)$	$((3, 1), \varnothing)$	$(0, 0)$	$(\varnothing, (1), \varnothing)$	$(0, -1, 1)$
$(2^3, 1)$	$(3, 2^2)$	$((1^2), (1))$	$(1, -1)$	$(\varnothing, (1), (1))$	$(1, 0, -1)$
$(2^2, 1^3)$	$(3, 2, 1^2)$	$((1^3), \varnothing)$	$(1, -1)$	$(\varnothing, (1^2), \varnothing)$	$(1, 0, -1)$
$(2, 1^5)$	$(3, 1^5)$	$((2, 1^2), \varnothing)$	$(0, 0)$	$((1), \varnothing, \varnothing)$	$(-1, 1, 0)$
(1^7)	(1^7)	$(\varnothing, (1^3))$	$(1, -1)$	$(\varnothing, \varnothing, (1^2))$	$(1, 0, -1)$

Table 1.

4.2 Actions of quantum affine algebras on $\Lambda^{s+\frac{\infty}{2}}$. Taking the action (17)–(20) of $U_q(\widehat{\mathfrak{sl}}_n)$ and the action (21)–(24) of $U_p(\widehat{\mathfrak{sl}}_l)$ as the input, we shall define actions of $U_q(\widehat{\mathfrak{sl}}_n)$ and $U_p(\widehat{\mathfrak{sl}}_l)$ on the semiinfinite wedge product $\Lambda^{s+\frac{\infty}{2}}$. First, we assign to each vector u of $\Lambda^{*+\frac{\infty}{2}} = \oplus_{s \in \mathbf{Z}} \Lambda^{s+\frac{\infty}{2}}$ a weight $\mathrm{wt}(u)$ of $\widehat{\mathfrak{sl}}_n$ and a weight $\dot{\mathrm{w}}\mathrm{t}(v)$ of $\widehat{\mathfrak{sl}}_l$. We shall write $\mathrm{Wt}(u)$ for the sum of $\mathrm{wt}(u)$ and $\dot{\mathrm{w}}\mathrm{t}(u)$. For $s \in \mathbf{Z}$, set

$|s\rangle := u_s \wedge u_{s-1} \wedge \cdots$. Let $\mathrm{wt}(|0\rangle) := l\Lambda_0$ and $\dot{\mathrm{wt}}(|0\rangle) := n\dot{\Lambda}_0$. Here and in what follows we put dots over the fundamental weights, fundamental roots, and the null root of $\widehat{\mathfrak{sl}}_l$ in order to distinguish them from those of $\widehat{\mathfrak{sl}}_n$. For each nonzero integer s, let

$$\mathrm{Wt}(|s\rangle) := \mathrm{Wt}(|0\rangle) + \begin{cases} -\mathrm{Wt}(u_0 \wedge u_{-1} \wedge \cdots \wedge u_{s+1}) & \text{if } s < 0, \\ \mathrm{Wt}(u_s \wedge u_{s-1} \wedge \cdots \wedge u_1) & \text{if } s > 0. \end{cases}$$

Here $\mathrm{Wt}(u_0 \wedge u_{-1} \wedge \cdots \wedge u_{s+1})$ and $\mathrm{Wt}(u_s \wedge u_{s-1} \wedge \cdots \wedge u_1)$ are defined by (19)–(20) and (23)–(24). Then for $r \leq t$ and $v \in \Lambda^r$, the expression

$$\mathrm{Wt}(v \wedge u_{s-r} \wedge u_{s-r-1} \wedge \cdots \wedge u_{s-t+1}) + \mathrm{Wt}(|s-t\rangle)$$

is independent of the choice of t. Hence the assignment

$$\mathrm{Wt}(v \wedge |s-r) := \mathrm{Wt}(v) + \mathrm{Wt}(|s-r\rangle)$$

gives a well-defined weight for the vector $v \wedge |s-r\rangle$ of $\Lambda^{s+\frac{\infty}{2}}$. Thus we have obtained a weight decomposition of $\Lambda^{s+\frac{\infty}{2}}$. It is straightforward to verify that in terms of the basis $\{|\boldsymbol{\lambda}_l, \boldsymbol{s}_l\rangle \mid \boldsymbol{\lambda}_l \in \Pi^l,\ \boldsymbol{s}_l \in \mathbf{Z}^l(s)\} = \{|\boldsymbol{\lambda}_n, \boldsymbol{s}_n\rangle \mid \boldsymbol{\lambda}_n \in \Pi^n,\ \boldsymbol{s}_n \in \mathbf{Z}^n(s)\}$ this decomposition looks as follows (here we use the notation of Section 2.1):

$$\mathrm{wt}(|\boldsymbol{\lambda}_l, \boldsymbol{s}_l\rangle) = -\Delta(\boldsymbol{s}_l, n)\,\delta + \Lambda_{s_1} + \cdots + \Lambda_{s_l} - \sum_{i=0}^{n-1} M_i(\boldsymbol{\lambda}_l|\boldsymbol{s}_l, n)\alpha_i, \tag{27}$$

$$\begin{aligned}\dot{\mathrm{wt}}(|\boldsymbol{\lambda}_l, \boldsymbol{s}_l\rangle) &= -(\Delta(\boldsymbol{s}_l, n) + M_0(\boldsymbol{\lambda}_l|\boldsymbol{s}_l, n))\,\dot{\delta} \\ &\quad + (n - s_1 + s_l)\dot{\Lambda}_0 + (s_1 - s_2)\dot{\Lambda}_1 + \cdots + (s_{l-1} - s_l)\dot{\Lambda}_{l-1},\end{aligned} \tag{28}$$

$$\dot{\mathrm{wt}}(|\boldsymbol{\lambda}_n, \boldsymbol{s}_n\rangle) = -\Delta(\boldsymbol{s}_n, l)\,\dot{\delta} + \dot{\Lambda}_{s_1} + \cdots + \dot{\Lambda}_{s_n} - \sum_{i=0}^{l-1} M_i(\boldsymbol{\lambda}_n|\boldsymbol{s}_n, l)\dot{\alpha}_i, \tag{29}$$

$$\begin{aligned}\mathrm{wt}(|\boldsymbol{\lambda}_n, \boldsymbol{s}_n\rangle) &= -(\Delta(\boldsymbol{s}_n, l) + M_0(\boldsymbol{\lambda}_n|\boldsymbol{s}_n, l))\,\delta \\ &\quad + (l - s_1 + s_n)\Lambda_0 + (s_1 - s_2)\Lambda_1 + \cdots + (s_{n-1} - s_n)\Lambda_{n-1}.\end{aligned} \tag{30}$$

Now we define actions of the Cartan parts of $U_q(\widehat{\mathfrak{sl}}_n)$ and $U_p(\widehat{\mathfrak{sl}}_l)$ by

$$t_i\, u = q^{\langle \mathrm{wt}(u), h_i\rangle} u, \qquad \partial\, u = \langle \mathrm{wt}(u), \partial\rangle u \quad (i = 0, 1, \ldots, n-1), \tag{31}$$

$$\dot{t}_j\, u = p^{\langle \dot{\mathrm{wt}}(u), \dot{h}_j\rangle} u, \qquad \dot{\partial}\, u = \langle \dot{\mathrm{wt}}(u), \dot{\partial}\rangle u \quad (j = 0, 1, \ldots, l-1). \tag{32}$$

Next, we define on $\Lambda^{s+\frac{\infty}{2}}$ actions of the raising generators of $U_q(\widehat{\mathfrak{sl}}_n)$ and $U_p(\widehat{\mathfrak{sl}}_l)$. Let $v \in \Lambda^r$ for some $r \in \mathbf{N}$. Lemma 3.18(i), (17), and (21) imply that the expressions

$$e_i(v \wedge u_{s-r} \wedge u_{s-r-1} \wedge \cdots \wedge u_{s-t+1}) \wedge (t_i)^{-1}|s-t\rangle \quad (i = 0, \ldots, n-1),$$

$$\dot{e}_j(v \wedge u_{s-r} \wedge u_{s-r-1} \wedge \cdots \wedge u_{s-t+1}) \wedge (\dot{t}_j)^{-1}|s-t\rangle \quad (j = 0, \ldots, l-1)$$

are independent of t for $t \geq r$. Hence the assignments

$$e_i(v \wedge |s-r\rangle) := e_i(v) \wedge (t_i)^{-1}|s-r\rangle \quad (i=0,\ldots,n-1),$$
$$\dot{e}_j(v \wedge |s-r\rangle) := \dot{e}_j(v) \wedge (\dot{t}_j)^{-1}|s-r\rangle \quad (j=0,\ldots,l-1)$$

determine well-defined endomorphisms of $\Lambda^{s+\frac{\infty}{2}}$.

Finally, we define actions of the lowering generators of $U_q(\widehat{\mathfrak{sl}}_n)$ and $U_p(\widehat{\mathfrak{sl}}_l)$ on $\Lambda^{s+\frac{\infty}{2}}$. Now using Lemma 3.18(ii), (18), and (22), we can check that the expressions

$$f_i(v \wedge u_{s-r} \wedge u_{s-r-1} \wedge \cdots \wedge u_{s-t+1}) \wedge u_{s-t} \wedge u_{s-t+1} \wedge \cdots \wedge u_{s-m+1},$$
$$\dot{f}_j(v \wedge u_{s-r} \wedge u_{s-r-1} \wedge \cdots \wedge u_{s-t+1}) \wedge u_{s-t} \wedge u_{s-t+1} \wedge \cdots \wedge u_{s-m+1}$$
$$(i=0,\ldots,n-1,\ j=0,\ldots,l-1)$$

are independent of $t \leq m$ provided $t > r$ and t is sufficiently large. Hence we obtain well-defined endomorphisms of $\Lambda^{s+\frac{\infty}{2}}$ by setting

$$f_i(v \wedge |s-r\rangle) := f_i(v \wedge u_{s-r} \wedge u_{s-r-1} \wedge \cdots \wedge u_{s-t+1}) \wedge |s-t\rangle,$$
$$\dot{f}_j(v \wedge |s-r\rangle) := \dot{f}_j(v \wedge u_{s-r} \wedge u_{s-r-1} \wedge \cdots \wedge u_{s-t+1}) \wedge |s-t\rangle,$$

where t is arbitrary such that $t \gg r$ $(i=0,\ldots,n-1,\ j=0,\ldots,l-1)$.

The actions of f_i and e_i defined above are easily described in terms of the basis $B(s) = \{|\boldsymbol{\lambda}_l, \boldsymbol{s}_l\rangle \mid \boldsymbol{\lambda}_l \in \Pi^l,\ \boldsymbol{s}_l \in \mathbf{Z}^l(s)\}$. Using the notation of Section 2.1, we have

$$f_i|\boldsymbol{\lambda}_l, \boldsymbol{s}_l\rangle = \sum_{\mathrm{res}_n(\boldsymbol{\mu}_l/\boldsymbol{\lambda}_l)=i} q^{N_i^{>}(\boldsymbol{\lambda}_l, \boldsymbol{\mu}_l|\boldsymbol{s}_l, n)} |\boldsymbol{\mu}_l, \boldsymbol{s}_l\rangle, \tag{33}$$
$$e_i|\boldsymbol{\mu}_l, \boldsymbol{s}_l\rangle = \sum_{\mathrm{res}_n(\boldsymbol{\mu}_l/\boldsymbol{\lambda}_l)=i} q^{-N_i^{<}(\boldsymbol{\lambda}_l, \boldsymbol{\mu}_l|\boldsymbol{s}_l, n)} |\boldsymbol{\lambda}_l, \boldsymbol{s}_l\rangle. \tag{34}$$

These actions and the actions of the generators t_i and ∂ defined by (27) are identical to those defined on the combinatorial Fock space $\mathbf{F}_q[\boldsymbol{s}_l]$. It follows that e_i, f_i, t_i, and ∂ satisfy the defining relations of $U_q(\widehat{\mathfrak{sl}}_n)$. Moreover, we see that for each $\boldsymbol{s}_l = (s_1, \ldots, s_l) \in \mathbf{Z}^l$, the Fock space $\mathbf{F}_q[\boldsymbol{s}_l]$ is realized inside $\Lambda^{s+\frac{\infty}{2}}$ with $s = s_1 + \cdots + s_l$ as the subspace spanned by $\{|\boldsymbol{\lambda}_l, \boldsymbol{s}_l\rangle \mid \boldsymbol{\lambda}_l \in \Pi^l\}$. Note that by (28), the Fock space $\mathbf{F}_q[\boldsymbol{s}_l]$ is the set of all vectors $u \in \Lambda^{s+\frac{\infty}{2}}$ such that $\dot{\mathrm{wt}}(u)$ is congruent to $(n - s_1 + s_l)\dot{\Lambda}_0 + (s_1 - s_2)\dot{\Lambda}_1 + \cdots + (s_{l-1} - s_l)\dot{\Lambda}_{l-1}$ modulo $\mathbf{Z}\dot{\delta}$.

Similarly, one can describe the actions of $\dot{f}_j$ and $\dot{e}_j$ in terms of the basis $B(s)$. Now the formulas acquire a simple form if we use the indexation of $B(s)$ by $(\boldsymbol{\lambda}_n, \boldsymbol{s}_n)$ $(\boldsymbol{\lambda}_n \in \Pi^n, \boldsymbol{s}_n \in \mathbf{Z}^n(s))$. We have

$$\dot{f}_j|\boldsymbol{\lambda}_n, \boldsymbol{s}_n\rangle = \sum_{\mathrm{res}_l(\boldsymbol{\mu}_n/\boldsymbol{\lambda}_n)=j} p^{N_j^{>}(\boldsymbol{\lambda}_n, \boldsymbol{\mu}_n|\boldsymbol{s}_n, l)} |\boldsymbol{\mu}_n, \boldsymbol{s}_n\rangle, \tag{35}$$
$$\dot{e}_j|\boldsymbol{\mu}_n, \boldsymbol{s}_n\rangle = \sum_{\mathrm{res}_l(\boldsymbol{\mu}_n/\boldsymbol{\lambda}_n)=j} p^{-N_j^{<}(\boldsymbol{\lambda}_n, \boldsymbol{\mu}_n|\boldsymbol{s}_n, l)} |\boldsymbol{\lambda}_n, \boldsymbol{s}_n\rangle. \tag{36}$$

These actions and the actions of the generators $\dot{t}_j$ and $\dot{\partial}$ defined by (29) are identical to those defined on the combinatorial Fock space $\mathbf{F}_q[s_n]$. It follows that $\dot{e}_j$, $\dot{f}_j$, $\dot{t}_j$, and $\dot{\partial}$ satisfy the defining relations of $U_p(\widehat{\mathfrak{sl}}_l)$. Moreover, for each $s_n = (s_1, \dots, s_n) \in \mathbf{Z}^n$, the Fock space $\mathbf{F}_q[s_n]$ is realized inside $\Lambda^{s+\frac{\infty}{2}}$ with $s = s_1 + \cdots + s_n$ as the subspace spanned by $\{|\boldsymbol{\lambda}_n, s_n\rangle \mid \boldsymbol{\lambda}_n \in \Pi^n\}$. Note that by (30), the Fock space $\mathbf{F}_q[s_n]$ is the set of all vectors $u \in \Lambda^{s+\frac{\infty}{2}}$ such that $\mathrm{wt}(u)$ is congruent to $(l - s_1 + s_n)\Lambda_0 + (s_1 - s_2)\Lambda_1 + \cdots + (s_{n-1} - s_n)\Lambda_{n-1}$ modulo $\mathbf{Z}\delta$.

4.3 Action of the Heisenberg algebra on $\Lambda^{s+\frac{\infty}{2}}$. Let $v \in \Lambda^r$ for some $r \in \mathbf{N}$. Lemma 3.18(i) and (25) imply that for $m > 0$ the expression

$$B_m(v \wedge u_{s-r} \wedge u_{s-r-1} \wedge \cdots \wedge u_{s-t+1}) \wedge |s - t\rangle$$

is independent of t for $t \geq r$. Hence the assignment

$$B_m(v \wedge |s - r\rangle) := B_m(v) \wedge |s - r\rangle \quad (m > 0)$$

determines a well-defined endomorphism of $\Lambda^{s+\frac{\infty}{2}}$. Now using Lemma 3.18(ii), one can check that for $m < 0$ the expression

$$B_m(v \wedge u_{s-r} \wedge u_{s-r-1} \wedge \cdots \wedge u_{s-t+1}) \wedge u_{s-t} \wedge u_{s-t+1} \wedge \cdots \wedge u_{s-k+1}$$

is independent of $t \leq k$ provided that $t > r$ and t is sufficiently large. Hence we obtain a well-defined endomorphism of $\Lambda^{s+\frac{\infty}{2}}$ by setting

$$B_m(v \wedge |s - r\rangle) := B_m(v \wedge u_{s-r} \wedge u_{s-r-1} \wedge \cdots \wedge u_{s-t+1}) \wedge |s - t\rangle \qquad (m < 0),$$

where t is arbitrary such that $t \gg r$. It is clear that B_m $(m \in \mathbf{N})$ has weight $m\delta + m\dot{\delta}$ with respect to the weight decomposition of $\Lambda^{s+\frac{\infty}{2}}$ defined in the previous section. This implies that the subspaces $\mathbf{F}_q[s_l]$ and $\mathbf{F}_q[s_n]$ are preserved by B_m. The next proposition shows that B_m generates a Heisenberg algebra $\mathcal{H}$.

Proposition 4.4. *There are nonzero $\gamma_m \in \mathbf{K}$ (independent of s) such that*

$$[B_m, B_{m'}] = \delta_{m+m',0}\gamma_m.$$

Proof. Since for each r the actions of B_m on Λ^r commute, for any $v \in \Lambda^r$ we have

$$[B_m, B_{m'}](v \wedge |s - r\rangle) = v \wedge [B_m, B_{m'}]|s - r\rangle. \tag{37}$$

It is therefore enough to show that the statement of the proposition holds when $[B_m, B_{m'}]$ is applied to $|s\rangle$ with arbitrary $s \in \mathbf{Z}$. Let us first assume that $m + m' > 0$. Then it is easy to see from (27)–(30) that $\mathrm{Wt}(|s\rangle) + (m + m')(\delta + \dot{\delta})$ is not a weight of $\Lambda^{s+\frac{\infty}{2}}$. Hence $[B_m, B_{m'}]|s\rangle = 0$ in this case. Next, let $m + m' < 0$. Write $[B_m, B_{m'}]|s\rangle$ as a linear combination of ordered wedges

$$[B_m, B_{m'}]|s\rangle = \sum_\nu c_\nu u_{k_1^\nu} \wedge u_{k_2^\nu} \wedge \cdots,$$

where c_ν are nonzero coefficients. Since $\mathrm{Wt}(|s\rangle) + (m+m')(\delta+\dot{\delta})$ is distinct from the weight of $|s\rangle$, we have $k_1^\nu > s$ for all ν. For any $t > 0$, (37) gives

$$[B_m, B_{m'}]|s\rangle = u_s \wedge u_{s-1} \wedge \cdots \wedge u_{s-nlt+1} \wedge [B_m, B_{m'}]|s-nlt\rangle.$$

From the structure of the ordering rules of Proposition 3.16, it follows that

$$[B_m, B_{m'}]|s-nlt\rangle = \sum_\nu c_\nu u_{k_1^\nu - nlt} \wedge u_{k_2^\nu - nlt} \wedge \cdots .$$

Choosing t sufficiently large so that $s \geq k_1^\nu - nlt$ holds for all ν and taking into account the inequality $k_1^\nu - nlt > s - nlt$, we obtain by Lemma 3.18(ii) that $u_s \wedge u_{s-1} \wedge \cdots \wedge u_{s-nlt+1} \wedge u_{k_1^\nu} \wedge u_{k_2^\nu} \wedge \cdots$ vanishes for all ν. Hence $[B_m, B_{m'}]|s\rangle$ is zero.

Finally, let $m + m' = 0$. Then the weight of $[B_m, B_{m'}]|s\rangle$ equals $\mathrm{Wt}(|s\rangle)$. It is easy to see that the vector $|s\rangle$ coincides with $|\varnothing_l, \mathbf{s}_l\rangle$ for a certain $\mathbf{s}_l \in \mathbf{Z}^l(s)$. However, from (27)–(30), it is clear that the weight subspace of $\mathbf{F}_q[\mathbf{s}_l]$ with the weight $\mathrm{Wt}(|\varnothing_l, \mathbf{s}_l\rangle)$ is one dimensional. Since B_m preserves $\mathbf{F}_q[\mathbf{s}_l]$, it follows that $[B_m, B_{m'}]|s\rangle = \gamma_m|s\rangle$ for some $\gamma_m \in \mathbf{K}$. Using $[B_m, B_{m'}]|s\rangle = u_s \wedge [B_m, B_{m'}]|s-1\rangle$ shows that γ_m does not depend on s. Specializing to $q = 1$, we obtain $\gamma_m|_{q=1} = mnl$. Hence $\gamma_m \neq 0$. □

Proposition 4.5. *For $m > 0$, we have*

$$\gamma_m = m\frac{1-q^{-2mn}}{1-q^{-2m}}\frac{1-q^{2ml}}{1-q^{2m}}.$$

For $l = 1$, a proof of this proposition is given in [13]. We give a proof for all $l \in \mathbf{N}$ in Section 5.1.1.

Following the same reasoning that was used above to show that B_m commutes with $B_{m'}$ unless $m + m' = 0$, we obtain the following.

Proposition 4.6. *The actions of $U_q'(\widehat{\mathfrak{sl}}_n)$, $U_p'(\widehat{\mathfrak{sl}}_l)$, and $\mathcal{H}$ on $\Lambda^{s+\frac{\infty}{2}}$ are pairwise mutually commutative.*

Taking into account the weight decomposition of $\Lambda^{s+\frac{\infty}{2}}$ given by (27)–(30) and the fact that weight of B_m equals $m(\delta+\dot{\delta})$, we see that vectors $|\varnothing_l, \mathbf{s}_l\rangle$ ($\mathbf{s}_l \in \mathbf{Z}^l(s)$) are singular vectors of $\mathcal{H}$, i.e., are annihilated by B_m with positive m. Clearly, these vectors are also singular vectors for $U_q(\widehat{\mathfrak{sl}}_n)$. Likewise, the vectors $|\varnothing_n, \mathbf{s}_n\rangle$ ($\mathbf{s}_n \in \mathbf{Z}^n(s)$) are singular for both $\mathcal{H}$ and $U_p(\widehat{\mathfrak{sl}}_l)$. Define the sets $A_l^n(s)$ and $A_n^l(s)$ by

$$A_l^n(s) := \{\mathbf{s}_l = (s_1, \ldots, s_l) \in \mathbf{Z}^l(s) \mid s_1 \geq s_2 \geq \cdots \geq s_l,\ s_1 - s_l \leq n\},$$
$$A_n^l(s) := \{\mathbf{s}_n = (s_1, \ldots, s_n) \in \mathbf{Z}^n(s) \mid s_1 \geq s_2 \geq \cdots \geq s_n,\ s_1 - s_n \leq l\}.$$

The definitions of $|\boldsymbol{\lambda}_l, \mathbf{s}_l\rangle$ and $|\boldsymbol{\lambda}_n, \mathbf{s}_n\rangle$ given in Section 4.1 imply that

$$\{|\varnothing_l, \mathbf{s}_l\rangle \mid \mathbf{s}_l \in A_l^n(s)\} = \{|\varnothing_n, \mathbf{s}_n\rangle \mid \mathbf{s}_n \in A_n^l(s)\}. \tag{38}$$

Hence $|\emptyset_l, s_l\rangle$ ($s_l \in A_l^n(s)$) or, equivalently, $|\emptyset_n, s_n\rangle$ ($s_n \in A_n^l(s)$) are the only vectors of the basis $B(s)$ that are simultaneously singular for $U_q(\widehat{\mathfrak{sl}}_n)$, $U_p(\widehat{\mathfrak{sl}}_l)$ and $\mathcal{H}$. Equality (38) shows that we have a bijection $A_l^n(s) \to A_n^l(s)$ such that

$$s_l \mapsto s_n \quad \text{if and only if} \quad |\emptyset_l, s_l\rangle = |\emptyset_n, s_n\rangle.$$

This bijection is completely determined by comparing the weights of $|\emptyset_l, s_l\rangle$ and $|\emptyset_n, s_n\rangle$ according to (27)–(30). Namely, $(t_1, \ldots, t_n) \in A_n^l(s)$ is the image of $(s_1, \ldots, s_l) \in A_l^n(s)$ if and only if

$$\Lambda_{s_1} + \cdots + \Lambda_{s_l} = (l - t_1 + t_n)\Lambda_0 + (t_1 - t_2)\Lambda_1 + \cdots + (t_{n-1} - t_n)\Lambda_{n-1}$$

or, equivalently, if and only if

$$\dot{\Lambda}_{t_1} + \cdots + \dot{\Lambda}_{t_n} = (n - s_1 + s_l)\dot{\Lambda}_0 + (s_1 - s_2)\dot{\Lambda}_1 + \cdots + (s_{l-1} - s_l)\dot{\Lambda}_{l-1}.$$

Example 4.7. Let $n = 5$, $l = 2$, and $s = 11$. Then $A_2^5(11)$ contains three elements: $(6, 5)$, $(7, 4)$, and $(8, 3)$. On the other hand, $A_5^2(11)$ is formed by the elements

$$(3, 2, 2, 2, 2), (3, 3, 2, 2, 1), (3, 3, 3, 1, 1).$$

The bijective correspondence between $A_2^5(11)$ and $A_5^2(11)$ is given by

$$\begin{aligned}
|\emptyset_2, (6,5)\rangle &= |\emptyset_5, (3,2,2,2,2)\rangle, && -(\delta + \dot{\delta}) + \Lambda_0 + \Lambda_1 + 4\dot{\Lambda}_0 + \dot{\Lambda}_1,\\
|\emptyset_2, (7,4)\rangle &= |\emptyset_5, (3,3,2,2,1)\rangle, && -2(\delta + \dot{\delta}) + \Lambda_2 + \Lambda_4 + 2\dot{\Lambda}_0 + 3\dot{\Lambda}_1,\\
|\emptyset_2, (8,3)\rangle &= |\emptyset_5, (3,3,3,1,1)\rangle, && -3(\delta + \dot{\delta}) + 2\Lambda_3 + 5\dot{\Lambda}_1.
\end{aligned}$$

Here we have listed the weights of the corresponding vectors in the right column.

The next theorem shows that $\{|\emptyset_l, s_l\rangle \mid s_l \in A_l^n(s)\} = \{|\emptyset_n, s_n\rangle \mid s_n \in A_n^l(s)\}$ is the complete set of singular vectors in $\Lambda^{s+\frac{\infty}{2}}$. A proof follows immediately from [6, Theorem 1.6] (see also [7, Theorem 3.2]).

Theorem 4.8.

$$\Lambda^{s+\frac{\infty}{2}} = \bigoplus_{s_l \in A_l^n(s)} U_q'(\widehat{\mathfrak{sl}}_n) \cdot \mathcal{H} \cdot U_p'(\widehat{\mathfrak{sl}}_l)\, |\emptyset_l, s_l\rangle;$$

equivalently,

$$\Lambda^{s+\frac{\infty}{2}} = \bigoplus_{s_n \in A_n^l(s)} U_q'(\widehat{\mathfrak{sl}}_n) \cdot \mathcal{H} \cdot U_p'(\widehat{\mathfrak{sl}}_l)\, |\emptyset_n, s_n\rangle.$$

Corollary 4.9. (i) *For $t_l = (t_1, \ldots, t_l) \in A_l^n(s)$, let $s_l = (s_1, \ldots, s_l)$ be any element of $\mathbf{Z}^l(s)$ such that $\Lambda_{s_1} + \cdots + \Lambda_{s_l} = \Lambda_{t_1} + \cdots + \Lambda_{t_l}$. Then $|\emptyset_l, t_l\rangle \in U_p'(\widehat{\mathfrak{sl}}_l)|\emptyset_l, s_l\rangle$.*

(ii) *For $t_n = (t_1, \ldots, t_n) \in A_n^l(s)$, let $s_n = (s_1, \ldots, s_n)$ be any element of $\mathbf{Z}^n(s)$ such that $\dot{\Lambda}_{s_1} + \cdots + \dot{\Lambda}_{s_n} = \dot{\Lambda}_{t_1} + \cdots + \dot{\Lambda}_{t_n}$.*

Then $|\emptyset_n, t_n\rangle \in U_q'(\widehat{\mathfrak{sl}}_n)|\emptyset_n, s_n\rangle$.

PROOF. Since the proofs of (i) and (ii) are almost identical, we show only (i). Since $|\varnothing_l, s_l\rangle$ is a singular vector for $U_q(\widehat{\mathfrak{sl}}_n) \cdot \mathcal{H}$, we have, by Theorem 4.8,

$$|\varnothing_l, s_l\rangle \in \bigoplus_{r_l \in A_l^n(s)} U_p'(\widehat{\mathfrak{sl}}_l)\, |\varnothing_l, r_l\rangle.$$

Observe that for two distinct elements r_l and t_l of $A_l^n(s)$, we have $\Lambda_{r_1} + \cdots + \Lambda_{r_l} \neq \Lambda_{t_1} + \cdots + \Lambda_{t_l}$. Therefore, comparing $\widehat{\mathfrak{sl}}_n$-weights, we have

$$|\varnothing_l, s_l\rangle \in U_p'(\widehat{\mathfrak{sl}}_l)\, |\varnothing_l, t_l\rangle.$$

Since $U_p'(\widehat{\mathfrak{sl}}_l)\, |\varnothing_l, t_l\rangle$ is an irreducible representation of $U_p'(\widehat{\mathfrak{sl}}_l)$, the claim follows. □

4.4 Canonical bases of the q-deformed Fock spaces. Fix an arbitrary integer s and define a gradation of the semiinfinite wedge product $\Lambda^{s+\frac{\infty}{2}}$ by setting $\deg |\lambda, s\rangle = |\lambda|$.

Lemma 4.10. *Let $k = (k_1, k_2, \dots) \in \mathbf{P}^{++}(s)$. Then for any $t, r \in \mathbf{N}$ such that $t > r \geq \deg u_k$, we have*

$$\overline{u_{k_1} \wedge u_{k_2} \wedge \cdots \wedge u_{k_r}} \wedge u_{k_{r+1}} \wedge \cdots \wedge u_{k_t} = \overline{u_{k_1} \wedge u_{k_2} \wedge \cdots \wedge u_{k_r} \wedge u_{k_{r+1}} \wedge \cdots \wedge u_{k_t}}.$$

For $l = 1$, a proof of this lemma is given in [18, proof of Lemma 7.7]. For arbitrary l, a proof is virtually identical and will be omitted.

From this lemma, it follows that for $k = (k_1, k_2, \dots) \in \mathbf{P}^{++}(s)$, the assignment

$$\overline{u_k} := \overline{u_{k_1} \wedge u_{k_2} \wedge \cdots \wedge u_{k_r}} \wedge u_{k_{r+1}} \wedge u_{k_{r+2}} \wedge \cdots \qquad (r \geq \deg u_k) \tag{39}$$

determines a well-defined semilinear involution $u \mapsto \overline{u}$ of $\Lambda^{s+\frac{\infty}{2}}$. It is easily seen from the weight decomposition of $\Lambda^{s+\frac{\infty}{2}}$ defined in Section 4.2 that $\mathrm{Wt}(\overline{u}) = \mathrm{Wt}(u)$ for any weight vector u of $\Lambda^{s+\frac{\infty}{2}}$. Hence for $s_l \in \mathbf{Z}^l(s)$ (resp., $s_n \in \mathbf{Z}^n(s)$), the Fock space $\mathbf{F}_q[s_l]$ (resp., $\mathbf{F}_q[s_n]$) is invariant with respect to the bar-involution. Therefore, in particular, we have

$$\overline{|\lambda_l, s_l\rangle} = \sum_{\mu_l \in \Pi^l} R_{\lambda_l, \mu_l}(s_l | q)\, |\mu_l, s_l\rangle,$$

where $R_{\lambda_l, \mu_l}(s_l | q)$ is a Laurent polynomial in q with integral coefficients. From (27) and the fact that the involution preserves the weight subspaces of $\Lambda^{s+\frac{\infty}{2}}$, it follows that $R_{\lambda_l, \mu_l}(s_l | q)$ is nonzero only if $|\lambda_l| = |\mu_l|$.

For $\lambda_l, \mu_l \in \Pi^l$ and $s_l \in \mathbf{Z}^l(s)$, let $k = (k_1, k_2, \dots)$ and $l = (l_1, l_2, \dots)$ be the unique elements of $\mathbf{P}^{++}(s)$ such that $|\lambda_l, s_l\rangle = u_k$ and $|\mu_l, s_l\rangle = u_l$. Then (39) implies that

$$R_{\lambda_l, \mu_l}(s_l | q) = R_{(k)_r, (l)_r}(q), \tag{40}$$

where $(\boldsymbol{k})_r := (k_1, k_2, \ldots, k_r)$, $(\boldsymbol{l})_r := (l_1, l_2, \ldots, l_r)$, and r is an arbitrary integer satisfying $r \geq \deg u_{\boldsymbol{k}}, \deg u_{\boldsymbol{l}}$. Here the coefficient $R_{(\boldsymbol{k})_r,(\boldsymbol{l})_r}(q)$ is defined in Section 3.4. The unitriangularity of the matrix $\|R_{\boldsymbol{k},\boldsymbol{l}}(q)\|$ $(\boldsymbol{k}, \boldsymbol{l} \in \mathbf{P}^{++})$ described in that section immediately leads to the following.

Proposition 4.11. *For $\boldsymbol{\lambda}_l, \boldsymbol{\mu}_l \in \Pi^l$ and $\boldsymbol{s}_l \in \mathbf{Z}^l(s)$, the coefficient $R_{\boldsymbol{\lambda}_l,\boldsymbol{\mu}_l}(\boldsymbol{s}_l|q)$ is zero unless the partition $\lambda = (\tau_l^s)^{-1}(\boldsymbol{\lambda}_l, \boldsymbol{s}_l)$ is greater than or equal to the partition $\mu = (\tau_l^s)^{-1}(\boldsymbol{\mu}_l, \boldsymbol{s}_l)$ with respect to the dominance order on partitions. Moreover, $R_{\boldsymbol{\lambda}_l,\boldsymbol{\lambda}_l}(\boldsymbol{s}_l|q) = 1$.*

The unitriangularity of the involution matrix $\|R_{\boldsymbol{\lambda}_l,\boldsymbol{\mu}_l}(\boldsymbol{s}_l|q)\|$ allows us to define canonical bases $\{\mathcal{G}^+(\boldsymbol{\lambda}_l, \boldsymbol{s}_l) \mid \boldsymbol{\lambda}_l \in \Pi^l\}$ and $\{\mathcal{G}^-(\boldsymbol{\lambda}_l, \boldsymbol{s}_l) \mid \boldsymbol{\lambda}_l \in \Pi^l\}$ of the Fock space $\mathbf{F}_q[\boldsymbol{s}_l]$ for arbitrary $\boldsymbol{s}_l \in \mathbf{Z}^l(s)$. These bases are characterized by

(i) $\overline{\mathcal{G}^+(\boldsymbol{\lambda}_l, \boldsymbol{s}_l)} = \mathcal{G}^+(\boldsymbol{\lambda}_l, \boldsymbol{s}_l)$, $\quad \overline{\mathcal{G}^-(\boldsymbol{\lambda}_l, \boldsymbol{s}_l)} = \mathcal{G}^-(\boldsymbol{\lambda}_l, \boldsymbol{s}_l)$,

(ii) $\mathcal{G}^+(\boldsymbol{\lambda}_l, \boldsymbol{s}_l) \equiv |\boldsymbol{\lambda}_l, \boldsymbol{s}_l\rangle \bmod q\mathcal{L}^+(s)$, $\quad \mathcal{G}^-(\boldsymbol{\lambda}_l, \boldsymbol{s}_l) \equiv |\boldsymbol{\lambda}_l, \boldsymbol{s}_l\rangle \bmod q^{-1}\mathcal{L}^-(s)$,

where $\mathcal{L}^+(s)$ (resp., $\mathcal{L}^-(s)$) is the $\mathbf{Q}[q]$-lattice (resp., $\mathbf{Q}[q^{-1}]$-lattice) of $\Lambda^{s+\frac{\infty}{2}}$ generated by the basis

$$\begin{aligned} B(s) &= \{|\lambda, s\rangle \mid \lambda \in \Pi\} \\ &= \{|\boldsymbol{\lambda}_l, \boldsymbol{s}_l\rangle \mid \boldsymbol{\lambda}_l \in \Pi^l,\ \boldsymbol{s}_l \in \mathbf{Z}^l(s)\} \\ &= \{|\boldsymbol{\lambda}_n, \boldsymbol{s}_n\rangle \mid \boldsymbol{\lambda}_n \in \Pi^n,\ \boldsymbol{s}_n \in \mathbf{Z}^n(s)\}. \end{aligned}$$

Set

$$\begin{aligned} \mathcal{G}^+(\boldsymbol{\lambda}_l, \boldsymbol{s}_l) &= \sum_{\boldsymbol{\mu}_l \in \Pi^l} \Delta^+_{\boldsymbol{\lambda}_l,\boldsymbol{\mu}_l}(\boldsymbol{s}_l|q)\, |\boldsymbol{\mu}_l, \boldsymbol{s}_l\rangle, \\ \mathcal{G}^-(\boldsymbol{\lambda}_l, \boldsymbol{s}_l) &= \sum_{\boldsymbol{\mu}_l \in \Pi^l} \Delta^-_{\boldsymbol{\lambda}_l,\boldsymbol{\mu}_l}(\boldsymbol{s}_l|q)\, |\boldsymbol{\mu}_l, \boldsymbol{s}_l\rangle. \end{aligned}$$

Then, keeping notation as in (40), we have

$$\Delta^+_{\boldsymbol{\lambda}_l,\boldsymbol{\mu}_l}(\boldsymbol{s}_l|q) = \Delta^+_{(\boldsymbol{k})_r,(\boldsymbol{l})_r}(q), \qquad \Delta^-_{\boldsymbol{\lambda}_l,\boldsymbol{\mu}_l}(\boldsymbol{s}_l|q) = \Delta^-_{(\boldsymbol{k})_r,(\boldsymbol{l})_r}(q),$$

where the matrices $\|\Delta^{\pm}_{\boldsymbol{k},\boldsymbol{l}}(q)\|$ $(\boldsymbol{k}, \boldsymbol{l} \in \mathbf{P}^{++})$ are defined in Section 3.4. Hence Theorem 3.26 shows that $\Delta^{\pm}_{\boldsymbol{\lambda}_l,\boldsymbol{\mu}_l}(\boldsymbol{s}_l|q)$ are parabolic Kazhdan–Lusztig polynomials. Note that $R_{\boldsymbol{\lambda}_l,\boldsymbol{\mu}_l}(\boldsymbol{s}_l|q) \neq 0$ only if $|\boldsymbol{\lambda}_l| = |\boldsymbol{\mu}_l|$ implies that $\Delta^{\pm}_{\boldsymbol{\lambda}_l,\boldsymbol{\mu}_l}(\boldsymbol{s}_l|q) \neq 0$ only if $|\boldsymbol{\lambda}_l| = |\boldsymbol{\mu}_l|$. For each nonnegative integer k, let us set

$$\|\Delta^{\pm}_{\boldsymbol{\lambda}_l,\boldsymbol{\mu}_l}(\boldsymbol{s}_l|q)\|_k = \|\Delta^{\pm}_{\boldsymbol{\lambda}_l,\boldsymbol{\mu}_l}(\boldsymbol{s}_l|q)\| \quad (|\boldsymbol{\lambda}_l| = |\boldsymbol{\mu}_l| = k).$$

A proof of the next proposition in the special case $l = 1$ is given in [18]. A proof of the general case is similar and will be omitted.

Proposition 4.12. *For each* $u \in \Lambda^{s+\frac{\infty}{2}}$, *one has*

$$\overline{e_i u} = e_i \overline{u}, \qquad \overline{f_i u} = f_i \overline{u} \qquad (i \in \{0, 1, \ldots, n-1\}),$$
$$\overline{\dot{e}_j u} = \dot{e}_j \overline{u}, \qquad \overline{\dot{f}_j u} = \dot{f}_j \overline{u} \qquad (j \in \{0, 1, \ldots, l-1\}),$$
$$\overline{B_{-m} u} = B_{-m} \overline{u}, \qquad \overline{B_m u} = q^{2m(n-l)} B_m \overline{u} \quad (m > 0).$$

Let us now describe how the canonical bases relate to the global crystal bases of Kashiwara. As in Section 2.2, let $\mathcal{L}[s_l] = \oplus_{\lambda_l \in \Pi^l} A\, |\lambda_l, s_l\rangle$ be the lower crystal lattice of $\mathbf{F}_q[s_l]$ at $q = 0$. Proposition 4.12 then implies that $\overline{\mathcal{L}[s_l]} = \oplus_{\lambda_l \in \Pi^l} \overline{A}\, \overline{|\lambda_l, s_l\rangle}$, where $\overline{A} \subset \mathbf{Q}(q)$ is the subring of rational functions regular at $q = \infty$, is a lower crystal lattice of $\mathbf{F}_q[s_l]$ at $q = \infty$ (cf. [11, 12]). Let $U_q(\widehat{\mathfrak{sl}}_n)_{\mathbf{Q}}$ be the $\mathbf{Q}[q, q^{-1}]$-subalgebra of $U_q(\widehat{\mathfrak{sl}}_n)$ generated by the q-divided differences $e_i^{(m)} and f_i^{(m)}$ and

$$\prod_{k=1}^{m} \frac{q^{1-k} t_i - (t_i)^{-1} q^{k-1}}{q^k - q^{-k}}$$

with $m \in \mathbf{N}$. One can show [3, Lemma 2.7] that

$$\mathbf{F}_q[s_l]_{\mathbf{Q}} = \oplus_{\lambda_l \in \Pi^l} \mathbf{Q}[q, q^{-1}]\, |\lambda_l, s_l\rangle$$

is invariant with respect to the action of $U_q(\widehat{\mathfrak{sl}}_n)_{\mathbf{Q}}$ on $\mathbf{F}_q[s_l]$.

The existence and uniqueness of the basis $\{\mathcal{G}^+(\lambda_l, s_l) \mid \lambda_l \in \Pi^l\}$ can, by using the unitriangularity of the bar-involution, be reformulated as the existence of an isomorphism

$$\mathbf{F}_q[s_l]_{\mathbf{Q}} \cap \mathcal{L}[s_l] \cap \overline{\mathcal{L}[s_l]} \xrightarrow{\sim} \mathcal{L}[s_l]/q\mathcal{L}[s_l]$$

such that the preimage of $|\lambda_l, s_l\rangle$ mod $q\mathcal{L}[s_l]$ is $\mathcal{G}^+(\lambda_l, s_l)$. Therefore, in the terminology of [11, 12],

$$\{\mathcal{G}^+(\lambda_l, s_l) \mid \lambda_l \in \Pi^l\}$$

is a lower global crystal basis of the integrable $U_q(\widehat{\mathfrak{sl}}_n)$-module $\mathbf{F}_q[s_l]$.

Now let us use the indexation of the basis $B(s)$ by pairs (λ_n, s_n) with $\lambda_n \in \Pi^n$ and $s_n \in \mathbf{Z}^n(s)$. Certainly, we may label the canonical bases by these pairs as well so that

$$\{\mathcal{G}^\pm(\lambda_n, s_n) \mid \lambda_n \in \Pi^n,\ s_n \in \mathbf{Z}^n(s)\} = \{\mathcal{G}^\pm(\lambda_l, s_l) \mid \lambda_l \in \Pi^l,\ s_l \in \mathbf{Z}^l(s)\}$$

and $\mathcal{G}^\pm(\lambda_n, s_n)$ are congruent to $|\lambda_n, s_n\rangle$ modulo $q^{\pm 1}\mathcal{L}^\pm(s)$. Comparing (35)–(36) with (33)–(34) and taking into account Theorem 2.4, we see that $\mathcal{L}[s_n] = \oplus_{\lambda_n \in \Pi^n} \overline{A}\, |\lambda_n, s_n\rangle$ is a lower crystal lattice of the $U_p(\widehat{\mathfrak{sl}}_l)$-module $\mathbf{F}_q[s_n]$ at $p := -q^{-1} = 0$. Then by Proposition 4.12 again, $\overline{\mathcal{L}[s_n]} = \oplus_{\lambda_n \in \Pi^n} A\, \overline{|\lambda_n, s_n\rangle}$ is a lower crystal lattice of $\mathbf{F}_q[s_n]$ at $p = \infty$, and the existence and uniqueness of the basis $\{\mathcal{G}^-(\lambda_n, s_n) \mid \lambda_n \in \Pi^n\}$ imply that there is an isomorphism

$$\mathbf{F}_q[s_n]_{\mathbf{Q}} \cap \mathcal{L}[s_n] \cap \overline{\mathcal{L}[s_n]} \xrightarrow{\sim} \mathcal{L}[s_n]/p\mathcal{L}[s_n]$$

taking $\mathcal{G}^-(\boldsymbol{\lambda}_n, \boldsymbol{s}_n)$ to $|\boldsymbol{\lambda}_n, \boldsymbol{s}_n\rangle$ mod $p\mathcal{L}[\boldsymbol{s}_n]$. Therefore,

$$\{\mathcal{G}^-(\boldsymbol{\lambda}_n, \boldsymbol{s}_n) \mid \boldsymbol{\lambda}_n \in \Pi^n\}$$

is a lower global crystal basis of the integrable $U_p(\widehat{\mathfrak{sl}}_l)$-module $\mathbf{F}_q[\boldsymbol{s}_n]$.

Let us now comment on how the canonical basis $\{\mathcal{G}^+(\boldsymbol{\lambda}_l, \boldsymbol{s}_l) \mid \boldsymbol{\lambda}_l \in \Pi^l\}$ is related to the lower global crystal basis $\{\mathcal{G}(\boldsymbol{\lambda}_l, \boldsymbol{s}_l) \mid \boldsymbol{\lambda}_l \in \Pi^l(\boldsymbol{s}_l)\}$ of the irreducible $U_q(\widehat{\mathfrak{sl}}_n)$-submodule $\mathbf{F}_q[\boldsymbol{s}_l]$ generated by $|\varnothing_l, \boldsymbol{s}_l\rangle$ (cf. Theorem 2.5). Using [3, Lemma 2.7], one can show that the rational form $\mathbf{F}_q[\boldsymbol{s}_l]_{\mathbf{Q}}$ of $\mathbf{F}_q[\boldsymbol{s}_l]$ belongs to $\mathbf{F}_q[\boldsymbol{s}_l]_{\mathbf{Q}}$. From the definition of $\mathcal{G}(\boldsymbol{\lambda}_l, \boldsymbol{s}_l)$, it now follows that $\mathcal{G}(\boldsymbol{\lambda}_l, \boldsymbol{s}_l)$ belongs to $\mathcal{L}^+(s)$ and hence has the same congruence property with respect to $\mathcal{L}^+(s)$ as does $\mathcal{G}^+(\boldsymbol{\lambda}_l, \boldsymbol{s}_l)$. On the other hand, by Proposition 4.12, the restriction of the bar-involution on $\mathbf{F}_q[\boldsymbol{s}_l]$ coincides with the involution of $\mathbf{F}_q[\boldsymbol{s}_l]$ defined in Section 2.2. By the uniqueness of $\mathcal{G}^+(\boldsymbol{\lambda}_l, \boldsymbol{s}_l)$, it now follows that $\mathcal{G}^+(\boldsymbol{\lambda}_l, \boldsymbol{s}_l) = \mathcal{G}(\boldsymbol{\lambda}_l, \boldsymbol{s}_l)$ for all $\boldsymbol{\lambda}_l \in \Pi^l(\boldsymbol{s}_l)$. Hence

$$\{\mathcal{G}^+(\boldsymbol{\lambda}_l, \boldsymbol{s}_l) \mid \boldsymbol{\lambda}_l \in \Pi^l(\boldsymbol{s}_l)\}$$

is the lower global crystal basis of the irreducible $U_q(\widehat{\mathfrak{sl}}_n)$-submodule $\mathbf{F}_q[\boldsymbol{s}_l]$.

For $\boldsymbol{s}_n \in \mathbf{Z}^n$, let $\mathbf{F}_q[\boldsymbol{s}_n]$ be the irreducible $U_p(\widehat{\mathfrak{sl}}_l)$-submodule of $\mathbf{F}_q[\boldsymbol{s}_n]$ generated by the highest-weight vector $|\varnothing_n, \boldsymbol{s}_n\rangle$. By the same argument as above, we conclude that

$$\{\mathcal{G}^-(\boldsymbol{\lambda}_n, \boldsymbol{s}_n) \mid \boldsymbol{\lambda}_n \in \Pi^n(\boldsymbol{s}_n)\}$$

is the lower global crystal basis of $\mathbf{F}_q[\boldsymbol{s}_n]$.

Note that the involution matrix $R_{\boldsymbol{\lambda}_l,\boldsymbol{\mu}_l}(\boldsymbol{s}_l|q)$, because of its unitriangularity, can be computed by using the ordering rules of Proposition 3.16. Therefore, we have an algorithm for computation of the transition matrices $\|\Delta^{\pm}_{\boldsymbol{\lambda}_l,\boldsymbol{\mu}_l}(\boldsymbol{s}_l|q)\|$. By the deep result of [2], the coefficients $\Delta^{+}_{\boldsymbol{\lambda}_l,\boldsymbol{\mu}_l}(\boldsymbol{s}_l|1)$ for $\boldsymbol{\lambda}_l \in \Pi^l(\boldsymbol{s}_l)$, $\boldsymbol{\mu}_l \in \Pi^l$ are identified with the decomposition numbers of Specht modules for an Ariki–Koike algebra and hence are nonnegative integers. Tables of the transition matrices suggest that for all $\boldsymbol{\lambda}_l, \boldsymbol{\mu}_l \in \Pi^l$ the entries $\Delta^{+}_{\boldsymbol{\lambda}_l,\boldsymbol{\mu}_l}(\boldsymbol{s}_l|q)$ are in $\mathbf{Z}_{\geq 0}[q]$ (and those of $\|\Delta^{-}_{\boldsymbol{\lambda}_l,\boldsymbol{\mu}_l}(\boldsymbol{s}_l|q)\|$ are in $\mathbf{Z}_{\geq 0}[p]$).

5 An inversion formula for Kazhdan–Lusztig polynomials

The aim of this section is to prove Theorem 5.15, which gives an inversion formula relating the matrices $\|\Delta^{+}_{\boldsymbol{\lambda}_l,\boldsymbol{\mu}_l}(\boldsymbol{s}_l|q)\|$ with $\|\Delta^{-}_{\boldsymbol{\lambda}_l,\boldsymbol{\mu}_l}(\boldsymbol{s}_l|q)\|$. In the $l = 1$ case, this formula has already been proved by Leclerc and Thibon in [18].

5.1 Some properties of the Heisenberg algebra action on $\Lambda^{s+\frac{\infty}{2}}$.

Definition 5.1. Let $m \in \mathbf{Z}_{\geq 0}$. We shall say that a pair $(\boldsymbol{\lambda}_r = (\lambda^{(1)}, \dots, \lambda^{(r)}), \boldsymbol{s}_r = (s_1, \dots, s_r)) \in \Pi^r \times \mathbf{Z}^r$ is *m-dominant* if for all $a = 1, 2, \dots, r-1$ we have the inequalities

$$s_a - s_{a+1} \geq m + |\boldsymbol{\lambda}_r|,$$

where $|\boldsymbol{\lambda}_r| = |\lambda^{(1)}| + \cdots + |\lambda^{(r)}|$.

Also, we shall say that a basis vector $|\boldsymbol{\lambda}_l, \mathbf{s}_l\rangle$ (resp., $|\boldsymbol{\lambda}_n, \mathbf{s}_n\rangle$) is m-dominant if the pair $(\boldsymbol{\lambda}_l, \mathbf{s}_l)$ (resp., $(\boldsymbol{\lambda}_n, \mathbf{s}_n)$) is. To explain the reason for introducing this definition, we need to present some notation. Let $n \in \mathbf{N}$, $l = 1$. Let x be a linear operator on $\Lambda^{s+\frac{\infty}{2}}$ acting on the elements of the basis $B(s)$ by

$$x\,|\lambda, s\rangle = \sum_{\mu\in\Pi} x_{\lambda,\mu}(s)\,|\mu, s\rangle,$$

where $x_{\lambda,\mu}(s)$ are coefficients in $\mathbf{K}$. Now let $n \in \mathbf{N}$, $l \in \mathbf{N}$. For each $b = 1, 2, \ldots, l$, we define an endomorphism $x^{(b)}[n, 1]$ of $\Lambda^{s+\frac{\infty}{2}}$ by

$$x^{(b)}[n, 1]\,|\boldsymbol{\lambda}_l, \mathbf{s}_l\rangle = \sum_{\mu\in\Pi} x_{\lambda^{(b)},\mu}(s_b)\,|(\lambda^{(1)}, \ldots, \lambda^{(b-1)}, \mu, \lambda^{(b+1)}, \ldots, \lambda^{(l)}), \mathbf{s}_l\rangle.$$

Similarly, let $n = 1$ and $l \in \mathbf{N}$. For an endomorphism y of $\Lambda^{s+\frac{\infty}{2}}$, we introduce the corresponding matrix elements $y_{\lambda,\mu}(s)$ on the basis $B(s)$ by

$$y\,|\lambda, s\rangle = \sum_{\mu\in\Pi} y_{\lambda,\mu}(s)\,|\mu, s\rangle.$$

Again, for arbitrary $n \in \mathbf{N}$ and $l \in \mathbf{N}$, we define for each $a = 1, 2, \ldots, n$ an endomorphism $y^{(a)}[1, l]$ of $\Lambda^{s+\frac{\infty}{2}}$ by

$$y^{(a)}[1, l]\,|\boldsymbol{\lambda}_n, \mathbf{s}_n\rangle = \sum_{\mu\in\Pi} y_{\lambda^{(a)},\mu}(s_a)\,|(\lambda^{(1)}, \ldots, \lambda^{(a-1)}, \mu, \lambda^{(a+1)}, \ldots, \lambda^{(n)}), \mathbf{s}_n\rangle.$$

Example 5.2. For $n = 2$, $l = 1$, and $s \in 2\mathbf{Z}$, one finds using the ordering rules of Proposition 3.16 that

$$\begin{aligned} B_{-2}|\varnothing, s\rangle &= |(4), s\rangle - q^{-1}|(3, 1), s\rangle + (q^{-2} - 1)|(2^2), s\rangle \\ &\quad + q^{-1}|(2, 1^2), s\rangle - q^{-2}|(1^4), s\rangle. \end{aligned}$$

Hence for $n = 2$ and $l = 2$, taking $\mathbf{s}_l = (s_1, s_2)$ such that $s_1, s_2 \in 2\mathbf{Z}$, we have

$$\begin{aligned} B^{(1)}_{-2}[2, 1]\,|\varnothing_l, \mathbf{s}_l\rangle &= |((4), \varnothing), \mathbf{s}_l\rangle - q^{-1}|((3, 1), \varnothing), \mathbf{s}_l\rangle \\ &\quad + (q^{-2} - 1)|((2^2), \varnothing), \mathbf{s}_l\rangle \\ &\quad + q^{-1}|((2, 1^2), \varnothing), \mathbf{s}_l\rangle - q^{-2}|((1^4), \varnothing), \mathbf{s}_l\rangle, \\ B^{(2)}_{-2}[2, 1]\,|\varnothing_l, \mathbf{s}_l\rangle &= |(\varnothing, (4)), \mathbf{s}_l\rangle - q^{-1}|(\varnothing, (3, 1)), \mathbf{s}_l\rangle \\ &\quad + (q^{-2} - 1)|(\varnothing, (2^2)), \mathbf{s}_l\rangle \\ &\quad + q^{-1}|(\varnothing, (2, 1^2)), \mathbf{s}_l\rangle - q^{-2}|(\varnothing, (1^4)), \mathbf{s}_l\rangle. \end{aligned}$$

Proposition 5.3.

(i) *Let* $(\boldsymbol{\lambda}_l, s_l) \in \Pi^l \times \mathbf{Z}^l$ *be* nm*-dominant for some* $m \in \mathbf{N}$. *Then*

$$B_{-m}\,|\boldsymbol{\lambda}_l, s_l\rangle = \sum_{b=1}^{l} q^{(b-1)m} B_{-m}^{(b)}[n,1]\,|\boldsymbol{\lambda}_l, s_l\rangle.$$

(ii) *Let* $(\boldsymbol{\lambda}_l, s_l) \in \Pi^l \times \mathbf{Z}^l$ *be* 0*-dominant. Then for any* $m \in \mathbf{N}$,

$$B_{m}\,|\boldsymbol{\lambda}_l, s_l\rangle = \sum_{b=1}^{l} q^{(b-1)m} B_{m}^{(b)}[n,1]\,|\boldsymbol{\lambda}_l, s_l\rangle.$$

(i′) *Let* $(\boldsymbol{\lambda}_n, s_n) \in \Pi^n \times \mathbf{Z}^n$ *be* lm*-dominant for some* $m \in \mathbf{N}$. *Then*

$$B_{-m}\,|\boldsymbol{\lambda}_n, s_n\rangle = \sum_{a=1}^{n} p^{(a-1)m} B_{-m}^{(a)}[1,l]\,|\boldsymbol{\lambda}_n, s_n\rangle.$$

(ii′) *Let* $(\boldsymbol{\lambda}_n, s_n) \in \Pi^n \times \mathbf{Z}^n$ *be* 0*-dominant. Then for any* $m \in \mathbf{N}$,

$$B_{m}\,|\boldsymbol{\lambda}_n, s_n\rangle = \sum_{a=1}^{n} p^{(a-1)m} B_{m}^{(a)}[1,l]\,|\boldsymbol{\lambda}_n, s_n\rangle.$$

A proof of this proposition is given in Section 5.4.

Example 5.4. To illustrate Proposition 5.3, take $n = 2$, $l = 2$, and $s_l = (2, -2)$. A straightforward computation using Proposition 3.16 gives

$$\begin{aligned} B_{-2}\,|\varnothing_l, s_l\rangle \;=\; & |((4),\varnothing), s_l\rangle - q^{-1}|((3,1),\varnothing), s_l\rangle + (q^{-2}-1)|((2^2),\varnothing), s_l\rangle \\ & + q^{-1}|((2,1^2),\varnothing), s_l\rangle - q^{-2}|((1^4),\varnothing), s_l\rangle \\ & + q^2|(\varnothing,(4)), s_l\rangle - q|(\varnothing,(3,1)), s_l\rangle + (1-q^2)|(\varnothing,(2^2)), s_l\rangle \\ & + q|(\varnothing,(2,1^2)), s_l\rangle - |(\varnothing,(1^4)), s_l\rangle. \end{aligned}$$

The pair $(\varnothing_l, s_l)$ is 4-dominant. Taking into account the formulas of Example 5.2, we see that the relation

$$B_{-2}\,|\varnothing_l, s_l\rangle = B_{-2}^{(1)}[2,1]\,|\varnothing_l, s_l\rangle + q^2 B_{-2}^{(2)}[2,1]\,|\varnothing_l, s_l\rangle$$

is indeed satisfied.

Remark 5.5. Simple decompositions for the actions of the bosons described in Proposition 5.3 fail to hold in general when bosons are applied to vectors that are not dominant. For example, let $n = l = 2$ and $s_l = (0, 0)$. Then the pair $(\varnothing_l, s_l)$ is not m-dominant for any $m \in \mathbf{N}$. In this case, an explicit computation yields

$$\begin{aligned} B_{-2}\,|\varnothing_l, s_l\rangle \;=\; & q|((4),\varnothing), s_l\rangle - |((3,1),\varnothing), s_l\rangle + (1-q^2)|((2^2),\varnothing), s_l\rangle \\ & + q|((2,1^2),\varnothing), s_l\rangle - |((1^4),\varnothing), s_l\rangle \\ & + |(\varnothing,(4)), s_l\rangle - q^{-1}|(\varnothing,(3,1)), s_l\rangle + (q^{-2}-1)|(\varnothing,(2^2)), s_l\rangle \\ & + |(\varnothing,(2,1^2)), s_l\rangle - q^{-1}|(\varnothing,(1^4)), s_l\rangle \\ & + (q^2-1)|((2),(2)), s_l\rangle + (q^{-1}-q)|((1),(2,1)), s_l\rangle \\ & + (q^{-1}-q)|((2,1),(1)), s_l\rangle + (1-q^{-2})|((1^2),(1^2)), s_l\rangle. \end{aligned}$$

For $m \in \mathbf{Z}$, let e_m and h_m be, respectively, the elementary symmetric function and the complete symmetric function (cf. [19]). In terms of the power-sum basis of the ring of symmetric functions, one has

$$e_m = e_m(p_1, p_2, \dots) = \sum_{\nu \in \Pi, |\nu| = m} a_{m,\nu}\, p_\nu,$$
$$h_m = h_m(p_1, p_2, \dots) = \sum_{\nu \in \Pi, |\nu| = m} b_{m,\nu}\, p_\nu,$$

where, as usual, we set $p_\nu = p_{\nu_1} p_{\nu_2} \cdots$ for a partition $\nu = (\nu_1, \nu_2, \dots)$. It will be understood that $e_0 = h_0 = 1$ and $e_m = h_m = 0$ for $m < 0$. Now, for all integers m, we define

$$E_m := e_m(B_1, B_2, \dots), \qquad \widetilde{E}_m := e_m(B_{-1}, B_{-2}, \dots),$$
$$H_m := h_m(B_1, B_2, \dots), \qquad \widetilde{H}_m := h_m(B_{-1}, B_{-2}, \dots).$$

Corollary 5.6.

(i) *Let $(\boldsymbol{\lambda}_l, s_l) \in \Pi^l \times \mathbf{Z}^l$ be nm-dominant for some $m \in \mathbf{N}$. Then*

$$\widetilde{E}_m\, |\boldsymbol{\lambda}_l, s_l\rangle = \sum_{m_1 + \cdots + m_l = m} \prod_{b=1}^{l} q^{(b-1)m_b} \widetilde{E}^{(b)}_{m_b}[n, 1]\, |\boldsymbol{\lambda}_l, s_l\rangle,$$
$$\widetilde{H}_m\, |\boldsymbol{\lambda}_l, s_l\rangle = \sum_{m_1 + \cdots + m_l = m} \prod_{b=1}^{l} q^{(b-1)m_b} \widetilde{H}^{(b)}_{m_b}[n, 1]\, |\boldsymbol{\lambda}_l, s_l\rangle.$$

(ii) *Let $(\boldsymbol{\lambda}_l, s_l) \in \Pi^l \times \mathbf{Z}^l$ be 0-dominant. Then for any $m \in \mathbf{N}$,*

$$E_m\, |\boldsymbol{\lambda}_l, s_l\rangle = \sum_{m_1 + \cdots + m_l = m} \prod_{b=1}^{l} q^{(b-1)m_b} E^{(b)}_{m_b}[n, 1]\, |\boldsymbol{\lambda}_l, s_l\rangle,$$
$$H_m\, |\boldsymbol{\lambda}_l, s_l\rangle = \sum_{m_1 + \cdots + m_l = m} \prod_{b=1}^{l} q^{(b-1)m_b} H^{(b)}_{m_b}[n, 1]\, |\boldsymbol{\lambda}_l, s_l\rangle.$$

(i′) *Let $(\boldsymbol{\lambda}_n, s_n) \in \Pi^n \times \mathbf{Z}^n$ be lm-dominant for some $m \in \mathbf{N}$. Then*

$$\widetilde{E}_m\, |\boldsymbol{\lambda}_n, s_n\rangle = \sum_{m_1 + \cdots + m_n = m} \prod_{a=1}^{n} p^{(a-1)m_a} \widetilde{E}^{(a)}_{m_a}[1, l]\, |\boldsymbol{\lambda}_n, s_n\rangle,$$
$$\widetilde{H}_m\, |\boldsymbol{\lambda}_n, s_n\rangle = \sum_{m_1 + \cdots + m_n = m} \prod_{a=1}^{n} p^{(a-1)m_a} \widetilde{H}^{(a)}_{m_a}[1, l]\, |\boldsymbol{\lambda}_n, s_n\rangle.$$

(ii′) *Let* $(\boldsymbol{\lambda}_n, \boldsymbol{s}_n) \in \Pi^n \times \mathbf{Z}^n$ *be* 0*-dominant. Then for any* $m \in \mathbf{N}$,

$$E_m\,|\boldsymbol{\lambda}_n, \boldsymbol{s}_n\rangle = \sum_{m_1+\cdots+m_n=m} \prod_{a=1}^{n} p^{(a-1)m_a} E^{(a)}_{m_a}[1, l]\,|\boldsymbol{\lambda}_n, \boldsymbol{s}_n\rangle,$$

$$H_m\,|\boldsymbol{\lambda}_n, \boldsymbol{s}_n\rangle = \sum_{m_1+\cdots+m_n=m} \prod_{a=1}^{n} p^{(a-1)m_a} H^{(a)}_{m_a}[1, l]\,|\boldsymbol{\lambda}_n, \boldsymbol{s}_n\rangle.$$

PROOF. It follows from (27)–(30) that for each $m \in \mathbf{Z}^*$, a vector $B_m|\boldsymbol{\lambda}_l, \boldsymbol{s}_l\rangle$ (resp., a vector $B_m|\boldsymbol{\lambda}_n, \boldsymbol{s}_n\rangle$) is a linear combination of $|\boldsymbol{\mu}_l, \boldsymbol{s}_l\rangle$ (resp., $|\boldsymbol{\mu}_n, \boldsymbol{s}_n\rangle$) with $|\boldsymbol{\mu}_l| = |\boldsymbol{\lambda}_l| - nm$ (resp., with $|\boldsymbol{\mu}_n| = |\boldsymbol{\lambda}_n| - lm$). Also, if Δ is the comultiplication on the ring of symmetric functions defined by $\Delta p_m = p_m \otimes 1 + 1 \otimes p_m$, then $\Delta e_m = \sum_{r+s=m} e_r \otimes e_s$, $\Delta h_m = \sum_{r+s=m} h_r \otimes h_s$ (cf. [19]). These facts and Proposition 5.3 imply the claims. □

5.1.1 *Proof of Proposition* 4.5. To emphasize the dependence on n and l, let us set, in the notation of Proposition 4.5, $\gamma_m[n, l] := \gamma_m$. First, let $n = l = 1$. In this case, the ordering rules of Proposition 3.16 reduce to $u_{k_1} \wedge u_{k_2} = -u_{k_2} \wedge u_{k_1}$ for all $k_1, k_2 \in \mathbf{Z}$. This makes it easy to verify that $\gamma_m[1, 1] = m$. Next, let $n > 1$, $l = 1$. It is clear that there is $\boldsymbol{s}_n \in \mathbf{Z}^n$ such that the pair $(\varnothing_n, \boldsymbol{s}_n)$ is m-dominant. Hence applying Proposition 5.3(i′), we obtain

$$\gamma_m[n, 1]|\varnothing_n, \boldsymbol{s}_n\rangle = B_m B_{-m}|\varnothing_n, \boldsymbol{s}_n\rangle = B_m \sum_{a=1}^{n} p^{(a-1)m} B^{(a)}_{-m}[1, 1]|\varnothing_n, \boldsymbol{s}_n\rangle.$$

However, $B_{-m}|\varnothing_n, \boldsymbol{s}_n\rangle$ is a linear combination of $|\boldsymbol{\lambda}_n, \boldsymbol{s}_n\rangle$ with $|\boldsymbol{\lambda}_n| = m$ and hence a linear combination of 0-dominant vectors. Therefore, one may apply Proposition 5.3(ii′) and get

$$B_m \sum_{a=1}^{n} p^{(a-1)m} B^{(a)}_{-m}[1, 1]|\varnothing_n, \boldsymbol{s}_n\rangle = \sum_{a=1}^{n} p^{2(a-1)m} B^{(a)}_{m}[1, 1] B^{(a)}_{-m}[1, 1]|\varnothing_n, \boldsymbol{s}_n\rangle.$$

This implies

$$\gamma_m[n, 1]|\varnothing_n, \boldsymbol{s}_n\rangle = \left(\sum_{a=1}^{n} p^{2(a-1)m} \gamma_m[1, 1]\right)|\varnothing_n, \boldsymbol{s}_n\rangle.$$

Thus

$$\gamma_m[n, 1] = m\frac{1 - p^{2mn}}{1 - p^{2m}}.$$

Finally, let $n \geq 1$ and $l > 1$. Obviously, there is $\boldsymbol{s}_l \in \mathbf{Z}^l$ such that the pair $(\varnothing_l, \boldsymbol{s}_l)$ is nm-dominant. Therefore, by Proposition 5.3(i), we have

$$\gamma_m[n, l]|\varnothing_l, \boldsymbol{s}_l\rangle = B_m B_{-m}|\varnothing_l, \boldsymbol{s}_l\rangle = B_m \sum_{b=1}^{n} q^{(b-1)m} B^{(b)}_{-m}[n, 1]|\varnothing_l, \boldsymbol{s}_l\rangle.$$

Again, it is clear that $B_{-m}|\varnothing_l, s_l\rangle$ is a linear combination of 0-dominant vectors. Hence using Proposition 5.3(ii), we obtain

$$\gamma_m[n,l]|\varnothing_l, s_l\rangle = \left(\sum_{b=1}^{l} q^{2(b-1)m}\gamma_m[n,1]\right)|\varnothing_l, s_l\rangle.$$

It follows that

$$\gamma_m[n,l] = m\frac{1-p^{2mn}}{1-p^{2m}}\frac{1-q^{2ml}}{1-q^{2m}}.$$

Recalling that $p := -q^{-1}$, we get the desired result. □

5.2 A scalar product of $\Lambda^{s+\frac{\infty}{2}}$. For each $s \in \mathbf{Z}$, we define on the semiinfinite wedge product $\Lambda^{s+\frac{\infty}{2}}$ a $\mathbf{K}$-bilinear scalar product by $\langle b, b'\rangle = \delta_{b,b'}$, where b and b' are any two elements of the basis

$$\begin{aligned} B(s) &= \{|\lambda, s\rangle \mid \lambda \in \Pi\} \\ &= \{|\boldsymbol{\lambda}_l, \boldsymbol{s}_l\rangle \mid \boldsymbol{\lambda}_l \in \Pi^l,\ \boldsymbol{s}_l \in \mathbf{Z}^l(s)\} \\ &= \{|\boldsymbol{\lambda}_n, \boldsymbol{s}_n\rangle \mid \boldsymbol{\lambda}_n \in \Pi^n,\ \boldsymbol{s}_n \in \mathbf{Z}^n(s)\}. \end{aligned}$$

It is clear that this scalar product is symmetric, and that for two weight vectors u and v, $\langle u, v\rangle$ is nonzero only if $\mathrm{Wt}(u) = \mathrm{Wt}(v)$.

Proposition 5.7. *For $u, v \in \Lambda^{s+\frac{\infty}{2}}$, one has*

$$\begin{aligned} &\langle e_i u, v\rangle = \langle u, q^{-1}(t_i)^{-1} f_i v\rangle, \quad \langle f_i u, v\rangle = \langle u, q^{-1} t_i e_i v\rangle \quad (i = 0, 1, \dots, n-1), \\ &\langle \dot{e}_j u, v\rangle = \langle u, p^{-1}(\dot{t}_j)^{-1} \dot{f}_j v\rangle, \quad \langle \dot{f}_j u, v\rangle = \langle u, p^{-1} \dot{t}_j \dot{e}_j v\rangle \quad (j = 0, 1, \dots, l-1). \end{aligned}$$

PROOF. This follows immediately from (33)–(36). □

Proposition 5.8. *For $m \in \mathbf{N}$ and $u, v \in\in \Lambda^{s+\frac{\infty}{2}}$, one has*

$$\langle B_{-m}u, v\rangle = \langle u, B_m v\rangle.$$

To prove this proposition, we use the following lemma.

Lemma 5.9. (i) *Assume that the statement of Proposition* 5.8 *is valid for some $n \in \mathbf{N}$ and $l = 1$. Then it is also valid for the same n and all $l \in \mathbf{N}_{>1}$.*

(ii) *Assume that the statement of Proposition* 5.8 *is valid for some $l \in \mathbf{N}$ and $n = 1$. Then it is also valid for the same l and all $n \in \mathbf{N}_{>1}$.*

PROOF. Since the proofs of (i) and (ii) are similar, we only give a proof of (i). Let us keep notation as in Proposition 5.8. Using Theorem 4.8, we assume without loss of generality that

$$u = \sum_k x_k y_k |\varnothing_l, \boldsymbol{t}_l\rangle,$$

where x_k is an element of $U_q(\widehat{\mathfrak{sl}}_n)^- \cdot \mathcal{H}^-$, y_k is an element of $U_p(\widehat{\mathfrak{sl}}_l)^-$ and $\boldsymbol{t}_l = (t_1, \dots, t_l)$ is an element of $A_l^n(s)$. By Corollary 4.9, for any $\boldsymbol{s}_l = (s_1, \dots, s_l) \in \mathbf{Z}^l(s)$ such that $\Lambda_{s_1} + \cdots + \Lambda_{s_l} = \Lambda_{t_1} + \cdots + \Lambda_{t_l}$, we have $|\varnothing_l, \boldsymbol{t}_l\rangle = Y(\boldsymbol{s}_l)|\varnothing_l, \boldsymbol{s}_l\rangle$ for some $Y(\boldsymbol{s}_l) \in U'_p(\widehat{\mathfrak{sl}}_l)$. Hence

$$\langle B_m u, v\rangle = \sum_k \langle B_{-m} x_k|\varnothing_l, \boldsymbol{s}_l\rangle, Y_k^*(\boldsymbol{s}_l)v\rangle,$$

where $Y_k^*(\boldsymbol{s}_l)$ is the adjoint of $Y_k(\boldsymbol{s}_l) := y_k Y(\boldsymbol{s}_l)$. Note that by Proposition 5.7, $Y_k^*(\boldsymbol{s}_l) \in U'_p(\widehat{\mathfrak{sl}}_l)$. We may and do assume that for each k the element $x_k \in U_q(\widehat{\mathfrak{sl}}_n)^- \cdot \mathcal{H}^-$ has a definite weight $\mathrm{wt}(x_k) = -(r_{k,0}\alpha_0 + \cdots + r_{k,n-1}\alpha_{n-1})$, where $r_{k,i}$ are some nonnegative integers. For all k, we have $x_k|\varnothing_l, \boldsymbol{s}_l\rangle \in \mathbf{F}_q[\boldsymbol{s}_l]$, and using (27), we see that $x_k|\varnothing_l, \boldsymbol{s}_l\rangle$ is a linear combination of $|\boldsymbol{\lambda}_l, \boldsymbol{s}_l\rangle$ with $|\boldsymbol{\lambda}_l| = r_k := r_{k,0} + \cdots + r_{k,n-1}$. Now let $r := \max_k\{r_k\}$ and choose $\boldsymbol{s}_l$ so that

$$s_b - s_{b+1} \geq r + nm \qquad (b = 1, \dots, l-1). \tag{41}$$

Then for each k, the vector $x_k|\varnothing_l, \boldsymbol{s}_l\rangle$ is a linear combination of nm-dominant vectors. Hence we may apply Proposition 5.3(i) and get

$$\langle B_m u, v\rangle = \sum_k \sum_{b=1}^{l} q^{(b-1)m} \langle B_{-m}^{(b)}[n, 1] x_k|\varnothing_l, \boldsymbol{s}_l\rangle, Y_k^*(\boldsymbol{s}_l)v\rangle.$$

Now using the assumption in statement (i) of the lemma, we obtain

$$\langle B_m u, v\rangle = \sum_k \sum_{b=1}^{l} q^{(b-1)m} \langle x_k|\varnothing_l, \boldsymbol{s}_l\rangle, B_m^{(b)}[n, 1] Y_k^*(\boldsymbol{s}_l)v\rangle.$$

For each k, the scalar product $\langle B_{-m} x_k|\varnothing_l, \boldsymbol{s}_l\rangle, Y_k^*(\boldsymbol{s}_l)v\rangle$ is nonzero only if $Y_k^*(\boldsymbol{s}_l)v \in \mathbf{F}_q[\boldsymbol{s}_l]$ and

$$\mathrm{wt}(Y_k^*(\boldsymbol{s}_l)v) = \mathrm{wt}(B_{-m} x_k|\varnothing_l, \boldsymbol{s}_l\rangle) = \mathrm{wt}|\varnothing_l, \boldsymbol{s}_l\rangle - \sum_{i=0}^{n-1} (r_{k,i} + m)\alpha_i.$$

It follows that $Y_k^*(\boldsymbol{s}_l)v$ is a linear combination of $|\boldsymbol{\mu}_l, \boldsymbol{s}_l\rangle$ with $|\boldsymbol{\mu}_l| = r_k + nm$, whence by (41) it is a linear combination of 0-dominant vectors. Therefore, we may apply Proposition 5.3(ii) and obtain

$$\langle B_{-m} u, v\rangle = \sum_k \langle x_k|\varnothing_n, \boldsymbol{s}_n\rangle, B_m Y_k(\boldsymbol{s}_n)^* v\rangle = \langle u, B_m v\rangle. \qquad \square$$

Proof of Proposition 5.8. In view of Lemma 5.9, it is sufficient to show that the statement of the proposition is valid for $n = l = 1$. However, in this case, the relation $\langle B_{-m} u, v\rangle = \langle u, B_m v\rangle$ is just a restatement of the fact that the endomorphism $m\partial/\partial p_m$ of the ring of symmetric functions is adjoint to the multiplication by p_m with respect to the scalar product orthonormalizing the basis of Schur functions. $\square$

Proposition 5.7 implies that for $m \in \mathbf{Z}_{\geq 0}$ and $u, v \in \Lambda^{s+\frac{\infty}{2}}$, one has

$$\langle \widetilde{E}_m u, v\rangle = \langle u, E_m v\rangle, \qquad \langle \widetilde{H}_m u, v\rangle = \langle u, H_m v\rangle. \tag{42}$$

5.3 A symmetry of the bar-involution. Define a semilinear involution $u \mapsto u'$ of $\Lambda^{*+\frac{\infty}{2}} = \oplus_{s\in\mathbf{Z}}\Lambda^{s+\frac{\infty}{2}}$ by

$$|\lambda, s\rangle' = |\lambda', -s\rangle, \qquad q' = q^{-1}.$$

Here λ' stands for the conjugate partition of λ. The description of the indexations of $|\lambda, s\rangle$ given in Remark 4.2 implies that for an l-multipartition $\boldsymbol{\lambda}_l = (\lambda^{(1)}, \dots, \lambda^{(l)})$ and $\boldsymbol{s}_l = (s_1, \dots, s_l) \in \mathbf{Z}^l$, we have

$$|\boldsymbol{\lambda}_l, \boldsymbol{s}_l\rangle' = |\boldsymbol{\lambda}_l', \boldsymbol{s}_l'\rangle,$$

where $\boldsymbol{\lambda}_l' = (\lambda^{(l)'}, \dots, \lambda^{(1)'})$ and $\boldsymbol{s}_l' = (-s_l, \dots, -s_1)$. Likewise, for an n-multipartition $\boldsymbol{\lambda}_n$ and $\boldsymbol{s}_n \in \mathbf{Z}^n$, we have

$$|\boldsymbol{\lambda}_n, \boldsymbol{s}_n\rangle' = |\boldsymbol{\lambda}_n', \boldsymbol{s}_n'\rangle.$$

Proposition 5.10. *For $u \in \Lambda^{s+\frac{\infty}{2}}$ $(s \in \mathbf{Z})$, we have*

$$(e_i u)' = q^{-1} t_{-i} e_{-i} u', \qquad (f_i u)' = q^{-1}(t_{-i})^{-1} f_{-i} u' \quad (i = 0, 1, \dots, n-1),$$
$$(\dot{e}_j u)' = p^{-1} \dot{t}_{-j} \dot{e}_{-j} u', \qquad (\dot{f}_j u)' = p^{-1}(\dot{t}_{-j})^{-1} \dot{f}_{-j} u' \quad (j = 0, 1, \dots, l-1).$$

PROOF. This follows from (33)–(36). □

Proposition 5.11. *For $u \in \Lambda^{s+\frac{\infty}{2}}$ $(s \in \mathbf{Z})$ and $m \in \mathbf{Z}_{\geq 0}$, we have*

(i) $$(\widetilde{E}_m u)' = (-q)^{m(n-1)}(-p)^{m(l-1)} \widetilde{H}_m u',$$

(ii) $$(E_m u)' = (-q)^{m(n-1)}(-p)^{m(l-1)} H_m u'.$$

To prove this proposition, we use the following lemma.

Lemma 5.12. (i) *Assume that statement* (i) *of Proposition* 5.11 *is valid for some $n \in \mathbf{N}$ and $l = 1$. Then it is also valid for the same n and all $l \in \mathbf{N}_{>1}$.*

(ii) *Assume that statement* (i) *of Proposition* 5.11 *is valid for some $l \in \mathbf{N}$ and $n = 1$. Then it is also valid for the same l and all $n \in \mathbf{N}_{>1}$.*

PROOF. Since the proofs of (i) and (ii) are virtually identical, we give only a proof of (i). Let us keep notation as in Proposition 5.11. Taking into account Theorem 4.8, we may and do assume without loss of generality that

$$u = \sum_k x_k y_k |\varnothing_l, \boldsymbol{t}_l\rangle,$$

where $\boldsymbol{t}_l = (t_1, \dots, t_l)$ is an element of $A_l^n(s)$, x_k is an element of $U_q(\widehat{\mathfrak{sl}}_n)^- \cdot \mathcal{H}^-$, and y_k is an element of $U_p(\widehat{\mathfrak{sl}}_l)^-$. Choose any sequence $\boldsymbol{s}_l = (s_1, \dots, s_l) \in \mathbf{Z}^l(s)$ such that the relation $\Lambda_{s_1} + \dots + \Lambda_{s_l} = \Lambda_{t_1} + \dots + \Lambda_{t_l}$ is satisfied. Corollary 4.9(i)

It is clear that we may assume all x_k to be weight vectors of $U_q(\widehat{\mathfrak{sl}}_n)^- \cdot \mathcal{H}^-$. Then $\mathrm{wt}(x_k) = -(r_{k,0}\alpha_0 + \cdots + r_{k,n-1}\alpha_{n-1})$ for some nonnegative integers $r_{k,i}$. Moreover, $x_k \in U_q(\widehat{\mathfrak{sl}}_n) \cdot \mathcal{H}$ implies that $x_k|\varnothing_l, s_l\rangle$ belongs to $\mathbf{F}_q[s_l]$. Hence $x_k|\varnothing_l, s_l\rangle$ is a linear combination of vectors $|\boldsymbol{\lambda}_l, s_l\rangle$ ($\boldsymbol{\lambda}_l \in \Pi^l$), and from formula (27) for the weight of $|\boldsymbol{\lambda}_l, s_l\rangle$, we see that for all these vectors $|\boldsymbol{\lambda}_l| = r_k := r_{k,0} + \cdots + r_{k,n-1}$. Now let $r := \max_k\{r_k\}$ and choose the sequence s_l so that the inequalities

$$s_b - s_{b+1} \geq r + nm \tag{43}$$

are satisfied for all $b = 1, \ldots, l-1$. Then for each k, the vector $x_k|\varnothing_l, s_l\rangle$ is a linear combination of nm-dominant $|\boldsymbol{\lambda}_l, s_l\rangle$, whereupon Corollary 5.6(i) gives

$$\widetilde{E}_m u = \sum_k Y_k(s_l) \sum_{m_1+\cdots+m_l=m} \prod_{b=1}^{l} q^{(b-1)m_b} \widetilde{E}^{(b)}_{m_b}[n,1]\, x_k|\varnothing_l, s_l\rangle.$$

Now we use the assumption in statement (i) of the lemma and obtain

$$(\widetilde{E}_m u)' = q^{-m(l-1)}(-q)^{m(n-1)} \times \sum_k Y_k(s_l)' \sum_{m_1+\cdots+m_l=m} \prod_{b=1}^{l} q^{(b-1)m_b} \widetilde{H}^{(b)}_{m_b}[n,1]\, (x_k|\varnothing_l, s_l\rangle)',$$

where $Y_k(s_l)'$ is the element of $U_p'(\widehat{\mathfrak{sl}}_l)$ defined by $Y_k(s_l)'v' = (Y_k(s_l)v)'$ (cf. Proposition 5.10). Next, observe that if a vector $|\boldsymbol{\lambda}_l, s_l\rangle$ is nm-dominant, then so is $|\boldsymbol{\lambda}_l, s_l\rangle'$. Hence from (43), it follows that for each k the vector $(x_k|\varnothing_l, s_l\rangle)'$ is a linear combination of nm-dominant $|\boldsymbol{\lambda}_l, s_l'\rangle$. Therefore, we may again apply Corollary 5.6(i) and obtain

$$\begin{aligned}(\widetilde{E}_m u)' &= q^{-m(l-1)}(-q)^{m(n-1)} \sum_k Y_k(s_l)' \widetilde{H}_m (x_k|\varnothing_l, s_l\rangle)' \\ &= (-p)^{m(l-1)}(-q)^{m(n-1)} \widetilde{H}_m u'.\end{aligned}$$

Thus (i) is proved. □

Proof of Proposition 5.11. By Lemma 5.12, the statement (i) of the proposition will be proved once it is shown to hold for $n = l = 1$. However, in this case, (i) is just a restatement of the relation $\omega(e_m) = h_m$ for the standard involution of the ring of symmetric functions defined in terms of the Schur functions by $\omega(s_\lambda) = s_{\lambda'}$.

It remains to observe that, the scalar product being nondegenerate, relation (ii) follows from (i), relations (42), and the easily checked formula $\langle u', v\rangle = \overline{\langle u, v'\rangle}$. □

Proposition 5.13. *For $u, v \in \Lambda^{s+\frac{\infty}{2}}$ ($s \in \mathbf{Z}$), one has*

$$\langle \overline{u}, v\rangle = \langle u', \overline{v'}\rangle.$$

PROOF. Using the decomposition of $\Lambda^{s+\frac{\infty}{2}}$ described in Theorem 4.8 we define a gradation of $\Lambda^{s+\frac{\infty}{2}}$ in the following way. We set the degrees of all the singular vectors $|\varnothing_l, s_l\rangle$ ($s_l \in A_l^n(s)$) to be zero, and we require that with respect to our gradation the operators f_i, $\dot{f}_j$, and B_{-m} ($m \in \mathbf{N}$) be homogeneous of degrees 1, 1, and m, respectively. Then the operators e_i, $\dot{e}_j$ are homogeneous of degree -1 and the operators $\widetilde{E}_m$, $\widetilde{H}_m$, E_m, and H_m are homogeneous of respective degrees m, m, $-m$, and $-m$.

Now we show the claim by induction. In degree zero, we have

$$\langle \overline{|\varnothing_l, s_l\rangle}, |\varnothing_l, t_l\rangle\rangle = \langle |\varnothing_l, s_l\rangle, |\varnothing_l, t_l\rangle\rangle = \delta_{s_l,t_l},$$
$$\langle |\varnothing_l, s_l'\rangle, \overline{|\varnothing_l, t_l'\rangle}\rangle = \langle |\varnothing_l, s_l'\rangle, |\varnothing_l, t_l'\rangle\rangle = \delta_{s_l,t_l}$$

for all $s_l, t_l \in A_l^n(s)$. Hence the claim holds in this case.

Assume that the proposition is proved for all u, v with degrees $< k$. By Theorem 4.8, to prove the proposition for all u, v, it is enough to show that

$$\langle \overline{(f_i u)}, v\rangle = \langle (f_i u)', \overline{v'}\rangle, \tag{44}$$
$$\langle \overline{(\dot{f}_j u)}, v\rangle = \langle (\dot{f}_j u)', \overline{v'}\rangle, \tag{45}$$
$$\langle \overline{(\widetilde{E}_m w)}, v\rangle = \langle (\widetilde{E}_m w)', \overline{v'}\rangle \tag{46}$$

for u, v, and w with degrees $k-1$, k, and $k-m$, respectively.

Let us show (44). We have

$$\langle \overline{(f_i u)}, v\rangle = \langle f_i \overline{u}, v\rangle = \langle \overline{u}, q^{-1} t_i e_i v\rangle = \langle u', \overline{(q^{-1} t_i e_i v)'}\rangle.$$

Here the first equality follows from Proposition 4.12, the second follows from Proposition 5.7, and the third follows from the induction assumption. Further,

$$\langle u', \overline{(q^{-1} t_i e_i v)'}\rangle = \langle u', \overline{e_{-i} v'}\rangle = \langle u', e_{-i}\overline{v'}\rangle = \langle q^{-1}(t_{-i})^{-1} f_{-i} u', \overline{v'}\rangle = \langle (f_i u)', \overline{v'}\rangle.$$

Here we used Propositions 5.10, 4.12, and 5.7. Thus (44) is established. A proof of (45) is similar.

Finally, let us show (46). We have

$$\langle \overline{(\widetilde{E}_m w)}, v\rangle = \langle \widetilde{E}_m \overline{w}, v\rangle = \langle \overline{w}, E_m v\rangle = \langle w', \overline{(E_m v)'}\rangle.$$

Here the first equality follows from Proposition 4.12, the second follows from (42), and the third follows from the induction assumption. Continuing, we have

$$\begin{aligned}\langle w', \overline{(E_m v)'}\rangle &= \langle w', \overline{(-q)^{m(n-1)}(-p)^{m(l-1)} H_m v'}\rangle \\ &= \langle w', (-q)^{m(n-1)}(-p)^{m(l-1)} H_m \overline{v'}\rangle \\ &= \langle (-q)^{m(n-1)}(-p)^{m(l-1)} \widetilde{H}_m w', \overline{v'}\rangle = \langle (\widetilde{E}_m w)', \overline{v'}\rangle.\end{aligned}$$

Here we used Propositions 5.11 and 4.12 and (42). Thus (46) is proved. □

For $s_l = (s_1, \dots, s_l) \in \mathbf{Z}^l$, let $\{\mathcal{G}^*(\boldsymbol{\lambda}_l, s_l) \mid \boldsymbol{\lambda}_l \in \Pi^l\}$ be the basis of $\mathbf{F}_q[s_l]$ dual to $\{\mathcal{G}^+(\boldsymbol{\lambda}_l, s_l) \mid \boldsymbol{\lambda}_l \in \Pi^l\}$ with respect to the scalar product $\langle u, v\rangle$. Write

$$\mathcal{G}^*(\boldsymbol{\lambda}_l, s_l) = \sum_{\boldsymbol{\mu}_l \in \Pi^l} \Delta^*_{\boldsymbol{\lambda}_l, \boldsymbol{\mu}_l}(s_l|q)\, |\boldsymbol{\mu}_l, s_l\rangle.$$

Since the basis $\{|\boldsymbol{\lambda}_l, s_l\rangle \mid \boldsymbol{\lambda}_l \in \Pi^l\}$ of $\mathbf{F}_q[s_l]$ is orthonormal relative to the scalar product, the matrix $\|\Delta^*_{\boldsymbol{\lambda}_l, \boldsymbol{\mu}_l}(s_l|q)\|$ is the transposed inverse of the matrix $\|\Delta^+_{\boldsymbol{\lambda}_l, \boldsymbol{\mu}_l}(s_l|q)\|$.

Proposition 5.14. *For $s_l \in \mathbf{Z}^l$ and $\boldsymbol{\lambda}_l \in \Pi^l$, one has*

$$\mathcal{G}^*(\boldsymbol{\lambda}_l, s_l)' = \mathcal{G}^-(\boldsymbol{\lambda}_l', s_l').$$

PROOF. Since the matrix $\|\Delta^+_{\boldsymbol{\lambda}_l, \boldsymbol{\mu}_l}(s_l|q)\|$ is unitriangular with off-diagonal entries in $q\mathbf{Z}[q]$, the same is true for its transposed inverse $\|\Delta^*_{\boldsymbol{\lambda}_l, \boldsymbol{\mu}_l}(s_l|q)\|$. It follows that

$$\mathcal{G}^*(\boldsymbol{\lambda}_l, s_l)' \equiv |\boldsymbol{\lambda}_l', s_l'\rangle \bmod q^{-1}\mathcal{L}^-(s),$$

where $s = s_1 + \cdots + s_l$. Hence $\mathcal{G}^*(\boldsymbol{\lambda}_l, s_l)'$ has the required congruence property relative to the basis $\{|\boldsymbol{\lambda}_l, s_l'\rangle\}$.

It remains to show that $\mathcal{G}^*(\boldsymbol{\lambda}_l, s_l)'$ is invariant with respect to the bar-involution. Since $\mathcal{G}^*(\boldsymbol{\lambda}_l, s_l)$ is dual to $\mathcal{G}^+(\boldsymbol{\lambda}_l, s_l)$, this is equivalent to

$$\langle \overline{\mathcal{G}^*(\boldsymbol{\lambda}_l, s_l)'}, \mathcal{G}^+(\boldsymbol{\mu}_l, s_l)'\rangle = \delta_{\boldsymbol{\lambda}_l, \boldsymbol{\mu}_l}.$$

Using Proposition 5.13, we obtain

$$\begin{aligned}\langle \overline{\mathcal{G}^*(\boldsymbol{\lambda}_l, s_l)'}, \mathcal{G}^+(\boldsymbol{\mu}_l, s_l)'\rangle &= \langle \mathcal{G}^*(\boldsymbol{\lambda}_l, s_l), \overline{\mathcal{G}^+(\boldsymbol{\mu}_l, s_l)}\rangle \\ &= \langle \mathcal{G}^*(\boldsymbol{\lambda}_l, s_l), \mathcal{G}^+(\boldsymbol{\mu}_l, s_l)\rangle = \delta_{\boldsymbol{\lambda}_l, \boldsymbol{\mu}_l}.\end{aligned}$$ □

This proposition immediately implies the following.

Theorem 5.15 (Inversion formula). *For $s_l \in \mathbf{Z}^l$, $\boldsymbol{\lambda}_l, \boldsymbol{\mu}_l \in \Pi^l$, one has*

$$\sum_{\boldsymbol{\nu}_l \in \Pi^l} \Delta^-_{\boldsymbol{\lambda}_l', \boldsymbol{\nu}_l'}(s_l'|q^{-1})\, \Delta^+_{\boldsymbol{\mu}_l, \boldsymbol{\nu}_l}(s_l|q) = \delta_{\boldsymbol{\lambda}_l, \boldsymbol{\mu}_l}.$$

5.4 Proof of Proposition 5.3. Let us prove part (i) of Proposition 5.3. First, we introduce some notation. Let s be an integer. For any pair $(\boldsymbol{\lambda}_l, s_l) \in \Pi^l \times \mathbf{Z}^l(s)$, where

$$\boldsymbol{\lambda}_l = (\lambda^{(1)}, \dots, \lambda^{(l)}), \qquad s_l = (s_1, \dots, s_l),$$

let $\boldsymbol{k} = (k_1, k_2, \dots)$ be the unique sequence from $\mathbf{P}^{++}(s)$ (cf. Section 4.1) such that

$$|\boldsymbol{\lambda}_l, s_l\rangle = u_{\boldsymbol{k}}.$$

As in Section 4.1, we write $k_i = a_i + n(b_i - 1) - nlm_i$, where $a_i \in \{1, \dots, n\}$, $b_i \in \{1, \dots, l\}$, and $m_i \in \mathbf{Z}$.

For any natural number r, set $(\boldsymbol{k})_r = (k_1, k_2, \dots, k_r)$. Then $(\boldsymbol{k})_r \in \mathbf{P}^{++}$ and $u_{\boldsymbol{k}} = u_{(\boldsymbol{k})_r} \wedge u_{k_{r+1}} \wedge u_{k_{r+2}} \wedge \cdots$, where $u_{(\boldsymbol{k})_r} = u_{k_1} \wedge \cdots \wedge u_{k_r}$ is an element of Λ^r.

We define $(\boldsymbol{k})_r^+ = (k_1^+, \dots, k_r^+)$ to be the unique permutation of the sequence $(\boldsymbol{k})_r$ characterized by the following two conditions:

$$\begin{aligned} b_i^+ &\le b_j^+ \quad \text{for all } i < j, \\ k_i^+ &> k_j^+ \quad \text{for all } i < j \text{ such that } b_i^+ = b_j^+. \end{aligned}$$

Here we set $k_i^+ = a_i^+ + n(b_i^+ - 1) - nlm_i^+$, where $a_i^+ \in \{1, \dots, n\}$, $b_i^+ \in \{1, \dots, l\}$, and $m_i^+ \in \mathbf{Z}$.

Example 5.16. Let $n = 2$, $l = 3$, and $s = -2$. Take $\boldsymbol{\lambda}_l = ((1^2), (1), (2))$ and $\boldsymbol{s}_l = (5, 0, -7)$. Let $r = 25$. In this case,

$$\begin{aligned} (\boldsymbol{k})_r = (&14, 13, 7, 3, 2, 1, -3, -4, -5, -8, -9, -10, -11, -13, -14, \\ &-15, -16, -17, -20, -21, -22, -23, -24, -25, -26), \end{aligned}$$

and

$$\begin{aligned} (\boldsymbol{k})_r^+ = (&14, 13, 7, 2, 1, -4, -5, -10, -11, -16, -17, -22, -23, 3, \\ &-3, -8, -9, -14, -15, -20, -21, -26, -13, -24, -25). \end{aligned}$$

Recall that in Section 4.1 we associated with $\boldsymbol{k}$ the semiinfinite sequences

$$\boldsymbol{k}^{(b)} = (k_1^{(b)}, k_2^{(b)}, \dots) \quad (b = 1, 2, \dots, l)$$

such that $k_i^{(b)} = s_b + i - 1 + \lambda_i^{(b)}$ for all $i \in \mathbf{N}$. The wedge $u_{(\boldsymbol{k})_r^+} = u_{k_1^+} \wedge \cdots \wedge u_{k_r^+}$ may be expressed in terms of these sequences in the following way. For $a \in \{1, \dots, n\}$, $b \in \{1, \dots, l\}$, and $m \in \mathbf{Z}$, set $u_{a-nm}^{(b)} := u_{a+n(b-1)-nlm}$. Then

$$\begin{aligned} u_{(\boldsymbol{k})_r^+} = u_{k_1^{(1)}}^{(1)} \wedge u_{k_2^{(1)}}^{(1)} \wedge \cdots \wedge u_{k_{r_1}^{(1)}}^{(1)} \wedge u_{k_1^{(2)}}^{(2)} \wedge u_{k_2^{(2)}}^{(2)} \wedge \cdots \wedge u_{k_{r_2}^{(2)}}^{(2)} \wedge \cdots \\ \cdots \\ \cdots \wedge u_{k_1^{(l)}}^{(l)} \wedge u_{k_2^{(l)}}^{(l)} \wedge \cdots \wedge u_{k_{r_l}^{(l)}}^{(l)}, \end{aligned} \tag{47}$$

where for each $b \in \{1, \dots, l\}$ we set $r_b := \sharp\{1 \le i \le r \mid b_i = b\}$.

Note that in general the wedge $u_{(\boldsymbol{k})_r^+}$ is not ordered, and using the ordering rules of Proposition 3.16 to straighten $u_{(\boldsymbol{k})_r^+}$ as a linear combination of ordered wedges, one typically obtains a linear combination with many terms. The first step towards the proof of the proposition is to show that if the pair $(\boldsymbol{\lambda}_l, \boldsymbol{s}_l)$ we started with is 0-dominant, then for any $r \in \mathbf{N}$, the straightening of $u_{(\boldsymbol{k})_r^+}$ produces only one term, which, up to a power of q, coincides with $u_{(\boldsymbol{k})_r}$.

Lemma 5.17. *Let $b_1, b_2 \in \{1, \dots, l\}$ and $a_1, a_2 \in \{1, \dots, n\}$ satisfy the inequalities $b_1 < b_2$, $a_1 \ge a_2$. Let $m \in \mathbf{Z}$. Then for any $t \in \mathbf{Z}_{\ge 0}$, one has the following relation:*

$$\begin{aligned} &u_{a_1-nm}^{(b_1)} \wedge u_{a_1-nm-1}^{(b_1)} \wedge u_{a_1-nm-2}^{(b_1)} \wedge \cdots \wedge u_{a_1-nm-t}^{(b_1)} \wedge u_{a_2-nm}^{(b_2)} \\ &= q^{\sum_{k=0}^{t} \delta_{(a_1-k) \equiv a_2 \bmod n}} u_{a_2-nm}^{(b_2)} \wedge u_{a_1-nm}^{(b_1)} \wedge u_{a_1-nm-1}^{(b_1)} \wedge u_{a_1-nm-2}^{(b_1)} \wedge \cdots \wedge u_{a_1-nm-t}^{(b_1)}. \end{aligned}$$

PROOF. This is shown by induction on t using relations (R3) and (R4) of Proposition 3.16 and Lemma 3.18. □

Keeping $(\boldsymbol{\lambda}_l, \boldsymbol{s}_l)$, $\boldsymbol{k}$, r, $(\boldsymbol{k})_r$, and $(\boldsymbol{k})_r^+$ as above, let us define

$$c_r(\boldsymbol{k}) := \sharp\{1 \leq i < j \leq r \mid b_i > b_j,\ a_i = a_j\},$$

and

$$|\boldsymbol{\lambda}_l, \boldsymbol{s}_l\rangle_r^+ := u_{(\boldsymbol{k})_r^+} \wedge u_{k_{r+1}} \wedge u_{k_{r+2}} \wedge \cdots .$$

Lemma 5.18. *Suppose that $(\boldsymbol{\lambda}_l, \boldsymbol{s}_l)$ is 0-dominant. Then*

$$|\boldsymbol{\lambda}_l, \boldsymbol{s}_l\rangle_r^+ = q^{c_r(\boldsymbol{k})} |\boldsymbol{\lambda}_l, \boldsymbol{s}_l\rangle.$$

PROOF. Since $|\boldsymbol{\lambda}_l, \boldsymbol{s}_l\rangle = u_{\boldsymbol{k}} = u_{(\boldsymbol{k})_r} \wedge u_{k_{r+1}} \wedge u_{k_{r+2}} \wedge \cdots$, we must prove that

$$u_{(\boldsymbol{k})_r^+} = q^{c_r(\boldsymbol{k})} u_{(\boldsymbol{k})_r}.$$

First of all, let us examine what the 0-dominance of $(\boldsymbol{\lambda}_l, \boldsymbol{s}_l)$ implies for the semiinfinite sequence $\boldsymbol{k}$. For each $b \in \{1, \ldots, l\}$, let p_b be the minimal number such that $k_i^{(b)} = s_b - i + 1$ for all $i \geq p_b$. Then $p_b = l(\lambda^{(b)}) + 1$, where $l(\lambda)$ denotes the length of a partition λ, and we have $k_{p_b}^{(b)} = s_b - l(\lambda^{(b)})$. On the other hand, $k_1^{(b)} = s_b + \lambda_1^{(b)}$. Hence using the assumption that $(\boldsymbol{\lambda}_l, \boldsymbol{s}_l)$ is 0-dominant (cf. Definition 5.1), we find that for all $b = 1, 2, \ldots, l-1$,

$$k_{p_b}^{(b)} - k_1^{(b+1)} = s_b - s_{b+1} - l(\lambda^{(b)}) - \lambda_1^{(b+1)} \geq s_b - s_{b+1} - |\boldsymbol{\lambda}_l| \geq 0.$$

The fact that we have the inequalities $k_{p_b}^{(b)} \geq k_1^{(b+1)}$ for all $b = 1, 2, \ldots, l-1$, implies that to straighten $u_{(\boldsymbol{k})_r^+}$ on the basis of ordered wedges, we need only to repeatedly apply Lemma 5.17. The result follows. □

Example 5.19. Let us illustrate the proof of Lemma 5.18 for $|\boldsymbol{\lambda}_l, \boldsymbol{s}_l\rangle_r^+$, where $(\boldsymbol{\lambda}_l, \boldsymbol{s}_l)$ and r are the same as in Example 5.16. Note that the pair $(\boldsymbol{\lambda}_l, \boldsymbol{s}_l)$ is 0-dominant.

In this case, $u_{(\boldsymbol{k})_r}$ is given by the following expression:

$$\begin{aligned} & u_{2+2\cdot 2}^{(1)} \wedge u_{1+2\cdot 2}^{(1)} \wedge u_{1+2}^{(1)} \wedge u_{1}^{(2)} \wedge u_{2}^{(1)} \wedge u_{1}^{(1)} \wedge u_{1-2\cdot 1}^{(2)} \wedge u_{2-2\cdot 1}^{(1)} \wedge u_{1-2\cdot 1}^{(1)} \wedge u_{2-2\cdot 2}^{(2)} \\ & \quad \wedge u_{1-2\cdot 2}^{(2)} \wedge u_{2-2\cdot 2}^{(1)} \wedge u_{1-2\cdot 2}^{(1)} \wedge u_{1-2\cdot 3}^{(3)} \wedge u_{2-2\cdot 3}^{(2)} \wedge u_{1-2\cdot 3}^{(2)} \wedge u_{2-2\cdot 3}^{(1)} \wedge u_{1-2\cdot 3}^{(1)} \\ & \quad \wedge u_{2-2\cdot 4}^{(2)} \wedge u_{1-2\cdot 4}^{(2)} \wedge u_{2-2\cdot 4}^{(1)} \wedge u_{1-2\cdot 4}^{(1)} \wedge u_{2-2\cdot 5}^{(3)} \wedge u_{1-2\cdot 5}^{(3)} \wedge u_{2-2\cdot 5}^{(2)}. \end{aligned}$$

Now let us rearrange (taking care of powers of q) the factors in this wedge by repeatedly applying Lemma 5.17. The rearrangement involves the following seven steps:

$$
\begin{aligned}
u_{(\boldsymbol{k})_r} &= u^{(1)}_{2+2\cdot 2} \wedge u^{(1)}_{1+2\cdot 2} \wedge u^{(1)}_{1+2} \wedge u^{(2)}_{1} \wedge u^{(1)}_{2} \wedge u^{(1)}_{1} \wedge u^{(2)}_{1-2\cdot 1} \wedge u^{(1)}_{2-2\cdot 1} \wedge u^{(1)}_{1-2\cdot 1} \\
&\quad \wedge u^{(2)}_{2-2\cdot 2} \wedge u^{(2)}_{1-2\cdot 2} \wedge u^{(1)}_{2-2\cdot 2} \wedge u^{(1)}_{1-2\cdot 2} \wedge u^{(3)}_{1-2\cdot 3} \wedge u^{(2)}_{2-2\cdot 3} \wedge u^{(2)}_{1-2\cdot 3} \wedge u^{(1)}_{2-2\cdot 3} \\
&\quad \wedge u^{(1)}_{1-2\cdot 3} \wedge u^{(2)}_{2-2\cdot 4} \wedge u^{(2)}_{1-2\cdot 4} \wedge u^{(1)}_{2-2\cdot 4} \wedge u^{(1)}_{1-2\cdot 4} \wedge \underline{u^{(3)}_{2-2\cdot 5} \wedge u^{(3)}_{1-2\cdot 5} \wedge u^{(2)}_{2-2\cdot 5}} \\
&= q^{-1} u^{(1)}_{2+2\cdot 2} \wedge u^{(1)}_{1+2\cdot 2} \wedge u^{(1)}_{1+2} \wedge u^{(2)}_{1} \wedge u^{(1)}_{2} \wedge u^{(1)}_{1} \wedge u^{(2)}_{1-2\cdot 1} \wedge u^{(1)}_{2-2\cdot 1} \wedge u^{(1)}_{1-2\cdot 1} \\
&\quad \wedge u^{(2)}_{2-2\cdot 2} \wedge u^{(2)}_{1-2\cdot 2} \wedge u^{(1)}_{2-2\cdot 2} \wedge u^{(1)}_{1-2\cdot 2} \wedge u^{(3)}_{1-2\cdot 3} \wedge u^{(2)}_{2-2\cdot 3} \wedge u^{(2)}_{1-2\cdot 3} \wedge u^{(1)}_{2-2\cdot 3} \\
&\quad \wedge u^{(1)}_{1-2\cdot 3} \wedge \underline{u^{(2)}_{2-2\cdot 4} \wedge u^{(2)}_{1-2\cdot 4} \wedge u^{(1)}_{2-2\cdot 4} \wedge u^{(1)}_{1-2\cdot 4}} \wedge u^{(2)}_{2-2\cdot 5} \wedge u^{(3)}_{2-2\cdot 5} \wedge u^{(3)}_{1-2\cdot 5} \\
&= q^{-3} u^{(1)}_{2+2\cdot 2} \wedge u^{(1)}_{1+2\cdot 2} \wedge u^{(1)}_{1+2} \wedge u^{(2)}_{1} \wedge u^{(1)}_{2} \wedge u^{(1)}_{1} \wedge u^{(2)}_{1-2\cdot 1} \wedge u^{(1)}_{2-2\cdot 1} \wedge u^{(1)}_{1-2\cdot 1} \\
&\quad \wedge u^{(2)}_{2-2\cdot 2} \wedge u^{(2)}_{1-2\cdot 2} \wedge u^{(1)}_{2-2\cdot 2} \wedge u^{(1)}_{1-2\cdot 2} \wedge u^{(3)}_{1-2\cdot 3} \wedge \underline{u^{(2)}_{2-2\cdot 3} \wedge u^{(2)}_{1-2\cdot 3} \wedge u^{(1)}_{2-2\cdot 3}} \\
&\quad \underline{\wedge u^{(1)}_{1-2\cdot 3} \wedge u^{(1)}_{2-2\cdot 4} \wedge u^{(1)}_{1-2\cdot 4}} \wedge u^{(2)}_{2-2\cdot 4} \wedge u^{(2)}_{1-2\cdot 4} \wedge u^{(2)}_{2-2\cdot 5} \wedge u^{(3)}_{2-2\cdot 5} \wedge u^{(3)}_{1-2\cdot 5} \\
&= q^{-7} u^{(1)}_{2+2\cdot 2} \wedge u^{(1)}_{1+2\cdot 2} \wedge u^{(1)}_{1+2} \wedge u^{(2)}_{1} \wedge u^{(1)}_{2} \wedge u^{(1)}_{1} \wedge u^{(2)}_{1-2\cdot 1} \wedge u^{(1)}_{2-2\cdot 1} \wedge u^{(1)}_{1-2\cdot 1} \\
&\quad \wedge u^{(2)}_{2-2\cdot 2} \wedge u^{(2)}_{1-2\cdot 2} \wedge u^{(1)}_{2-2\cdot 2} \wedge u^{(1)}_{1-2\cdot 2} \wedge \underline{u^{(3)}_{1-2\cdot 3} \wedge u^{(1)}_{2-2\cdot 3} \wedge u^{(1)}_{1-2\cdot 3} \wedge u^{(1)}_{2-2\cdot 4}} \\
&\quad \underline{\wedge u^{(1)}_{1-2\cdot 4} \wedge u^{(2)}_{2-2\cdot 3} \wedge u^{(2)}_{1-2\cdot 3} \wedge u^{(2)}_{2-2\cdot 4} \wedge u^{(2)}_{1-2\cdot 4} \wedge u^{(2)}_{2-2\cdot 5}} \wedge u^{(3)}_{2-2\cdot 5} \wedge u^{(3)}_{1-2\cdot 5} \\
&= q^{-11} u^{(1)}_{2+2\cdot 2} \wedge u^{(1)}_{1+2\cdot 2} \wedge u^{(1)}_{1+2} \wedge u^{(2)}_{1} \wedge u^{(1)}_{2} \wedge u^{(1)}_{1} \wedge u^{(2)}_{1-2\cdot 1} \wedge u^{(1)}_{2-2\cdot 1} \wedge u^{(1)}_{1-2\cdot 1} \\
&\quad \wedge \underline{u^{(2)}_{2-2\cdot 2} \wedge u^{(2)}_{1-2\cdot 2} \wedge u^{(1)}_{2-2\cdot 2} \wedge u^{(1)}_{1-2\cdot 2} \wedge u^{(1)}_{2-2\cdot 3} \wedge u^{(1)}_{1-2\cdot 3} \wedge u^{(1)}_{2-2\cdot 4} \wedge u^{(1)}_{1-2\cdot 4}} \\
&\quad \wedge u^{(2)}_{2-2\cdot 3} \wedge u^{(2)}_{1-2\cdot 3} \wedge u^{(2)}_{2-2\cdot 4} \wedge u^{(2)}_{1-2\cdot 4} \wedge u^{(2)}_{2-2\cdot 5} \wedge u^{(3)}_{1-2\cdot 3} \wedge u^{(3)}_{2-2\cdot 5} \wedge u^{(3)}_{1-2\cdot 5} \\
&= q^{-17} u^{(1)}_{2+2\cdot 2} \wedge u^{(1)}_{1+2\cdot 2} \wedge u^{(1)}_{1+2} \wedge u^{(2)}_{1} \wedge u^{(1)}_{2} \wedge u^{(1)}_{1} \wedge \underline{u^{(2)}_{1-2\cdot 1} \wedge u^{(1)}_{2-2\cdot 1} \wedge u^{(1)}_{1-2\cdot 1}} \\
&\quad \underline{\wedge u^{(1)}_{2-2\cdot 2} \wedge u^{(1)}_{1-2\cdot 2} \wedge u^{(1)}_{2-2\cdot 3} \wedge u^{(1)}_{1-2\cdot 3} \wedge u^{(1)}_{2-2\cdot 4} \wedge u^{(1)}_{1-2\cdot 4}} \wedge u^{(2)}_{2-2\cdot 2} \wedge u^{(2)}_{1-2\cdot 2} \\
&\quad \wedge u^{(2)}_{2-2\cdot 3} \wedge u^{(2)}_{1-2\cdot 3} \wedge u^{(2)}_{2-2\cdot 4} \wedge u^{(2)}_{1-2\cdot 4} \wedge u^{(2)}_{2-2\cdot 5} \wedge u^{(3)}_{1-2\cdot 3} \wedge u^{(3)}_{2-2\cdot 5} \wedge u^{(3)}_{1-2\cdot 5} \\
&= q^{-21} u^{(1)}_{2+2\cdot 2} \wedge u^{(1)}_{1+2\cdot 2} \wedge u^{(1)}_{1+2} \wedge \underline{u^{(2)}_{1} \wedge u^{(1)}_{2} \wedge u^{(1)}_{1} \wedge u^{(1)}_{2-2\cdot 1} \wedge u^{(1)}_{1-2\cdot 1} \wedge u^{(1)}_{2-2\cdot 2}} \\
&\quad \underline{\wedge u^{(1)}_{1-2\cdot 2} \wedge u^{(1)}_{2-2\cdot 3} \wedge u^{(1)}_{1-2\cdot 3} \wedge u^{(1)}_{2-2\cdot 4} \wedge u^{(1)}_{1-2\cdot 4}} \wedge u^{(2)}_{1-2\cdot 1} \wedge u^{(2)}_{2-2\cdot 2} \wedge u^{(2)}_{1-2\cdot 2} \\
&\quad \wedge u^{(2)}_{2-2\cdot 3} \wedge u^{(2)}_{1-2\cdot 3} \wedge u^{(2)}_{2-2\cdot 4} \wedge u^{(2)}_{1-2\cdot 4} \wedge u^{(2)}_{2-2\cdot 5} \wedge u^{(3)}_{1-2\cdot 3} \wedge u^{(3)}_{2-2\cdot 5} \wedge u^{(3)}_{1-2\cdot 5} \\
&= q^{-26} u^{(1)}_{2+2\cdot 2} \wedge u^{(1)}_{1+2\cdot 2} \wedge u^{(1)}_{1+2} \wedge u^{(1)}_{2} \wedge u^{(1)}_{1} \wedge u^{(1)}_{2-2\cdot 1} \wedge u^{(1)}_{1-2\cdot 1} \wedge u^{(1)}_{2-2\cdot 2} \wedge u^{(1)}_{1-2\cdot 2} \\
&\quad \wedge u^{(1)}_{2-2\cdot 3} \wedge u^{(1)}_{1-2\cdot 3} \wedge u^{(1)}_{2-2\cdot 4} \wedge u^{(1)}_{1-2\cdot 4} \wedge u^{(2)}_{1} \wedge u^{(2)}_{1-2\cdot 1} \wedge u^{(2)}_{2-2\cdot 2} \wedge u^{(2)}_{1-2\cdot 2} \\
&\quad \wedge u^{(2)}_{2-2\cdot 3} \wedge u^{(2)}_{1-2\cdot 3} \wedge u^{(2)}_{2-2\cdot 4} \wedge u^{(2)}_{1-2\cdot 4} \wedge u^{(2)}_{2-2\cdot 5} \wedge u^{(3)}_{1-2\cdot 3} \wedge u^{(3)}_{2-2\cdot 5} \wedge u^{(3)}_{1-2\cdot 5} \\
&= q^{-26} u_{(\boldsymbol{k})_r^+}.
\end{aligned}
$$

Here at each step we underline the part to which we apply Lemma 5.17. Note that we use this lemma twice at steps 1, 2, 3, 4, 5 and we use it once at steps 6 and 7.

Let $l = 1$ temporarily and for partitions λ and μ, define the matrix elements $(B_{-m})_{\lambda,\mu}(s)$ by

$$B_{-m}|\lambda, s\rangle = \sum_{\mu\in\Pi}(B_{-m})_{\lambda,\mu}(s)\,|\mu, s\rangle.$$

Now we proceed with the proof of the proposition. Assume that the pair $(\boldsymbol{\lambda}_l, s_l)$ is nm-dominant for some $m \in \mathbf{N}$. Then

$$B_{-m}|\boldsymbol{\lambda}_l, s_l\rangle = B_{-m}u_{\boldsymbol{k}} = (B_{-m}u_{(\boldsymbol{k})_r}) \wedge |s-r\rangle = q^{-c_r(\boldsymbol{k})}(B_{-m}u_{(\boldsymbol{k})_r^+}) \wedge |s-r\rangle,$$

where the second equality is obtained by taking r sufficiently large and the third equality follows from Lemma 5.18. Using (47), we have

$$B_{-m}u_{(\boldsymbol{k})_r^+} = \sum_{b=1}^{l}\sum_{i=1}^{r_b} u^{(1)}_{(\boldsymbol{k}^{(1)})_{r_1}} \wedge u^{(2)}_{(\boldsymbol{k}^{(2)})_{r_2}} \wedge\cdots\wedge u^{(b)}_{(\boldsymbol{k}^{(b)})_{r_b}-\epsilon_i nm} \wedge\cdots\wedge u^{(l)}_{(\boldsymbol{k}^{(l)})_{r_l}},$$

where we set $(\boldsymbol{k}^{(b)})_{r_b} = (k_1^{(b)}, \ldots, k_{r_b}^{(b)})$ and $\epsilon_i = (0, \ldots, 0, 1, 0 \ldots, 0)$ with 1 on the ith position. Let us now straighten the expression

$$\sum_{i=1}^{r_b} u^{(b)}_{(\boldsymbol{k}^{(b)})_{r_b}-\epsilon_i nm}$$

on the basis of ordered wedges. It is clear that to do so we only need to use relations (R1) and (R2) of Proposition 3.16. However, these two relations are the same as in the $l = 1$ case. Hence assuming (as we may by choosing large enough r) that $r_1, r_2, \ldots, r_l$ are sufficiently large, we get

$$B_{-m}|\boldsymbol{\lambda}_l, s_l\rangle$$
$$= q^{-c_r(\boldsymbol{k})}\sum_{b=1}^{l}\sum_{\mu\in\Pi}(B_{-m})_{\lambda^{(b)},\mu}(s_b)\,|(\lambda^{(1)}, \ldots, \lambda^{(b-1)}, \mu, \lambda^{(b+1)}, \ldots, \lambda^{(l)}), s_l\rangle_r^+.$$

Observe now that in the above sum we have for all b and μ,

$$|\mu| = |\lambda^{(b)}| + nm.$$

Hence the nm-dominance of $(\boldsymbol{\lambda}_l, s_l)$ implies the 0-dominance of each pair

$$((\lambda^{(1)}, \ldots, \lambda^{(b-1)}, \mu, \lambda^{(b+1)}, \ldots, \lambda^{(l)}), s_l). \tag{48}$$

It follows that we may apply Lemma 5.18 in order to straighten each wedge under the sum. Doing so, we get

$$B_{-m}|\boldsymbol{\lambda}_l, s_l\rangle = \sum_{b=1}^{l}\sum_{\mu\in\Pi} q^{c_r(\boldsymbol{l})-c_r(\boldsymbol{k})}$$
$$\times (B_{-m})_{\lambda^{(b)},\mu}(s_b)\,|(\lambda^{(1)}, \ldots, \lambda^{(b-1)}, \mu, \lambda^{(b+1)}, \ldots, \lambda^{(l)}), s_l\rangle,$$

where $\boldsymbol{l}$ is the unique element of $\mathbf{P}^{++}(s)$ such that

$$|(\lambda^{(1)}, \dots, \lambda^{(b-1)}, \mu, \lambda^{(b+1)}, \dots, \lambda^{(l)}), s_l\rangle = u_{\boldsymbol{l}}.$$

Finally, using the 0-dominance of (48) and the 0-dominance of $(\boldsymbol{\lambda}_l, s_l)$, it is not difficult to see that $c_r(\boldsymbol{l}) - c_r(\boldsymbol{k}) = (b-1)m$ for all large enough r. Proposition 5.3(i) follows.

A proof of Proposition 5.3 (i′)is obtained by following the same steps as above but interchanging everywhere the roles of n and l and the roles of p and q. The proofs of (ii) and (ii′) are similar to the proofs of (i) and (i′) and will be omitted. □

Acknowledgments. I would like to thank S. Ariki, T. Baker, M. Kashiwara, B. Leclerc, T. Miwa and, J.-Y. Thibon for illuminating discussions. I am especially indebted to B. Leclerc for explanations concerning the papers [5], [17], and [18]. The influence of [17] and [18] on the present article will be obvious.

References

[1] T. Arakawa, T. Suzuki, and A. Tsuchiya, Degenerate double affine hecke algebra and conformal field theory, preprint q-alg/9710031, 1997.

[2] S. Ariki, On the decomposition numbers of the Hecke algebra of $G(m, 1, n)$, *J. Math. Kyoto Univ.*, **36** (1996), 789–808.

[3] S. Ariki and A. Mathas, The number of simple modules of the Hecke algebra of $G(m, 1, n)$, preprint, 1998.

[4] V. V. Deodhar, On some geometric aspects of Bruhat orderings II, *J. Algebra*, **111** (1987), 483–506.

[5] O. Foda, B. Leclerc, M. Okado, J.-Y. Thibon, and T. Welsh, Branching functions of $A^{(1)}_{n-1}$ and Jantzen-Seitz problem for Ariki-Koike algebras, *Adv. Math.*, **141** (1999), 322–365.

[6] I. Frenkel, Representations of affine Lie algebras, Hecke modular forms and Korteweg–de Vries type equations, *Lecture Notes in Math.*, **933** (1982), 71–110.

[7] I. Frenkel, Representations of affine Kac-Moody algebras and dual resonance models, *Lectures in Appl. Math.*, **21** (1985), 325–353.

[8] V. Ginzburg, N. Reshetikhin, and E. Vasserot, Quantum groups and flag varieties, *Contemp. Math.*, **175** (1994), 101–130.

[9] T. Hayashi, q-analogues of Clifford and Weyl algebras: Spinor and oscillator representations of quantum enveloping algebras, *Comm. Math. Phys.*, **127** (1990), 129–144.

[10] M. Jimbo, K. Misra, T. Miwa, and M. Okado, Combinatorics of representations of $U_q(\widehat{\mathfrak{sl}}_n)$ at $q = 0$, *Comm. Math. Phys.*, **136** (1991), 543-566.

[11] M. Kashiwara, On crystal bases of the q-analogue of universal enveloping algebras, *Duke Math. J.*, **63** (1991), 465–516.

[12] M. Kashiwara, Global crystal bases of quantum groups, *Duke Math. J.*, **69** (1993), 455–485.

[13] M. Kashiwara, T. Miwa, and E. Stern, Decomposition of q-deformed Fock spaces, *Selecta Math. N.S.*, **1** (1995), 787–805.
[14] M. Kashiwara, T. Miwa, J.-U. H. Petersen, and C. M. Yung, Perfect crystals and q-deformed Fock spaces, *Selecta Math. N.S.*, **2** (1996), 415–499.
[15] A. A. Kirillov, Jr., Lectures on the affine Hecke algebras and Macdonald conjectures, preprint math/9501219, 1995.
[16] A. Lascoux B. Leclerc, and J.-Y. Thibon, Hecke algebras at roots of unity and crystal bases of quantum affine algebras, *Comm. Math. Phys.*, **181** (1996), 205–263.
[17] B. Leclerc and J.-Y. Thibon, Canonical bases of q-deformed Fock spaces, *Internat. Math. Res. Notices*, **9** (1996), 447–456.
[18] B. Leclerc and J.-Y. Thibon, Littelwood-Richardson coefficients and Kazhdan-Lusztig polynomials, preprint math.QA/9809122, 1998.
[19] I. G. Macdonald, *Symmetric Functions and Hall Polynomials*, Clarendon Press, Oxford, 1979, .
[20] K. C. Misra and T. Miwa, Crystal base of the basic representation of $U_q(\widehat{\mathfrak{sl}}_n)$, *Comm. Math. Phys.*, **134** (1990), 79–88.
[21] W. Soergel, Kazhdan-Lusztig-Polynome und eine Kombinatorik für Kipp-Moduln, *Represent. Theory*, **1** (1997), 115–132.
[22] E. Stern, Semi-infinite wedges and vertex operators, *Internat. Math. Res. Notices*, **4** (1995), 201–220.
[23] K. Takemura and D. Uglov, Representations of the quantum toroidal algebra on highest weight modules of the quantum affine algebra of type $\mathfrak{gl}_n$, preprint math.QA/9806134, 1998.
[24] D. Uglov, Canonical bases of higher-level q-deformed Fock spaces, preprint math.QA/9901032, 1999.
[25] M. Varagnolo and E. Vasserot, On the decomposition matrices of the quantized Schur algebra, preprint math.QA/9803023, 1998.

Denis Uglov†
Research Institute for Mathematical Sciences
Kyoto University
606 Kyoto
Japan

Finite-Gap Difference Operators with Elliptic Coefficients and Their Spectral Curves

A. Zabrodin

Abstract. We review recent results on the finite-gap properties of difference operators with elliptic coefficients and give an explicit characterization of spectral curves for difference analogues of higher Lamé operators. This curve parametrizes double-Bloch solutions to the difference Lamé equation. The curve depends on a positive integer number ℓ—related to its genus g by $g = 2\ell$—and two continuous parameters: the lattice spacing η and the modular parameter τ. Isospectral deformations of the difference Lamé operator under Volterra flows are also discussed.

1 Introduction

The spectrum of the Schrödinger operator $-\partial_x^2 + u(x)$ with a periodic potential $u(x) = u(x+T)$ has a band structure; there are stable energy bands separated by gaps. For smooth potentials, the width of gaps rapidly decreases as energy becomes higher. However, gaps generically occur at arbitrarily high energies, so there are infinitely many of them.

Of particular interest are the exceptional cases when for sufficiently high energies there are no longer gaps and their number is therefore finite. Such operators are usually referred to as algebraically integrable or finite-gap operators. Their study goes back to the classical works of the last century. The renewed interest in the theory of finite-gap operators is due to their role in constructing quasi-periodic exact solutions to nonlinear integrable equations.

Among examples of finite-gap operators, the first and most familiar is the classical Lamé operator

$$\mathcal{L} = -\frac{d^2}{dx^2} + \ell(\ell+1)\wp(x+\omega'|\omega,\omega'), \tag{1.1}$$

where $\wp(x|\omega, \omega')$ is the Weierstrass $\wp$-function and ℓ is a parameter. The potential is a doubly-periodic function on the complex plane with periods 2ω and $2\omega'$, where $\mathrm{Im}\,(\omega'/\omega) = \tau > 0$. If ω is real while ω' is purely imaginary, the spectral problem is self-adjoint. The finite-gap property of higher Lamé operators for integer values of ℓ was established in [11]. If ℓ is a positive integer, then the Lamé operator has exactly ℓ gaps in the spectrum. Such a remarkable spectral property is a sign of a hidden algebraic symmetry, which, in turn, leads to an intimate connection with integrable systems.

The finite-gap property becomes even more striking for difference operators. A natural difference analogue of the Schrödinger equation has the form

$$a(x)\Psi(x+\eta) + b(x)\Psi(x) + c(x)\Psi(x-\eta) = E\Psi(x), \tag{1.2}$$

where the parameter η is the lattice spacing. Let us assume that η is real, the coefficients are real functions of x, and $c(x) = a(x-\eta)$; then the problem is self-adjoint. Let the coefficient functions be periodic with a common period T: $a(x+T) = a(x)$, $b(x+T) = b(x)$. The difference Schrödinger operators with periodic coefficients exhibit much richer spectral properties because the problem has two competing periods (T and η) rather than one. Nevertheless, the class of finite-gap operators survives.

The structure of the spectrum of a typical difference operator crucially depends on whether the ratio T/η of the two periods is a rational or irrational number. In the former (commensurate) case, one can always set $T/\eta = Q \in \mathbf{Z}$ without loss of generality. Then there are no more than Q stable bands in the spectrum. Indeed, set $\Psi(x_0 + n\eta) = \Psi_n$, $a(x_0 + n\eta) = a_n$, etc. and rewrite (1.2) in the form

$$a_n\Psi_{n+1} + b_n\Psi_n + c_n\Psi_{n-1} = E\Psi_{n,},$$

where $a_{n+Q} = a_n$, $b_{n+Q} = b_n$, and $c_{n+Q} = c_n$. Since the coefficients are periodic, one may look for solutions in the Bloch form $\Psi_n = e^{ik\eta n}\chi_n$, where χ_n is Q-periodic and k is the Bloch momentum. Therefore, the spectral problem is reduced to the eigenvalue problem for a Hermitian $Q \times Q$ matrix. For each real value of the Bloch momentum k, the secular equation has Q real solutions $E = E_i(k)$. As k sweeps over the Brillouin zone, $E_i(k)$ sweep over the stable bands labeled by i. Several neighboring bands can merge, so the total number of stable bands can be less than or equal to Q.

The latter, incommensurate case can be practically realized as a proper limit of the former when both the numerator and denominator of the fraction $T/\eta = Q/P$ tend to infinity. The resulting spectra can be (and usually are) extremely complicated chaotic generations of the Cantor set type. Some of them, like those in the Azbel–Hofstadter problem [3, 10], though of a multifractal nature, nevertheless keep a good deal of hidden regularity revealed in terms of string solutions to Bethe equations [1]. Very little is known about the spectra of generic type; they seem to be completely irregular. In this paper, we discuss the opposite case of the utmost regular spectra in the sense that the number of bands is finite although T/η is irrational. Moreover, the number of bands does not really depend on this ratio, being determined by another (integer) parameter. The operators with this type of spectra are true difference analogues of the finite-gap operators.

In [15], the following difference analogue of the Lamé operator (1.1) was proposed:

$$L = \frac{\theta_1(x-\ell\eta)}{\theta_1(x)} e^{\eta\partial_x} + \frac{\theta_1(x+\ell\eta)}{\theta_1(x)} e^{-\eta\partial_x}. \tag{1.3}$$

Here $\theta_1(x) \equiv \theta_1(x|\tau)$ is the odd Jacobi θ-function and ℓ is a nonnegative integer. The coefficients are periodic functions with period 1. This operator can be made self-adjoint by the similarity transformation $L \to g^{-1}(x)Lg(x)$ with a function $g(x)$ such that $g(x+1) = g(x)$, so the spectrum is real. The operator (1.3) first appeared in a completely different context of representations of the Sklyanin algebra as early as in 1983 [18, 19]. Namely, L coincides with one of the four generators of the Sklyanin algebra in the functional realization found by Sklyanin. Remarkably, for positive integer values of ℓ and *arbitrary* generic η, the operator L has $2\ell+1$ stable bands (and 2ℓ gaps) in the spectrum. The finite-gap property of this operator for integer ℓ was proved in [15]. It was also shown [15, 23] that the Sklyanin algebra provides a natural algebraic framework for analyzing the spectral properties of the operator L. (A different algebraic approach to the difference analogues of the Lamé operators was proposed in [8].)

Another similarity transformation, $\tilde{L} = f^{-1}Lf$, where

$$f(x) = \prod_{j=1}^{\ell} \theta_1(x - j\eta), \tag{1.4}$$

makes coefficients of the difference operator

$$\tilde{L} = e^{\eta\partial_x} + \frac{\theta_1(x+\ell\eta)\theta_1(x-(\ell+1)\eta)}{\theta_1(x)\theta_1(x-\eta)} e^{-\eta\partial_x} \tag{1.5}$$

double-periodic functions of x with periods 1 and τ. The limit $\eta \to 0$ gives the Lamé operator (1.1). Indeed, replacing x by $x + \frac{1}{2}\tau$ in (1.5), we obtain

$$\tilde{L} = 2 - \eta^2(\mathcal{L} + \text{const}) + O(\eta^3), \quad \eta \to 0, \tag{1.6}$$

where the $\wp$-function in the $\mathcal{L}$ has periods 1 and τ (see (A.3)).

Let us mention that spectral curves of the classical Lamé operator (1.1) and its Treibich–Verdier generalizations [21] for small values of ℓ were studied in [7, 20]. A detailed analysis of solutions to the difference Lamé equation at $\ell = 1$ was recently carried out in [17].

The paper is organized as follows. Section 2 is a continuation of the introduction. To present the problem in a broader context, we discuss the general notion of the finite-gap operator. In Section 3, a family of Bloch eigenfunctions of the operator (1.5) is constructed. These eigenfunctions are parametrized by points of the spectral curve. Section 4 contains equations for the edges of bands and some examples. In Section 5, we work out an explicit relation between the Bloch multipliers. The form of the result suggests that some hypothetical combinatorial identities for "elliptic numbers" may be relevant. Finally, Section 6 contains some remarks on the isospectral deformations of the difference Lamé operator. In this case, the coefficient in (1.5) has more poles. The location of the poles, however, is not arbitrary; they are constrained by locus equations.

2 A general view of finite-gap operators

The key idea of the modern approach to spectra of differential or difference operators is to regard the solutions $\Psi(x, E)$ to the spectral problem (say (1.2)) as functions of E for any complex values of E and to study their analytic properties in E. In so doing, it is not necessary to assume that the problem is self-adjoint, so the parameter η and the coefficients may be complex numbers.

In practice, one may try to construct a family of eigenfunctions $\Psi = \Psi(x, E, p_1, p_2, \dots)$ depending, apart from E, on a finite number of additional parameters p_i. For instance, one of these could be the Bloch momentum k: $\Psi(x) = e^{ikx}\chi(x, E, k)$, where $\chi(x+T, E, k) = \chi(x, E, k)$. Suppose that such a family does exist. Then the spectral parameters appear to be constrained by some relations $F_i(E, p_1, p_2, \dots) = 0$ so that only one of the parameters is independent. These relations define a complex curve (a Riemann surface) in the parameter space called the *spectral curve*. The true spectral parameter is a point of the curve. This is the proper mathematical formulation of the dispersion law $E = E(k)$. Usually, this function is multivalued. It becomes single-valued on the spectral curve (when the latter is well defined). Moreover, the solution $\Psi(x, E)$ of the spectral problem also becomes a single-valued function on the spectral curve.

In the case of second-order difference operators, the spectral problem (1.2) has no more than two linearly independent solutions. In other words, the function E can take any of its values at most twice. The existence of such a function implies that the spectral curve is a hyperelliptic curve. Any hyperelliptic curve can be represented in the form of a two-sheet covering of the complex plane of the variable E:

$$w^2 = \prod_i (E - E_i), \tag{2.1}$$

where E_i are called *branch points*. For $E = E_i$, (1.2) has only one linearly independent solution. The curve (2.1) is well defined if the set of branch points is finite. Then the curve is algebraic and has finite genus. Equivalently, this means that there exists a difference operator W such that W cannot be represented as a polynomial function of the difference operator in the right-hand side of (1.2) but rather commutes with this operator. In this case, they have a set of common eigenfunctions. Equation

(2.1) is then lifted to the operator relation. The parameter w is the eigenvalue of the operator W on the common eigenfunction $\Psi(x, E)$: $W\Psi(x, E) = w\Psi(x, E)$.

For self-adjoint spectral problems, the branch points E_i are real numbers $E_1 < E_2 < E_3 < \dots$. In the stable bands, the Bloch momentum takes real values. The stable bands are segments of the real line $[E_{2i+1}, E_{2i+2}], i = 0, 1, \dots$, so the branch points are just edges of bands. At the edges of bands, the Bloch solutions $\Psi(x, E_i)$ are periodic or antiperiodic.

The notion of the spectral curve is really useful in the exceptional case of the spectra of regular type mentioned in the introduction. Recall that in the typical case, the set of branch points may even be uncountable, so the spectral curve defined above does not make sense. The very existence of a well-defined spectral curve of finite genus is the precise characterization of the finite-gap operators.

Then it is natural to address the inverse problem: given a hyperelliptic curve of finite genus regarded as a spectral curve of some difference operator, find coefficients of this operator, i.e., the functions $a(x)$ and $b(x)$ in (1.2). In this way, one is able to construct a representative family of finite-gap operators [6, 14]. For difference operators, this was done in [5, 16, 12]. The coefficients of the operators can be expressed through Riemann's theta functions associated with the curve.

The curve itself does not determine the operator uniquely. There is a remaining finite-parametric freedom that can be fixed by some additional data on the curve (essentially, a number of marked points). In other words, any finite-gap operator admits a class of isospectral deformations. The coefficients of the operator with respect to the isospectral flows obey certain nonlinear integrable equations.

Looking for formulas more effective than a bunch of Riemann's theta functions, one may inquire whether they can be expressed through simpler functions, for instance, elliptic functions. For a particular class of curves, namely, for special coverings of elliptic curves, this is indeed possible (see, e.g., [4]). The Riemann theta function associated with such a curve factorizes into a product of Jacobi θ-functions, so the coefficients of the operator become elliptic functions. Moreover, the whole family of isospectral deformations of the operator enjoys the same property.

An important example of this phenomenon in the differential setup is provided by the Lamé operator (1.1) for $\ell \in \mathbf{Z}_+$ and its isospectral deformations $-\partial_x^2 + u(x)$ with

$$u(x) = 2 \sum_{j=1}^{\ell(\ell+1)/2} \wp(x - x_j) + \text{const.} \tag{2.2}$$

The Lamé operator itself corresponds to a very degenerate configuration when all the poles sit in one and the same point. The isospectral flows are the flows of the KdV hierarchy [6] for the potential $u(x)$. Solving, say, the KdV equation $\dot{u} = 6uu' - u'''$ for $u = u(x, t)$ with the initial condition $u(x, 0) = \ell(\ell + 1)\wp(x - x_0)$, we get a family of Schrödinger operators with elliptic potential that have the same spectral curve as the Lamé operator. The poles x_j (and the constant term) in (2.2) become

t-dependent. By direct substitution in the KdV equation, it can be shown [2] that they are constrained by the conditions

$$\sum_{j=1,\neq i}^{\ell(\ell+1)/2} \wp'(x_i - x_j) = 0, \quad i = 1, 2, \ldots, \ell(\ell+1)/2, \tag{2.3}$$

and obey the differential equations

$$\dot{x}_j = -12 \sum_{k=1,\neq j}^{\ell(\ell+1)/2} \wp(x_j - x_k). \tag{2.4}$$

Equations (2.3) are the famous equations defining the equilibrium locus of the elliptic Calogero–Moser system of particles. From the general theory that connects the pole dynamics of elliptic solutions of nonlinear integrable equations with systems of Calogero–Moser type [2, 13], it follows that the connected component of the locus is parametrized by the Jacobian of the spectral curve of the Lamé operator. Therefore, it is an ℓ-dimensional submanifold, spanned by higher Calogero–Moser flows, in the $\frac{1}{2}\ell(\ell+1)$-dimensional configuration space with coordinates x_j.

In Section 6, we present analogues of equations (2.2), (2.3), and (2.4) in the difference setup. The isospectral flows are connected with elliptic solutions to the Volterra hierarchy.

3 Double-Bloch eigenfunctions of the difference Lamé operator and the spectral curve

In this section, we study Bloch eigenfunctions of the difference Lamé operator. Following the general scheme outlined at the beginning of Section 2, we construct a family of eigenfunctions depending on E and two additional spectral parameters. All three parameters are constrained by two equations that define the spectral curve.

Consider the eigenvalue equation for the operator $\tilde{L}$ (1.5):

$$\psi(x+\eta) + \frac{\theta_1(x+\ell\eta)\theta_1(x-(\ell+1)\eta)}{\theta_1(x)\theta_1(x-\eta)} \psi(x-\eta) = E\psi(x). \tag{3.1}$$

The coefficient function is double-periodic. Therefore, it is natural to look for solutions in the class of *double-Bloch functions* [15], i.e., such that $\psi(x+1) = B_1\psi(x)$ and $\psi(x+\tau) = B_\tau\psi(x)$ with some constants B_1 and B_τ. These are going to be the additional parameters p_i from the general scheme of Section 2.

Consider the function

$$\Phi(x,\zeta) = \frac{\theta_1(\zeta+x)}{\theta_1(x)\theta_1(\zeta)}. \tag{3.2}$$

Its monodromy properties in x are $\Phi(x+1,\zeta) = \Phi(x,\zeta)$, $\Phi(x+\tau,\zeta) = e^{-2\pi i\zeta}\Phi(x,\zeta)$, i.e., it is a double-Bloch function. Moreover, it is the simplest nontrivial (i.e., different from the exponential function) double-Bloch function since it has only one pole. This function serves as a building block for more general double-Bloch functions.

Let ℓ be a positive integer. We employ the following double-Bloch ansatz for the ψ:

$$\psi(x) = K^{x/\eta} \sum_{j=1}^{\ell} s_j(\zeta, K, E)\Phi(x - j\eta, \zeta), \tag{3.3}$$

where ζ and K parametrize the Bloch multipliers of the function $\psi(x)$: $B_1 = K^{\frac{1}{\eta}}$, $B_\tau = K^{\frac{\tau}{\eta}} e^{-2\pi i \zeta}$. The coefficients s_j depend on the indicated parameters alone.

Substituting (3.3) into (3.1) and computing the residues at the points $x = j\eta$, $j = 0, \ldots, \ell$, we get $\ell + 1$ linear equations

$$\sum_{j=1}^{\ell} M_{ij} s_j = 0, \quad i = 0, 1, \ldots, \ell, \tag{3.4}$$

for ℓ unknowns s_j. The matrix elements M_{ij} of this system are given by the formula

$$\begin{aligned} M_{ij} &= K\delta_{i,j-1} - E\delta_{i,j} + K^{-1}\frac{\theta_1((j+\ell+1)\eta)\theta_1((j-\ell)\eta)}{\theta_1((j+1)\eta)\theta_1(j\eta)}\delta_{i,j+1} \\ &+ K^{-1}\frac{\theta_1(\zeta-(j-i+1)\eta)}{\theta_1(\zeta)}\frac{\theta_1((i+\ell)\eta)\theta_1((i-\ell-1)\eta)}{\theta_1(\eta)\theta_1((j-i+1)\eta)}(\delta_{i,0}-\delta_{i,1}). \end{aligned} \tag{3.5}$$

Here $i = 0, 1, \ldots, \ell$ and $j = 1, 2, \ldots, \ell$. The overdetermined system (3.5) has nontrivial solutions if and only if the rank of the rectangular matrix M_{ij} is less than ℓ. Let $M^{(0)}$ and $M^{(1)}$ be $\ell \times \ell$ matrices obtained from M by deleting the rows with $i = 0$ and $i = 1$, respectively. Then the values of three parameters ζ, K, and E for which (3.5) has solutions of the form (3.3) are determined by the system of two equations: $\det M^{(0)} = \det M^{(1)} = 0$. They indeed define a curve.

To obtain the explicit form of these equations, we expand the determinants with respect to the first row. This yields an explicit characterization of the spectral curve summarized in the theorem below. Henceforth, the "elliptic factorial" and "elliptic binomial" notation is convenient:

$$\begin{aligned} [n]! &= \prod_{j=1}^{n} [j], \quad [j] \equiv \theta_1(j\eta)/\theta_1(\eta), \\ \begin{bmatrix} n \\ m \end{bmatrix} &\equiv \frac{[n]!}{[m]![n-m]!}. \end{aligned} \tag{3.6}$$

Theorem 3.1. *The difference Lamé equation* (3.1) *has double-Bloch solutions of the form* (3.3) *if and only if the spectral parameters* ζ, K, *and* E *obey the equations*

$$\sum_{j=0}^{\ell}(-1)^j K^{-j}\theta_1(\zeta - j\eta)\begin{bmatrix}\ell\\ j\end{bmatrix} A_j^{(\ell)}(E) = 0,$$
$$\sum_{j=0}^{\ell+1}(-1)^j K^{-j}\theta_1(\zeta - j\eta)[j-1]\begin{bmatrix}\ell+1\\ j\end{bmatrix} A_{|j-1|}^{(\ell)}(E) = 0, \tag{3.7}$$

where $A_j^{(\ell)}(E)$ are polynomials of $(\ell - j)$th degree explicitly given by the determinant formula

$$A_{\ell-s}^{(\ell)}(E) \tag{3.8}$$
$$= \begin{bmatrix}\ell\\ s\end{bmatrix}\begin{bmatrix}2\ell\\ s\end{bmatrix}^{-1} \det\left(E\delta_{i,j} + \frac{[-i]}{[\ell+1-i]}\delta_{i,j-1} + \frac{[2\ell+2-i]}{[\ell+1-i]}\delta_{i,j+1}\right)_{1\le i,j\le s}$$

(here $0 \le s \le \ell$), $A_\ell^{(\ell)} = 1$.

Let us list some useful properties of the polynomials $A_j^{(\ell)}(E)$. First, they obey the recurrence relation

$$A_{\ell-s-1}^{(\ell)}(E) = \frac{[\ell-s]}{[2\ell-s]} E A_{\ell-s}^{(\ell)}(E) + \frac{[s]}{[2\ell-s]} A_{\ell-s+1}^{(\ell)}(E) \tag{3.9}$$

with the initial condition $A_\ell^{(\ell)}(E) = 1$, $A_{\ell-1}^{(\ell)}(E) = ([\ell]/[2\ell])E$. Next, it is clear from (3.9) that

$$A_{\ell-s}^{(\ell)}(-E) = (-1)^s A_{\ell-s}^{(\ell)}(E), \quad 0 \le s \le \ell. \tag{3.10}$$

Equations (3.7) define a Riemann surface $\tilde{\Gamma}$ that covers the complex plane. The monodromy properties of the θ-function (see (A.4) in the appendix) make it clear that this surface is invariant under the transformation

$$\zeta \longmapsto \zeta + \tau, \qquad K \longmapsto K e^{2\pi i \eta}. \tag{3.11}$$

The factor of the $\tilde{\Gamma}$ over this transformation is an algebraic curve Γ that is a ramified covering of the elliptic curve with periods 1 and τ. It is clear from (3.7) and (3.10) that the curve admits the involution

$$(\zeta, K, E) \longmapsto (\zeta, -K, -E), \tag{3.12}$$

so the spectrum is symmetric with respect to the reflection $E \to -E$. Another result of [15], which is not so easy to see from (3.7), is that the curve Γ is at the same time a hyperelliptic curve.

Theorem 3.2. *The curve Γ is a hyperelliptic curve of genus $g = 2\ell$. The hyperelliptic involution is given by*

$$(\zeta, K, E) \longmapsto (2N\eta - \zeta, K^{-1}, E), \quad N = \frac{1}{2}\ell(\ell+1). \tag{3.13}$$

The points $P = (\zeta, K, E) \in \Gamma$ *of the curve parametrize double-Bloch solutions* $\psi(x) = \psi(x, P)$ *to* (3.1), *and the solution* $\psi(x, P)$ *corresponding to each point* $P \in \Gamma$ *is unique up to a constant multiplier.*

For the proof of Theorem 3.2, see [15]. Here we give a few remarks. The involution (3.13) looks best in terms of the function $\Psi(x) = \psi(x)\prod_{j=1}^{\ell}\theta_1(x - j\eta)$ that satisfies the eigenvalue equation $L\Psi = E\Psi$ with L as in (1.3) (cf. (1.4)). Then the hyperelliptic involution simply takes $\Psi(x)$ to $\Psi(-x)$. The genus of the curve can be found by counting the number of fixed points of this involution. At the fixed points, the two solutions $\Psi(x)$ and $\Psi(-x)$ are linearly dependent: $\Psi(-x) = r\Psi(x)$. Writing out the eigenvalue equation at $x = 0$, we obtain the necessary condition $\Psi(-\eta) = \Psi(\eta)$, so $r = 1$. The ansatz (3.3) for ψ is equivalent to the ansatz

$$\Psi(x) = K^{x/\eta}\prod_{j=1}^{\ell}\theta_1(x + y_j)$$

with $\sum_j y_j = \zeta$. At $K = 1$, the dimension of the linear space of even functions of this form is known to be $2\ell + 1$. Adding images of the fixed points under the involution (3.12), we eventually get $4\ell + 2$ fixed points, so by the Riemann–Hurwitz formula, the genus is equal to 2ℓ.

Taking into account the symmetry $E \to -E$, we can write the equation of the hyperelliptic curve in the standard form

$$w^2 = \prod_{i=1}^{2\ell+1}(E^2 - E_i^2). \tag{3.14}$$

The hyperelliptic involution takes (w, E) to $(-w, E)$.

In (3.14), w is the eigenvalue of a nontrivial operator W commuting with L on their common eigenfunction. The explicit form of the operator W was found in [9]:

$$W = \varphi_\ell(x)\sum_{k=0}^{2\ell+1}(-1)^k\begin{bmatrix}2\ell+1\\ k\end{bmatrix}\frac{\theta_1(x + (2\ell - 2k + 1)\eta)}{\prod_{j=0}^{2\ell-k+1}\theta_1(x + j\eta)\prod_{j'=1}^{k}\theta_1(x - j'\eta)}e^{(2\ell-2k+1)\eta\partial_x},$$

where $\varphi_\ell(x) = \prod_{j=0}^{2\ell}\theta_1(x + (j - \ell)\eta)$.

Let us conclude this section by examining the behavior of the spectral curve in the vicinity of its "infinite points," i.e., the points at which the function E has poles. From (3.7), we conclude that there are two such points: $\infty_+ = (\zeta \to 0,\ K \to \infty,\ E \to \infty)$ and $\infty_- = (\zeta \to 2N\eta,\ K \to 0,\ E \to \infty)$. In the neighborhood of $\infty_\pm$, $E = K^{\pm 1} + o(K^{\pm 1})$. In terms of the variables (w, E), these points are $\infty_\pm = (w \to \pm\infty,\ E \to \infty)$.

4 Edges of bands

The edges of bands $\pm E_i$, i.e., the branch points of the two-sheet covering (3.14), are values of the function $E = E(P)$ at the fixed points of the hyperelliptic involution.

As is clear from (3.13), the fixed points lie above the points $\zeta = N\eta + \omega_a$, where ω_a are the half-periods: $\omega_1 = 0$, $\omega_2 = \frac{1}{2}$, $\omega_3 = \frac{1}{2}(1+\tau)$, $\omega_4 = \frac{1}{2}\tau$. The corresponding values of K are determined from (3.11). Then the set of the branch points E_i is fixed by Theorem 3.1.

Corollary 4.1. *Let $\mathcal{E}_a$, $a = 1, \ldots, 4$, be the set of common roots of the polynomial equations*

$$\begin{aligned} \sum_{j=0}^{\ell}(-1)^j\theta_a((N-j)\eta)\begin{bmatrix} \ell \\ j \end{bmatrix} A_j^{(\ell)}(E) = 0, \\ \sum_{j=0}^{\ell+1}(-1)^j\theta_a((N-j)\eta)[j-1]\begin{bmatrix} \ell+1 \\ j \end{bmatrix} A_{|j-1|}^{(\ell)}(E) = 0, \end{aligned} \tag{4.1}$$

where θ_a are Jacobi θ-functions.[1] *Then the set of the edges of bands $\pm E_i$ is the union of $\bigcup_{a=1}^4 \mathcal{E}_a$ and its image under the reflection $E \to -E$.*

Let us give two examples. At $\ell = 1$, the set $\mathcal{E}_1$ is empty while $\mathcal{E}_a$ for $a = 2, 3, 4$ contains one point. From (4.1) we find that

$$E_\alpha = 2\frac{\theta_{\beta+1}(\eta)\theta_{\gamma+1}(\eta)}{\theta_{\beta+1}(0)\theta_{\gamma+1}(0)},$$

where $\{\alpha, \beta, \gamma\}$ is any cyclic permutation of $\{1, 2, 3\}$. At $\ell = 2$, the set $\mathcal{E}_1$ has two elements E_1 and E_2 obtained as solutions of the quadratic equation $[2]E^2 + [2]^3E + 2[4] = 0$ so that

$$E_{1,2} = \frac{1}{2}\left(\frac{\theta_1^2(2\eta)}{\theta_1^2(\eta)} \pm \sqrt{\frac{\theta_1^4(2\eta)}{\theta_1^4(\eta)} - 8\frac{\theta_1(4\eta)}{\theta_1(2\eta)}}\right).$$

For $a = 2, 3, 4$, each set $\mathcal{E}_a$ has one element E_{a+1}:

$$E_{a+1} = \frac{\theta_1(2\eta)\theta_a(2\eta)}{\theta_1(\eta)\theta_a(\eta)}.$$

In general, it is possible to prove [15] that

$$\#(\mathcal{E}_1) = \begin{cases} \frac{1}{2}(\ell-1), & \ell \text{ odd}, \\ \frac{1}{2}\ell+1, & \ell \text{ even}, \end{cases} \qquad \#(\mathcal{E}_{2,3,4}) = \begin{cases} \frac{1}{2}(\ell+1), & \ell \text{ odd}, \\ \frac{1}{2}\ell, & \ell \text{ even}. \end{cases} \tag{4.2}$$

Note that $\#(\cup_{a=1}^4\mathcal{E}_a) = 2\ell + 1$, which agrees with (3.14).

---

[1]The definitions and transformation properties of the Jacobi θ-functions $\theta_a(x|\tau)$, $a = 1, 2, 3, 4$, are listed in the appendix. For brevity, we write $\theta_a(x|\tau) \equiv \theta_a(x)$.

5 Relation between the Bloch multipliers

To simplify equations of the curve (3.7), one can try to eliminate one of the variables and obtain a single equation for the other two. Here we show how to eliminate E. This leads to a closed relation between the two Bloch multipliers of the function (3.3) (parametrized through ζ and K). Its form (see (5.6) and (5.8) below) suggests an interpretation in terms of hypothetical combinatorial identities for elliptic numbers.

At the first glance, the elimination of E from (3.7) is hardly possible. Nevertheless, there is an alternative argument leading directly to the relation between the Bloch multipliers. Here it is more convenient to deal with the difference Lamé operator in the gauge equivalent form (1.3). Our construction is based on the following simple lemma [23].

Lemma 5.1. *Let $\Psi(x)$ be any solution to the equation*

$$\frac{\theta_1(x-\ell\eta)}{\theta_1(x)}\Psi(x+\eta)+\frac{\theta_1(x+\ell\eta)}{\theta_1(x)}\Psi(x-\eta)=E\Psi(x) \tag{5.1}$$

in the class of entire functions on the complex plane of the variable x. Then

$$\Psi(j\eta)=\Psi(-j\eta),\quad j=1,2,\ldots,\ell. \tag{5.2}$$

This assertion follows from the specific location of zeros and poles of the coefficients of (5.1). Indeed, setting $x=0$ in (5.1), we have $\Psi(\eta)=\Psi(-\eta)$. The proof can be completed by induction. At $x=\pm\ell\eta$, one of the coefficients in the left-hand side of (5.1) vanishes, so the chain of relations (5.2) truncates at $j=\ell$.

Remarkably, the conditions (5.2) and the ansatz

$$\Psi(x)=K^{x/\eta}\left(\prod_{j=1}^{\ell}\theta_1(x-j\eta)\right)\sum_{m=1}^{\ell}s_m(K,\zeta)\Phi(x-m\eta,\,\zeta) \tag{5.3}$$

for Ψ (equivalent to the ansatz (3.3) for ψ) with the same function $\Phi(x,z)$ given by (3.2) allow one to find the relation between the Bloch multipliers even without explicit use of the difference Lamé equation (5.1). Plugging (5.3) into (5.2), we obtain ℓ equalities (for $m=1,2,\ldots,\ell$):

$$K^m s_m=(-1)^{\ell}K^{-m}\theta_1(2m\eta)\left(\prod_{j=1,\neq m}^{\ell}\frac{\theta_1((m+j)\eta)}{\theta_1((m-j)\eta)}\right)\sum_{n=1}^{\ell}\Phi(-(m+n)\eta,\,\zeta)s_n. \tag{5.4}$$

This is a system of linear homogeneous equations for s_n. It has nontrivial solutions if and only if its determinant is equal to zero, whence we obtain the equation connecting ζ and K:

$$\det\left(K^{2m}\delta_{mn}+G_{mn}(\zeta)\right)_{1\le m,n\le\ell}=0, \tag{5.5}$$

where

$$G_{mn}(\zeta) = (-1)^{\ell+1}[2m]\left(\prod_{j=1,\neq m}^{\ell} \frac{[m+j]}{[m-j]}\right)\Phi\big(-(m+n)\eta,\, \zeta\big).$$

This equation defines a curve that is the image of the spectral curve Γ under the projection that takes (ζ, K, E) to (ζ, K).

The equation of the spectral curve (5.5) can be represented in the form

$$\sum_{j=0}^{N}(-1)^j C_j^{(\ell)}(\eta)\theta_1(\zeta - 2j\eta)K^{2(N-j)} = 0, \tag{5.6}$$

where $N = \frac{1}{2}\ell(\ell+1)$ and $C_j^{(\ell)}(\eta)$ are some coefficients depending only on η and τ such that $C_j^{(\ell)}(\eta) = C_{N-j}^{(\ell)}(\eta)$, $C_0^{(\ell)}(\eta) = 1$. To see this, we expand the determinant (5.5) in powers of K with the help of the identity

$$\det\left(\frac{\theta_1(x_i+x_j+\zeta)}{\theta_1(x_i+x_j)}\right)_{1\le i,j\le n} = \frac{\theta_1^{n-1}(\zeta)\theta_1(\zeta + 2\sum_{i=1}^n x_i)}{\prod_{i=1}^n \theta_1(2x_i)} \prod_{i<j}^{n} \frac{\theta_1^2(x_i-x_j)}{\theta_1^2(x_i+x_j)}$$

(a particular case of the formula for the elliptic Cauchy determinant). Let Λ be the set $\{1, 2, \ldots, \ell\}$. For any subset $J \subseteq \Lambda$, $\Lambda \setminus J$ is its complement and $\sigma(J) = \sum_{j\in J} j$. Evaluating the elliptic Cauchy determinant at $x_n = -2\eta n$, we get

$$\begin{aligned} &\det\left(K^{2m}\delta_{mn} + G_{mn}(\zeta)\right)_{1\le m,n\le\ell} \\ &\quad= \sum_{J\subseteq\Lambda} \frac{\theta_1(\zeta - 2\sigma(J)\eta)}{\theta_1(\zeta)} K^{2N-2\sigma(J)}(-1)^{\sigma(J)} \prod_{k\in J}\prod_{k'\in\Lambda\setminus J} \frac{[k+k']}{[|k-k'|]}. \end{aligned} \tag{5.7}$$

Thus the coefficient $C_j^{(\ell)}$ reads

$$C_j^{(\ell)} = \sum_{J\subseteq\Lambda,\sigma(J)=j}\ \prod_{k\in J}\prod_{k'\in\Lambda\setminus J} \frac{[k+k']}{[|k-k'|]}. \tag{5.8}$$

The symmetry $j \leftrightarrow N-j$ is now transparent. Note that the sum in (5.8) runs over partitions of the number j into *distinct* parts not exceeding ℓ.

We note that for any elliptic module τ, we have that

$$\lim_{\eta\to 0} C_j^{(\ell)} = \binom{N}{j} \tag{5.9}$$

(the usual binomial coefficient). To see this, consider the limiting case $\tau \to i\infty$. We have $\exp(-\pi i\tau/4)\theta_1(x|\tau) \to 2\sin(\pi x)$ as $\tau \to i\infty$ (see (A.2)), so

$$[j] \longrightarrow \frac{\sin(\pi\eta j)}{\sin(\pi\eta)} \equiv (j)_q, \quad q = e^{2\pi i\eta}.$$

Then (5.9) follows from the identity [22]

$$\sum_{J\subseteq\Lambda} z^{\sigma(J)} \prod_{k\in J} \prod_{k'\in\Lambda\setminus J} \frac{(k+k')_q}{(|k-k'|)_q} = \prod_{1\le j\le k\le \ell} (1+zq^{j+k-\ell-1}), \tag{5.10}$$

which is a specialization of the C_ℓ-type Weyl denominator formula.[2] A detailed combinatorial analysis of the limiting cases $\tau \to i\infty$ or $\tau \to 0$ of the difference Lamé operators can be found in [22].

6 Isospectral deformations of the difference Lamé operator and locus equations

Finally, let us comment on isospectral deformations of the difference Lamé operator. We are going to present difference analogues of the operator (2.2) and of the locus equations (2.3).

Instead of (3.1), consider the equation

$$\psi(x+\eta) + c(x)\psi(x-\eta) = E\psi(x) \tag{6.1}$$

with a more general coefficient $c(x)$, which is an elliptic function represented in the form

$$c(x) = \frac{\rho(x+\eta)\rho(x-2\eta)}{\rho(x)\rho(x-\eta)}, \quad \rho(x) = \prod_{j=1}^{\ell(\ell+1)/2} \theta_1(x-x_j). \tag{6.2}$$

Note that in the case of the difference Lamé operator, the configuration of zeros of the $\rho(x)$ is very specific,

$$\rho(x) = \prod_{1\le j\le k\le\ell} \theta_1(x+(j+k-\ell-1)\eta), \tag{6.3}$$

so all but two cancel in the $c(x)$.

The isospectral flows are the flows of the Volterra hierarchy for the $c(x)$. The first equation of the hierarchy,

$$\frac{\partial c(x)}{\partial t} = -c(x)(c(x+\eta) - c(x-\eta)), \tag{6.4}$$

is the compatibility condition of the spectral problem (6.1) and the linear problem

$$\frac{\partial\psi(x)}{\partial t} = c(x)c(x-\eta)\psi(x-2\eta).$$

[2] I am grateful to A. N. Kirillov for pointing out this identity and drawing my attention to the paper [22].

Recall that changing the variables as $t \to \frac{1}{3}\eta^3 t$ and $x \to x - 2\eta t$ and setting $c(x) = 1 - \eta^2 u(x)$, one gets the KdV equation for u as $\eta \to 0$.

Substituting the pole ansatz (6.2) into the Volterra equation and requiring the residues at the poles to be zero, we get the following two systems of equations ($j = 1, 2, \ldots, \frac{1}{2}\ell(\ell+1)$):

$$\begin{aligned} \frac{\theta_1'(0)}{\theta_1(2\eta)} \dot{x}_j &= \prod_{k=1,\neq j}^{\ell(\ell+1)/2} \frac{\theta_1(x_j - x_k + 2\eta)\theta_1(x_j - x_k - \eta)}{\theta_1(x_j - x_k + \eta)\theta_1(x_j - x_k)}, \\ \frac{\theta_1'(0)}{\theta_1(2\eta)} \dot{x}_j &= \prod_{k=1,\neq j}^{\ell(\ell+1)/2} \frac{\theta_1(x_j - x_k - 2\eta)\theta_1(x_j - x_k + \eta)}{\theta_1(x_j - x_k - \eta)\theta_1(x_j - x_k)}, \end{aligned} \tag{6.5}$$

where $\dot{x}_j = \partial_t x_j$. Solving these equations or, equivalently, the Volterra equation with initial condition (6.3), one arrives at a family of isospectral deformations of the difference Lamé operator. The two systems (6.5) must be satisfied simultaneously. Therefore, the right-hand sides are identical, whence we obtain the necessary conditions for solutions to exist:

$$\prod_{k=1,\neq j}^{\ell(\ell+1)/2} \frac{\theta_1(x_j - x_k + 2\eta)\theta_1^2(x_j - x_k - \eta)}{\theta_1(x_j - x_k - 2\eta)\theta_1^2(x_j - x_k + \eta)} = 1, \tag{6.6}$$

which are the difference analogues of (2.3). The results of [15] imply that these equations define an equilibrium locus of the Ruijsenaars–Schneider model. As in the differential case, the locus is not compact. Its closure contains, in particular, the point corresponding to the degenerate configuration (6.3). From the general arguments of [15], it follows that the connected component of the locus that contains this point at the boundary is ℓ-dimensional. Expanding (6.6) in $\eta \to 0$, we get (2.3).

Appendix Theta functions

We use the following definition of the Jacobi θ-functions:

$$\begin{aligned} \theta_1(x|\tau) &= -\sum_{k\in\mathbf{Z}} \exp\left(\pi i \tau \left(k + \frac{1}{2}\right)^2 + 2\pi i \left(x + \frac{1}{2}\right)\left(k + \frac{1}{2}\right)\right), \\ \theta_2(x|\tau) &= \sum_{k\in\mathbf{Z}} \exp\left(\pi i \tau \left(k + \frac{1}{2}\right)^2 + 2\pi i x \left(k + \frac{1}{2}\right)\right), \\ \theta_3(x|\tau) &= \sum_{k\in\mathbf{Z}} \exp\left(\pi i \tau k^2 + 2\pi i x k\right), \\ \theta_4(x|\tau) &= \sum_{k\in\mathbf{Z}} \exp\left(\pi i \tau k^2 + 2\pi i \left(x + \frac{1}{2}\right) k\right). \end{aligned} \tag{A.1}$$

They can be represented as infinite products:

$$\theta_1(x|\tau) = 2\sin(\pi x)e^{\pi i\tau/4}\prod_{k=1}^{\infty}\big(1-e^{2\pi ik\tau}\big)\big(1-e^{2\pi i(k\tau+x)}\big)\big(1-e^{2\pi i(k\tau-x)}\big),$$
$$\theta_2(x|\tau) = 2\cos(\pi x)e^{\pi i\tau/4}\prod_{k=1}^{\infty}\big(1-e^{2\pi ik\tau}\big)\big(1+e^{2\pi i(k\tau+x)}\big)\big(1+e^{2\pi i(k\tau-x)}\big),$$
$$\theta_3(x|\tau) = \prod_{k=1}^{\infty}\big(1-e^{2\pi ik\tau}\big)\big(1+e^{\pi i(2k\tau-\tau+2x)}\big)\big(1+e^{\pi i(2k\tau-\tau-2x)}\big),$$
$$\theta_4(x|\tau) = \prod_{k=1}^{\infty}\big(1-e^{2\pi ik\tau}\big)\big(1-e^{\pi i(2k\tau-\tau+2x)}\big)\big(1-e^{\pi i(2k\tau-\tau-2x)}\big). \tag{A.2}$$

The $\wp$-function is related to the θ_1 as follows:

$$\wp(x|1/2,\tau/2) = -\frac{d^2}{dx^2}\log\theta_1(x|\tau) + \text{const.} \tag{A.3}$$

Throughout the paper, we write $\theta_a(x|\tau) = \theta_a(x)$.

The transformation properties of the theta functions for shifts by (half-)periods are

$$\begin{aligned}\theta_a(x\pm 1) &= (-1)^{\delta_{a.1}+\delta_{a.2}}\theta_a(x),\\ \theta_a(x\pm\tau) &= (-1)^{\delta_{a.1}+\delta_{a.4}}e^{-\pi i\tau\mp 2\pi ix}\theta_a(x),\end{aligned} \tag{A.4}$$

$$\begin{aligned}\theta_1\left(x\pm\frac{1}{2}\right) &= \pm\theta_2(x),\\ \theta_1\left(x\pm\frac{\tau}{2}\right) &= \pm ie^{-\frac{1}{4}\pi i\tau\mp\pi ix}\theta_4(x),\\ \theta_1\left(x\pm\frac{1+\tau}{2}\right) &= \pm e^{-\frac{1}{4}\pi i\tau\mp\pi ix}\theta_3(x).\end{aligned} \tag{A.5}$$

Acknowledgments. The author would like to thank Professors M. Kashiwara and T. Miwa for the opportunity to present these results at the "Physical Combinatorics" workshop. Discussions with I. M. Krichever, A. D. Mironov, T. Takebe, and P. B. Wiegmann are gratefully acknowledged.

This work was supported in part by RFBR grant 98-01-00344 and by a grant in support of scientific schools.

References

[1] A. Abanov, J. Talstra, and P. Wiegmann, Hierarchical structure of Azbel-Hofstadter problem: Strings and loose ends of Bethe ansatz, *Nuclear Phys. B*, **525** (1998), 571–596.

[2] H. Airault, H. McKean, and J. Moser, Rational and elliptic solutions of the KdV equation and related many-body problem, *Comm. Pure Appl. Math.*, **30** (1977), 95–125.

[3] M. Azbel, The energy spectrum of conducting electron in magnetic field, *Zh. Èksper. Teor. Fiz.*, **46** (1964), 929–946.

[4] E. Belokolos, A. Bobenko, V. Enolskii, A. Its, and V. Matveev, *Algebraic-Geometrical Approach to Nonlinear Integrable Equations*, Springer-Verlag, Berlin, 1994.

[5] E. Date and S. Tanaka, Exact solutions of the periodic Toda lattice *Progr. Theoret. Phys.*, **5** (1976), 457–465.

[6] B. Dubrovin, V. Matveev, and S. Novikov, Non-linear equations of Korteweg-de Vries type, finite zone linear operators and Abelian varieties, *Uspekhi Mat. Nauk*, **31**(1) (1976), 55–136.

[7] V. Enolskii and J. Eilbeck, On the two-gap locus for the elliptic Calogero-Moser model, *J. Phys. A*, **28** (1995), 1069–1088.

[8] G. Felder and A. Varchenko, Algebraic Bethe ansatz for the elliptic quantum group $E_{\tau,\eta}(sl_2)$, *Nuclear Phys. B*, **480** (1996), 485–503.

[9] G. Felder and A. Varchenko, Algebraic integrability of the two-body Ruijsenaars operator, preprint q-alg/9610024, 1996.

[10] D. R. Hofstadter, Energy levels and wave functions for Bloch electrons in rational and irrational magnetic fields, *Phys. Rev. B*, **14** (1976), 2239–2249.

[11] E. L. Ince, Further investigations into the periodic Lamé functions, *Proc. Roy. Soc. Edinburgh*, **60** (1940), 83–99.

[12] I. M. Krichever, Algebraic curves and non-linear difference equations, *Uspekhi Mat. Nauk*, **33**(4) (1978), 215–216.

[13] I. M. Krichever, Elliptic solutions of Kadomtsev-Petviashvilii equation and integrable systems of particles, *Functional Anal. Appl.*, **14**(4) (1980), 282–290.

[14] I. M. Krichever, Nonlinear equations and elliptic curves, *Itogi Nauki i Tekhniki*, **23** (1983).

[15] I. Krichever and A. Zabrodin, Spin generalization of the Ruijsenaars-Schneider model, non-abelian 2D Toda chain and representations of Sklyanin algebra, *Uspekhi Mat. Nauk*, **50**(6) (1995), 3–56.

[16] D. Mumford, Algebro-geometric construction of commuting operators and of solutions to the Toda lattice equation, Korteweg-de Vries equation and related non-linear equations, in *Proceedings of the International Symposium on Algebraic Geometry, Kyoto,* 1977, Kinokuniya Book Store, Tokyo, 1978, 115–153.

[17] S. N. M. Ruijsenaars, Relativistic Lamé functions: The special case $g = 2$, *J. Phys. A*, **32** (1999), 1737–1772.

[18] E. K. Sklyanin, On some algebraic structures related to the Yang-Baxter equation, *Funktsional Anal. i Prilozhen*, **16**(4) (1982), 27–34.

[19] E. K. Sklyanin, On some algebraic structures related to the Yang-Baxter equation: Representations of the quantum algebra, *Funktsional Anal. i Prilozhen*, **17**(4) (1983), 34–48.

[20] A. O. Smirnov, Elliptic solutions of the Korteweg-De Vries equation, *Mat. Zametki*, **45**(6) (1989), 66–73.

[21] A. Treibich and J.-L. Verdier, Solitons elliptiques, in P. Cartier, L. Illusie, N. M. Katz, G. Laumon, Y. Manin, and K. A. Ribet, eds., *Grothendieck Festschrift*, Progress in Mathematics 88, Birkhäuser, Boston, 1990

[22] J. F. Van Diejen and A. N. Kirillov, *Formulas for q-Spherical Functions Using Inverse Scattering Theory of Reflectionless Jacobi Operators*, Hokkaido University Preprint Series in Mathematics 430, Hokkaido University, Hokkaido, 1998.

[23] A. Zabrodin, On the spectral curve of the difference Lamé operator, *Internat. Math. Res. Notices*, **11** (1999), 589–614.

Joint Institute of Chemical Physics
Kosygina Street 4
117334 Moscow
Russia

and

ITEP
117259 Moscow
Russia
zabrodin@heron.itep.ru